MAJOR FUNGAL DISEASES OF RICE

RECENT ADVANCES

Major Fungal Diseases of Rice

Recent Advances

Professor K. Manibhushan Rao Festschrift

by

S. Sreenivasaprasad

and

R. Johnson

KLUWER ACADEMIC PUBLISHERS
DORDRECHT / BOSTON / LONDON

A C.I.P. Catalogue record for this book is available from the Library of Congress.

ISBN 1-4020-0050-2

Published by Kluwer Academic Publishers,
P.O. Box 17, 3300 AA Dordrecht, The Netherlands.

Sold and distributed in North, Central and South America
by Kluwer Academic Publishers,
101 Philip Drive, Norwell, MA 02061, U.S.A.

In all other countries, sold and distributed
by Kluwer Academic Publishers,
P.O. Box 322, 3300 AH Dordrecht, The Netherlands.

Printed on acid-free paper

Printed in the Netherlands.

CONTENTS

Professor K. Manibhushan Rao - Reminiscence

We are pleased and privileged to bring out this festschrift as a commemoration of the 60th birthday of Prof. K. Manibhushan Rao, which he attained on 19th November, 1997. The board range of topics presented in this festschrift is a fitting tribute to his enormous contributions to both basic and applied aspects of rice fungal pathology. He is held among his colleagues and students with great esteem and affection. For more than three decades, Prof. Rao has extensively contributed to our understanding of major fungal diseases viz blast, sheath blight and sheath rot of rice. It is by virtue of application of laboratory findings with field based experiments and observations as well as statistical and computer simulation models he was able to carve a niche among the rice pathologists. Writing a traditional profile of a scientist, quite often would turn out to be a catalogue of scientific achievements and *curriculum vitae*. For us writing these reminiscences has proved an extraordinary experience by way of including lot of personal annotations.

Kalla Manibhushan Rao, was born in a agricultural family at Jeypore, Koraput Dist., Orissa State, India on 19th of November 1937 as the second son of Late Kalla Nagabhushan Rao and Achiyamma. After graduating from Maharaja's college. Vizianagram, AP, he joined Ravenshaw college, Cuttack and obtained his M.Sc., degree in 1962 with an impressive first class University 2nd rank.

Immediately after post-graduation, he was employed for a short while as a Research Assistant in the division of Plant pathology at Central Rice Research Institute (CRRI), Cuttack headed by late Dr. S.Y. Padmanabhan, a distinguished Rice Pathologist. Then began Prof. Rao's long and untiring journey in the rice fungal pathology and he never looked back. Even during his short period of stay at CRRI, he could publish a paper on "false - smut incidence in rice" which is well quoted by most rice pathologists. In 1963, he moved to University Botany Laboratory (UBL), University of Madras and started working on rice blast disease under the guidance of Prof. T.S. Sadasivan, an eminent plant pathologist of international repute and obtained his Ph.D. degree in 1968. Even before obtaining his doctorate degree he was offered a lecturer post in July 1968 at the Centre for Advanced Studies (CAS) in Botany as UBL was recognized as CAS by UGC (University Grants Commission), India. Subsequently, Dr. Rao was elevated as reader in 1978 and as professor in 1987 at the Centre. Further, in 1990 University of Madras awarded him the prestigious D.Sc. (Doctor of Science) for his outstanding research "Contributions on blast and sheath-blight diseases of rice in India".

Recently, in view of his outstanding qualifications and research experience, University of Madras appointed him as Professor Emeritus (August 2000 onwards) at CAS in Botany. His expertise is being used in both teaching and research and it is really a great honour for him in rejoining the Centre. Prof. Rao is active in attending symposia at National and International level and also sharing his experience in rice pathology to Academic staff college participants of the neighboring / sister universities.

GENETIC AND PATHOGENIC DIVERSITY OF THE RICE BLAST PATHOGEN

R. SRIDHAR and U. D. SINGH
Molecular Plant pathology Laboratory, Department of Plant pathology
Central Rice Research Institute, Cuttack 753 006, Orissa, India.

1. Introduction

Rice blast caused by *Pyricularia grisea* (Rossman *et al.*, 1990) (perfect state *Magnaporthe grisea*, Barr, 1977) continues to be a serious constraint to rice production in both tropical and temperate regions (Ou, 1980, 1985). Occurrence of blast was recorded for the first time in the Thanjavur delta of Tamil Nadu in 1918 (McRae, 1920). The disease has occurred from time to time in the country causing severe damage to rice production (Sundararaman, 1927; Thomas, 1930). Blast is common in all the three main ecosystems in which rice is grown namely, irrigated, and rainfed uplands and lowlands although the incidence and severity is greater in upland situations. The existence of strains of the blast pathogen *Pyricularia grisea* differing in pathogenicity was first noticed by Sasaki (1922). However, intensive studies on pathogenic races of this fungus was initiated in Japan only after some blast resistant cultivars such as Futaba, developed by hybridisation, became susceptible to this disease after cultivation for nearly a decade. Goto (1960, 1965) identified 13 races and placed them into three groups. Following this, studies on pathogenic races of *P. grisea* were intensified in several countries resulting in the identification of many races. However, the differentials used by different workers were not identical.

2. Pathogenic variability

In a co-operative effort, a set of eight cultivars consisting of Raminad Str. 3, Zenith, NP-125, Usen, Dular, Kanto 51, Sha-tiao-tsao (s) (CI 8970-S) and Caloro were selected to serve as international differential set for identifying races of blast fungus (Atkins *et al.*, 1967). This set of differentials continues to be in use in India and elsewhere, despite the fact that the genetic analysis of it is incomplete. However, in many places local sets of differentials capable of providing better differential reaction are in use (Ou, 1985). Japan has registered 34 races (Matsumoto *et al.*, 1969), the USA 16 (Latterell *et al.*, 1960; Atkins, 1962), Taiwan 27 (Chien, 1967), Korea 18 (Lee and Matsumoto, 1966), and more than 250 were reported in the Philippines (IRRI, 1975). Variability in the blast fungus was first noticed in India in the extensive multi-location testing of several resistant varieties during 1955-1958 in several parts of the country by the Central Rice Research Institute (Padmanabhan, 1965a). By 1964, the existence of 22

S. Sreenivasaprasad and R. Johnson (eds.), Major Fungal diseases of Rice Recent Advances, 1-7.
Prof. K. Manibhushan Rao Festschrift

Our relationship with Prof. Rao is more than two decades old. We know his reputation as an excellent researcher and a gifted biologist in general. Our lasting impressions of him are his keen perception and intelligence and his open and generous spirit. As a scientist, Prof. Rao was highly productive, publishing more than 100 research papers in reputed national and international journals, author of several chapters in symposia proceedings, books, 3 monographs and 2 books. Earlier in 1971, Dr. Rao was deputed to go abroad with an UNESCO Post-Doctoral research fellowship for specialised training in cytology and genetics of the blast pathogen with Dr. P.R. Day a well reputed fungal geneticist at Connecticut Agricultural Experimental Station, New Haven, USA. After completing his assignment in USA he moved to the University of Dundee, Dundee, Scotland to work with Dr. A.C. Hastie, in the Genetics Dept. Further, during 1975 - 76, Dr. Rao visited International Rice Research Institute (IRRI), Philippines as a senior post-doctoral research fellow and worked with Dr. S.H. Ou, the then principal rice pathologist.

Behind every successful man there is always a woman and in Prof. Rao's case, his wife Mrs. Annapurna Rao is the person who always looked after him so well and but for her co-operation the enormous amount of work done by Dr. Rao would not have been possible. They have two sons and a daughter who are all well settled.

Though Dr. Rao was primarily wedded to the rice fungal pathology, he was also very much interested in horticulture, landscaping and gardening. His first acquaintance with gardening was during Indo-Chinese war in 1962. To raise the war fund, staff at the CRRI, Cuttack were engaged after office hours in vegetable cultivation by utilising the available waste lands. Enthused Rao started collecting literature on the cultivation of vegetable crops and the valuable information enabled him to write a series of popular articles under the column "Gardening News" in "Eastern Times", the only English daily then published from Cuttack. Later, he moved to University of Madras and after a lapse of almost two decades he once again resumed his writings on "Gardening" in "The Hindu" and completed the 100th article in May 1994 and the Centre could celebrate the occasion. At the suggestion of his well wishers and friends he wrote a comprehensive book as a compilation of the published articles over years. M/s. Macmillan India Ltd., published the book as "Text Book of Horticulture" in 1990 and reprinted twice in 1995 and 2000. This book is very popular in most agricultural universities as it is prescribed as a text book for horticultural students. Prof. Rao's articles on "Gardening" series are also published in popular Telugu daily "Vartha" from Hyderabad.

Although this is not the correct forum to describe in detail all his credentials, we want to recount some of Rao's academic achievements. Prof. Rao has more than 30 years of post-graduate teaching experience and has guided six Ph. D. and 21 M. Phil. Scholars. One may not fail to notice the excellence of his students which bears testimony to his remarkable legacy. He has successfully carried out five major research projects funded by UGC / ICAR (Indian Council of Agricultural Research), New Delhi. It may be recalled that Prof. Rao was the first Indian rice researcher to apply computer simulation models on the forecasting of tropical rice blast disease. He was the only invited speaker from India to present his findings at the International Rice Conference (IRC) held at Seoul, South Korea in 1990. Prof. Rao has served in the evaluation committees of several South Indian Universities as well as in the Public Service Commissions of AP, Karnataka, Tamil Nadu and Orissa. Currently, he is serving as an expert committee member of UGC to evaluate the research work done under SAP, DRS and COSIST Programmes. He has delivered a number of lectures to professional bodies including few coveted endowment lectures such as Dr. Todla Ekambaram Endowment lecture, University of Madras (1984 - 85) and Prof. M. Chinnaveeriah Endowment lecture, Karnataka University, Dharwad (1989 - 90) to mention a few. He is a life member / member of a number of professional societies and has organised many symposia / seminars at various levels and edited the proceedings. Prof. Rao is a recipient of the prestigious ICAR sponsored Hari Om Ashram Trust award for the year 1991 - 92 for the development of "EPIBLA" - a simulator for forecasting tropical rice blast disease. He coined the term from "EPIdemiology of BLAst". In 1998 he was elected as a fellow of the National Academy of Agriculture Sciences (FNA AS), New Delhi. Most of his students are in high esteemed academic positions at International and National level.

As the first research student of Prof. Rao, I (S.M.) have been privileged to know him very closely. It was the summer of 1976 and after obtaining my M.Sc., degree in Botany from St. Joseph College, Trichy, I enrolled for a Ph.D. degree in 1976 with Dr. Rao, who just then returned from IRRI, Philippines. With the facilities available at the Centre, he established a dedicated team of workers, but he is critical in choosing his students. Despite some hard moments, I fondly recall my private discussions and collection trips with him. Prof. Rao's criticisms were inevitably friendly and constructive. He seems to have given his students the impression that he was learning through them. He succeeded in nurturing a "burning yearning for publications" in many individuals, including myself. It is indeed a pleasure and honour to contribute this profile and I wish him many more years of purposeful life and bliss.

As the convener of Prof. Rao's festschrift, I had the privilege of inviting a number of eminent agricultural scientists, with whom Prof. Rao is closely associated, to be in the Advisory committee and it's gratifying to mention that almost all readily agreed without any reservations. The list of contributors amply reflects the standard of this festschrift as most of them are well reputed scientists in respective fields of research. Moreover, one of the editors (S.S.) joined our team at the Centre much later and could bring out this prestigious volume along with the co-editor (R.J.) with whom Prof. Rao has close association. Despite their commitments, the editors managed to bring out this volume successfully.

I am (N.M.) going to take the liberty of making a few personal comments. I first met Prof. Rao at Dr. Peter R. Day's (former Head of Genetics Dept., Connecticut Agricultural Experiment Station, New Haven) residence as the later hosted a dinner in honour of Dr. Rao joining his laboratory as an UNESCO post doctoral research fellow. Since we worked on different aspects of rice blast in two different laboratories our friendship developed from day one, to be precise from October 6, 1971. We used to meet at weekends and my two children were very much attached to Dr. Rao. After a gap of 10 years I met Prof. Rao for the 2nd time, this time at his own place in Madras as I visited India. This led to Dr. Zuber, one of Prof. Rao's students joining our department for post-doctoral work. Later, Prof. Rao attended the 5th International Congress of Phytopathology (ICPP) held at Kyoto in August 1988 and after the congress he came as a visiting Professor to my laboratory at Kyushu University, Fukuoka. He stayed with us mostly and acquainted himself with the Japanese way of living too, sleeping on a traditional mat and Japanese way of eating with chop- sticks in addition to academic work. Later, in 1995, he visited us with his spouse under the prestigious INSA (India) - JSPS (Japan) Senior Professorship and we had a very fruitful time at Fukuoka, twenty-six years after our first meeting at Connecticut, USA. It is now my privilege to recognise him as one of the authorities on major fungal diseases of rice as his work is very well known to rice researchers all over the world. It gives me great pleasure to contribute this profile by wishing Professor and Mrs. Rao many more happy years of successful life.

N. Matsuyama
S. Manian.

Professor K. Manibhushan Rao - Publications

(a) Full research papers

1964 Environmental conditions and false-smut incidence in rice. Indian Phytopath. **22** : 110-114.

1971 Studies on the cultures of *Pyricularia oryzae* Cav.-I. Morphology and pathogenicity. Riso **20** : 283-297.

1971 Studies on the cultures of *Pyricularia oryzae* Cav.-II. Utilization of nitrogen and carbon compounds. Phytopath. Z. **72** : 173-183.

1971 Studies on the cultures of *Pyricularia oryzae* Cav.-III. *In vitro* production of pectolytic and cellulolytic enzymes. Phytopath. Z. **70** : 203-212.

1971 (with R. Kalyanasundaram and S. Suryanarayanan) Antigenic comparison of cultures of *Pyricularia oryzae* Cav. Phytopath. Z. **70** : 121-126.

1972 (with P.R. Day) Low night temperature and blast disease development on rice. Phytopathology **62** : 1005-1007.

1972 (with A. Narayanarao and S. Suryanarayanan) Sporulation of *Pyricularia* spp. in culture. Effect of some aromatic compounds. Proc. Indian Acad. Sci. **45B** : 105-110.

1972 (with A. Narayanarao and S. Suryanarayanan) Effect of some phenolic compounds on sporulation and growth of *Pyricularia* spp. Proc. Indian Acad. Sci. **45B** : 140-144.

1972 (with A. Narayanarao and S. Suryanarayanan) Sporulation of *Pyricularia* spp. on detached leaves of Graminae and in media supplemented with leaf extracts. Proc. Indian Acad. Sci. **45B** : 145-151.

1976 (with S.H. Ou) Multinucleate state following conidial germination in *Pyricularia oryzae* Cav. Philippine Phytopathology **12** : 4-6.

1978 (with C.V. Subramanian and C. Rajendran) Development of race ID-1 of *Pyricularia oryzae* Cav. Phytopath. Z. **91**: 340-435.

1979 (with S.H. Ou) Genetic recombination in the blast pathogen *Pyricularia oryzae* Cav. Philippine Phytopathology **15**: 122-126.

1980 (with S. Manian) The efficacy of the scale by the normal distribution method in screening rice cultivars for sheath blight disease. Risò **29** : 313-316.

1980 (with S.H. Ou) Karyological studies in *Pyricularia oryzae* Cav. Riso **29** : 305-308.

1982 (with S. Manian) Histopathological studies in rice sheath blight disease incited by *Rhizoctonia solani* Kühn. Z. Pflkrankh. Pflschutz. **89** : 523-521.

1983 (with M. Zuber) Physiology of host-pathogen interaction in rice-sheath blight disease with reference to changes in carbohydrate and nitrogen contents. Ind. J. Bot. **6** : 117-124.

1984 (with M.L. Jesudhas) Conidium ontogeny in *Pyricularia*. Phytopath. Z. **110** : 235-244.

1984 (with M. Zuber) Changes in phenolics and enzymes of phenol metabolism in sheath blight disease of rice caused by *Rhizoctonia solani*. Acta Phytopath. Hung. **19** : 251-262.

1986 (with S. Sreenivasaprasad and V. Parvathi) Changes in phenolic constituents of rice in response to *Rhizoctonia solani* infection. Ind. J. Bot. **9** : 104-111.

1989 (with U.I. Baby and Y. Joe, 1989) Effect of organic amendments on the saprophytic survival of rice sheath blight pathogen and the soil microflora. Oryza **26** : 70-78.

1990 (with Y. Joe and P. Madathiammal) Elicitation of resistance in rice to sheath blight disease. Int. J. Trop. Pl. Dis. **8** : 193-197.

1990 (with S. Sreenivasaprasad) Biological control of plant pathogens. Potential of antagonistic fungi. Encology **4** : 14-18.

1990 (with S. Sreenivasaprasad) Antagonistic potential of *Gliocladium virens* and *Trichoderma longibrachiatum* to phytopathogenic fungi. Mycopathologia **109** : 19-26.

1990 (with S. Sreenivasaprasad) Biocontrol potential of fungal antagonists *Gliocladium virens* and *Trichoderma longibrachiatum*. Z. Pflkrankh. Pflschutz. **97** : 570-579.

1991 (with S. Kumaresan) Studies on the biological control of rice sheath blight disease. Indian J. Pl. Pathol. **9** : 64-70.

1992 (with U.I. Baby) Vesicular arbuscular mycorrhizal fungi of rice field soils. Acta Botanica Indica **20** : 10-15.

1992 (with P. Krishnan, T.T. Muralidhar and B.C. Reddy) Epidemiological studies of rice blast disease-I. Spore releasing capacity of various types of blast lesions and spore release pattern against weather parameters. Intl. J. Top. Pl. Dis. **10** : 153-165.

1993 (with U.I. Baby) Control of rice sheath blight through the integration of fungal antagonists and organic amendments. Trop. Agric. (Trinidad) **70** : 240-244.

1993 (with T. Mani) Application of glucosamine assay as a technique to detect fungal mass in rice - Sheath blight system. Z. Pflkrankh. Pflschutz. **100** : 285-290.

1993 (with P. Ramalingam) Studies on the phytotoxic metabolites of *Rhizoctonia solani*. 1. Toxin production, biosynthesis and phytotoxicity. Ind. J. Microbial Ecol. **3** : 47-61.

1993 (with G. Viji and U.I. Baby) Induction of fungicidal resistance in *Tricoderma* spp. through UV irradiation. Indian J.Microbiology **33** : 125-129.

1993 (with S. Sreenivasaprasad) Efficacy of *Gliocladium virens* and *Trichoderma longibrachiatum* biocontrol agents of ground-nut, root-rot and stem-rot diseases. Intl. J. Pest Management **29** : 167-171.

1995 (with Y. Joe) Biocontrol constituents of differentially virulent isolates of *Sarocladium* spp. causing sheath rot disease of rice. Z. Pflkrankh. Pflschutz. **102** : 291-198.

1995 (with S. Manian) Changes in the phenol metabolism of rice cultivars with reference to nitrogen fertilization and sheath blight disease. Acta Phytopath. et. Entamol. Hung. **30** : 111-203.

1995 (with D.M. Chelladorai and S. Sreenivasaprasad) Effect of phenylacetic acid on the phenolics and enzymes of phenol metabolism in rice. Acta Phytopath. et. Entamol. Hung. **30** : 205-213.

1996 (with U.I. Baby) Influence of organic soil amendments on vesicular arbuscular mycorrhizal fungi in relation to rice sheath blight disease. Mycorrhizae **6** : 201-206.

1997 (with G. Viji and U.I. Baby) Factors affecting the parasitic activity of *Gliocladium virens* and *Trichoderma longibrachiatum* and *T. harzianum* on *Rhizoctonia solani*. Indian Phytopath. **50** : 491-495.

(b) Short research papers

1977 (with P. Ramalingam and S. Suryanarayanan) Isolation of diauxotrophic mutants of *Pyricularia oryzae* Cav. with N-methyl-N'-nitro-nitrosoguanidine (MNNG). Microbial Genet. Bull. **42** : 1.

1979 (with S. Manian) Resistance to sheath blight disease in India. Int. Rice Res. Newsl. **4** : 5-6.

1979 (with M. Zuber) Phytoalexin from the germinating seeds of rice (*Oryza sativa*). Curr. Sci. **48**: 497-498.

1981 (with S. Manian and M. Zuber) The phytotoxic metabolites of *Rhizoctonia solani*. J. Sci. & Indstr. Res. **40** : 502-607.

1982 (with K. Kalaiselvi and S. Sreenivasaprasad) Acquired resistance of rice leaves to *Rhizoctonia solani*. Int. Rice. Newsl. **11** : 16.

1982 (with S. Manian) Screening rice germplasm for tolerance to sheath blight incited by *Rhizoctonia solani* Kühn. Phytopath. Medit. **21** : 37-38.

1987 (with M. Zuber and S. Sreenivasaprasad) Mycelial amino acid composition of differently virulent isolates of *Rhizoctonia solani*. Curr. Sci. **56** : 619-621.

1987 (with A. Waheeta and S. Sreenivasaprasad) Induced resistance in rice to sheath blight disease. Curr. Sci. **56** : 486-489.

1988 (with S. Sreenivasaprasad, R.S. Chitralekha and K. Kalaiselvi) Cross-protection in rice sheath blight disease. Ind. Bot. Soc. **67** : 97-100.

1989 (with S. Sreenivasaprasad, U.I. Baby and Y. Joe) Susceptibility of rice sheath blight pathogen to mycoparasites. - Curr. Sci., **58** : 515-518.

1990 (with S. Manian) Influence of some factors on the survival of *Rhizoctonia solani* in soil. Trop. Agric. **67** : 207-208.

1991 (with U.I. Baby) Biological control of rice sheath blight with *Gliocladium virens* and *Trichoderma longibrachiatum*. Petria **1** : 151-153.

1991 (with S. Sreenivasaprasad) Potential of *Glicocladium virens* and *Trichoderma longibrachiatum* as biocontrol agents of fungal pathogens. Petria **1** : 145.

1997 (with G. Viji and U.I. Baby) Non-target effect of systemic fugicides against *Rhizoctonia solani*. Indian Phytopath. **50** : 324-328.

(c) Scientific reviews

1978 (with M. Zuber) Disease resistance in cereals. Acata phytopath. Hung. **13** : 313-335.

1979 (with S. Manian and M. Zuber) Sheath blight disease of rice in South East Asia. Z. .Pflkrankh. Pflschutz. **86** : 499-509.

1980 Rice diseases-I. Fungal diseases of rice. Biology **2** : 14-19.

1980 Rice diseases-II. Bacterial diseases of rice. Biology **2** : 20-22.

1980 Rice diseases-III. Virus diseases of rice. Biology **2** : 23-25.

1986 (with S. Sreenivasaprasad and R.S. Chitralekha) Sheath rot disease or rice. Z. Pflkrankh. Pflschutz. **93** : 319-329.

1988 (with M. Zuber and N. Matsuyama) Phenol metabolism and plant disease resistance. Acta Phytopath. et. Entamol. Hung. **23** : 103-114.

1988 (with R. Sridhar and A.K. Sinha) Innovative approaches to management of some major disease of rice. J. Sci. & Indstr. Res. **48** : 181-185.

1989 Sheath blight disease threatening rice cultivation in South East Asia. J. Ind. Bot. Soc. **68** : 10-18.

(d) Articles published in symposia proceedings

1986 (with S. Sreenivasaprasad) The principal diseases of rice in India. In : 14th Annual Meeting .Mycol. Soc. India. Thanjavur, Tamil Nadu, India. p.17-22.

1990 (with P. Krishnan, K.N. Ponnuswamy, S. Natarajan and B. Chenchu Reddy 1990) Development of computerised simulation model (EPIBA) for rice leaf bast epidemic. In : *Intl. Symp. Rice Res. & New Frontiers*, DRR, Hyderabad, p.219-220.

1993 (with U.I. Baby) Control of rice sheath blight disease by organic manuring. In : *Proc. 5th Kerala Sci. Cong.* held at Kottayam, India, p.158-161.

1997 (with U.I. Baby) Integrated disease management of sheath blight in rice. In : *Proc. Intl. Symp. on Integrated disease management for sustainable agriculture* held at Golden Jubilee Celebrations of Ind. Phytopathol. Soc., IARI, New Delhi , p.191-194.

2000 Integrated management of rice – seed borne fungal pathogens with special reference to sheath blight disease. In : *Proc. National Seminar on Seed Science and Technology.* (eds. H. Shekar Shetty and H.S. Prakash) Department of Applied Botany and Seed Pathology, University of Mysore, Mysore, p.125-129.

2000 Simulation models in epidemiology of rice blast disease. In : *Proc. Intl. Symp. on Rainfed rice for sustainable food security* held at the Golden Jubilee Celebration of CRRI, Cuttack, India. (In press).

(e) Chapters contributed to books

1977 Low temperature on the susceptibility of International rice differentials to races of *Pyricularia oryzae.* In : *Physiology of host-pathogen interaction.* (ed. A.Mahadevan). Today and Tomorrow's Printers and Publishers, New Delhi, p.301-312.

1977 (with M. Zuber) *Pyricularia* sp. on ragi seeds. In : *Physiology of host-pathogen interaction* (ed. A.Mahadevan). Today and Tomorrow's Printers and Publishers, New Delhi. p.473-478.

1989 (with U.I. Baby and Y. Joe) Basic research on rice sheath blight disease management. In : *Basic research for crop disease management.* (ed. P. Vidhyasekaran). Daya Publishing House. New Delhi, p.376-381.

1991 (with P. Krishnan) Epidemiology of blast (EPIBLA), a simulation model and forecasting system for tropical rice in India. In : *Rice blast modelling and forecasting.* Int. Rice Res. Inst., (IRRI) P.O.Box. 933, Manila, Philippines, p.31-38.

1991 (with S. Sreenivasaprasad and Y. Joe) Host-sensitization as an approach to plant disease control. In : *Genetic engineering and molecular biology and tissue culture* (ed. P. Vidhyasekaran). Daya Publishing House, Delhi, p.187-207.

1989 (with Y. Joe and P. Krishnan) Rice blast epidemiology and application of its techniques in disease management. (eds. S.P. Raychoudhuri and J.P.Verma). *Review of Tropical Plant Pathology* **6** : 119-149.

1996 (with U.I. Baby) Fungal antagonists and VA mycorrhizal fungi for biocontrol of *Rhizoctonia solani,* the rice sheath blight pathogen, In : *Recent developments in biocontrol of plant of pathogens.* (eds. K. Manibhushanarao and A. Mahadevan). Today and Tomorrow's Printers and Publishers, New Delhi, p.1-9.

1996 (with K. Natarajan) Fungi as biocontrol agents against fungal plant pathogens. In : *Recent developments in biocontrol of plant pathogens* (eds. K. Manibhushanrao and A.Mahadevan). Today and Tomorrow's Printers and Publishers, New Delhi, p.87-95.

1998 (With P. Raghuveer Rao, M. Venkaiah and J. Saraswathi) Biopesticides. In : *Plan Products* (eds. K.V. Madhava Rao and U.Sanjeevi) Telugu Academy, Hyderabad. Ecomax, Offset Printers, Hyderabad, p. 260.

1999 (with S. Natarajan *et al.*) The power of climate components on the flare-up diseases and pests in rice. In : *Rice ina variable climate* (eds. Y.P.Abral and Sulochana Gadgil). APC Publications Pvt. Ltd., Box 2679, New Delhi, p.167-195.

2000 (With U.I. Baby) Innovative approaches in rice sheath blight disease management. In : *Biocontrol potential and its exploitation in sustainable agriculture* vol.1 (eds. Upadhayay et al.). Kluwer Academy / Plenum Publishers, N.Y. p.279-286.

2001 (With U.I. Baby) Major fungal diseases of rice. In : *Plant Pathology in India* (50 years 1994-1997) Festscherift volume of Professor S.B.Chattopathyay's 75th birth Anniversary. Dept. of Plant Pathology, Bidhan Chandra Krishi Viswavidyalaya, Kalyani, Nadia, West Bengal (In press).

2001 Sheath rot disease of rice. In : *Fifty years of Plant Pathology in India*. In : Indian Phytopathological Society, Golden Jubilee Celebration, IARI, New Delhi (In Press).

2001 (With T.R.K.Reddy and J.Saraswathi) Mycorrhizae as biofertilizers. In : *Biofertilizers and Manures* (eds. K.V. Madhava Rao and U.Sanjeevi). Telugu Academy, Hyderabad. Ecomax Offset Printers, Hyderabad (In press).

(f) Books / Monographs - authored or edited

1991 (Reprinted 1995 & 2000). *Textbook of Horticulture*. Macmillan Ltd., New Delhi, India, 342 pp.

1994 *Rice blast disease*. Daya Publishing House, Delhi, 180 pp.

1995 *Sheath blight disease of rice*. Daya Publishing House, Delhi, 101 pp.

1996 *Sheath rot disease of rice*. Daya Publishing House, Delhi, 71 pp.

1996 (with A. Mahadevan) *Recent developments in biocontrol of plant pathogens*. Today and Tomorrow's Printers and Publishers, New Delhi, 160 pp.

(g) Other Publications (Scientific interest)

1979 National Symposium on blast and sheath blight disease of rice in India. Int. Rice Res. Newsl. **4** : 16.

1979 Fungal disease of Paddy. The Hindu **102(362)** : 28.

1986 Orchids. Biology Education **3** : 5-10.

1986 Plant Taxonomy. Biology Education **3** : 34-38.

1988 (with S. Sreenivasaprasad and R.S. Chitralekha) Cross-protection in rice against sheath blight disease. The Hindu **111(5)** : 28.

1988 (with Y. Joe and U.I. Baby) Soil amendments and control of rice sheath blight. The Hindu **111(294)** : 24.

1989 (with K. Murugasan, P.T. Kalaichelvan and N. Raman) First National Seminar on molecular Plant Pathology. Curr. Sci. **58** : 537-538.

1989 (with Y. Joe) Sheath rot-A serious rice disease. Kisan World, April 1989, p. 27.

1993 (with Y. Joe) Controlling sheath-rot of rice. The Hindu **166(249)** : 24.

2001 Orchids as botanical curiosities. In : *Encyclopaedia of Cultivated Plants* (eds. M.S. Swaminathan and S.L. Kochhar (In Press).

EDITORS' NOTE

We are pleased to bring out this volume and would like to warmly thank the members of Professor K. Manibhushan Rao Festschrift Advisory and Organising Committees for their help and guidance.

Advisory Committee
Prof. T.S. Sadasivan, Chennai, India
Prof. M.S. Swaminathan, Chennai, India
Prof. A. Appa Rao, Hyderabad, India
Prof. E. A. Siddiq, Hyderabad, India
Dr. P.R. Day, New Jersey, USA
Prof. N. Matsuyama, Fukuoka, Japan
Dr. C.K. Kim, Suweon, S. Korea
Dr. P.S. Teng, Manila, Philippines.

Organising Committee
Prof. S. Manian, Coimbatore, India
Prof. Louis Jesudhos, Coimbatore, India.

Prof. Manian in particular shared a lot of the effort involved in initiating and organising the process and we very much appreciate his efforts. The papers in this volume cover most of the major fungal diseases of rice, in the order of their importance in general and bring out many of the recent advances made in better understanding the pathogens as well as the host. We are grateful to all our contributors for their efforts and patience and would like to thank them all, on behalf of the Festschrift Committees. Our special thanks to Ms. Marilyn Nicholl, Horticulture Research International, U.K. for preparing the camera ready manuscripts and the editorial and production staff at Kluwer, the Netherlands for seeing the volume through to publication and the excellent finish. We join the members of the Festschrift Advisory and Organising Committees and Prof. Rao's other friends and colleagues to wish Professor and Mrs. Manibhushan Rao many more years of happy life.

S. Sreenivasaprasad
R. Johnson.

races of the blast fungus, classified using the international set of differentials, was established. Subsequently, a total of 31 races of the fungus were identified (Padmanabhan, 1965b; Padmanabhan *et al.*, 1970). In contrast, existence of a few relatively stable races of the fungus has also been reported from India (Veeraraghavan, 1975a) and China (Luo *et al.*, 1979; Sun *et al.*, 1981). Although the rice-blast pathosystem is believed to follow a gene-for-gene relationship (Silue *et al.*, 1992; Valent and Chumley, 1994), virulence analysis has led to conflicting opinions on the degree of variability of *P. grisea* (Veeraraghavan, 1975a, 1975b; Veeraraghavan and Dath, 1976; Ou, 1985; Latterell and Rossi, 1986; Bonnman *et al.*, 1987). Biochemical studies describing isozyme polymorphisms (Leung and Williams, 1986) and double-stranded RNA content (Hunst *et al.*, 1986) detected some variation, but meaningful correlation with pathotype could not be established.

3. Molecular analysis of pathogen populations

In the 1980's, a range of molecular techniques became available for population genetics, offering new tools to describe and understand the genetic variation and evolution. Data on DNA polymorphisms can now be used to describe the genetic variation and its partitioning, thus expanding our understanding of the evolution of the pathogen populations. Restriction fragment length polymorphism (RFLP) analysis has been used to study genetic relationships of bacterial and fungal genera, species, and pathovars/races.

Hamer *et al.* (1989) identified a family of dispersed repetitive DNA sequences, called MGR (named after the perfect stage of the blast fungus, *Magnaporthe grisea*), that was conserved in *P. grisea* genomes. Several features of MGR sequences suggested that they could be useful for population level genetic analysis. MGR sequences occurred in approximately 50 copies per haploid genome and were dispersed among all chromosomes. DNA gel blot analysis using an MGR sequence (MGR586, previously known as pCB 586) as hybridisation probe yielded mototically stable isolate specific *Eco*RI RFLP profiles, containing 60 - 80 resolvable fragments of 0.7 to 20.0 kb in size with all *P. grisea* isolates that infect rice, but only one or a few fragments in *P. grisea* isolates limited to other grass hosts. World wide conservation of MGR586 sequences in rice blast pathogens indicates that they descend from a common ancestral source, genetically isolated from other host-limited forms of *P. grisea.* Consequently, MGR-RFLP profiles could serve as genotype-specific DNA fragments (in the general sense of Jeffreys *et al.*, 1985) that would also identify the distribution of clonal lineages (statistically similar group of individuals inferred to be related via common ancestry) within and among pathotypes of the rice blast fungus. DNA fingerprinting has been widely used for studying the population structure of *P. grisea* occurring in Philippines (Borromeo *et al.*, 1993; Chen *et al.*,1995; Zeigler *et al.*,1995), Korea (Han *et al.*, 1993, 1995), USA (Levy *et al.*, 1991a), Columbia (Levy *et al* 1991b, 1993), China (Zhu *et al.*, 1993), Thailand (Mekwatanakarn *et al.*, 1996), Vietnam (Nghia *et al.*, 1996) and in India (Gnanamanickam *et al.*, 1993; Sivaraj *et al.*, 1995, 1996; Sridhar *et al.*, 1996).

MGR-DNA fingerprinting was first used in the United States using a sample consisting of 79 isolates of *P. grisea* collected from southern United States belt over a period of 30 years (Levy *et al.*,1991a). This sample consisting of eight most frequently observed pahotypes were partitioned into eight discretely distinct lineages. Each lineage was associated with only one or two pathotypes, most of which were lineage specific. However, the relationship between lineage and pathotype was soon found to be quite complex (Levy *et al.*, 1991b), as an analysis of isolates from a rice germplasm evaluation site in Columbia showed that 50 pathotypes were grouped into seven lineages. Each lineage was associated with a specific group of pathotypes. However, an understanding of the pathotypic evolution by detailed analysis of lineage and pathotype slowly started to emerge. Apparently, in the rice blast fungus, pathotypic evolution proceeds rapidly relative to the molecular markers used to define genetic lineages of the pathogen (Leung *et al.*, 1993). Subsequently, based on MGR586-RFLP profile, Levy *et al.* (1993) differentiated six distinct genetic lineages from among 151 isolates of the blast fungus collected from 15 rice cultivars in Columbia while, virulence analysis on international differentials revealed the presence of 39 pathotypes among these isolates. However, each lineage was associated with a specific subset of cultivars and expressed a generally non-overlapping subset of pathotypes. This led to the conclusion that MGR-DNA fingerprint analysis may provide the genealogical framework for evaluating pathogen variation that is essential for understanding disease epidemiology and pathotype evolution, useful for selecting germplasm sources for breeding more durable blast resistance.

Analysing the population structure of the rice blast fungus occurring at two field sites used for evaluating blast resistance during two seasons involving over 1500 isolates collected from 38 cultivars in the Philippines. Chen *et al.* (1995) observed that the populations were highly differentiated both geographically and temporally based on MGR-DNA fingerprinting with *Eco*RI restriction endonuclease.

Using MGR586 fingerprints, 36 lineages of the blast fungus occurring in southern India were so far identified by the Madras group of workers (Gnanamanickam *et al.*, 1993; Sivaraj *et al.*, 1995, 1996). At the Central Rice Research Institute the DNA of 85 isolates blast fungus was fingerprinted employing RFLP technique using MGR586 as a hybridisation probe. The population of Jagadalpur in Madhya Pradesh (a hot spot for blast) was very diverse as it consisted of six lineages of the seven detected by us (Sridhar *et al.*, 1996). Two lineages were detected among the isolates (small number examined so far) from Sikkim, which were also detected in Madhya Pradesh and in Orissa.

4. Host specificity

Many weed species occurring in and around rice fields have been reported to be hosts of *P. grisea* (Mackill and Bonnman, 1986). Whether or not these weeds serve as source of rice blast epidemics has long been an issue for debate. Genetic and molecular analyses of the blast fungus have revealed the genetic differences between blast fungus isolates from different hosts. DNA hybridisation analysis has shown that genomes of rice infecting isolates contain many copies of repetitive DNA elements (MGR sequences) that are absent or present only in low copy number in non-rice infecting isolates (Hamer *et al.*, 1989; Hamer, 1991). Further intensive study in the Philippines employing restriction analysis of mitochondrial DNA with five different endonucleases, and DNA blot hybridisation using ribosomal, single copy, and repetitive DNA probes involving *P. grisea* isolates from 17 host species including rice (Borromeo *et al.*, 1993) revealed that isolates from rice and weeds obtained from the same locations in the Philippines are genetically differentiated on the basis of MGR sequence conservation. Dissimilarity between the MGR 586 hybridisation profiles of blast fungus isolates from rice and wheat demonstrated in Brazil illustrates the genetic differentiation between *P. grisea* isolates infecting different hosts (Valent, 1990). Despite this evidence, the occurrence of *P. grisea* in diverse ecologies with different local flora does not completely rule out the possibility of some *P. grisea* strains cross infecting weeds and rice (Borromeo *et al.*, 1993).

Recent observations lend support to this speculation. Both dissimilar and similar banding patterns were observed in MGR fingerprints between rice and non-rice isolates from Korea (Han *et al.*, 1995) and cross infectivity tests of rice isolates and non-rice isolates revealed that rice isolates infect some of the grasses and *vice versa* (Han *et al.*, 1995; Kumar and Singh, 1995).

Dissimilarity between populations of blast fungus infecting different ecotypes of rice has also been reported. Pathogen populations infecting *japonica* and *indica* cultivars have been reported to differ (Han *et al.*, 1993; Correa-Victoria *et al.*, 1994; Zeigler *et al.*, 1994, 1995).

Likewise, conflicting reports exist about the similarity of neck-infecting and leaf-infecting strains of *P. grisea.* If the same strains cause both the disease syndromes, then the genes governing resistance may be the same. If not different genes might confer resistance to neck and leaf infections. In that case, the breeding strategies should be tailored suitably for incorporating the necessary resistance genes for developing varieties resistant to both leaf and neck blast infections.

5. Conclusion

Although fungicides are available for controlling blast of rice, it is quite expensive for Asian farmers. Growing resistant varieties is not only economical for minimising the losses caused by the diseases but it is also an environmentally friendly method. Blast resistance tends to be unreliable with resistance often failing or 'breakingdown' under field conditions (Kiyosawa, 1972). This has been attributed to extreme variability of virulence in the pathogen, inadequate resistance screening and evaluation methods, inappropriate use of resistance genes, or a combination of these (Zeigler *et al.*, 1994). Molecular marker technology offers opportunities to overcome these problems. It is essential to have a clear understanding of the structure of pathogen population prevailing in specific areas , which would lead to the identification of useful resistance genes that can be meaningfully deployed.

It has been difficult to achieve long-lasting resistance to blast, although the desirability of combining multiple genes conferring resistance in rice breeding has long been recognised (Ikehashi and Khush, 1979; Jenning, 1979). Molecular markers can improve the efficiency and resolution of genetic analysis, particularly when multiple resistance genes are present in a single cultivar (Paterson *et al.*, 1991). Thus, marker technology might facilitate future efforts for achieving durable resistance in rice against blast.

6. References

Atkins, J.G. (1962) Prevalence and distribution of pathogenic races of *Piricularia oryzae* in the US, *Phytopathology* **52**, 2 (abstract).

Atkins, J.G., Robert, A.L, Adair, C.R., Goto, K., Kozaka, T., Yanagida, R., Yamada, M. and Matsumoto, S. (1967) An international set of rice varieties for differentiating races of *Piricularia oryzae*, *Phytopathology* **57**, 297-301.

Barr, M.E. (1977) *Magnaporthe*, *Telimenella* and *Hyponectria* (Physosporellaceae), *Mycologia* **69**, 952-966.

Bonnman, J.M., Vergel de Dios, T.L., Bandong, J.M. and Lee, E.J. (1987) Pathogenic variability of monoconidial isolates of *Pyricularia oryzae* in Korea and in the Philippines, *Plant Disease* **71**, 127-130.

Borromeo, J.M., Vergel de Dios, T.L., Bandong, J.M. and Lee, E.J. (1993) Genetic differentiation among isolates of *Pyricularia* infecting rice and weed hosts, *Phytopathology* **83**, 393-399.

Chen, D., Zeigler, R.S., Leung, H. and Nelson, R.J. (1995) Population structure of *Pyricularia grisea* at two screening sites in the Philippines, *Phytopathology* **85**, 1011-1020.

Chien, C.C. (1967) Studies on the physiologic races of the rice blast fungus, Piricularia oryzae, Cav. *Bulletin of the Taiwan Agriculture Research Institute* **26**, 1-63.

Correa-Victoria, F.J., Zeigler, R.S. and Levy, M. (1994) Virulence characteristics of genetic families of *Pyricularia grisea* in Colombia, in R.S. Zeigler, S.A. Leong and P.S. Teng (eds.), *Rice Blast Disease*, CAB International, Wallingford, U.K., pp.211-230.

Goto, K. (1960) Progress report of the joint research on the rice blast fungus, *Piricularia oryzae* (Fascicle 1), *Special Bulletin of the Plant Disease and Insect Forecast Service* **No. 5**, 1-89.

Goto, K. (1965) Physiological races of *Piricularia oryzae* in Japan, in *The Rice Blast Disease*, John Hopkins Press, Baltimore, U.S.A., pp. 237-242.

Gnanamanickam, R., Sivaraj, R., Levy, M., Renganathan, N. and Hamer, J.E. (1993) Genetic organization of the rice blast fungus in India (Abstract), in *6th Annual Meeting, International Programme on Rice Biotechnology*, China National Rice Research Institute and Purdue University, west Lafayette, IN, U.S.A., p. 232.

Hamer, J.E. (1991) Molecular probes for rice blast disease, *Science* **252**, 632-633.

Hamer, J.E., Farrall, L., Orbach, M.J., Valent, B. and Chumley, F.G. (1989) Host species-specific conservation of a family of repeated DNA sequences in the genome of a fungal plant pathogen, *Proceedings of the National Academy of Sciences USA* **86**, 9981-9985.

Han, S.S., Ra, D.S. and Nelson, R.J. (1993) Comparison of RFLP-based phylogenetic trees and pathotypes of *Pyricularia oryzae* in Korea, *RDA Journal of Agricultural Science* **35**, 315-323.

Han, S.S., Ra, D.S. and Nelson, R.J. (1995)Relationship between DNA fingerprints and virulence of *Pyricularia grisea* from rice and other hosts in Korea, *International Rice Research Notes* **20**, 26-27.

Hunst, P.L., Latterell, F.M. and Rossi, A.E. (1986) Variation in double-stranded RNA from isolates of *Pyricularia oryzae, Phytopathology* **76**, 674-678.

IRRI (1975) *The International Rice Research Institute Annual Report for* **1975**, Philippines.

Ikehashi, I. And Khush, G.S. (1979) Breeding for blast resistance at IRRI, in *Proceedings of the Rice Blast Workshop*, International Rice Research Institute, Philippines, pp. 69-80.

Jeffreys, A.J., Wilson, V. and Thien, S.L. (1985) Individual specific 'fingerprints' of human DNA, *Nature* **316**, 76-79.

Jennings, P.R. (1979) Concluding remarks, in *Proceedings of the Rice Blast Workshop*, International Rice Research Institute, Philippines, pp 217-222.

Kiyosawa, S. (1972) Genetics of blast resistance, in *Rice Breeding*, International Rice Research Institute, Philippines, pp. 203-225.

Kumar, A. and Singh, R.A. (1995) Host range of some isolates of *Magnaporthe grisea* and their grouping into *forma-specialis, Indian Phytopathology* **48**, 210-212.

Latterell, F.M. and Rossi, A.E. (1986) Longevity and pathogenic stability of *Pyricularia oryzae*, *Phytopathology* **76**, 231-235.

Latterell, F.M., Tullis, E.C. and Collier, J.W. (1960) Physiologic races of *Pyricularia oryzae* Cav., *Plant Disease Reporter* **44**, 679-683.

Lee, S.C. and Matsumoto, S. (1966) Studies on the physiologic races of rice blast fungus in Korea during the period of 1962-1963, *Annals of the Phytopathological Society of Japan* **32**, 40-45.

Leung, H. and Williams, P. H. (1986) Enzyme polymorphism and genetic differentiation among geographic isolates of the rice blast fungus, *Phytopathology* **76**, 778-783.

Leung, H., Nelson, R.J. and Leach, J.E. (1993) Population structure of plant pathogenic fungi and bacteria, *Advances in Plant pathology* **10**, 157-205.

Levy, M., Romao, J., Marchetti, M.A. and Hamer, J.E. (1991b) DNA fingerprinting with a dispersed repeated sequence resolves pathotype diversity in the rice blast fungus, *The Plant Cell* **3**, 95-102.

Levy, M., Correa-Victoria, F.J., Zeigler, R.S., Xu, S. and Hamer, J.E. (1991b) Organization of genetic and pathotype variation in the rice blast fungus at a Colombian 'hot spot', *Phytopathology* **81**, 1236-1237.

Levy, M., Correa-Victoria, F.J., Zeigler, R.S., Xu, S. and Hamer, J.E. (1993) Genetic diversity of the rice blast fungus in a disease nursery in Columbia, *Phytopathology* **83**, 1427-1433.

Luo, K., Huang, S.Y., Xiao, Q.M. and Wang, B.X. (1979) Preliminary assay on virulence of rice blast fungus in Hunan province, *Acta Phytopathologica Sinica* **9**, 83-92.

Mackill, A.O. and Bonnman, J.M. (1986) New hosts of *Pyricularia oryzae*, *Plant Disease* **70**, 125-127.

Matsumoto, S., Kozaka, T. and Yamada, M. (1969) Pathogenic races of *Pyricularia oryzae* Cav. in Asia and some other countries, *Bulletin of National Institute of Agricultural Sciences* (Tokyo) **C23**, 1-36.

McRae, W. (1920) *Administrative report of Mycologist, Madras*, **1919-1920**.

Mekwatanakarn, P., Kositratana, W., Zeigler, R.S. and Levy, M. (1996)Characterisation of the *Pyricularia grisea* population by lineage using MGR-DNA fingerprinting in Thailand, in *Abstracts of the Third Asia-Pacific Conference on Agricultural Biotechnology: Issues and Choices*, Nov. 10-15, 1996, Prachuapkhirikhan, Thailand, p.66.

Nghia, L.T., Ha, N.T., Chinh, D.Q., Quang, V.D. and Quy, T.D. (1996) Preliminary study on molecular analysis of rice blast fungus population in Vietnam, in *Abstracts of the Third Asia-Pacific Conference on Agricultural Biotechnology: Issues and Choices*, Nov. 10-15, 1996, Prachuapkhirikhan, Thailand, p. 108.

Ou, S.H. (1980) A look at world-wide rice blast disease control, *Plant Disease* **64**, 439-445.

Ou, S.H. (1985) *Rice Diseases*, 2nd ed., Commonwealth Mycological Institute, CABI, Surrey, England.

Padmanabhan, S.Y. (1965a) Breeding for blast resistance in India, , in *Rice Blast Disease*, John Hopkins Press, Baltimore, U.S.A., pp. 343-359.

Padmanabhan, S.Y. (1965b) Physiologic specialization of *Piricularia oryzae* Cav., the causal organism of blast disease of rice, *Current Science* **34**, 307-308.

Padmanabhan, S.Y., Chakrabarti, N.K., Mathur, S.C. and Veeraraghavan, J. (1970) Identification of pathogenic races of *Pyricularia oryzae* in India, *Phytopathology* **60**, 1574-1577.

Paterson, A.H., Tanksley, S.D. and Sorrells, M.E. (1991) DNA markers in plant improvement, *Advances in Agronomy* **46**, 39-90.

Rossman, A.Y., Howard, R. J. and Valent, B. (1990) *Pyricularia grisea*, the correct name of the rice blast disease fungus, *Mycologia* **82**, 509-512.

Sasaki, R. (1922) Existence of strains in rice blast fungus, *International Journal of Plant Protection* (Tokyo) **9**, 631-644.

Silue, D., Nottegham, J.L. and Tharreau, D. (1992) Evidence of a gene-for-gene relationship in the *Oryza sativa-Magnaporthe grisea* pathosystem, *Phytopathology* **82**, 577-580.

Sivaraj, R., Gnanamanickam, S.S. and Levy, M. (1995) Genetic and pathotype diversity in Pyricularia grisea populations in India (abstract), , in *Proceedings of the 4th Annual Meeting, National Rice Biotechnology Network*, July 4-6, 1995, National Chemical Laboratory, Pune and Directorate of Rice Research, Hyderabad, India, pp. 50-52.

Sivaraj, R., Krishnamurthi, K., Viji, G., Dayakar, B. V., Gnanamanickam, S.S. and Levy, M. (1996) Approaches for the management of rice blast: Biological control with bacterial antagonists and pyramiding of blast resistance genes, in *Proceedings of the 5th Annual Meeting, National Rice Biotechnology Network*, Nov. 13-16, 1996, Indian Agricultural Research Institute, New Delhi, India, pp. 101-104.

Sridhar, R., Sahnti, M.L., Singh, U.D., Variar, M., Maiti, D., Reddy, J.N., Sinha, P.K., George, M.L.C., Bernardo, M.A., Baraoidan, M.R. and Nelson, R.J. (1996) Molecular characterization of bacterial blight and blast pathogen populations of India and breeding for cultivars with bacterial blight and blast resistance genes, in *Proceedings of the 5th Annual Meeting, National Rice Biotechnology Network*, Nov. 13-16, 1996, Indian Agricultural Research Institute, New Delhi, India, pp. 95-97.

Sun, S., Jin, M., Zhang, S., Tao, X. and Fang, D. (1981) Rice blast disease and its control, *Shangai Science Technical Publication* p.182.

Sundararaman, S. (1927) Administrative report of the Government Mycologist, Coimbatore for 1926-1927, *Department report, Madras Presidency for the official year 1926-1927*, pp. 326-344.

Thomas, K.M. (1930) Some aspects of the control of blast disease of paddy, *Madras Agriculture Journal* **18**, 596-604.

Valent, B. (1990) Rice blast as a model system for plant pathology, *Phytopathology* **80**, 33-36.

Valent, B. and Chumley, F.G. (1994) Avirulence genes and mechanisms of genetic instability in the rice blast fungus, in R.S.Zeigler, S.A.Leong, and P.S.Teng (eds.), *Rice Blast Disease*, CABI, Wallingford, U.K., pp. 111-134.

Veeraraghavan, J. (1975a) Stability in *Pyricularia oryzae* Cav., *Proceedings of the Indian National Academy of Sciences* **82**, 127-131.

Veeraraghavan, J. (1975b) A new method of classification and nomenclature of physiological races of *Pyricularia oryzae* Cav., *International Rice Communications Newsletter* **24**, 128-138.

Veeraraghavan, J. and Dath, A.P. (1976) Host pathogen equilibrium between rice (*Oryza sativa*) and its pathogen *Pyricularia oryzae* Cav., *Current Science* **45**, 333-334.

Zeigler, R.S., Tohme, J., Nelson, R., Levy, M. and Correa-Victoria, F. (1994) Lineage exclusion: A proposal linking blast population analysis to resistance breeding, in R.S.Zeigler, S.A.Leong, and P.S.Teng (eds.), *Rice Blast Disease*, CABI, Wallingford, U.K., pp. 267-292.

Zeigler, R.S., Cuoc, L.X., Scott, R.P., Bernardo, M.A. Chen, D.H., Valent, B. and Nelson, R.J. (1995) The relationship between lineage and virulence in *Pyricularia grisea* in the Philippines, *Phytopathology* **85**, 443-451.

Zhu, P., Shen, Y., Yuan, X., Levy, M., Manry, J. and Hamer, J.E. (1993) Genetic organization of rice blast fungus in China (abstract), in *Proceedings of the 6th Annual Meeting of the International Rice Biotechnology Programme*, China National Rice Research Institute and Purdue University, Lafayette, IN, U.S.A., p. 111.

CHARACTERISATION OF THE BLAST PATHOGEN POPULATIONS AT RICE SCREENING SITES IN WEST AFRICA

J. CHIPILI [1], S. SREENIVASAPRASAD*[1], A. E. BROWN[1], N.J. TALBOT[2], M. HOLDERNESS[3], Y. SERE[4], S.K. NUTSUGAH[5], J. TWUMASI[6] and K. DARTEY[6]
[1]*Horticulture Research International, Wellesbourne, UK.*
[2]*University of Exeter, Exeter, UK*
[3] *CABI Bioscience, International Mycological Institute' UK*
[4]*West Africa Rice Development Association, Bouaké, Côte d'Ivoire*
[5]*Savanna Agricultural Research Institute, Ghana*
[6]*Crops Research Institute, Ghana.*

1. Introduction

In West Africa, the demand for rice is growing faster than any other major staple food. Rice constitutes a major source of calories for the rural and urban poor and is grown in approximately 4.3 million ha in 17 West African countries – member states of West Africa Rice Development Association (WARDA). Annual production in the region is approximately 7.4 million tonnes of paddy. FAO estimates for current annual rice imports in West Africa is 4 million tonnes, equivalent to more than one billion US$. The average yield is 1.7 tonnes per ha, lowest compared to the rest of the world and a wide range of biophysical constraints reduces the yield potential of the varieties (WARDA Annual Report, 1998; WARDA MTP, 200-2002).

Blast is a primary constraint affecting rice production in the region and is widespread and most serious in all rice growing ecologies (Notteghem and Baudin, 1981; (Fomba and Taylor, 1994; Singh *et al.*, 1998), causing 3.2 - 77% yield losses. Deployment of resistant varieties integrating good cultural practices is the most effective and economical way to combat the blast disease. However, breakdown of resistance is common due to the dynamic nature of the pathogen in responding to the host genotype and environment. Understanding the diversity and dynamics of the pathogen populations and identification of resistance sources based on this knowledge is critical to the development of blast resistance that is stable over space and durable over time.

Breeding for stable blast resistance has been extremely difficult and there have been conflicting opinions on the diversity and stability of blast pathotypes (Zeigler *et al.*, 1995). Hamer *et al.* (1989) discovered repetitive DNA sequences in the *M. grisea* genome and one such element, the MGR586 has been widely used to assess the genetic variability in *M. grisea* in various countries (Levy *et al.*, 1991, 1993; Zeigler *et al.*,

S. Sreenivasaprasad and R. Johnson (eds.), Major Fungal Diseases of Rice Recent Advances, 9-20.
Prof. K. Manibhushan Rao Festschrift

1995; Roumen *et al.*, 1997). MGR586 is a dispersed repetitive DNA element found in 30-50 copies in rice pathogens, but only one to a few copies exist in other host-limited forms of *M. grisea*. In general, sub-populations of the pathogen attacking rice and weeds were found to be genetically differentiated (Borromeo *et al.*, 1993), although a number of instances of *M. grisea* isolates cross-infecting different hosts is known.

With the aid of MGR586 fingerprinting, Levy *et al.* (1991) demonstrated the genetic structure of *M. grisea* representing the 8 major pathotypes observed throughout the southern rice belt of the USA. In a similar study in Colombia, six distinct clonal lineages (Levy *et al.*, 1993) each associated with a subset of cultivars and pathotypes were identified. These results, together with those obtained in China, Bangladesh and the Philippines (Hamer *et al.*, 1991), show that in each country, isolates pathogenic on rice correspond to a limited number of clonal lineages. Globally the majority of these lineages show characteristics of a country or appear to be indigenous, although there are at least four lineages that are widespread in the Americas and at least two other lineages that are present in Asia and the Americas (Hamer *et al.*, 1991). In places with long history of rice cultivation, lineage and pathotype relationships were found to be complex, although the clonality of the blast lineages and the stability of the pathotypes held true in general. The pathotype diversity results, together with the recent definition of lineages have allowed a higher resolution of the pathogen population structure and novel breeding strategies based on 'lineage-exclusion' are being tested in various rice growing regions with a view to prolonging the durability of resistant cultivars (Zeigler *et al.*, 1995; Gnanamanickam *et al.*, 1998; Sivaraj *et al.*, 1998; Mekwatanakarn *et al.*, 2000). The 'lineage-exclusion' method proposes bringing together into cultivars those genes that are each effective against an entire 'lineage', so that durable resistance can be achieved against various lineages prevalent in a region (Levy *et al.*, 1993; Zeigler *et al.*, 1995).

In West Africa, the current emphasis on blast control is developing rice cultivars with adequate levels of resistance/tolerance to the disease in conjunction with sound crop management practices. A key component within this programme is developing rice cultivars appropriate for a particular ecosystem and resistant to the local blast pathogen populations. In order to achieve the above, information on the genetic and pathotype diversity of the blast pathogen populations and characterisation and identification of key sites suitable for resistance screening programmes are essential. This resulted in initiating a collaborative programme between UK institutes and African partners, funded by the UK Department for International Development (DFID)-Crop Protection Programme (CPP), managed by the Natural Resources International, UK. Blast component of the project R6738 (September 1996 – September 1999) involved Horticulture Research International (HRI); CAB International; West Africa Rice Development Association (WARDA), Côte d'Ivoire; Savanna Agricultural Research Institute (SARI) and Crops Research Institute (CRI), Ghana. Further strategic and adaptive activites are being carried out as part of a CPP project R7552 'Strategies for the development and deployment of durable blast resistance in West Africa' (December 1999 – November 2002) led by HRI in collaboration with WARDA, SARI and CRI.

2. Materials and methods

More than 200 *M. grisea* isolates were obtained from blast samples collected from WARDA and the National Agticultural Research System (NARS) rice screening sites in Burkina Faso, Côte d'Ivoire, Ghana and Nigeria (Chipili, 2000).

DNA was extracted from mycelial powder by the CTAB method (Valent *et al.*, 1986; Hamer and Givan, 1990). DNA digestion and Southern hybridisation with the MGR586 probe were carried out following standard protocols (Levy *et al.*, 1993). Virulence spectrum of the *M. grisea* isolates was determined on the international set of rice cultivars, following the scale of Valent *et al.* (1991), and pathotype designations were assigned based on Ling and Ou (1969).

3. Results

3.1. MGR586 FINGERPRINT GROUPS (LINEAGES) OF *MAGNAPORTHE GRISEA* ISOLATES FROM BURKINA FASO, CÔTE D'IVOIRE, GHANA AND NIGERIA

MGR586 fingerprints of *M. grisea* isolates obtained from Burkina Faso, Côte d'Ivoire, Ghana and Nigeria, along with a reference strain R (4375.R.26), were scored for the presence and absence of restriction fragments in the 0.8-20 kb size range and subjected to cluster analysis to identify genetic groups/lineages. Based on the similarity of the fingerprint patterns obtained for sixty-four *M. grisea* isolates from Burkina Faso, seven lineages designated BF-1 to BF-7 were identified. These isolates were collected in 1996 and 1997 growing seasons and were from two main screening sites, Farako-Ba and Banfora, with a few isolates collected from three other sites, Sideradougou, Vallée du Kou and Labola. BF-1 was the dominant lineage including approximately 30% of the isolates collected and comprised isolates collected in both seasons. Isolates within this lineage were from four of the five sites that were sampled and were collected from ten cultivars (including a local variety) and some non-rice hosts (*Paspalum*, *Brachiaria* and *O. longistaminata*). However, the bulk of isolates in lineage BF-1 were from one site, Farako-Ba, from eight cultivars.

Lineage BF-2 comprised 20% of isolates from seven cultivars collected in 1996 from a single site, Farako-Ba. About 10% of the isolates from Burkina Faso belonged to lineage BF-3 and again most were collected from Farako-Ba in 1996 and from five different cultivars. Lineage BF-4 comprised four isolates collected from three different rice cultivars from Farako-Ba and two isolates from *O. longistaminata* each collected from Banfora and Sideradougou. One of the *O. longistaminata* isolates, S572, was from wild rice that was growing close to cultivar Usen. Nine *M. grisea* isolates collected from Farako-Ba, Sideradougou and Banfora comprised BF-5 and BF-6. Two non-rice *M. grisea* isolates (*O. longistaminata* and *Setaria*) from Banfora and Sideradougou, respectively formed lineage BF-7. The remaining eight isolates were distinct and were not grouped into any of the lineages described above.

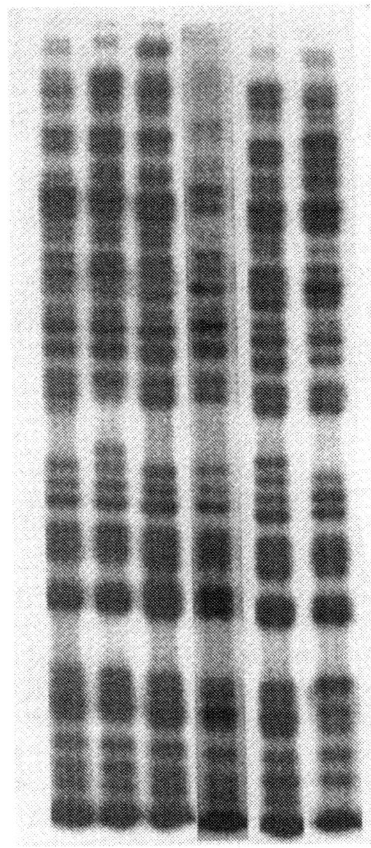

Figure 1. Magnaporthe grisea isolates from six rice varieties at two sites in Nigeria show similar MGR586 fingerprints and belong to a dominant lineage in West Africa.

M. grisea isolates from Côte d'Ivoire were from the main screening sites at Man, M'be and Korhogo, with a small number of samples also obtained from Boundiali, Koumadougou and Sakassou. These isolates were grouped into five lineages designated CD-1 to CD-5. CD-1 present on at least eighteen different rice cultivars was the dominant lineage (50% of isolates) and was present at three main sites that were sampled. However, 50% of isolates in lineage CD-1 (from nine different cultivars) were from a single site, Man; 25% of the isolates were from another site, Korhogo, and were recovered from five different cultivars. Lineage CD-2 consisted of five isolates from four different cultivars collected in 1996 from Man and Korhogo. Five isolates from five different cultivars collected in 1996 from site Man, formed Lineage CD-3. Lineage CD-4 comprised three isolates from different cultivars; two of these were from Man but the origin of the third sample was not known. Lineage CD-5 comprised two isolates from different cultivars at Man. Eleven isolates in this sample set from Côte d'Ivoire were not grouped into any of the lineages and based on the low similarity values to other isolates, could form distinct lineages.

M. grisea isolates from Ghana (collected from four main blast-screening sites Bolgatanga, Hohoe, Kwadaso and Nyankpala) were grouped into three distinct lineages designated as GH-1, GH-2 and GH-3. GH-1 was the major lineage comprising 50% of all the isolates and was present at all the sites sampled except Nyankpala (Tamale). This lineage (GH-1) occurred on at least 13 rice cultivars, some of which could have shared the same genetic background, for example Tox-related cultivars. Lineage GH-2 comprised 20% of the isolates sampled and all were from the same site, Hohoe, recovered from six different cultivars (two of which were Tox-related). Lineage GH-3 consisted of two isolates from the same cultivar (red rice) and both were from the same site (Bolgatanga). About 11 isolates from Ghana derived from four different sites and different cultivars were not grouped into any of the lineages above by the group average cluster analysis. One *M. grisea* isolate from rice at Kwadaso produced an 'atypical' fingerprint pattern with only four MGR586 hybridising bands.

Twenty-three *M. grisea* isolates from Nigeria were fingerprinted of which 20 formed a major lineage, NI-1 (example, Figure 1). One isolate, B58 from Badeggi, was quite distinct and showed only 36-40% similarity with any isolate in lineage NI-1 and was designated lineage NI-2. These isolates were collected from three main WARDA trial sites (Badeggi, Oyo and Uyo) and were from 23 rice cultivars. The majority of isolates in lineage NI-1 were from Oyo mainly from Tox-related cultivars. Lineage NI-1 was present at all three sites sampled. Some isolates in lineage NI-1 from Badeggi appeared to be closely related in their fingerprint pattern to isolates in lineage GH-1 from Ghana and isolates from Côte d'Ivoire and could belong to the same lineage.

3.2. PATHOTYPE DIVERSITY

The virulence characteristics of representatives of some prevailing lineages of the rice blast pathogen in Burkina Faso, Côte d'Ivoire, Ghana and Nigeria were determined on the international rice differentials. In Burkina Faso, 23 pathotypes from 16 different cultivars were identified from two main sites Banfora, Farako-Ba and two other sites Sideradougou, Vallée du Kou together with one pathotype each from Labola and Koumadougou sites. The most frequently observed pathotype group (around 40% of the sampling) was IC. Some pathotypes (for example IC-1, IC-9, IC-25, IF-1, IG-1 and IG-2) were represented by 2-3 isolates whereas the rest were represented by only one isolate. Up to 12 different pathotypes were present at Farako-Ba. Some isolates (S520, S576, S387, S389, S528, S529-2 and S567-1) from wild rice *O. longistaminata* (from Banfora, Sideradougou and Vallée du Kou), were found to be pathogenic on the two susceptible checks, B40 and CO39, and also on the international differentials. Similarly, isolates collected from non-rice hosts *Paspalum* (S326) and *Setaria* (S508) were observed to be pathogenic (pathotypes IA-105 and IF-2, respectively) when tested on the international differentials. One isolate from *O. longistaminata* (S386), however, was non-pathogenic to both the susceptible checks and the international differentials.

Eleven pathotypes were recovered from four sites Man, M'be, Korhogo and Sakassou in Côte d'Ivoire. These isolates were from at least nine rice cultivars and an unknown weed. IA and ID (30% each) were the most frequently occurring pathotype groups. One site, Man, expressed seven different pathotypes belonging to four different pathotype groups (IA, IB, IC and ID).

In Ghana, 21 pathotypes were detected from the four main sites Bolgatanga, Hohoe, Kwadaso and Nyankpala and originated from 17 known rice cultivars, four unknown rice cultivars and a wild rice. In most instances, each isolate showed a distinct virulence spectrum defined by a particular pathotype and there were only a few examples where different isolates had an identical virulence spectrum thus were of the same pathotype. The most frequently observed pathotype groups were IB and IC (50% and 33% of the isolates, respectively).

In Nigeria, 16 pathotypes were observed from three sites Badeggi, Oyo and Uyo and were from 14 different cultivars. IA and IB were the most prominent pathotype groups (30% each) and each of the pathotypes was represented by only one isolate. Two isolates, B12 and IKR10/2 recovered from rice were non-pathogenic on susceptible cultivars B40 and CO39 and also on the international differentials.

3.3. SITE CHARACTERISATION

In Burkina Faso, Farako-Ba and Banfora with all the major lineages recorded so far and a wide range of virulence diversities are key sites suitable for blast resistance screening. However, major lineage BF-1 has not been recorded so far at Sideradougou and it would be interesting to monitor whether BF-1 appears at this site.

In Côte d'Ivoire, Man displayed high blast pathogen diversity with all five lineages recorded (CD-1 to CD-5) and four international pathotype groups (IA, IB, IC and ID) and is a key site suitable for blast resistance screening. M'be site showed low pathogen diversity and even the dominant lineage CD-1 was not recorded.

In Ghana, Hohoe with two lineages (GH-1 and GH-2) and four pathotype groups (IA, IB, IC and IG) and Bolgatanga with two lineages (GH-1 and GH-3) and three pathotype groups (IA, IB and IC) are key blast resistance screening sites (Figure 2). At Nyankpala (Tamale) several ungrouped genotypes were identified and whether any of these develop into dominant lineage(s) merits further investigation.

In Nigeria, the dominant lineage NI-1 was present at Badeggi, Oyo and Uyo and all three sites displayed overlapping pathotypes. However, further characterisation of *M. grisea* population is necessary to provide improved assessment of blast pathogen diversity at these sites. Man and M'be sites in Côte d'Ivoire (WARDA) as well as Bolgatanga and Hohoe sites (NARES–CRI and SARI) in Ghana are being used for follow on activities in R7552.

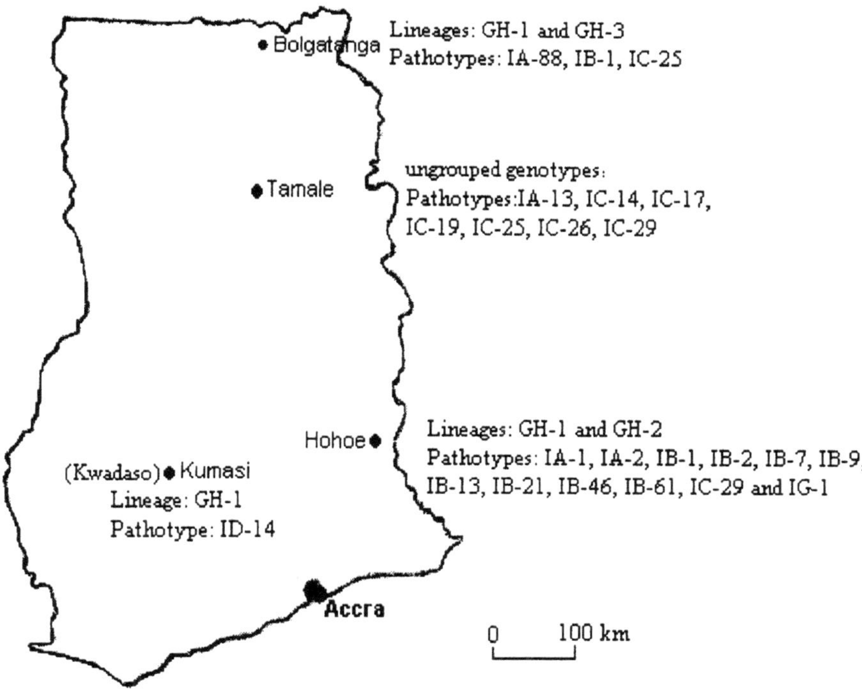

Figure 2. Diversity and distribution of the rice blast pathogen *Magnaporthe grisea* lineages and pathotypes in Ghana.
Reprinted from Chipili *et al.*, Ghana Science Association Biennial Conference Proceedings, August 2001.

3.4. LINEAGE-PATHOTYPE RELATIONSHIPS OF *MAGNAPORTHE GRISEA* ISOLATES

Combined analysis of the lineage-pathoytpe data has provided some understanding of the lineage-pathotype relationships. In Nigeria, lineage NI-1 represented 12 pathotypes. Although these pathotypes originated from 12 different cultivars, a number of these isolates showed related virulence patterns, for example, pathotypes IA-8 and IA-65; IB-1, IB-3 and IB-42; ID-6 and ID-16. *M. grisea* isolates originating from the same site showed lesser degrees of genetic diversity, although their virulence patterns varied. For instance, seven isolates collected from Oyo from different cultivars expressed seven different pathotypes, but all belonged to lineage NI-1 with very high MGR586 fingerprint similarity.

In Ghana, the lineage/pathotype relationships showed a similar trend to that in Nigeria. Pathotypes represented in lineage GH-1 were mostly IB group and some of these were closely related (for example IB-1, IB-2 and IB-7) differing by a few compatibility reactions on the international differentials. Some of the other isolates in this lineage originating from Tox-related cultivars at Hohoe, also expressed related pathotypes (for example IA-1 and IA-2).

Eleven pathotypes identified in Burkina Faso were represented by lineages BF-1, BF-2, BF-5, and BF-7. The lineage/pathotype relationships observed in Burkina Faso were more complex compared to Nigeria and Ghana. Lineage BF-1 represented diverse pathotypes IA-105, IB-64 and IC-13, which originated from diverse hosts and different sites. Some isolates in lineage BF-2, however, expressed either closely related pathotypes (e.g. IC-25 and IC-27) or the same pathotype (e.g. IC-9 from cultivars K3 and Yashiro-mochi at the Farako-Ba site).

In Côte d'Ivoire, pathotypes IA-13, IB-45 and ID-14 identified from Man, belonged to CD-1, indicating high virulence diversity in this lineage. Results obtained so far generally suggest that each lineage comprised isolates that infected different cultivars and a range of pathotypes. The degree of pathotype diversity varied between lineages with a limited number of pathotypes belonging to some lineages, whilst other lineages expressed a much broader range of pathotypes. In some cases, pathotypes belonging to a lineage expressed relatively similar virulence spectra. Further understanding of the virulence diversities of the blast lineages in the region is essential to test the 'lineage-exclusion' model for breeding and deployment of resistance.

4. Discussion

This paper presents data from four major rice growing countries in West Africa on: (i) the genetic diversity (lineages) of *M. grisea* populations and their distribution pattern, (ii) diversity and distribution of the *M. grisea* pathotypes prevalent in the region (iii) the lineage-pathotype relationships of representative isolates and (iv) characterisation of resistance screening sites.

Based on MGR586-DNA fingerprint analysis, the diversity observed among the *M. grisea* populations, mainly from screening sites, comprised at least seven lineages in Burkina Faso, five lineages in Côte d'Ivoire, three lineages in Ghana and two lineages in Nigeria. The relatively small number of lineages identified in each country supports the widely held view that rice blast pathogen, *M. grisea* populations are predominantly clonal (Levy *et al.*, 1991, 1993; Zeigler *et al.*, 1994, 1995). The isolates analysed were recovered from more than 20 rice cultivars and breeding lines in each of the four countries. These lineages did not, however, include several individual isolates with distinct fingerprints, which if considered as distinct lineages, could increase the number of lineages in each of the countries.

The number of lineages present per site varied both within and across different countries. In some instances different lineages were recovered from a single cultivar collected from the same site in the same year. In other instances multiple samples from the same cultivar collected but from different sites expressed the same lineage. For example, *M. grisea* isolates from WITA 11 from two different sites M'be and Sakassou in Côte d'Ivoire expressed the same lineage, CD-1. It was also observed that the same cultivar sampled in two consecutive growing seasons expressed different lineages, in some cases. These results indicate that shifts in lineage diversity and distribution are likely to occur in this region. Some lineages were detected across different countries, for example, lineage NI-1 from Nigeria was also present in Ghana (lineage GH-1). Similarly, lineages CD-1, CD-2 and CD-4 from Côte d'Ivoire were also present in Burkina Faso (BF-1, BF-4 and BF-2, respectively). This finding indicates a recent international movement of the pathogen due to local trade and/or movement of seed infested with *M. grisea*. In this context, a rice cultivar resistant to lineage NI-1 from Nigeria could also be resistant to lineage GH-1 in Ghana. This suggests that identification and development of cultivars that are resistant to the common dominant lineages by a regional organisation such as WARDA in collaboration with the National Agricultural Research and Extension Systems (NARES) could be useful for deployment of these varieties in a number of countries in the region. In this context, close linkages between the project activities and the IPM and Breeding Task Forces, under the ROCARIZ network, are envisaged.

Analysis of the virulence spectrum of the *M. grisea* isolates indicated a high degree of pathotype diversity within populations in each of the four countries, and generally, the range of pathotypes differed between countries and at different sites within a country. However, some of the pathotype groups were found to be predominant in more than one country i.e. in Ghana they were IB and IC, groups IA and IB predominated in Nigeria; IC in Burkina Faso and IA and ID in Côte d'Ivoire. In some instances, the same pathotype was recovered from different cultivars both within a country and in different countries. For example, IB-13 recovered from two different Tox-related rice cultivars from Hohoe in Ghana. Similarly pathotype IA-1 was recovered from Ghana, Nigeria and Burkina Faso from Tox 4004-43-1-2-1, CO39 and *O. longistaminata*, respectively.

Among the *M. grisea* isolates analysed, 7% were classified as 'atypical' based on the MGR586 fingerprint patterns and their host origin. A number of isolates from wild rice, with rice pathogen-like and grass pathogen-like fingerprints, were pathogenic on rice cultivars tested and developed typical blast lesions. One isolate each from *Paspalum* and *Setaria* (with rice pathogen-like fingerprints) also produced blast symptoms in inoculation assays. However, two 'atypical' isolates recovered from rice from Nigeria, B12 and IKR10/2, with nine and seven MGR586 hybridising bands, respectively, were found to be non-pathogenic, although similar isolates with few MGR586 bands from wild rice and an unknown weed (with a single band) were pathogenic on rice in inoculation tests.

Baseline data, new to West Africa, on the diversity and distribution pattern of the pathogen populations has been established and key sites identified. The outputs generated have been disseminated to WARDA and the NARS collaborators as well as to other NARS pathologists and breeders through the Integrated Pest Management-Task Force (IPM-TF) of the West and Central Africa Rice Research and Development Network (ROCARIZ) for up take and adoption in their programmes. Recent research by the International Rice Research Institute (IRRI), Philippines, International Centre for Tropical Agriculture (CIAT), Colombia and Purdue University, USA highlights the importance of blast pathogen characterisation to develop novel resistance gene deployment technologies.

Project R7552, managed by HRI, builds on the outputs and collaborations from R6738. The collaborating institutes are undertaking strategic and adaptive research to understand the dynamics of the pathogen populations, develop technologies suitable for long-term local monitoring of blast, identify sources of resistance to characterised pathogen groups and develop appropriate blast management strategies. This project also complements work on rice weed management and participatory varietal evaluation in West Africa funded by CPP as well as other DFID programmes. The pathogen characterisation data generated in R6738 and R7552 will also contribute to filling the existing knowledge gap for West Africa in the Rice Blast Database – an international initiative to provide a global atlas to track the inter-continental diversity and distribution of the blast pathogen populations.

The IPM-TF, Rice Breeding-TF and the Technology Transfer-TF of ROCARIZ, closely co-ordinated by WARDA, provide important linkages for uptake pathways in West Africa. The focus is also on promotion of technology to and capability strengthening at WARDA and NARS to achieve sustainable disease management. Linkages have also been established with NARS in India undertaking complementary rice blast research and development work funded by the Asia Rice Biotechnology Network managed by IRRI and other donors. This provides the opportunity for exchange and effective utilisation of knowledge between Africa and Asia to develop resistance gene rotation, gene pooling and/ or gene pyramiding technologies for blast management and achieve a wider impact. Amelioration of the constraint posed by blast to rice production will benefit the resource poor farmers and consumers in West Africa.

Furthermore, the above area of research provides good complementarity to another DFID-CPP pipeline project 'Finger millet blast in East Africa: Pathogen diversity and disease management strategies'. The collaborative project links HRI with The Queen's University Belfast, the International Crops Research Institute for the Semi-Arid Tropics (a CG Centre in India and Africa) and Serere Agricultural and Animal Production Research Institute (NARS in Uganda). The molecular and pathological tools and the resistance gene deployment technologies developed for rice blast in West Africa could be utilised to develop appropriate management methods to tackle finger millet blast in East Africa enhancing the value and impact of the developmental research funded by the programme.

5. Summary

Characterisation of the genetic and pathogenic diversity of the rice blast pathogen, *Magnaporthe grisea*, populations in four West African countries (Burkina Faso, Côte d'Ivoire, Ghana and Nigeria) was undertaken. More than 200 *M. grisea* isolates from a range of cultivars/breeding lines were collected from key screening sites across the region. Based on MGR586 fingerprinting, seven lineages were identified from Burkina Faso, five from Côte d'Ivoire, three from Ghana and two from Nigeria. Some lineages were common between countries. The number of lineages per site varied both within and across different countries but, in general, genetically similar cultivars expressed the same lineage. Analysis of the virulence spectrum of representative *M. grisea* isolates, assessed on international rice differentials indicated a high degree of pathotype diversity within populations in each of the countries studied; 28 pathotypes in Burkina Faso, 11 in Côte d'Ivoire, 21 in Ghana and 16 in Nigeria. The number of pathotypes at a site and in a country varied considerably. Relationships between the lineages identified and the pathotypes prevalent in the four West African countries was complex, pathotype heterogeneity being common in most lineages. This study has provided the baseline data on the *M. grisea* population structure in the region and paves the way for further monitoring and the development of blast control measures appropriate to the pathogen diversity and dynamics.

6. Acknowledgements

This publication is an output from research projects funded by the United Kingdom Department for International Development (DFID) for the benefit of developing countries. The views expressed are not necessarily those of DFID. Projects R6738 and R7552, DFID-Crop Protection Programme, managed by the Natural Resources International, UK.

7. References

Borromeo, E.S., Nelson, R.J., Bonman, J.M. and Leung, H. (1993) Genetic differentiation among isolates of *Pyricularia* infecting rices and weed hosts, *Phytopathology* **83**, 393-399.

Chipili, J. (2000) *Characterisation of populations of Magnaporthe grisea, the rice blast fungus, in some of the West African countries*, PhD thesis, University of Exeter, UK, pp. 130.

Fomba, S.N. and Taylor, D.R. (1994) Rice blast in West Africa: Its Nature and Control, in R.S. Zeigler, S.A. Leong and P.S. Teng (eds.), *Rice Blast Disease*, CAB International, Wallingford, Oxon, UK and International Rice Research Institute, Manilla, The Philippines, pp. 343-355.

Gnanamanickam, S.S., Lavanya, B., Sivaraj, R. and Levy, M. (1998) Lineage-exclusion resistance breeding: Pyramiding of blast resistance genes for management of rice blast in India. (Abstract), *2nd International Rice Blast Conference* 4-7 August, Ecole Nationale Supèrieure Agronomique de Montpellier, France.

Hamer, J.E., Farrall, L., Orbach, M.J., Valent, B. and Chumley, F.G. (1989) Host species-specific conservation of a family of repeated DNA sequences in the genome of a fungal plant pathogen, *Proceedings of the National Academy of Sciences of the USA* **86**, 9981-9985.

Hamer, J.E. and Givan, S. (1990) Genetic-mapping with dispersed repeated sequences in the rice blast fungus-mapping the *Smo* locus, *Molecular and General Genetics* **223**, 487-495.

Hamer, J.E., Talbot, N.J. and Levy, M. (1991) Genome Dynamics and Pathotype Evolution in the rice blast fungus, *6th International Symposium on Molecular Plant-Microbe Interactions*, Seattle, Washington, USA.

Levy, M., Romao, J., Marchetti, M.A. and Hamer, J.E. (1991) DNA fingerprinting with a dispersed repeated sequence resolves pathogen diversity in the rice blast fungus, *The Plant Cell* 3, **95**-102.

Levy, M., Correa-Victoria, F.J., Zeigler, R., Xu, S. and Hamer, J. (1993) Genetic diversity of the rice blast fungus in a disease nursery in Colombia, *Phytopathology* **83**, 1427-1433.

Ling, K.C. and Ou, S.H. (1969) Standardization of the international race numbers of *Pyricularia oryzae*, *Phytopathology* **59**, 339-342.

Mekwatanakarn, P., Kositratana, W., Levy, M. and Zeigler, R.S. (2000) Pathotype and Avirulence Gene Diversity of *Pyricularia grisea* in Thailand as determined by Rice Lines Near-Isogenic for Major Resistance Genes, *Plant Disease* **84**, 60-70.

Notteghem, J.L. and Baudin, P. (1981) *Main Rice Diseases in West Africa*, West Africa Rice Development Association, Monrovia, Liberia.

Roumen, E., Levy, M. and Notteghem, J.L. (1997) Characterisation of the European pathogen population of *Magnaporthe grisea* by DNA fingerprinting and pathotype analysis, *European Journal of Plant Pathology* **103**, 363-371.

Singh, B.N., Fomba, S.N., Sere, Y., Sy, A.A., Akator, K., Ngninbeyie, P. and Ahn, S.W. (1998) Breeding for blast resistance in lowland rice in West Africa. (Abstract), *2nd International Rice Blast Conference* 4-7 August, Ecole Nationale Supèrieure Agronomique de Montpellier, France.

Sivaraj, R., Gnanamanickam, S.S. and Levy, M. (1998) " Lineage-exclusion" tests for blast resistance in southern India. (Abstract), *2nd International Rice Blast Conference* 4-7 August, Ecole Nationale Supèrieure Agronomique de Montpellier, France.

Valent, B., Crawford, M.S., Weaver, C.G. and Chumley, F.G. (1986) Genetic studies of fertility and pathogenicity in *Magnaporthe grisea* (*Pyricularia grisea*), *Iowa State Journal of Research* **60**, 569-594.

Valent, B., Farrall, L. and Chumley, F.G. (1991) *Magnaporthe grisea* genes for pathogenicity and virulence identified through a series of backcrosses, *Genetics* **127**, 87-101.

WARDA Annual Report, 1998.

WARDA Medium Term Plan, 2000-2002 .

Zeigler, R.S., Tohme, J., Nelson, R., Levy, M. and Correa-Victoria, F.J. (1994) Lineage exclusion: A proposal for linking blast population analysis to resistance breeding, in R.S. Zeigler, S.A. Leong and P.S. Teng (eds.), *Rice Blast Disease*, CAB International, Wallingford, Oxon, UK and International Rice Research Institute, Manilla, The Philippines, pp. 267-292.

Zeigler, R.S., Cuoc, L.X., Scott, R.P., Bernado, M.A., Chen, D.H., Valent, B. and Nelson, R.J. (1995) The relationship between lineage and virulence in *Pyricularia grisea* in the Philippines, *Phytopathology* **85**, 443-451.

IDENTIFICATION OF PATHOGENICITY DETERMINANTS IN THE RICE BLAST FUNGUS *MAGNAPORTHE GRISEA*

N.J. TALBOT, H.R.K. MCCAFFERTY, M.J. KERSHAW and K.P. DIXON
Department of Biological Sciences,
University of Exeter,
Washington Singer Laboratories,
Perry Road, Exeter, EX4 4QG, UK.

1. Introduction

Magnaporthe grisea, (Hebert) Barr [anamorph, *Pyricularia grisea* Sacc.], is a heterothallic ascomycete fungus (Rossman *et al.*, 1990) capable of causing disease in more than fifty members of the Gramineae (Ou, 1985). These include many forage grasses and weeds, and a number of important crops such as wheat (*Triticum aestivum*), barley (*Hordeum vulgare*), millet (*Pennisetum americanum*), maize (*Zea mays*) and rice (*Oryza sativa*). *Magnaporthe grisea* is most widely known, however, as the causal agent of rice blast, the most devastating disease of cultivated rice. Rice blast is a persistent problem and in the past 20 years has caused average annual losses from 11 to 30 % of the world rice harvest (Baker *et al.*, 1997), equivalent to 157 million tons of rice.

This chapter describes progress made towards understanding the molecular basis of pathogenicity in the rice blast fungus. Rapid advances have recently been made in identifying pathogenicity factors in *M. grisea* and characterising the infection process in some detail. This has required development of a large number of experimental techniques, including DNA-mediated transformation, gene replacement, insertional mutagenesis and differential cDNA screens in addition to isolation of fertile rice pathogenic strains of the fungus. The accompanying Chapter by Bhargava and Hamer describes these processes in detail and explores the strategies used to identify pathogenicity genes in *M. grisea*. This chapter focuses instead on our current understanding of pathogenicity and likely future avenues for research. A number of excellent reviews on aspects of pathogenicity in *M. grisea* have appeared in recent years (Valent and Chumley, 1991; Holden and Hamer, 1996; Howard and Valent, 1996) and the reader is referred to these for complementary, and occasionally alternative, perspectives.

S. Sreenivasaprasad and R. Johnson (eds.), Major Fungal diseases of Rice Recent Advances, 21-33.
Prof. K. Manibhushan Rao Festschrift

2. Defining pathogenicity

The pathogenicity phenotype is multifaceted, encompassing all aspects of the development of a fungus in relation to its host. In order to understand what allows a pathogen to cause disease it is therefore necessary to determine, at the molecular level, what separates plant pathogens from their saprophytic relatives. The most straight forward way is to produce a mutation in a given gene and then see if the resulting mutant can cause disease. If it cannot, then this obviously provides the most conclusive evidence that the gene encodes a pathogenicity determinant. However, if the mutation causes pleiotropic effects on fungal growth and development, or in general fitness, then it becomes a matter of conjecture as to whether loss of pathogenicity is due to specific loss of the pathogenicity factor, or merely reduced fitness. This question has been debated frequently and has often divided plant pathologists (see Schafer, 1994; Van Etten *et al.,* 1994; Oliver and Osbourn, 1995). Classifications of genes as either 'pathogenicity genes' or 'virulence genes' have, for example, emerged depending on the relative reduction in disease symptoms produced by a corresponding gene mutation.

As more genes are identified and their roles in pathogenicity tested, it seems likely, however, that any system of classification will break down. This is because the pathogenicity phenotype is sufficiently complex that it invariably relies on the orchestrated action of many gene products. It is likely that mutations in many of these genes will have only a minor effect on pathogenicity and yet they may encode critical factors for the overall success of a pathogen. If fully understanding the molecular mechanisms by which a fungus gains entry to a plant, colonises its living tissue, derives nutrition and ultimately propagates itself, is the purpose of plant pathology, then it seems likely that any gene expressed during the disease may be pertinent to pathogenicity. Although this seems an unfocused view, it seems more likely (to us) to be closer to reality because it allows for the inherent complexity of the phenotype. For this reason fully understanding fungal pathogenicity is likely to be a mammoth task, but therefore offers a significant and exciting challenge.

A number of global regulators of pathogenicity are, however, likely to be rapidly identified with extremely specific roles in pathogenesis (for a discussion of this see Holden and Hamer, 1996) and understanding the interplay between these genes and their morphogenetic and physiological targets may, in the short term, prove the most powerful means of exploring pathogenesis.

3. The life cycle of *Magnaporthe grisea.*

3.1. INFECTION-RELATED DEVELOPMENT

M. grisea spreads from plant to plant in the form of three-celled, tear drop-shaped conidia (asexual spores). *M. grisea* conidia are not long-lived and the fungus does not overwinter in this form. They appear instead to be highly specialised structures designed simply to attach to rice leaves and quickly germinate, allowing the fungus to gain entry into the rice tissue as rapidly as possible. The conidium adheres to the hydrophobic leaf surface by means of a glue called spore tip mucilage (Hamer *et al.*, 1988) released from a compartment in the conidial apex. Spore tip mucilage is composed of protein, carbohydrate and lipid, and has been shown to bind to the lectin concanavalinA, suggesting that the carbohydrate moiety contains α-linked glucosyl or mannosyl residues (Howard and Valent, 1996). The adhesive binds strongly to hydrophobic substrates and is released as soon as the conidium becomes hydrated. Once attached to the leaf, the spore immediately germinates producing a short germ tube from one of its terminal cells (Howard, 1994). In the environment of the leaf surface, the only requirement for germination and consequent development is the presence of free water and neither surface contact, or an external nutrient supply are necessary (Xiao *et al.*, 1994; Howard, 1994). Cell division occurs soon after germination, with one nucleus remaining in the spore and the other migrating into the appressorium via the germ tube (Bourett and Howard, 1990). The germ tube is extremely sensitive and involved in perception of the topography and rigidity of a substrate, conducive to appressorium development (Bourett and Howard, 1990). A number of modifications of the germ tube occur to allow leaf surface recognition, including the positioning of apical vesicles close to the substrate and flattening of the germ tube when in contact with the leaf surface (Bourett and Howard, 1990). After 4 - 5 hours, an appressorium differentiates from the end of the germ tube. This process starts with the formation of a swollen hook, which tightly adheres to the surface. This swollen hyphal tip differentiates and after only 40 -50 minutes a septum forms separating the appressorium from the germ tube and conidium.

Surface contact and free water are undoubtedly required for successful appressorium formation (Howard, 1994; Xiao *et al.*, 1994). Rice leaves are coated in a thick waxy cuticle (Uchiyama and Okuyama, 1990) that makes them extremely hydrophobic (See Figure 1). Rice has, in fact, one of the most hydrophobic surfaces of all plants, producing contact angles of up to 170° with a 10 μL drop of water (Talbot, 1995). For comparison, artificial hydrophobic surfaces such as Teflon (PTFE, trademark of DuPont Co.) have contact angles of $90\text{-}130^{\circ}$ (Talbot, 1995). The ridged surface of the leaf contributes greatly to this extreme hydrophobicity, but even in the micro-environment of the leaf cuticle, where *M. grisea* develops, the extensive wax deposition is likely to lead to very hydrophobic conditions. High surface has been found to induce appressorium formation (Hamer *et al.*, 1988; Lee and Dean, 1994; Jelitto *et al.*, 1994) and appressoria can be formed *in vitro* on surfaces such as Teflon or wax-coated plastics (Hamer *et al.*, 1988; Lee and Dean, 1994). A cyclic AMP-dependent (cAMP) signal transduction pathway appears to be involved in this process as exogenous application of cAMP can induce appressorium formation on poorly inductive (hydrophilic) surfaces such as glass (Lee and Dean, 1993). It is, however, becoming increasingly apparent that a number of other inductive cues affect appressorium development in addition to surface hydrophobicity. Recently, for example, it has been shown that a number of cutin monomers, including *cis*-9,10-epoxy-18-hydroxyoctadecanoic acid and *cis*-9-octadecen-1-ol can induce appressorium formation on normally non-inductive surfaces, even when added in nanomolar concentrations (Gilbert *et al.*, 1996). Plant lipid and other wax compounds were also tested and 1,16-hexadecanediol was found to be a potent inducer of appressorium development (Uchiyama *et al.*, 1979; Gilbert *et al.*, 1996; Sweigard *et al.*, unpublished). A number of other factors also influence appressorium development such as starvation stress (Talbot *et al.*, 1993; Howard, 1994; Jelitto *et al.*, 1994) and light (Jellito *et al.*, 1994).

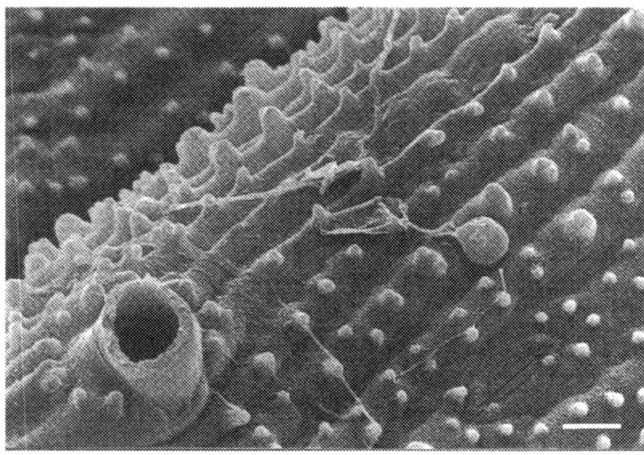

Figure 1. Cryo-scanning electron micrograph showing differentiation of a *M. grisea* appressorium on the surface of a rice leaf. In this case an appressorium has elaborated from a short germ tube and begun to generate turgor, approximately 24 h after conidial germination.
The three-celled conidium and germ tube have collapsed after full differentiation of the appressorium and separation of the cell by septation. Bar = 10 μm.

It would appear therefore that a number of signals can induce development of appressoria and that perhaps a threshold of multiple inducing signals normally elicits appressorium development on rice leaves. Some results that shed new light on developmental signalling in appressoria come from a study by Thines *et al.* (1997) who found that they could inhibit appressorium development on hydrophobic surfaces using Glisoprenin A, even in the presence of cAMP, phosphodiesterase inhibitors, or 1,16 hexadecanediol. The inhibition was, however, specific to appressorium development on hydrophobic substrates because Glisoprenin A failed to inhibit development on hydrophilic surfaces in the presence of these inducers. This suggests that the signal transduction pathway involved in the perception of hydrophobic surfaces may be independent of other pathways able to lead to appressorium morphogenesis and of cAMP.

The clearest evidence that multiple input signals can influence appressorium development, however, comes from the consistent observation that higher numbers of appressoria are present when observed directly on rice leaves than on artificial substrates (Hamer *et al.*, 1988; Lee and Dean, 1993; Jelitto *et al.*, 1994; Talbot *et al.*, 1996). This suggests that there are factors influencing development which are only partially simulated in these *in vitro* assays. This also accounts for the high level of variability in appressorium development assays frequently observed in many labs (Talbot, unpublished; J.A. Sweigard pers. comm.; R.J. Howard pers. comm.).

The relationship of the cell cycle to appressorium morphogenesis has been studied using DAPI staining (Bourett and Howard, 1990). Each of the three conidial cells contain a single nucleus. As the germ tube emerges from one of the terminal cells, the nucleus in that cell undergoes a mitotic division and one daughter nucleus remains in the conidium while the other moves through the germ tube and enters the developing appressorium. Once septation occurs, this nucleus becomes effectively separated in the appressorium. The germ tube and conidium then collapse as the appressorium develops turgor. The appressorium nucleus is therefore the 'mother-nucleus' for all subsequent cell divisions which occur during growth of the fungus after infection. Interestingly, the physical characteristics of surfaces appear to influence the length of germ tubes greatly and, as a consequence, the number of nuclear divisions which take place during the pre-penetration phase of development (Howard, 1994). How this occurs, however, is currently unknown and clearly requires further study.

The morphogenesis of appressoria clearly involves the perception of multiple input signals from the leaf surface and surrounding aqueous medium. The correct perception of the leaf surface as a conducive surface for infection appears to involve surface-acting cell wall proteins. The first of these to be identified was the product of the *MPG1* gene which is important in efficient development of appressoria and pathogenicity (Talbot *et al.*, 1993, 1996). The relative importance of surface-acting proteins like the MPG1 hydrophobin is not surprising because they are probably present at the interface between the fungal cell wall and leaf cuticle where surface perception must take place. They are also seldom likely to be absolute requirements for appressorium morphogenesis, because of the multiple input signals clearly inducing development of these cells. The physical attributes of hydrophobins are, however, remarkable and are described in detail in the accompanying chapter by Kershaw and Talbot.

Of greater fundamental importance to appressorium development appears to be the action of underlying signal transduction pathway. The cAMP-dependent morphogenesis of appressoria appears to act via a MAP kinase cascade containing homologs of the pheromone response pathway studied in detail in the budding yeast *Saccharomyces cerevisiae* (see Herskowitz, 1995). Evidence for this comes from two independent studies. First Xu and Hamer (1996) identified a MAP kinase-encoding gene called *PMK1* which is homologous, both in sequence and function, to the *FUS3* gene of *S. cerevisiae*. A gene replacement was carried out to create a *pmk1*$^-$ null mutant and this was found to be blocked in its ability to elaborate appressoria. This is the first mutation recorded which blocks appressorium development completely and therefore identifies *PMK1* as a central regulator of the morphogenetic programme which underlies their formation. Interestingly, *PMK1* also affects subsequent pathogenic development suggesting that the formation of appressoria and further colonisation of plant tissue are co-regulated processes.

The second piece of evidence implicating this signal transduction pathway comes from the study of Beckerman *et al.*, (1997) who observed that appressorium development in *M. grisea* could be inhibited in a mating type specific manner by the *S. cerevisiae* α-factor pheromone. This remarkable observation indicates that inappropriate interference in the analogous pathway in *M. grisea* blocks appressorium morphogenesis. Taken together, these studies suggest that elements of the *M. grisea* pheromone response pathway have been co-opted into regulating appressorium development and subsequent pathogenic growth of the fungus. Sexual development in fungi normally requires departure from the polarised growth form and cellular differentiation, prior to nuclear fusion and meiosis. As appressorium formation requires a similar differentiation and loss of polarity, it is logical that a similar signalling mechanism might underlie both processes, albeit with different input signals. The absence of sexual fertility in most *M. grisea* rice pathogens (Valent *et al.*, 1990), however, suggests that the primary function of the pheromone response pathway in this fungus may, indeed, be in infection-related development.

3.2. APPRESSORIUM-MEDIATED INFECTION

The structure of the *M. grisea* appressorium has been studied in detail by the pioneering ultrastructural analyses of Howard and co-workers. The outer cell wall has a high chitin content and during maturation of the appressorium its affinity to lectins changes, either due to the addition of new wall material or from the matrix becoming so dense that lectins no longer have access to corresponding haptens (Howard *et al.*,, 1991). Another major constituent of the appressorium is melanin which is synthesised via the polymerisation of 1,8-dihydroxynaphthalene. Melanin biosynthesis has long been recognised as a target for anti-fungal agents and tricyclazole, one of the most effective blast fungicides, inhibits polyhydroxy-naphthalene reductase preventing infection by the fungus. It therefore appears to have an essential role in appressorium-mediated infection, but until recently this has not been characterised. Three colour mutants deficient in melanin biosynthesis have helped to define the role of melanin in the appressorium (Howard and Ferrari, 1989; Chumley and Valent, 1990). These mutants, named *ALB1, RSY1* and *BUF1* due to their albino, rosy and buff appearance respectively, were shown to produce non-functional appressoria which failed to penetrate rice leaf cuticles and were consequently non-pathogenic (Howard and Ferarri, 1989). *ALB1* shows homology to genes encoding polyketide synthases (Howard and Valent, 1996), while *RSY1* encodes scytalone dehydratase (Lundqvist *et al.*, 1994) and *BUF1* encodes NADPH-dependent polyhydroxy-naphthalene reductase (Vidal-Cros *et al.*, 1994).

The first clue to the likely role of melanin in appressorium function came from experiments which showed that appressoria generate enormous turgor (Figure 2) required for their action. These experiments were carried out by incubation of appressoria in hyperosmotic solutions to a point where cytorrhysis (cell collapse) occurred. This lead to an estimate of intracellular hydrostatic pressure of up to 8 MPa (Howard *et al.*, 1991). It was also shown that artificially lowering turgor, by suspending appressoria in polyethylene glycol solutions of varying molarities, would prevent penetration, indicating that appressorial turgor is required for infection (Howard *et al.*, 1991).

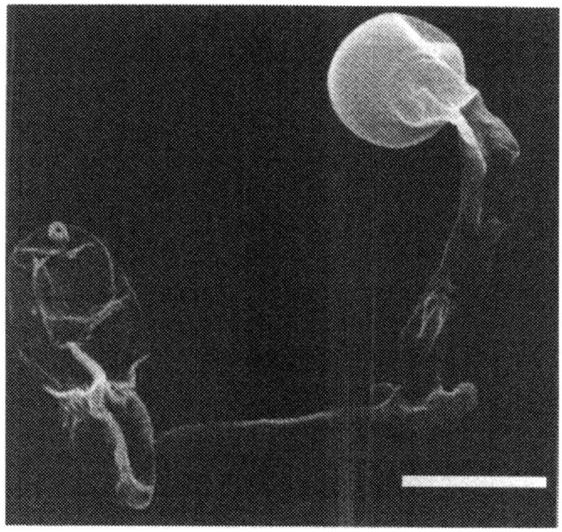

Figure 2. Cryo-scanning electron micrographs of an appressorium of *M. grisea* Detail of an appressorium at maximum turgor, 48 h after formation on a teflon membrane. Adapted from De Jong *et al.*, (1997). Notice the collapse of all structures except the appressorium. Bar = 10 μm

Remarkably, even a small reduction in turgor (0.5 MPa) was found to reduce infection efficiency (by 80%). This study suggested that *M. grisea* appressoria function primarily by generating mechanical force by the application of turgor pressure (Howard *et al.*, 1991). This was consistent with the observation that disruption of a cutinase gene, *CUT1*, from *M. grisea* failed to inhibit infection (Sweigard *et al.*, 1992). It was also found that melanin contributes greatly to the integrity of the cell wall, reducing its porosity to a likely pore size of less than 1 nm (Howard *et al.*, 1991). This suggested that one of its roles might be in maintenance of cellular turgor (Money and Howard, 1996).

Recently, we investigated this process by extracting the contents of appressoria during turgor generation (De Jong *et al.*, 1997). This showed that *M. grisea* appressoria accumulate up to 3.22 M glycerol. This is, in theory, sufficient to generate 8.7 MPa of turgor pressure assuming glycerol acts as an ideal solute. Because this is unlikely at such enormous concentrations, we carried out psychrometry of glycerol solutions and estimated that 3.22 M glycerol would be likely to generate 5.8MPa turgor pressure in an appressorium (De Jong *et al.*, 1997). To investigate the role of melanin in appressoria we incubated them in hyperosmotic solutions of glycerol and observed cytorrhysis of 52% of cells in a 1.75 M glycerol solution. This confirmed the likely internal molarity of glycerol as being in the order of 2-3 M, but also showed that the appressorial cell wall is impermeable to the solute. This was in marked contrast to non-melanin pigmented appressoria. Walls of these strains were fully permeable to glycerol showing recovery form cytorrhysis in less than 1 min, followed by plasmolysis of the appressorial protoplast (De Jong *et al.*, 1997).

These experiments indicate that appressorial turgor in *M. grisea* results from accumulation of molar concentrations of glycerol within a melanised appressorium designed to hold the solute in the cell during turgor generation. This provides a biochemical mechanism for turgor generation which will form the basis of future studies into the molecular control of appressorium function.

Consistent with the formation of large concentrations of glycerol in the appressorium, a reduction in the level of stored glycogen in appressoria has previously been observed by electron microscopy (Bourett and Howard, 1990). This suggests that carbohydrate, in the form of glycogen, is transported into the appressorium for subsequent metabolism to glycerol via glucose and the primary steps of glycolysis. It seems likely that the key enzymes in this process will be NADH-dependent glycerol-3-phosphate dehydrogenase (EC 1.1.1.8) which catalyses the reduction of dihydroxyacetone phosphate (DHAP) to glycerol-3-phosphate (G3P) and glycerol phosphatase which converts G3P to free glycerol. Investigations to determine whether activity of these enzymes is present in appressoria are currently underway.

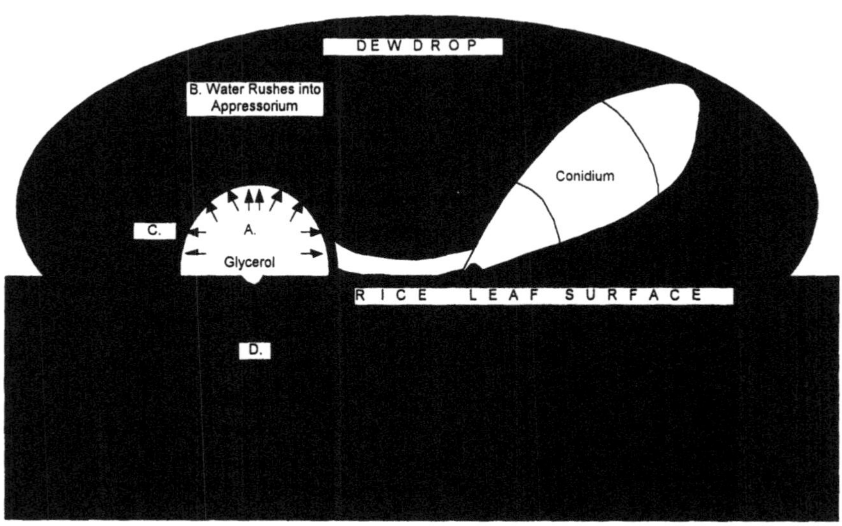

Figure 3. Model for Generation of Appressorial turgor in *M. grisea*. A. Glycerol is synthesised in the appressorium which is separated from the germ tube and conidium by a septum. The appressorium generates turgor due to a rapid influx of water (B.). Glycerol is prevented from leaving the appressorium due to the melanin pigmented wall (C) shown in black. The turgor pressure is focused on a point (D.) where localised cell wall dissolution occurs and a penetration peg ruptures the leaf cuticle. Adapted and up-dated from Talbot (1995) based on data presented in Howard and Valent (1996) and De Jong *et al.*, (1997).

The regulation of turgor generation by *M. grisea* appears to involve cAMP-dependent signalling and the cAMP-dependent protein kinase-encoding gene *CPKA*. This gene was identified by Mitchell and Dean (1995) and *cpka*⁻ null mutants, generated by gene replacement, were shown to be non-pathogenic. This appears to be due to a drastic reduction in the functional ability of appressoria. This was tested by determining the ability of *cpka*⁻ mutants to penetrate onion epidermis. Mutants were consistently reduced in their ability to penetrate epidermal cells, suggesting that they do not generate turgor (Xu *et al.,* 1997). The effect of mutations at the *CPKA* locus on glycerol accumulation in appressoria has yet to be determined, but should clarify the role of this essential regulatory gene further. The involvement of cAMP in signal transduction leading both to appressorium morphogenesis and appressorium function strongly indicates that a second cAMP-dependent protein kinase must be present in *M. grisea*, although it has not yet been identified. The degree of independence and redundancy in these parallel signal transduction pathways will undoubtedly prove illuminating in understanding the function of appressoria (Kronstad, 1997).

Hydrostatic pressure generated within the appressorium is focused on a point at the base of the appressorium, the penetration pore, which is free from both wall material and melanin (Bourett and Howard, 1990). It is surrounded by a ring-like structure which seals and strengthens the pore border. A penetration peg forms from this pore and is forced through the plant cuticle and into the epidermis. This process involves actin which has been localised to the penetration peg apex (Bourett and Howard, 1992).

3.3. COLONISATION OF RICE LEAF TISSUE BY *M. GRISEA*

Once within the rice epidermis the fungus grows in a determinate growth pattern with bulbous swollen hyphae. In this form it colonises the first 2-3 epidermal cells, growing intracellularly (Bourett and Howard 1990; Heath *et al.,* 1990). Rapid colonisation of the plant then occurs with infection hyphae ramifying throughout the epidermal and mesophyll cells. The hyphae narrow to a more conventional growth pattern as they traverse further from the initial site of infection and about 72 hours following initial infection, lesions become visible on the surface of the rice leaves. Very little is currently known about the physiology of invasive growth by *M. grisea.* It is clear that growth is very rapid because it has been estimated that in a heavy blast infection, up to 10% of the total leaf biomass is composed of fungal mycelium after 72 hours (Talbot *et al.,* 1993). Consistent with rapid growth of the fungus during this period, a number of genes have been found to be differentially regulated during early infection. These include *UEP1* which encodes a ubiquitin fusion protein and is highly expressed 48 hours after inoculation of rice seedlings. At this stage *M. grisea* has just entered the leaf epidermis and started to colonise the plant. Interestingly while *UEP1* was highly expressed at this time the transcript was not detectable subsequently, in spite of extensive fungal growth (McCafferty and Talbot, submitted). *UEP1* was unique in this expression pattern compared to other members of the ubiquitin gene family which were not highly expressed during pathogenesis. It is likely that *UEP1* is involved in ribosomal biogenesis and protein turnover which may therefore reach a peak at this time. Subsequent to this, the high level expression of the *M. grisea* mitochondrial small sub-unit ribosomal RNA has been observed, at a time just prior to symptom outbreak by the fungus. The regulated expression of this rRNA during pathogenesis indicates that a burst of mitochondrial biogenesis probably accompanies rapid growth of the fungus in plant tissue, but ceases almost completely before disease symptom formation (Talbot and Tongue, submitted). These observations suggest that plant colonisation is punctuated by the expression of specific genes at particular stages of development. They also suggest that the fungus grows extremely rapidly through leaf tissue but ceases growth abruptly prior to symptom outbreak.

3.4. DISEASE SYMPTOM DEVELOPMENT

Rice blast lesions are elliptical in shape with a pale chlorotic margin and necrotic centre. Some five to seven days following infection, aerial conidiophores differentiate from within the coalescing lesions and thousands of new conidia are produced to perpetuate the cycle (Ou, 1985; Bastiaans, 1990). During periods when the prevailing relative humidity exceeds 93%, lesions can produce 2000-6000 conidia each night for about two weeks (Ou, 1985).

Whether *M. grisea* lesions result due to fungal gene expression, or as a result of plant defence systems, remains unclear. A number of toxins are formed by *M. grisea* including tenuazonic acid, pyricularin, pyricularic acid, pyrichalasin H, pyricularol, tyrosol, picolinic acid. The requirement of these toxins for pathogenicity has, however, yet to be demonstrated and only tenuazonic acid appears to be present in diseased rice leaves at toxic levels (Lebrun *et al.*, 1990; Valent and Chumley, 1991).

A number of cell wall degrading enzymes are known to be produced by *M. grisea* including xylanases and other hemicellulases. The importance of these in tissue colonisation and symptom development is an area of increasing interest. A recent study by Wu *et al.* (1997), for example, showed that *M. grisea* has at least seven distinct endo-xylanase activities. Mutations at the *XYL1* and *XYL2* genes, encoding two endo-□-1,4,-D-xylanases, were generated by targeted deletion and a *xyl1/xyl2* double mutant then produced. None of the mutants were reduced in their ability to infect compatible rice cultivars, although growth defects were observed in the *xyl1/xyl2* double mutant in culture. Interestingly, they also discovered that four further xylanases could be detected in these mutants. This highlights the flexibility exhibited by *M. grisea* and suggests that cell-wall degrading enzyme activities specific to the pathogenic growth phase *in M. grisea*, may be present with particular roles in proliferation of the fungus in leaf tissue.

Disease symptom development may be influenced by starvation stress. It is clear, for example, that genes such as *MPG1* are starvation-stress induced and also highly expressed during disease symptoms (Talbot *et al.*, 1993; Lau and Hamer, 1996). Genes expressed early in infection, such as *UEP1* and the mito-ss rRNA, meanwhile are down-regulated by starvation stress and not highly expressed during disease symptom outbreak. The regulation of nitrogen starvation has been shown to be involved in *M. grisea* pathogenicity as two genes *NPR1* and *NPR2* which regulate nitrogen source utilisation and the expression of *MPG1*, are also required for full pathogenicity. Npr1 and Npr2 mutants were reduced in pathogenicity and their ability to express *MPG1* in response to starvation stress (Lau and Hamer, 1996). This is in contrast to the wide domain nitrogen regulator *NUT1* (homologous to *Aspergillus nidulans areA*) which regulates nitrogen source utilisation in the absence of ammonia. This gene was found not to be required for pathogenicity (Froehliger and Carpenter, 1996) indicating that in *M. grisea*, pathogenic growth may involve novel regulatory genes not found in related saprotrophic fungi. Nitrogen starvation of *M. grisea* has also recently been shown to elicit production of phytotoxic high molecular weight compounds which can cause rice leaf senescence. The production of this activity, thought to be a protein or glycoprotein-containing fraction, was nitrogen-starvation specific and was significantly reduced in *nut1*$^-$, *npr1*$^-$ or *npr2*$^-$ mutants. Taken together these results suggest that that starvation-stress of *M. grisea* may be one of the inductive cues for disease symptom development.

4. Conclusion

It is clear that the balance of studies of pathogenesis in *M. grisea* has so far been tipped heavily toward investigating appressorial morphogenesis. This is not surprising because it is a fascinating subject area and highly amenable to experimental manipulation. The key area of future research regarding appressorium development is likely to be identifying the receptors which mediate development of infection structures. This should prove possible now that signal transduction intermediates such as *PMK1* have been identified, by using the two-hybrid system (Xu and Hamer, 1996). The interplay with cAMP-dependent signalling will also prove to be highly informative as cAMP clearly mediates appressorium development and turgor generation perhaps through alternative forms of cAMP-dependent protein kinase (Xu *et al.,* 1997). Characterisation of genes encoding adenylate cyclase and ras will clearly be required, however, to implicate this signalling fully in the morphogenetic sequence (Gimeno *et al.,* 1992).

Appressorial turgor generation and the means by which turgor is translated to mechanical force are areas requiring extensive research. Although the biochemical basis of turgor appears to be centred around glycerol production, the exact route of its biosynthesis and the regulation of this process are unclear. The translation of turgor to mechanical force and alteration in the axis of cellular polarity during infection are also worthy of investigation. They are likely to require interplay of a large number of metabolic and biosynthetic pathways, including those controlling cell wall biosynthesis, polarity establishment and cell division.

The molecular biology of plant colonisation by *M. grisea* will also require a much more concerted effort if the pathogenic growth state is to be fully understood. A full temporal analysis of *M. grisea* gene expression would seem to be the most straightforward, albeit labour-intensive route, to understanding this process. The analysis of the increasing number of non-pathogenic mutants available will also be critical. Tipping the balance toward understanding the whole of pathogenesis, rather than simply primary leaf infection will take some time, but clearly presents the most exciting challenge of all.

5. References

Baker, B., Zambryski, P., Staskawitz, B. and Dinesh-Kumar, S.P. (1997) Signalling in plant-microbe interactions, *Science* **276**, 726-733.

Bastiaans, L. (1990) Ratio between virtual and visual lesion size as a measure to describe reduction in leaf photosynthesis of rice due to leaf blast, *Phytopathology* **81**, 611-615.

Beckerman, J.L., Naider, F. and Ebbole, D.J. (1997) Inhibition of pathogenicity of the rice blast fungus by *Saccharomyces cerevisiae* □-factor, *Science* **276**, 1116-1119.

Bourett T.M. and Howard, R.J. (1990) *In vitro* development of penetration structures in the rice blast fungus *Magnaporthe grisea, Canadian Journal of Botany* **68**, 329-342.

Bourett T.M.and Howard, R.J. (1992) Actin in penetration pegs of the rice blast fungus *Magnaporthe grisea, Protoplasma* **168**, 20-26.

Chumley F.G. and Valent B. (1990) Genetic analysis of melanin deficient non-pathogenic mutants of *Magnaporthe grisea, Molecular Plant Microbe Interactions* **3**, 135-143.

De Jong J.C., McCormack, B.J., Smirnoff, N. and Talbot, N.J. (1997) Glycerol generates turgor in rice blast, *Nature* **389**, 244-245.

Froeliger, E.H.and Carpenter, B.E. (1996) *NUT1*, a major nitrogen regulatory gene in *Magnaporthe grisea*, is dispensable for pathogenicity, *Molecular General Genetics* **251**, 647-656.

Gilbert, R. D., Johnson, A. M. and Dean, R. A. (1996). Chemical signals responsible for appressorium formation in the rice blast fungus, *Physiological and Molecular Plant Pathology* **48**, 335-346.

Gimeno, P. O., Ljungdahl, P. O., Styles, C.A. and Fink, G.R. (1992) Unipolar cell divisions in the yeast *Saccharomyces cerevisiae* lead to filamentous growth:regulation by starvation and RAS, *Cell* **68**, 1077-1090.

Hamer, J.E., Howard, R.J., Chumley, F.G. and Valent B. (1988) A mechanism for surface attachment of spores of a plant pathogenic fungus. *Science* **239**, 288-290.

Heath, M.C., Valent, B., Howard, R.J. and Chumley, F.G. (1990) Correlations between cytologically detected plant-fungal interactions and pathogenicity of *Magnaporthe grisea* toward weeping lovegrass, *Phytopathology* **80**, 1382-1386.

Herskowitz, I. (1995). MAP kinase pathways in yeast - for mating and more, *Cell* **80**, 187-197.

Holden, D. W. and Hamer, J.E. (1997). Linking approaches in the study of fungal pathogenesis: A commentary, *Fungal Genetics and Biology* **21**, 11-16.

Howard, R.J. (1994) Cell biology of pathogenesis, in R.S. Zeigler, S.A. Leong, P.S. Teng (eds.), *The Rice Blast Disease*, CABI, Wallingford, U.K. pp. 3-22.

Howard, R.J. and Ferrari, M.A. (1989) Role of melanin in appressorium formation, *Experimental Mycology* **13**, 403-418.

Howard R.J. and Valent B. (1996) Breaking and entering - host penetration by the fungal rice blast pathogen *Magnaporthe grisea, Annual Review of Microbiology* **50**, 491-512.

Howard, R.J., Ferrari, M.A., Roach, D.H. and Money, N.P. (1991) Penetration of hard substrates by fungu employing enormous turgor pressures. *Proceedings of the National Academy of Sciences USA* **88,** 11281-11284.

Jelitto, T.C., Page, H.A. and Read, N.D. (1994) Role of external signals in regulating the pre-penetration phase of infection by the rice blast fungus, *Magnaporthe grisea, Planta* **194**, 471-477.

Kronstad, J. W. (1997) Virulence and cAMP in smuts, blast and blights, *Trends in Plant Sciences* **2**, 193-199.

Lau, G.W. and Hamer, J.E. (1996) Regulatory genes controlling *MPG1* expression and pathogenicity in the rice blast fungus *Magnaporthe grisea The Plant Cell* **8,** 771-781.

Lebrun, M.H., Dutfoy, F., Gaudemer, F., Kunesch, G. and Gaubemer, A. (1990) Detection and quantification of the fungal phytotoxin tenuazonic acid produced by *Pyricularia oryzae,Phytochemistry* **29**, 3777-3783.

Lee Y-H. and Dean, R.A. (1993) cAMP Regulates Infection Structure Formation in the Plant Pathogenic Fungus *Magnaporthe grisea, The Plant Cell* **5**, 693-700.

Lee Y-H. and Dean, R.A. (1994) Hydrophobicity of contact surface induces appressorium formation in *Magnaporthe grisea, FEMS Microbiology Letters* **115**, 71-75.

Lundqvist, T., Rice, J., Hodge, C.N., Basarab, G.S., Pierce, J. and Lundqvist, Y. (1994) Crystal structure of scytalone dehydratase- a disease determinant of the rice pathogen *Magnaporthe grisea, Structure* **2**, 937-944.

Mitchell, T.K. and Dean, R.A. (1995) The cAMP-dependent protein kinase catalytic sub-unit is required for appressorium formation and pathogenesis by the rice blast fungus *Magnaporthe grisea, The Plant Cell* **7**, 1869-1878

Money, N. P. and Howard, R. J. (1996). Confirmation of a link between fungal pigmentation, turgor pressure, and pathogenicity using a new method of turgor measurement, *Fungal Genetics and Biology* **20**, 217-227.

Oliver, R.P. and Osbourn, A. (1995) Molecular dissection of fungal phytopathogenicity, *Microbiology* **141**, 1-9.

Ou, S.H. (1985) *Rice Diseases*, CABI, Surrey, U.K., pp. 109-201.

Rossman, A.Y., Howard R.J. and Valent B. (1990) *Pyricularia grisea*, the correct name for the rice blast fungus, *Mycologia* **82**, 509-512.

Schafer, W. (1994) Molecular mechanisms of fungal pathogenicity to plants, *Annual Review of Phytopathology* **32**, 461-477.

Sweigard, J.A., Chumley, F.G. and Valent, B. (1992) Disruption of a *Magnaporthe grisea* cutinase gene, *Molecular General Genetics* **232**, 183-190.

Talbot, N.J. (1995) Having a blast: exploring the pathogenicity of *Magnaporthe grisea, Trends in Microbiology* **3**, 9-16.

Talbot, N.J., Ebbole, D.J. and Hamer, J.E. (1993) Identification and characterization of *MPG1*, a gene involved in pathogenicity from the rice blast fungus *Magnaporthe grisea, The Plant Cell* **5**, 1575-1590.

Talbot, N.J., Kershaw, M.J., Wakley, G.E., de Vries, O.M.H., Wessels, J.G.H. and Hamer, J.E. (1996) *MPG1* encodes a fungal hydrophobin involved in surface interactions during infection-related development of *Magnaporthe grisea, The Plant Cell* **8**, 985-999.

Thines, E., Eilbert, F., Sterner, O. and Anke, H. (1997) Glisoprenin A, an inhibitor of the signal transduction pathway leading to appressorium formation in germinating conidia of *Magnaporthe grisea* on hydrophobic surfaces, *FEMS Microbiology Letters* **151**, 219-224.

Uchiyama, T., Ogasawara, N., Nanba, Y. and Ito, H. (1979) Conidial germination and appressorial formation on the coverglass or cellophane coated with various lipid components of plant leaf waxes, *Agriccutural Biological Chemistry* **43**, 383-384.

Uchiyama, T. and Okuyama, K. (1990) Participation of *Oryza sativa* leaf wax in appressorium formation by *Pyricularia oryzae, Phytochemistry* **29**, 91-92.

Valent, B. and Chumley, F.G. (1991) Molecular genetic analysis of the rice blast fungus *Magnaporthe grisea, Annual Revie of Phytopathology* **29**, 443-467.

Valent, B., Farrall, L. and Chumley, F.G. (1990) *Magnaporthe grisea* genes for pathogenicity and virulence identified through a series of backcrosses, *Genetics* **127**, 87-101.

Van Etten, H.D., Soby, S. Wasmann, C. and McCluskey, K. (1994) Pathogenicity genes in fungi, in M.J. Daniels, J.A. Downie and A.E. Osbourn (eds.). *Advances in Molecular Genetics of Plant Microbe Interaction, Vol. 3,* Kluwer, Dordrecht, The Netherlands.

Vidal-Cros, A., Viviani, F., Labesse, G., Boccara, M. and Gaudry, M. (1994) Polyhydroxynaphthalene reductase involved in melanin biosynthesis in *Magnaporthe grisea, European Journal of Biochemistry* **219**, 985-992.

Wu, S.C., Ham, K.S., Darvill, A.G. and Albersheim, P. (1997) Deletion of two endo-beta-1,4-xylanase genes reveals additional isozymes secreted by the rice blast fungus, *Molecular Plant Microbe Interactions* **10**, 700-708.

Xiao J-Z., Ohshima, A., Kamakura, T., Ishiyama, T. and Yamaguchi, I. (1994) Extracellular glycoprotein(s) associated with cellular differentiation in *Magnaporthe grisea, Molecular Plant Microbe Interactions* 7, 639-644.

Xu, J.R. and Hamer, J.E. (1996) MAP Kinase and cAMP signaling regulate infection structure formation and pathogenic growth in the rice blast fungus *Magnaporthe grisea, Genes and Development* **10**, 2696-2706.

Xu, J.R. Urban, M., Sweigard, J.A. and Hamer, JE (1997) The *CPKA* gene of *Magnaporthe grisea* is essential for appressorial penetration, *Molecular Plant Microbe Interactions* **10**, 187-194.

THE ROLE OF *MPG1* HYDROPHOBIN IN PATHOGENESIS OF THE RICE BLAST FUNGUS *MAGNAPORTHE GRISEA*

M.J. KERSHAW and N.J. TALBOT
Department of Biological Sciences,
University of Exeter,
Washington Singer Laboratories,
Perry Road, Exeter, EX4 4QG, UK.

1. Introduction

Fungal plant pathogens have evolved diverse mechanisms for gaining entry into host tissue, ranging from entry through natural plant openings to various mechanisms of direct penetration through the outer cuticle. In the bean rust pathogen *Uromyces appendiculatus*, for example, differentiation into the infective cell, or appressorium, is triggered when germ tubes are able to sense the height of the stomatal guard cells (Hoch *et al.*, 1987). This highlights the specialised nature of host perception by fungal pathogens and the likely complexity of the underlying biology.

The rice blast fungus *Magnaporthe grisea* (Hebert) Barr. has emerged as a tractable model organism in which to study plant infection processes. Very rapid progress has been made in understanding the mechanism of appressorium-mediated infection in this organism (see Howard and Valent, 1996). The infective cycle of *M. grisea* and the concerted efforts made to identify and characterise pathogenicity determinants of *M. grisea* are described in detail in the accompanying chapters by Talbot *et al.*,; Bhargava and Hamer. This chapter is focused on the pre-penetration phase of *M. grisea* development which begins when an asexual spore, or conidium, lands on the leaf surface and ends with successful infection through the cuticle. Recently, a gene called *MPG1* has been found to encode an important component of infection-related development and this chapter reviews the isolation of this gene, the characterisation of its putative product, and its likely function in *M. grisea* pathogenesis.

2. Infection-related Morphogenesis in *M. grisea*

From the time a conidium lands on the leaf surface, *M. grisea* begins a series of developmental and metabolic processes that enable it to penetrate directly into the host. Successful plant infection is dependent on a number of environmental cues and the completion of a defined morphogenetic sequence. A conidium quickly germinates to produce a short germ tube which differentiates at the tip into the infective cell, or

S. Sreenivasaprasad and R. Johnson (eds.), Major Fungal diseases of Rice Recent Advances, 35-48.
Prof. K. Manibhushan Rao Festschrift

appressorium (Howard *et al.*, 1991). Formation of the appressorium involves a switch from apical to radial growth and melanisation of the cell wall (Howard and Ferrari, 1989). This morphogenetic process only occurs on conducive surfaces which have to be hard and normally hydrophobic, conditions associated with the rice leaf surface (see Talbot, 1995).

3. Identification of the *MPG1* gene from *M. grisea*

A powerful strategy to identify genes playing important roles in pathogenicity is to isolate genes expressed preferentially during growth of the pathogen in the host plant. Using a differential cDNA screen, *M. grisea* genes abundantly expressed during infection of a susceptible rice cultivar were isolated, from among which the *MPG1* gene was identified (Talbot *et al.*, 1993). Total RNA was extracted from rice plants infected with the compatible *M. grisea* isolate Guy-11, and used to construct a directional cDNA library. This cDNA library was screened with radio-labelled cDNA probes derived from uninfected rice leaf poly$(A)^+$ RNA and *M. grisea*- infected rice leaf poly$(A)^+$ RNA.

A number of cDNAs were found that hybridised specifically to the cDNA probe from the infected tissue. Some of these clearly originated from the host, particularly as host plant gene expression is known to be induced in response to fungal invasion (Bowles, 1990; Yang *et al.*, 1997)), but a significant number represented novel *M. grisea* transcripts. The selected cDNAs were classified by their hybridisation patterns in RNA gel blots of poly $(A)^+$ RNA from uninfected rice leaves, infected rice leaves, or *M. grisea* grown in a complete medium. In this way one of the cDNAs, pNJT-15, was shown to be expressed abundantly during plant infection in comparison to its expression in *M. grisea* grown in axenic culture.

Southern blot analysis defined pNJT-15 as corresponding to a single copy gene conserved in all host-limited forms and pathotypes of *M. grisea* examined. The gene was RFLP mapped and defined a new locus, *MPG1*, near the end of linkage group H (designation of Romao and Hamer, 1992). This was confirmed by hybridisation of pNJT-15 to the corresponding chromosomal-sized DNA by pulse field gel electrophoresis.

The expression of *MPG1* was studied in detail by developing a method to quantify fungal biomass during rice infection. In this analysis plant and fungal DNA was extracted from rice plants 72 h after infection and probed with *ILV1*, a single copy gene encoding acetolactate synthase. The hybridisation signals obtained were compared to a dilution series of total genomic DNA from *M. grisea* also probed with *ILV1*. Using laser densitometry of the hybridisation signals it was calculated that 10% of the infected rice DNA at 72 h was composed of *M. grisea* genomic DNA. Because DNA extraction was mass-dependent this suggested that up to 10% of the rice leaf biomass was composed of colonising *M. grisea* mycelium. This value was used to determine that the *MPG1* transcript was 62 times more abundant in *M. grisea* mycelium growing *in planta* than when grown in liquid culture.

4. Inactivation of *MPG1* leads to a reduction in pathogenicity on rice.

The *MPG1* gene was inactivated using a one-step gene replacement (Talbot *et al.*, 1993). This involved the construction of a gene replacement vector containing flanking regions of the *MPG1* locus, with a selectable hygromycin B resistance gene cassette (*Hph*) inserted between them to replace the MPG1 open reading frame. The construct was linearised and transformed into *M. grisea* strain Guy-11. The use of the flanking regions allowed homologous integration of the vector at the *MPG1* locus and led to a direct replacement of the *MPG1* coding region with *Hph* (Talbot *et al.*, 1993).

Two separate hygromycin B resistant transformants, TM400-2 and TM400-5, showed the predicted gene replacement and were tested for pathogenicity. Significantly, the *mpg1::Hph* mutants showed a reduced ability to cause disease symptoms on rice leaves. The pathogenicity assays were carried out on susceptible rice plants by spraying a conidial suspension. The two transformants TM400-2 and TM400-5 caused very reduced symptoms with only a few visible lesions on rice leaves (Figure 1), in contrast to the wild type Guy-11 which caused large spreading lesions (Talbot *et al.*, 1993). The transformants were re-isolated from the resulting small number of lesions and used to re-infect rice seedlings. The lesion deficiency phenotype was consistently observed in these experiments confirming its association with the transformants (Talbot, unpublished). The mean lesion density was reduced by approximately 80% from 37.4 per 5-cm leaf tip in Guy-11 to 7.5 per 5 cm leaf tip in TM400-2.

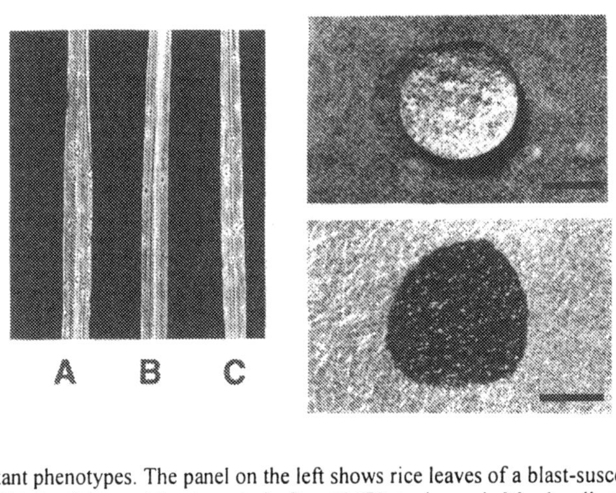

Figure 1. Mpg1 mutant phenotypes. The panel on the left shows rice leaves of a blast-susceptible cultivar, CO-39, infected with (A) A wild type *M. grisea* strain Guy-11 (B) An isogenic Mpg1 null mutant, TM400-2, (C) A wild type hygromycin-susceptible segregant from a cross of TM400-2 to TH3. All infections were carried out with a suspension of 1×10^4 conidia ml^{-1}. See Talbot *et al.* (1996) for details. The panels on the right show the Mpg1 'easily-wettable' phenotype. The top panel shows a water drop suspended on the surface of a conidiating culture of Guy-11. The bottom panel shows how the water droplet has soaked into the surface of a conidiating culture of the Mpg1 mutant, TM400-2. Bar = 1 mm

The reduced lesion density suggested that Mpg1 null mutants were defective in an early part of the infection process. *M. grisea* will attach, germinate and form appressoria on PTFE-teflon (Du Pont) cover-slips providing a convenient technique for assessing this process. Examination of conidial germination and appressorium development showed that the Mpg1 mutants TM400-2 and TM400-5 were able to germinate normally but had a significantly reduced ability to undergo appressorium development.

Germ tubes of Mpg1 null mutants undergo hooking and swelling, which has been shown to occur prior to appressorium formation (Bourett and Howard, 1990), but then often fail to differentiate into appressoria. Typically, a long germ tube forms which undergoes hooking and swelling at a number of points along the hypha but fails to produce a mature appressorium. Mpg1 mutants typically produced appressoria from 15-20% of germinated conidia, compared with 80-90% from the isogenic wild type strain Guy-11 (Talbot *et al.*, 1993). Another phenotype associated with Mpg1 mutants is their reduced ability to conidiate. Wild type strains such as Guy-11 will normally produce in the region of 7×10^6 conidia ml^{-1} from an agar plate culture flooded with water, whereas the Mpg1 mutants produced as few as 5×10^4 ml^{-1}. This indicated that *MPG1* may have a role not only in appressorium formation but also in conidial morphogenesis. The possibility that *MPG1* is involved in these two distinct developmental processes was supported by the detection of the *MPG1* transcript at two separate times during pathogenesis, first during appressorium differentiation, and subsequently at the time of symptom development. *MPG1* was also shown to be highly expressed in conidiating cultures (Talbot *et al.*, 1993; Beckerman and Ebbole, 1996).

5. MPG1 encodes a fungal hydrophobin

MPG1 was found to encode a 15kD secreted protein belonging to a family of fungal proteins designated hydrophobins (Talbot *et al.*, 1993, 1996). Fungal hydrophobins were first discovered as the products of genes expressed abundantly during the development of the basidiomycete *Schizophylllum commune* (Wessels *et al.*, 1991) . Four *S. commune* genes were identified encoding small moderately hydrophobic, secreted proteins with a conserved spacing of eight cysteine residues. Hydrophobins as a group, are defined by this characteristic spacing of cysteine residues in their amino acid sequence (Wessels, 1996). All hydrophobin genes have classic signal peptide sequences and in the case of the *S. commune* hydrophobin *Sc3*, secretion has been confirmed by immunolocalisation. (Wösten *et al.*, 1993). To date, twenty different hydrophobins have been reported from a range of fungal species, including ascomycetes, basidiomycetes and zygomycetes. The diversity of fungi identified with hydrophobin-encoding genes, suggests they may be ubiquitous among filamentous fungi. (Wessels, 1996)

Hydrophobins have been implicated in aerial morphogenesis in *S. commune* (Wessels, 1996; Talbot, 1997) and as spore wall proteins in *Aspergillus nidulans* (Stringer *et al.*, 1991) and *Neurospora crassa* (Bell-Pederson *et al.*, 1992). Hydrophobins have also been suggested to have a role as plant toxins. The Dutch Elm disease toxin Cerato-ulmin, for example, is a hydrophobin (Stringer and Timberlake,

1993), although it appears not to be required for full pathogenicity (Bowden *et al.*, 1996).

Hydrophobins have been divided into two distinct classes by the arrangement of hydrophobic and hydrophilic residues in their primary amino acid sequence (Wessels, 1994). These differences are reflected in their biochemical properties. Class I hydrophobins, which include the products of *MPG1*, *rodA* , *dewA* , *EAS* and the *S. commune* hydrophobins, form high molecular weight, SDS-insoluble, protein complexes which can only be broken into their constituent monomers by treatment with strong oxidising agents such as performic acid. (de Vries *et al.*, 1993). These characteristics are based on analysis of Sc3p, Sc4p and MPG1p but are likely to be shared by all members of the group. It is probably because of these unusual biochemical properties that most Class I hydrophobins have only been discovered subsequent to the identification of their corresponding genes (Wessels, 1996). Purification and characterisation of the Sc3 hydrophobin from *S. commune* lead to the discovery of inter-facial self assembly which describes the ability of hydrophobins to respond to interfaces between water and air, or between water and hydrophobic surfaces (Wösten *et al.*, 1993). The self-assembly process appears to involve spontaneous aggregation of hydrophobin monomers mediated by short-range hydrophobic interactions (Wösten *et al.*, 1993). The aggregates formed are extremely insoluble and, strikingly, are not dispersed, even by extraction with hot detergents. Only by treating these insoluble aggregates with trifluoroacetic acid (TFA), or formic acid, are the polymers dissociated into the component hydrophobin monomers (de Vries *et al.*, 1993).

Class II hydrophobins, which include cerato-ulmin (*Ophiostoma ulmi)* and cryparin (*Cryphonectria parasitica*), also polymerise into aggregates but these are much more unstable and can be dissolved in 60% ethanol (Takei and Richards, 1978). They have been studied in some detail due to the pioneering work done on the phytotoxin cerato-ulmin (Richards and Takai, 1993; Takai 1974), but were only found to be homologous to hydrophobins after the amino acid sequence of cerato-ulmin was fully derived (Bowden *et al.*, 1996).

The purification scheme established for *Sc3* was adapted to purify the Mpg1 hydrophobin from shake cultures of *M. grisea* grown under conditions of nitrogen starvation and labelled with $^{35}SO4^{2-}$ (Talbot *et al.*, 1996). This work established the *MPG1* gene product as a 15kD secreted protein with properties characteristic of a class I hydrophobin. The putative MPG1 protein monomer is soluble in TFA, and exhibits a mobility shift after treatment with performic acid due to oxidation of cysteine residues. This has been shown to be characterisitc of hydrophobins (Talbot *et al.*, 1996). One of the many roles attributed to fungal hydrophobins is the formation of a hydrophobic rodlet layer as the outermost layer of the cell wall of conidia. The hydrophobic properties this layer provides, enables spores to survive desiccation and also assists in their dispersal. Conidia of the *Eas* mutants in *N. crassa*, for example, differ from the wild type in that they are easily-wetted and do not disperse readily in air currents. The conidia of these mutants lack rodlets on their surface (Beever and Dempsey, 1978; Dempsey and Beever, 1979). The rodlet layer has been characterised biochemically and shown to be composed of *EAS* hydrophobin (Templeton *et al.,* 1995). Gene disruption of the *EAS* gene, meanwhile, resulted in loss of the rodlet layer (Bell-Pederson *et al.,* 1992; Lauter *et al.,* 1992). The *A. nidulans* hydrophobins *rodA* and *dewA* were cloned from a differential cDNA screen which identified sporulation-specific genes. Both appear to contribute to the hydrophobicity of the spore wall because spores of *dewA⁻ rodA⁻* double mutants were found to be less hydrophobic than those of either mutant alone (Stringer and Timberlake, 1995). Disruption of the *rodA* gene results in a phenotype in which the conidiophore cells and spores lack an external rodlet layer (Stringer *et al.,* 1991). *dewA* mutants were named for their detergent wettable phenotype because sporulating colonies were wetted rapidly in detergent solutions (Stringer and Timberlake, 1995). The *dewA* gene product was localised to conidia by epitope tagging using the influenza haemagglutinin epitope (Stringer and Timberlake, 1995). Localisation of dewAp was found on the outer conidial wall. Interestingly, *dewA* does not seem to encode rodlets directly because rodlets were still present in *dewA⁻* mutants, as long as they expressed *rodA* (Stringer and Timberlake, 1995).

The expression of *MPG1* during conidiation and its relatedness to hydrophobin-encoding genes suggested that it might encode a component of conidial cell walls. To test this hypothesis the spore surface of wild type and Mpg1 null mutants was examined using freeze fracture of conidial suspensions, which were then coated with carbon/platinum to produce replicas. These were examined by transmission electron microscopy to determine the surface topology (Figure 2).

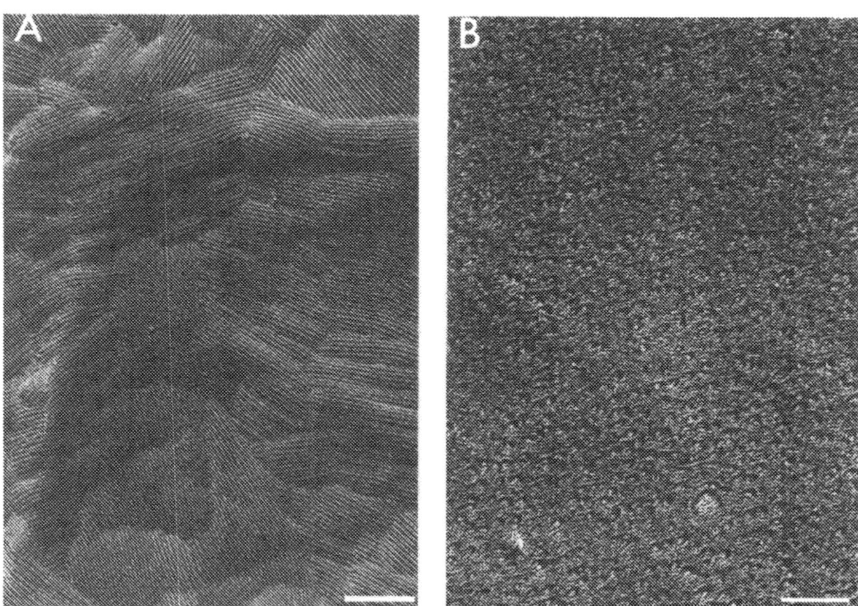

Figure 2. The *M. grisea* conidial rodlet layer is absent in Mpg1 mutants. Transmission electron micrographs of carbon-platinum replicas of freeze-fractured conidia to show surface topology. (A) Conidium of the wild type strain Guy-11 (B) Conidium of the isogenic Mpg1 mutant TM400-2. Bar = 100nm

These studies revealed the presence of a wall layer in *M. grisea* conidia composed of interwoven rodlets with an apparent diameter of 5 nm (Talbot *et al.,* 1996). The rodlet layer was completely absent from conidia of the *mpg1*⁻ mutant (TM400-2). Analysis of 20 wild type and mutant co-segregants showed that the presence of the rodlet layer completely co-segregated with the *MPG1* gene (Talbot *et al.,*. 1996). Interestingly, the *M. grisea* rodlets are small in comparison to those observed on conidia of *A. nidulans* (10 to 13 nm) *N. crassa* (11-13nm), and *S. commune* (~ 13 nm). The significance of this size difference is unknown, although it is raises the possibility that rodlet size is an indicator of the physical characteristics of the constituent hydrophobin.

Hydrophobin extractions were performed from *M. grisea* conidia harvested from aerial hyphae of wild type and *mpg1*⁻ mutants. Extractions were performed by dry harvesting conidia and selecting the TFA-extractable fraction. SDS-polyacrylamide gel electrophoresis showed a single 15 kD protein from conidial wall extracts of Guy-11. This protein was completely absent from *mpg1*⁻ mutant strains (Talbot *et al.,* 1996).

The absence of spore rodlets in *eas*⁻ mutants leads to a pronounced clumping of conidia with cultures becoming severely water-logged, hence the mutant's name "easy-wettable" (Bell-Pedersen *et al.,* 1992). In the case of *rodA*⁻ mutants this phenotype is slightly less pronounced, but still produces an obvious water-soaked appearance to cultures (Stringer *et al.,* 1991). These phenotypes are thought to be caused by the loss of surface hydrophobicity brought about by the absence of spore rodlets. The hydrophobicity of the Mpg1 mutants was therefore tested by applying drops of water to the surface of conidiating cultures of wild type and Mpg1 mutant strains. After a period of 12 hours the water drop was still suspended on the hydrophobic surface of the wild type strain but had soaked into the surface of the *mpg1*⁻ strain TM400-2, leaving a pronounced water-soaked mark. The *MPG1* hydrophobin must therefore contribute to cell surface hydrophobicity of aerial hyphae of *M. grisea* (Talbot *et al.,* 1993).

6. Regulated expression of *MPG1*

MPG1 is highly expressed during infection related development and at the time of symptom development. It is also known to be regulated by starvation stress, being induced by either carbon or nitrogen starvation, but repressed during growth in rich medium (Talbot *et al.,* 1993; Beckerman and Ebbole, 1996; Lau and Hamer, 1996).

In *A. nidulans* and *N. crassa* nitrogen utilisation pathways are activated by wide domain regulators (see Arst and Cove, 1973; Coddington, 1976). These are members of the GATA family of DNA-binding proteins and have been shown to be transcriptional activators encoded by *areA* in *A. nidulans* (Caddick *et al.,* 1996), and *nit2* in *N. crassa* (Fu and Marzluf, 1990). *MPG1* is regulated by the wide domain regulator of nitrogen source utilization, *NUT1* , which is homologous to *areA* and *nit2* of *A. nidulans* and *N. crassa* respectively. Analysis of the *MPG1* promoter has also revealed the presence of 4 GATA-like binding sites which may be recognised by *NUT1* (Lau and Hamer, 1996;

Talbot, unpublished). Most genes involved in regulating nitrogen metabolism, including *NUT1*, are dispensable for pathogenicity (Lau and Hamer, 1996; Froeliger and Carpenter, 1996). *MPG1* is, however, also regulated by two novel *trans*-acting regulators encoded by the *NPR1* and *NPR2* genes (Lau and Hamer, 1996). These genes regulate the response of *M. grisea* to nitrogen starvation and/or carbon starvation, so that corresponding mutants are unable to utilise a wide range of nitrogen sources. By screening for chlorate resistant mutants which could only grow on glutamine as a nitrogen source *npr1* and *npr2* mutants were selected. The dramatic loss in pathogenicity in *npr1*⁻ and *npr2*⁻ mutants is more severe than that observed in Mpg1 deletion mutants. It is therefore likely that *NPR1* and *NPR2* encode regulators of a number of other important genes involved in pathogenesis by *M. grisea*.

The late expression of *MPG1* during pathogenesis indicates an involvement in conidial morphogenesis consistent with its function in encoding the spore rodlet layer. *EAS* is similarly highly expressed during conidiation and its regulation has been studied in detail, acting as a template for future studies of *MPG1* expression. *EAS* was independently isolated as a clock-controlled (circadian) gene (ccg-2) (Loros *et al.*, 1989) and as a blue light inducible gene (Lauter *et al.*, 1992) and was only later shown to encode a fungal hydrophobin (Bell-Pedersen *et al.*, 1992). Deletion analysis of the *EAS* (*ccg*-2) promoter was carried out to localise *cis*-acting elements mediating clock, light, and developmental control (Bell-Pederson *et al.*, 1996). They identified a distinct activating clock element (ACE) residing within a 45-bp region, spacially distinct from other *cis*-acting elements mediating photo-induction and developmental regulation (Bell-Pedersen *et al.*, 1996). This study highlighted the likely complexity of expression of hydrophobin genes as *EAS* is subject to repression during mycelial growth of *N. crassa*, developmental regulation, circadian expression, photo-induction and probably starvation-stress induction. Each of these environmental influences requires specific *cis*-acting factors and corresponding *trans*-acting regulators to co-ordinate EAS expression during conidiation by the fungus (Bell-Pederson *et al.*, 1996).

The production of asexual conidia has been studied in great detail in *A. nidulans* (Clutterbuck, 1969; Timberlake, 1990). The regulation of this system has also been investigated and numerous regulatory genes identified by mutation (Clutterbuck, 1969). Morphogenetic target genes have also been isolated that direct the synthesis of spore-specific enzymes or structural proteins and *rodA* and *dewA* clearly belong to this class. These studies therefore also act as a model for the likely regulation of *MPG1* during fungal development and pathogenesis.

Mutant phenotypes of *A. nidulans* asexual development regulators result in arrested conidiophore development. This has led to the identification of a number of genes including *fluffy* (*fluG*), *bristle* (*brlA*), *abacus* (*aba*) and *wet-white* (*wetA*) which are central to the regulation of conidiophore development (Adams, 1995). Null *brlA* mutants, for example, make only conidiophore stalks and fail to produce any of differentiated conidiophore cells (Clutterbuck, 1969), *abaA* mutants produce metulae but not phialides or conidia, and *wetA* mutants produce normal conidiophores but the spores fail to undergo normal maturation and autolyse.

The best characterised regulatory gene, *brlA*, plays a central role in controlling conidiophore development in *A. nidulans. It* has been shown that *brlA* activates specific structural genes including *rodA* as well as the next regulatory gene *abaA* Therefore . *brlA* is required for the switch from polarised cell extension of the conidiophore stalk to non-polar growth, resulting in conidiophore vesicle formation. Consistent with this, forced expression of *brlA* in vegetative cells results in a cessation of polar growth (Adams *et al*., 1990) Using an *in vivo* approach, utilising *Saccharomyces cerevisiae* as a heterologous expression system, it has been shown that *brlA* is capable of activating transcription from a minimal yeast promoter to which upstream sequences from the *rodA* gene had been fused. Additional random DNA fragments mediating *brlA* - dependent gene activation were also identified and lead to the identification of a *cis*-acting sequence motif also present in the *rodA* fragment, designated *brlA* response elements (BREs).

Interestingly, a number of near-consensus bristle response elements (BREs) have been identified in the *MPG1* promoter implying that *MPG1* may be regulated by a *brlA*-homologous regulator during conidiation and perhaps appressorium development (Talbot, unpublished). This suggests that expression studies of *MPG1* may lead to a fuller understanding of the regulators of appressorium development and pathogenicity in *M. grisea.*

It is clear that *MPG1* is under the control of complex regulatory mechanisms involving multiple *trans*-acting factors, including the products of *NUT1*, *NPR1*, *NPR2* and perhaps a *brlA* homolog. The elaborate control of *MPG1* expression probably reflects the complex spatio-temporal interplay of environmental cues influencing appressorium development, conidial development and pathogenesis in the fungal mycelium.

7. The Role of *MPG1* in Pathogenesis

A gene replacement of *MPG1* resulted in reduction in pathogenicity, consistent with the gene serving a role in appressorium development (Talbot *et al.,* 1993). The re-introduction of *MPG1* fully complemented all mutant phenotypes including easy wettability of cultures, conidial production, the formation of a conidial rodlet layer and appressorium development. The transformants were also fully restored in pathogenicity on rice (Talbot *et al.,* 1996).

The induction and efficient development of appressorium formation is dependent upon a number of conditions being met, and perception of these factors, appears to be the most likely function for *MPG1*. These include the surface hardness (Xiao *et al.,* 1994) and hydrophobicity (Lee and Dean, 1994), as well as external factors including light (Jellito *et al.,* 1994), and nutrient limitation (Lau and Hamer, 1996). Interestingly, nutrient rich conditions which repress *MPG1* expression (Talbot *et al.,* 1993), also inhibit appressorium formation (Beckerman and Ebbole, 1996). This suggests that expression of *MPG1* during appressorium development may be partly induced by the lack of nutrients surrounding the germinating conidium. In the entomopathogenic fungus *Metarhizium anisopliae* nutrient limitation has similarly been associated with formation of appressoria and expression of a hydrophobin gene, *ssgA* (St. Leger *et al.*, 1992).

High level expression of *MPG1* during appressorium development would be likely to lead to secretion of the hydrophobin directly onto the rice leaf surface. Inter-facial self assembly of hydrophobins has been found to occur in response to hydrophobic surfaces (Wösten *et al.*, 1994a) and *sc3*$^-$ gene disruptants, for example, are reduced in their ability to adhere to hydrophobic surfaces. Inter-facial self assembly of the *MPG1* hydrophobin at the fungus-plant interface may therefore enable it to function in attachment and perception of the leaf surface (Talbot, 1995). To investigate this possible function for *MPG1*, a series of experiments was carried out by Talbot *et al.* (1996). First, a series of flow chamber experiments using water flow rates of up to 7.5 L min^{-1} failed to remove appressoria or attached germ tubes from hydrophobic surfaces. This showed the strength of attachment to hydrophobic surfaces, but surprisingly the same result was obtained with both wild-type and *mpg1*$^-$ strains. These findings indicated that attachment of fungal germ tubes and appressoria was due to multiple extracellular components.

To investigate the components of attachment, germinated conidia/appressoria were grown on teflon membranes and subjected to a series of washes in water, 60% ethanol, and 2% hot SDS. These washes were designed to disrupt any carbohydrate or protein-mediated attachment processes, but without affecting a hydrophobin-associated process.

After washing in hot SDS, it was found that wild type *M. grisea* germ tubes and appressoria were still predominantly attached to hydrophobic surfaces. However, *mpg1*$^-$ mutants had mainly been washed away. This strongly indicates that attachment of *M. grisea* germ tubes and appressoria to hydrophobic surfaces contains a hydrophobin component which is impervious to treatment by hot SDS. The fact that this component is absent from *mpg1*$^-$ mutants, co-segregates with *MPG1*, and is fully restored by re-introduction of the *MPG1* gene, suggests that *MPG1* encodes this activity (Talbot *et al.*, 1996).

Consistent with a function for MPG1p in surface perception, Beckerman and Ebbole (1996) found that *MPG1* is required for efficient induction of appressoria in response to surface hydrophobicity. They found that a number of artificial substrates could support efficient appressorium formation in *mpg1*$^-$ strains including relatively hydrophilic glass surfaces. This shows that in the absence of the *MPG1* hydrophobin other factors may induce development, particularly on non-hydrophobic surfaces. This is consistent with recent observations that independent signalling pathways may trigger appressorium development on hydrophobic and hydrophilic surfaces. Interestingly both Beckerman and Ebbole (1996) and Talbot *et al.* (1996) found that cAMP remediates the *mpg1*$^-$ appressorial deficiency phenotype. This suggests that *MPG1* acts upstream of the cAMP signalling pathway which controls appressorium development.

The main function of *MPG1* in *M. grisea* pathogenesis therefore appears to be as a sensor of the leaf surface. This is likely, however, simply to be a consequence of secreting a hydrophobin onto a hydrophobic surface. Interfacial self-assembly of the secreted hydrophobin would produce an amphipathic monolayer on the leaf surface with the hydrophobic side in contact with the rice leaf, and its hydrophilic side facing the germ tube. This is likely to aid attachment by presenting a hydrophilic surface for the action of mucilages and glycoprotein adhesives known to be involved in appressorium morphogenesis (Xiao *et al.*, 1994). It is, however, also likely to act as a conformational cue for appressorium formation. The absence of *MPG1* results in a reduced efficiency of this process, but not in its complete elimination. This is because all of the regulatory and structural components for appressorium development are still present and likely to be induced, albeit with reduced frequency.

8. Conclusion

A number of critical experiments need to be carried out to understand the function of *MPG1* completely. First and foremost, the localisation of the *MPG1* gene product has not yet been achieved due to the technical difficulty of isolating sufficient protein product. Over-expression of *MPG1* in prokaryotic hosts has not been successful due to the presumed toxicity of the hydrophobin product. Over-expression in a eukaryotic host is currently underway, in addition to epitope tagging experiments and should allow localisation of MPG1p during appressorium development (Kershaw, unpublished; Tenjo and Hamer, pers. comm.).

The large scale purification of MPG1p would also allow direct determination of its ability to self-assemble (Wösten *et al.*, 1993; 1994b) and the exact environmental conditions allowing this to occur. This would define its physical characteristics in greater detail and form the basis for a full structure-function study which would involve large scale targeted mutagenesis and structural studies. Perhaps the most powerful way of carrying out such a study, however, would be to examine the functional relatedness of the fungal hydrophobins using the genetic tractability of *M. grisea*. A large number of hydrophobin-encoding genes have now been described in ascomycetes and basidiomycetes and yet the functions of many of these proteins remain unclear, although they are expressed during a variety of developmental stages, or in response to environmental stress (Wessels, 1997).

MPG1 is known to direct formation of the spore wall protein but, more significantly, it also appears to function as a developmental sensor for appressorium formation (Talbot *et al.*, 1996). This raises the possibility that plant pathogenic fungi have recruited hydrophobins for critical surface interactions. This evolutionary development may have taken two possible routes: either hydrophobins have specific properties that enable them to self-assemble, dependent on the prevailing physical properties of a surface, or particular regulatory pathways have evolved to secrete generic hydrophobins at critical times.

These alternative hypotheses can be readily tested by expression of fungal hydrophobin in *mpg1*$^-$ deletion mutants, either expressed under their own promoters or under control of the *MPG1* promoter. Hydrophobin genes from *S. commune, N. crassa,*

A. nidulans and *M. anisopliae* have so far been transformed into an *mpg1*⁻ mutant background in these two alternative forms. Transformants with single copy insertions of the gene have recently been phenotypically characterised to determine whether any of these hydrophobins can complement any, or all, of the Mpg1 mutant phenotypes. Preliminary results from this investigation suggest that different hydrophobins show differential abilities in complementing Mpg1 mutant phenotypes (Kershaw and Talbot, unpublished). Interestingly, some show no complementation even when expressed under control of the *MPG1* promoter. These preliminary observations suggest that hydrophobins are diverse in function and *MPG1* may have been specifically adapted for its role in pathogenesis.

9. References

Adams, T.H. (1995) Asexual sporulation in higher fungi, in N.A.R Gow and G.M. Gadd (eds.) The Growing Fungus, Chapman and Hall, London, UK, pp.367-382.

Adams, T.H., Boylan, M.T. and Timberlake, W.E. (1990) brlA is necessary and sufficient to direct conidiophore development in Aspergillus nidulans, *Cell* **54**, 353-362.

Arst, H.N. and Cove, D.J. (1973) Nitrogen metabolite repression in Aspergillus nidulans, *Molecular General Genetics* **126**, 111-141.

Beckerman, J. L. and Ebbole, D. J. (1996) *MPG1*, a gene encoding a fungal hydrophobin of *Magnaporthe grisea*, is involved in surface recognition, *Molecular Plant Microbe Interactions* **9**, 450-456.

Beever, R. E. and Dempsey, G. P. (1978) Function of rodlets on the surface of fungal spores, *Nature* **272**, 608-610.

Bell-Pederson, D., Dunlap, J. C. and Loros, J. J. (1992) The *Neurospora* circadian clock-controlled gene, *ccg-2*, is allelic to *eas* and encodes a fungal hydrophobin required for formation of the conidial rodlet layer, *Genes and Development* **6**, 2382-2394.

Bell-Pederson, D., Dunlap, J. C. and Loros, J. J. (1996) Distinct *cis*-acting elements mediate clock, light, and developmental regulation of the *Neurospora crassa EAS* (ccg-2) gene, *Molecular Cell Biology* **16**, 513-521.

Bourett, T.M. and Howard, R.J. (1990) In vitro development of penetration structures in the rice blast fungus *Magnaportthe grisea, Canadian Journal of Botany* **68**, 329-342.

Bowden, C.G., Smalley, E., Guries, R.P., Hubbes, M. and Horgen, P.A. (1996) Lack of association between cerato-ulmin production and virulence in *Ophiostoma novo-ulmi, Molecular Plant Microbe Interactions* **9**, 556-564.

Bowles, D.J.(1990) Defense related proteins in higher plants, *Annual Review of Biochemistry* **59**, 873-907.

Caddick, M.X., Arst, H.N. Taylor, L.H., Johnson, R.I. and Brownlee, A.G. (1986). Cloning of the regulatory gene *areA* mediating nitrogen metabolite repression in *Aspergillus nidulans, EMBO Journal* **5**, 1087-1090.

Clutterbuck, A.J. (1969) A mutational analysis of conidial development in *Aspergillus nidulans, Genetics* **63**, 317-327.

Coddington, A. (1976) Biochemical studies on the nit mutants of *Neurospora crassa*, *Molecular General Genetics* **145**, 195-206

Dempsey, G.P. and Beever, R.E. (1979) Electron microscopy of the rodlet layer of *Neurospora crassa* conidia, *Journal of Bacteriology* **139**, 757-761.

de Vries, O. M. H., Fekkes, M. P., Wosten, H. A. B. and Wessels, J. G. H. (1993) Insoluble hydrophobin complexes in the walls of *Schizophyllum commune* and other filamentous fungi, *Archieves in Microbiol*ogy **159**, 330-335.

Froeliger, E. and Carpenter, B. (1996) *NUT-!*, a major nitrogen regulator in *Magnaporthe grisea* is dispensible for pathogenicity, *Molecular General Genetics* **251**, 647-656.

Fu, Y.-H. and Marzluf, G.A. (1990) *nit-2,* the major positive acting nitrogen regulating gene of *Neurospora crassa,* encodes a sequence-specific DNA binding protein, *Proceedings of the National Academy of Sciences USA.* **87**, 5331-5335.

Hoch, H.C., Staples, R.C., Whitehead, B., Comeau, J. and Wolf, E.D. (1987) Signaling for growth orientation and cell differentiation by surface topography in *Uromyces*, *Science* **235**, 1659-1662.

Howard, R.J. and Ferrari, M.A. (1989) Role of melanin in appressorium function, *Experimental Mycology* **13**, 403-418.

Howard R.J., and Valent B. (1996) Breaking and entering - host penetration by the fungal rice blast pathogen *Magnaporthe grisea, Annual Review of Microbiology* **50**, 491-512.

Howard, R.J., Ferrari, M.A., Roach, D.H. and Money, N.P. (1991) Penetration of hard substrates by a fungus employing enormous turgor pressures, *Proceedings of the National Academy of Sciences USA* **88**, 11281-11284.

Jelitto, T.C., Page, H.A. and Read, N.D. (1994) Role of external signals in regulating the pre-penetration phase of infection by the rice blast fungus, *Magnaporthe grisea, Planta* **b194**, 471-477.

Lau, G. and Hamer, J.E. (1996) Regulatory genes controlling *MPG1* expression and pathogenicity in the rice blast fungus *Magnaporthe grisea, The Plant Cell* **8**, 771-781.

Lauter, F. R., Russo, V. E. A. and Yanofsky, C. (1992) Developmental and light regulation of *Eas*, the structural gene for the rodlet protein of *Neurospora crassa, Genes and Development* **6**, 2373-2381.

Lee, Y-H. and Dean, R.A. (1994) Hydrophobicity of contact surface induces appressorium formation in *Magnaporthe grisea, FEMS Microbiology Letters* **115**, 71-76.

Loros, J.J., Denome, S.A. and Dunlap, J.C. (1989) Molecular cloning of genes under the control of the circadian clock in *Neurospora, Science* **243**, 385-388.

Richards W.C. and Takai, S. (1993) Amino acid sequence and spectroscopic studies of Dutch elm disease toxin, cerato-ulmin, in M.B. Sticklen and J.L. Sherald (eds), *Dutch Elm Disease Research, Cellular and Molecular Approaches*, Springer Verlag, New York, pp 152-170.

Romao, J. and Hamer, J.E. (1992) Genetic organization of a repeated DNA sequence family in the rice blast fungus, *Proceedings of the National Academy of Sciences USA* **89**, 5316-5320.

St. Leger, R. J., Staples, R. C. and Roberts, D. W. (1992) Cloning and regulatory analysis of starvation-stress gene, *ssgA*, encoding a hydrophobin-like protein from the entomopathogenic fungus *Metarhizium anisopliae, Gene* **120**, 119-124.

Stringer, M. A. and Timberlake, W. E. (1993) Cerato-ulmin, a toxin involved in Dutch Elm disease, is a fungal hydrophobin, *The Plant Cell* **5**, 145-146.

Stringer, M. A. and Timberlake, W. E. (1995) *dewA* encodes a fungal hydrophobin component of the *Aspergillus* spore wall, *Molecular Microbiology* **16**, 33-44.

Stringer, M. A., Dean, R. A., Sewell, T. C. and Timberlake, W. E. (1991) Rodletless, a new Aspergillus developmental mutant induced by directed gene inactivation, *Genes and Development* **5**, 1161-1171.

Takai, S. (1974) Pathogenicity and cerato-ulmin production in *Ceratocystis ulmi, Nature* **52**, 124-126.

Takai, S. and Richards, W. C. (1978) Cerato-ulmin, a wilting toxin of *Ceratocystis ulmi*: Isolation and some properties of cerato-ulmin from the culture of *C. ulmi, Journal of Phytopathology* **91**, 129-146.

Talbot, N. J. (1995) Having a blast: exploring the pathogenicity of *Magnaporthe grisea, Trends in Microbiology* **3**, 9-16.

Talbot, N.J. (1997) Fungal Biology: Growing into the air, *Current Biology* **7**, R78-82

Talbot, N. J., Ebbole, D. J. and Hamer, J. E. (1993) Identification and characterization of *MPG1*, a gene involved in pathogenicity from the rice blast fungus *Magnaporthe grisea, The Plant Cell* **5**, 1575-1590.

Talbot, N. J., Kershaw, M. J., Wakley, G. E., de Vries, O. M. H, Wessels, J. G. H. and Hamer, J. E. (1996) MPG1 encodes a fungal hydrophobin invovled in surface interactions during infection-related development by *Magnaporthe grisea, The Plant Cell* **8**, 985-989.

Templeton, M. D., Greenwood, D. R. and Beever, R. E. (1995) Solubilization of *Neurospora crassa* rodlet proteins and identification of the predominant protein as the proteolytically processed *eas* (*ccg-2*) gene product, *Experimental Mycology* **19**, 166-169.

Timberlake, W.E. (1990) Molecular genetics of *Aspergillus* development, *Annual Review of Genetics* **24**, 5-36.

Wessels, J. G. H. (1994) Developmental regulation of fungal cell wall formation, *Annual Review of Phytopathology* **32**, 413-437.

Wessels, J. G. H. (1996) Fungal hydrophobins. Proteins that function at an interface, *Trends in Plant Science* **1**, 9-15.

Wessels, J.G.H. (1997) Hydrophobins: Proteins that change the nature of the fungal surface, *Advances in Microbial Physiology* **38**, 1-45.

Wessels, J. G. H., de Vries, O. M. H., Ásgeirsdóttir, S. A. and Schuren, F. H. J. (1991a) Hydrophobin genes involved in formation of aerial hyphae and fruit bodies in *Schizophyllum*, The *Plant Cell* **3**, 793-799.

Wessels, J. G. H., de Vries, O. M. H., Ásgeirsdóttir, S. A. and Springer, J. (1991b) The *thn* mutation of *Schizophyllum commune*, which suppresses formation of aerial hyphae, affects expression of the *Sc3* hydrophobin gene, *Journal of General Microbiology* **137**, 2439-2445.

Wösten, H. A. B., de Vries, O. M. H. and Wessels, J. G. H. (1993) Interfacial self-assembly of a fungal hydrophobin into a hydrophobic rodlet layer, *The Plant Cell* **5**, 1567-1574.

Wösten, H. A. B., Ásgeirsdóttir, S. A., Krook, J. H., Drenth, J. H. H. and Wessels, J. G. H. (1994a) The fungal hydrophobin Sc3p self-assembles at the surface of aerial hyphae as a protein membrane constituting the hydrophobic rodlet layer, *European Journal of Cell Biology* **63**,122-129.

Wösten, H. A. B., Schuren, F. H. J. and Wessels, J. G. H. (1994b) Interfacial self-assembly of a hydrophobin into an amphipathic membrane mediates fungal attachment to hydrophobic surfaces, *EMBO Journal* **13**, 5848-5854.

Xiao, J.Z., Watanabe, T., Kamakura, T., Ohshima, A. and Yamaguchi, I. (1994) Studies on cellular differentiation of *Magnaporthe grisea.* Physio-chemical aspects of substratum surfaces in relation to appressorium formation, *Physiological and Molecular Plant Pathology* **44**, 227-236.

Yang, Y., Shah, J. and Klessig, D.F. (1997) Signal perception and transduction in plant defense responses, *Genes and Development* **11**, 1621-1639

MOLECULAR SIGNALING PATHWAYS INVOLVED IN APPRESSORIUM DEVELOPMENT BY *MAGNAPORTHE GRISEA*

G. C. FANG and R. A. DEAN
Department of Plant Pathology and Physiology,
Clemson University,
Clemson, South Carolina 29634 USA.

1. Introduction

The infection process of phytopathogenetic fungi involves a series of continuous interactions between the pathogenic agents and their hosts. Plants are equipped with a variety of physical and chemical strategies to restrain fungal invasion, including a waxy cuticle layer, preformed antimicrobial compounds, induction of the hypersensitive reaction, pathogenesis-related proteins (PR proteins) and phytoalexin biosynthesis, as well as the establishment of systemic acquired resistance (SAR). To establish a successful infection, pathogenic fungi must first be able to penetrate the epidermis, which is coated with a waxy cuticle and thick cell wall. Some biotrophic fungi such as rusts and powdery mildews bypass the plant physical barrier by entering through stomata. However, many fungal pathogens produce appressoria as the primary weapon of direct penetration. Prior to the development of the appressorium, a variety of biological events occur, including secretion of certain adhesive compounds, hydrophobins or other proteins which assist in attaching the germ tubes to the host plant surface (Hamer *et al.*, 1988; Talbot *et al.*, 1993; Beckerman and Ebbole, 1996; Nicholson, 1996). Hydrolytic enzymes may be also secreted to soften and change the surface of the cuticle layer facilitating attachment and penetration (Deising *et al.*, 1992; Pascholati *et al.*, 1993). An infection peg emerges beneath the appressorium and, often driven by hydrostatic pressure, penetrates the epidermal cell. After entering, the infection hyphae continues to develop and ramifies through the plant tissue.

Recent progress in molecular biology, biochemistry and cell biology has provided new opportunities for understanding the complex phenomenon of fungal development at the molecular level. *M. grisea,* the causal agent of rice blast disease is not only a serious problem world wide but also serves as a valuable model system for studying infection-related development and the molecular genetic basis of pathogenicity (Ou, 1985; Teng, 1994). This chapter summarises recent data on elucidating the development of the appressorium by this phytopathogenic fungus, with emphasis on the underlying signal transduction pathways.

S. Sreenivasaprasad and R. Johnson (eds.), Major Fungal diseases of Rice Recent Advances, 49-60.
Prof. K. Manibhushan Rao Festschrift

2. Environmental cues for the induction of the appressorium

How does *M grisea* recognise environmental signals in preparation for infection? And once recognised, how do these cues trigger the developmental infection process? These are fascinating questions and are being intensely addressed. It appears that both physical and biochemical induction are important for signaling infection development. Physical induction signals include surface hydrophobicity and hardness of the contact surface. Since the surface of rice leaves is highly hydrophobic, this led us to suspect that hydrophobicity activates appressorium development. In a series of studies, surfaces with different hydrophobic properties were experimentally created and tested for their effects on stimulating appressorium development. A strong correlation between the surface hydrophobicity and appressorium formation was demonstrated (Lee and Dean, 1993, 1994). These results have been supported by other studies (Hamer *et al.*, 1988; Howard *et al.*, 1991). Other reports suggest that hardness of the contact surface is an important environmental cue for appressorium development (Jelitto *et al.*, 1994; Xiao *et al.*, 1994). However, a certain degree of discrepancy exists in the literature regarding the importance of particular physical cues required to stimulate appressorium development. It is possible that fungal stains, material and methodology used by different research groups may account for these results. Hegde and Kolattukudy (1997) reported that the requirement of hydrophobicity is dependent on the population density of conidia.

Chemical components on the plant leaf surface may be also involved in the induction of appressorium formation. Treatment of glass slides with a variety of leaf wax extracts from rice have been shown to induce appressorium formation (Uchiyama *et al*, 1979). Hegde and Kolattukudy (1997) made similar observations and reported that both spore germination and appressorium formation can be induced by synthetic n-C_{22} fatty acid, alcohol or alkane as well as by plant surface waxes isolated from both rice and nonhost plants. They also proposed that compounds on the surface of conidia inhibited appressorium formation, and that plant surface waxes relieved this self-inhibition and promoted infection development. A lipid was isolated from the conidium surface which blocked germination and appressorium formation in a dose-dependent manner. How the plant surface waxes relieve this self-inhibition remains to be elucidated. Of particular interest may be the cutin monomer 1,16-hexadecanediol which has a backbone of 16 carbons and contains terminal hydroxyl groups (Gilbert *et al.*, 1996). This simple molecule is able to activate appressorium development at the nanomolar level. A change in the chemical structure, including removal of terminal hydroxyl group, changes in the chain length, and positioning of the hydroxyl group leads to a great reduction in activity. The origin of this compound or similar ones is believed to be hydrolytic products of the cuticle layer released by cutinases secreted from the emerging germ tubes.

3. The Role of melanisation and turgor pressure in directing penetration

Melanisation of the appressorium is crucial for penetration and pathogenicity. Mutants deficient in developing melanised appressoria are greatly affected in their ability to penetrate (review by Kubo and Furusawa, 1991). *Buf-* mutants of *M. grisea* (Chumley and Valent, 1990), and *Brm2-* mutants of *Alternaria alternata* (Kimura and Tsuge, 1993) which lack melanin were greatly impaired in their infectivity. Also, treatment of germinated conidia of *M. grisea* with tricyclazole inhibited the formation of melanin in the cell wall of the appressorium and blocked penetration (Woloshuk *et al.*, 1980). In *Colletotrichum graminicola*, the ability to infect corn leaves from albino mutants was greatly reduced, but was restored by the addition of scytalone, an intermediate product in melanin biosynthesis (Rasmussen and Hanau, 1989). Mutations in other genes involved in the melanin biosynthesis have confirmed the essential role of melanisation in penetration and pathogenicity (Kubo and Furusawa, 1991; Tanaka *et al.*, 1994; Perpetua et al., 1996). Melanin appears to provide the physical rigidity of the cell wall for sustaining high turgor pressure that is necessary for the infection peg to penetrate the underlying plant cuticle (Howard and Ferrari, 1989; Howard *et al.*, 1991). This is supported by direct evidence from the measurement of the turgor pressure in the appressorium of *M. grisea*. In melanised appressoria pressure increased to 1.92 MPa, while in appressoria of nonmelanised *buf-* mutants the pressure only reached 0.93 MPa (Money and Howard, 1996).

It has been discovered that a high concentration of glycerol, estimated to be at molar level, is present in the appressorium of *M. grisea* (de Jong *et al.*, 1997). Such concentrations of glycerol might contribute to the turgor pressure generated inside the appressorium. Melanisation of the appressorium appears to be essential for glycerol accumulation. First, the levels of intracellular glycerol in appressoria of non-melanised strains of *M. grisea* were estimated to be only 1/10 to 1/4 of that of the wild type. Second, treatment of spores with the melanin biosynthesis inhibitor, tricyclazole, caused a great reduction in glycerol accumulation in appressoria (de Jong *et al.*, 1997). It has been argued that melanisation reduces the permeability of the appressorium cell wall and allows glycerol to accumulate. In yeast the hydrolytic enzyme trehalase hydrolyses the storage carbohydrate trehalose to glucose, and the activity of trehalase can be activated by cAMP signaling (De Vergilio *et al.*, 1990; Uno *et al.*, 1983). Similarly, in *M. grisea* cAMP may mediate hydrolysis of the storage carbohydrate during appressorium development and facilitate the generation of turgor pressure for penetration. However, this hypothesis remains to be tested.

4. Signal transduction pathways involved in the development of the appressorium

It has been known for some time in higher eukaryotes that protein phosphorylation and dephosphorylation regulate a wide variety of cellular processes (Hubbard and Cohen, 1993; Karin, 1994; Walsh and van Patten, 1994); however, the notion that cyclic AMP acts as an intracellular signal molecule for appressorium development has been only demonstrated recently. Our first interest in the cAMP-mediated signal transduction pathway came from observations that exogenous cyclic AMP or cAMP derivatives induced appressorium development in noninducive environments such as a hydrophilic surface. Inhibition of phosphodiesterase activity by 3-isobutyl-1-methylxanthine (IBMX) also triggered appressorium development presumably by preventing the degradation of intracellular cAMP thus allowing it to accumulate (Lee and Dean, 1993). Because cAMP is commonly encountered in higher eukaryotes as a second messenger for transducing environmental signals for cellular development, we suspected that the induction of appressorium formation may be regulated through a cAMP kinase cascade. To further support the hypothesis we cloned and characterised the genes involved in cAMP-dependent signal transduction pathway including *cpk*A and *mac1* (Mitchell and Dean, 1995). Our results clearly demonstrate that cAMP plays a crucial role in infection related development, in which the cAMP signal is mediated by cAMP-dependent protein kinase (PKA). This in turn may phosphorylate and activate downstream metabolic proteins to carry out the developmental process. The following sections describe our current knowledge of this signal transduction pathway.

To determine if cAMP signals appressorium development through a PKA, our initial studies focused on disabling this protein. PKA is a tetrameric protein composed of two regulatory and two catalytic subunits in the inactive form. Upon binding of cAMP, two active catalytic subunits become dissociated from the dimer of regulatory subunits. To clone the gene encoding the catalytic subunit of PKA (PKA-C), we applied a PCR based strategy using degenerate primers designed from conserved regions in eukaryotic PKA-C proteins. A product of 189 bp was amplified from genomic DNA of *M. grisea*. A set of nested primers was then used to ensure the product was PKA-C. The resulting 60-bp fragment was labeled and used as probe to screen a *M. grisea* cosmid library. Positive clones were further characterised by restriction and hybridisation analysis and appropriate fragments subcloned. Sequence analysis of two independent clones revealed an open reading frame of 1894 bp, interrupted by three introns, encoding a polypeptide of 539 amino acids (Mitchell and Dean, 1995). The deduced amino acid of the *cpkA* gene shared 46 to 61% homology with the PKA-C of other organisms. It also contained a catalytic core, which shared 81% homology with analogous proteins in *S. cerevisiae*. The core region contained several functionally important motifs including an ATP binding motif, a catalytic loop motif and a magnesium ion-chelating loop. Two autophosphorylation sites were also conserved. The N-terminal of *cpkA* was glutamine rich, which is also present in certain regulatory proteins (Gerber *et al.*, 1994; Gill *et al.*, 1994; Tanaka *et al.*, 1994), providing a potential regulatory activity for *cpk*A. The gene was disrupted by gene replacement strategy in which the entire gene was replaced by a hygromycin resistance gene cassette. As expected, the *cpkA(-)* knock-out mutants failed to respond fully to cAMP and were unable to form normal appressoria even in the presence of cAMP or IBMX up to 12 hours after treatment. Strains lacking CPKA were nonpathogenic on susceptible rice. Further analysis of the *cpkA(-)* disruption mutants revealed a delayed response to cAMP -- appressoria that did form were smaller and germ tubes were longer than the wild type (Xu *et al.*, 1997, Brooks and Dean, unpublished data). However, vegetative growth, conidiation and the competency of sexual reproduction of the *cpkA(-)* disruption mutants remained unaffected.

In *Ustilago maydis*, the cAMP pathway is important in regulating dimorphic growth and gall formation and may be also involved in the b mating-type morphogenetic pathway (Gold *et al.*, 1994, 1997). In *N. crassa*, hyphal growth polarity appears to be regulated by a PKA pathway (Bruno *et al.*, 1996). A PKA-C gene has also been cloned from *Aspergillus niger*. Transformants overexpressing *pkaC* were phenotypically different from wild type in various respects including slower vegetative growth and reduced sporulation (Bencina *et al.*, 1997). Interestingly, these phenotypes were particularly pronounced on media containing trehalose and glycerol. Thus, this may also implicate a possible involvement of cAMP mediated signalling pathway in the metabolism of these carbohydrates in *A. niger*.

To further dissect the cAMP signaling pathway in *M. grisea*, we used a PCR based strategy to clone the *mac1* gene which encodes an adenylate cyclase, an enzyme that converts ATP to cAMP. PCR primers were synthesised from conserved region of adenylate cyclase genes from other filamentous fungi (Kore-eda *et al.*, 1991; Loubradou *et al.*, 1996). The amplified fragment showed high homology with other adenylate cyclase genes and was used as a probe to screen a *M. grisea* BAC library (Zhu *et al.*, 1997). To recover the entire gene, positive BAC clones were further subcloned and screened by probing with *nac1*, a gene which encodes adenylate cyclase in *N. crassa*. Results from sequence analyses revealed that *mac1* contained an ORF of 6483-bp, divided by three introns, and encoded a putative protein of 2160 amino acids. The deduced polypeptide was quite similar to other fungal adenylate cyclase enzymes and shared 62% and 56% identity with *Podospora anserina* and *N. crassa*, respectively. MAC1 also contained characteristics common to adenylate cyclase enzymes. A PEST domain was conserved in the N-terminal region. At least 9 leucine-rich repeats were present in the central region. A catalytic domain was also present close to the C-terminal. The C-terminal region was composed of two sub-domains, the CAP (cyclase-associated protein) binding site and the leucine zipper motif. All of these domains are crucial for adenylate cyclase activity.

To understand the functional role of *mac1*, a disruption construct composed of the hygromycin phosphotransferase gene flanked by the 5' and 3' ends of the *mac1* gene was designed for gene replacement. Several independent *mac1(-)* transformants were obtained. These mutants exhibited reduced vegetative growth as well as a reduction in conidiation and were deficient in sexual reproduction. Adenylate cyclase is known to regulate growth and morphogenesis in a variety of other fungi (Gold *et al.*, 1994; Kore-eda *et al.*, 1991; Loubradou et al., 1996; Uno and Ishikawa, 1973). *mac1* disruption mutants in *M. grisea* also showed delayed spore germination, and were not able to produce appressoria on inductive surfaces, including the hydrophobic surface of GelBond, rice leaves and onion skin. This deficiency in appressorium development was restored by exogenous cAMP, 1,16-hexadecanediol and cAMP derivatives such as monobutyryl-cAMP, but not by IBMX. *mac1(-)* strains were also unable to infect susceptible rice. Our results with *mac1* strongly support the important role that cAMP-dependent signalling pathway plays in appressorium development in *M. grisea*.

G proteins are ubiquitously involved in transducing signals from cell-surface receptors to the interior of the cell (reviewed by Simon *et al.*, 1991; Neer, 1995). Signal ligands bind to receptors and promote an exchange of GTP for GDP on the α subunit of the G protein followed by a conformational change in the α subunit and dissociation of α subunit from $\beta\gamma$ subunits. Activated G proteins activate downstream effectors, including adenylate cyclase, phospholipases, and multiple ion channels, which in turn direct intracellular responses. Several fungal Gα genes have been identified and alteration of these genes leads to a variety of phenotypical changes. In *N. crassa* disruption of *gna-1* produced female infertility and increased sensitivity to hyperosmotic media. A second Gα gene, *gna-2* appears to have overlapping functions (Baasiri *et al.*, 1997). In *A. nidulans*, G protein-dependent signaling pathway mediates sporulation and mycotoxin biosynthesis (Hicks *et al.*, 1997). In *U. maydis*, G proteins are coupled to pheromone receptors and are responsible for the transmission of the pheromone signal resulting in the development of infectious filamentous dikaryon (Regenfelder *et al.*, 1997). G proteins are also believed to be involved in the pathogenicity in *Cryphonectria parasitica*. Deletion or suppression of the expression of the Gα gene *CPG-1* conferred reduced pathogenicity similar to hypovirulent strains harboring RNA viruses (Choi *et al.*, 1995; Chen *et al.*, 1996).

We isolated three G protein α subunit genes, *magA*, *magB* and magC, using degenerate oligonucleotide primers, from *M.* grisea. These genes were less than 50% identical to each other, but are closely related to other G protein α subunit genes in fungi. *MagA* was very similar to *CPG-2* of *Cryphonectria parasitica; magB* was virtually identical to *CPG-1* of *C. parasitica*, to *gna1* of *N. crassa*, and to *fadA* of *Emericella nidulans*; and *magC* was most similar to *gna2* of *N. crassa* (Liu and Dean, 1997). Genomic hybridisation analysis indicated that each gene existed as a single copy in the *M. grisea* genome. To obtain insight into the roles of these genes in the signal transduction pathways of infection related development, each gene was disrupted and the phenotype analysed. Deletion of *magA* did not have significant phenotypic consequence (except as noted below). *MagC* is involved in conidiation but not in vegetative growth or appressorium development. However, disruption of *magB* produced a diverse effect ranging from reduced vegetative growth, conidiation, and appressorium formation. It is interesting to note that sexual development was perturbed by loss of any one of the 3 G protein α subunit genes, indicating the ubiquitous involvement of G proteins in certain aspects of cellular development in *M. grisea.* The ability of *magB(-)* transformants to infect susceptible rice leaves was also greatly attenuated. However, appressorium formation by *magB(-)* transformants was less reduced on older leaves than on younger ones. Appressorium formation by *magB(-)* transformants was restored by 1,16-hexadecanediol. Perhaps, as leaves mature certain stimulatory cutin components are produced and are deposited on the leaf surface (Liu and Dean, 1997). Exogenous cAMP also restored appressorium formation in *magB(-)* strains. These results also support our model that the second messenger molecule cAMP functions downstream of G proteins in the cAMP-dependent signal transduction pathway regulating appressorium development.

It is interesting to summarise the responses of *cpkA*, *mac1*, and *magB* to the environmental cues. *cpkA(-)* and *mac1(-)* mutants fail to respond to 1,16-hexadecanediol and hydrophobicity signal which suggests that both signals act on development prior to the protein kinase A cascade. *magB(-)* mutants can be restored by exogenous 1,16-hexadecanediol but hydrophobicity alone is unable to stimulate appressorium formation, suggesting two different mechanisms -- one for the thigmotropic stimulants and another for chemical stimulation. Putting these data together, we can fine-tune our model. At least two distinct signal sensing pathways exist for triggering appressorium development; the thigmotropic pathway is mediated through G proteins whereas the chemical stimulation pathway bypasses the requirement of G proteins. These two pathways might converge prior to or at least at the activation of the *mac1* gene (Figure 1).

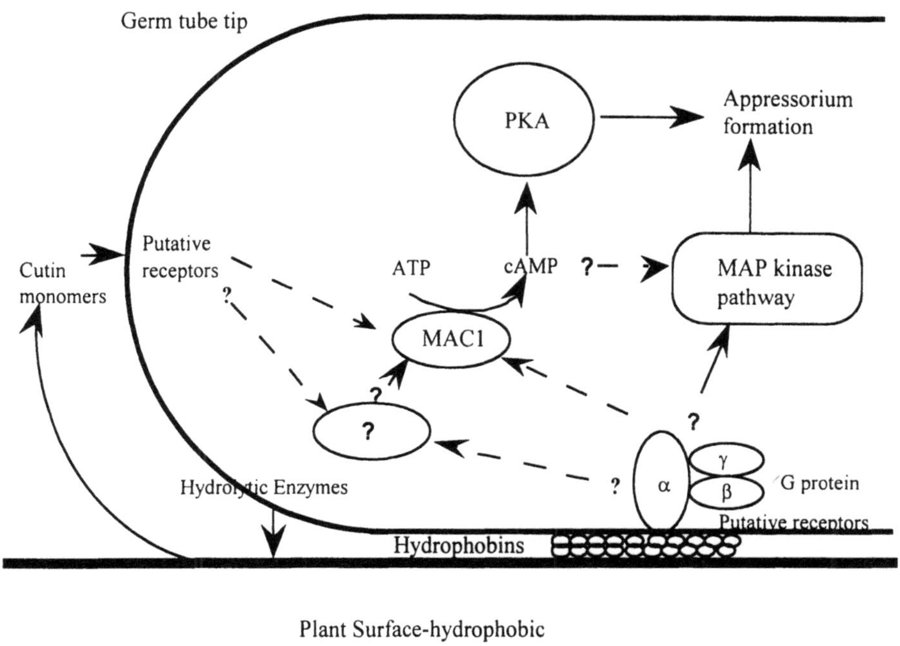

Figure 1. A model of signaling systems regulating appressorium formation in *M. grisea*. At least two distinct signal sensing pathways exist for triggering appressorium development. The thigmotropic pathway, which may be mediated by hydrophobin protein such as MPG1 (Beckerman and Ebbole, 1996), is relayed through G proteins. This leads to the activation of the adenylate cyclase MAC1, which converts ATP to cAMP and activates PKA. The activated PKA then initiates a protein phosphorylation/dephosphorylation cascade and directs appressorium development. The chemical induction pathway bypasses the requirement of G proteins. These two pathways might converge prior to or at least at the activation of the MAC1. Both PKA and MAP kinase pathways are required for the development of mature and functional appressoria.

A mitogen-activated protein (MAP) kinase pathway also appears to be involved in regulating appressorium development in *M. grisea*. A typical pathway comprises Raf-1 (a MAPKKK), Ras, MAPKK and MAPK and has been demonstrated to play an important role in the transduction of extracellular signals controlling cellular proliferation and differentiation in a number of eukaryotic systems (Neiman, 1993; Nishida and Gotoh, 1993). In *U. maydis*, a MEK/MAPKK homolog of *fuz7*, has been identified to control two distinct signaling pathways; the pheromone response and the invasive response (Banuett and Herskowitz, 1994). Xu and Hamer (1996) isolated *PMK1* from *M. grisea* which not only shares high homology with the MAP kinase *FUS3* and *KSS1* in *S. cerevisiae*, but also can functionally complement the *fus3kss1* double mutant for pheromone response. Disruption of the *PMK1* gene in *M. grisea* yielded mutants deficient in appressorium development and invasive growth.

Vegetative growth and sexual and asexual reproduction were not affected in the *pmk1(-)* mutants like the *cpkA* (-) mutants,. It appears that the *PMK1* gene may function downstream or be independent of the cAMP-dependent signal pathway because the mutants responded to cAMP to some extent. Appressoria failed to develop but the germ tube tips became hooked. However, the interactions of both signaling pathways remain to be resolved. Recently, a fungal metabolite, glisoprenin A, was isolated from *Gliocladium roseum* that appears to interfere with appressorium development. The compound inhibited appressorium development on a hydrophobic contact surface but failed to inhibit cAMP induction when tested on a hydrophilic surface (Thines *et al.*, 1997). This finding supports the possibility that two signal transduction pathways are involved in controlling appressorium formation in *M. grisea.* In *Colletotrichum trifolii*, the activity of calmodulin, protein kinase C and a serine-threonine kinase gene (*TB3*) were found to be elevated during or prior to the development of appressoria (Buhr and Dickman, 1997). Again this suggests integrated interactions of complex signaling pathways regulating appressorium development.

5. Identification of genes required for appressorium development

A series of appressorium deficient mutants were created by UV mutagenesis. To isolate these genes required for appressorium development, a map-based cloning strategy has been employed. The *app1* mutant failed to form appressoria and was non-pathogenic. This mutant was also unresponsive to cAMP and 1,16-hexadecanediol treatment. Genetic analysis revealed a single gene mutation event located in the central region of the chromosome 2 between RFLP markers A14B10 and cos94. The mutation also cosegregated with four other RFLP markers (Zhu *et al.*, 1996). However, physical mapping analysis of this region indicated a great reduction in recombination frequency in this region, as reflected by the physical/genetic distance ratio of 300 kbp/cM. This is 6 fold higher than the average 47kbp/cM for the *M. grisea* genome (Zhu and Dean, unpublished data). To facilitate chromosome walking, a BAC library was constructed (Zhu *et al.*, 1997). The 9216 clone BAC library contains 25 genomes equivalents with an average insert size of 130 kb. A BAC contig, which comprises 15 BAC clones and spans 1 cM with a coverage of ~1 Mb, was constructed to cover the region. These clones have been analyzed and a candidate BAC clone containing *app1* has been identified. Further analysis is under way to locate the *app1* gene in this clone.

The *M. grisea* BAC library is also facilitating isolation and characterisation of other genes involved in appressorium development. We have been able to isolate several genes from this library which play important roles in regulating appressorium development such as *mac1*, *cpk*A, *rpk*A, as well as spanning difficult regions in the genome that are prone to rearrangement such as *Buf.* This library is available to the *Magnaporthe* community and can be requested through our web site (http://genome.clemson.edu/index.html). The BAC library has been used to reconstruct chromosome 7. We have created a minimum tile of 41 BAC clones that covers the entire 4.2 Mb chromosome (Zhu and Dean, unpublished data). These clones will facilitate physical mapping, chromosome-walking and also provide immediate material for large-scale genomic sequencing.

Restriction enzyme-mediated integration (REMI) mutagenesis (Schiestl and Petes, 1991; Kuspa and Loomis, 1992) and similar techniques are useful for directly identifying genes affecting appressorium development. We have used a plasmid vector, pCB1179, containing a hygromycin resistance gene cassette linearised with *Pst* I to transform *M. grisea*. Hygromycin resistant transformants were further screened for impaired appressorium formation. We isolated a total of 1078 hygromycin resistant transformants, and ca. 4% of them were initially identified as appressorium deficient mutants. Genomic Southern hybridisation analysis revealed single plasmid integration events in 14 of the mutants. Genetic analysis is being conducted to evaluate cosegregation of the hygromycin resistance gene and the phenotypes. Sequences flanking the plasmid insertion sites have been rescued from several mutants and are being used to screen the BAC library to identify the genes.

6. Summary

Recent progress in pathogen-host recognition and the signal transduction pathways involved in appressorium development has provided significant insight into this unique developmental process. However, key receptors for sensing environmental cues that initiate the developmental process remain to be elucidated. The yeast two-hybrid or three-hybrid system may turn out to be a useful strategy to uncover candidate receptor proteins. Elucidating the interactions of the signal transduction pathways including the PKA pathway and the MAP kinase pathway and the downstream components of these pathways will provide opportunities for blocking the infection process and controlling this important disease. As these signal transduction pathways are universally involved in the regulation of cellular differentiation, it is plausible that these pathways also play a crucial role in the pathogenic development of other phytopathogenic fungi. Antifungal agents designed from the target genes involved in the signaling pathways in *M. grisea* may be also applied to control other fungal diseases.

7. References

Baasiri, R. A., Lu, X, Rowley, P. S., Turner, G. E. and Borkovich, K. A. (1997) Overlapping functions for two G protein a subunits in *Neurospora crassa*, *Genetics* **147**, 137-145.

Banuett, F. and Herskowitz, I. (1994) Identification of Fuz7, a *Ustilago maydis* MEK/MAPKK homolog required for a-locus-dependent and -independent steps in the fungal life cycle, *Genes and Development* **8**, 1367-1378.

Beckerman, J. L. and Ebbole, D. J. (1996) MPG1, a gene encoding a fungal hydrophobin of *Magnaporthe grisea*, is involved in surface recognition, *Molecular Plant-Microbe Interactions* **9**, 450-456.

Bencina, M., Panneman, H., Ruijter, G. J. G., Legisa, M. and Visser, J. (1997) Characterization and overexpression of the *Aspergillus niger* gene encoding the cAMP-dependent protein kinase catalytic subunit, *Microbiology* **143**, 1211-1220.

Bruno, K. S., Aramayo, R., Minke, R. F., Metzenberg, R. L. and Plamann, M. (1996) Loss of growth polarity and mislocalization of septa in a *Neurospora mutant* altered in the regulatory subunit of cAMP-dependent protein kinase, *The EMBO Journal* **15**, 5772-5782.

Buhr, T. L. and Dickman, M. B. (1997) Gene expression analysis during conidial germ tube and appressorium development in *Colletotrichum trifolii, Applied and Environmental Microbiology* **63**, 2378-2383.

Chen, B., Gso, S., Choi, G. H. and Nuss, D. L. (1996) Extensive alteration of fungal gene transcripts accumulation and elevation of G-protein-regulated cAMP levels by a virulence-attenuation hypovirus, *Proceedings of the National Academy of Sciences USA* **93**, 7999-8000.

Choi, G. H., Chen, B. and Nuss, D. L. (1995) Virus-mediated or transgenic suppression of a G-protein alpha subunit and attenuation of fungal virulence, *Proceedings of the National Academy of Sciences USA* **92**, 305-309.

Chumley, F. G. and Valent, B. (1990) Genetic analysis of melanin-deficient, nonpathogenic mutants of *Magnaporthe grisea, Molecular Plant-Microbe Interactions* **3**, 135-143.

Deising, H., Nicholson, R. L., Haug, M., Howard, R. J. and Mendgen, K. (1992) Adhesion pad formation and the involvement of cutinase and esterases in the attachment of uredospores to the host cuticle, *The Plant Cell* **4**, 1101-1111.

de Jong, J. C., McCormack, B. J., Smirnoff, N. and Talbot, N. J. (1997) Glycerol generates turgor in rice blast, *Nature* **389**, 244-245.

De Virgilio, C., Simmen, U., Hottiger, T., Boller, T. and Wiemken, A. (1990) Heat shock induces enzymes of trehalose metabolism, trehalose accumulation, and thermotolerance in *Schizosaccharomyces pombe*, even in the presence of cycloheximide, *The FEBS Letters* **273**, 107-110.

Gerber, H.-P., Seipel, K., Georgiev, O., Hofferer, M., Hug, M., Rusconi, S. and Schaffner, W. (1994) Transcriptional activation modulated by homopolymeric glutamine and proline stretches, *Science* **263**, 808-811.

Gilbert, R. D., Johnson, A. M. and Dean, R. A. (1996) Chemical signals responsible for appressorium formation in the rice blast fungus *Magnaporthe grisea, Physiological and Molecular Plant Pathology* **48**, 335-346.

Gill, G., Pascal, E., Tseng, Z. H. and Tjian, R. (1994) A glutamine-rich hydrophobic patch in transcription factor Sp1 contacts the dTAF$_{II}$110 component of the *Drosophila* TFIID complex and mediates transcriptional activation, *Proceedings of the National Academy of Sciences USA* **91**, 192-196.

Gold, S. E., Duncan, G. A., Barrett, K. J. and Kronstad, J. W. (1994) cAMP regulates morphogenesis in the fungal pathogen *Ustilago maydis, Genes and Development* **8**, 2805-2816.

Gold, S. E., Brogdon, S. M., Mayoraga, M. E. and Kronstad, J. W. (1997) The *Ustilago maydis* regulatory subunit of a cAMP-dependent protein kinase is required for gall formation in maize, *The Plant Cell* **9**, 1585-1594.

Hamer, J. E., Howard, R. J. Chumley, F. G. and Valent, B. (1988) A mechanism for surface attachment in spores of a plant pathogenic fungus, *Science* **239**, 288-290

Hegde, Y. and Kolattukudy, P. E. (1997) Cuticular waxes relieve self-inhibition of germination and appressorium formation by the conidia of *Magnaporthe grisea, Physiological and Molecular Plant Pathology* **51**, 75-84.

Hicks, J. K., Yu, J,-H. Keller, N. P. and Adams, T. H. (1997) *Aspergillus* sporulation and mycotoxin production both require inactivation of the *FadA* G□ protein-dependent signalling pathway, *The EMBO Journal* **16**, 4916-4923.

Howard, R., J. and Ferrari, M. A. (1989) Role of melanin in appressorium function, *Experimental Mycology* **13**, 403-418.

Howard, R. J., Ferrari,, M. A., Roach, D. H. and Money, N. P. (1991) Penetration of hard substrates by a fungus employing enormous turgor pressures, , *Proceedings of the National Academy of Sciences USA* **88**, 11281-11284.

Hubbard, M. J., and Cohen, P. (1993) On target with a new mechanism for the regulation of protein phosphorylation, *Trends in Biochemical Science* **18**, 172-177.

Jelitto, T. C., Page, H. A. and Read, N. D. (1994) Role of external signals in regulating the pre-penetration phase of infection by the rice blast fungus, *Magnaporthe grisea*, *Planta* **194**, 471-477.

Karin, M. (1994) Signal transduction from the cell surface to the nucleus through the phosphorylation of transcription factors, *Current Opinion in Cell Biology* **6**, 415-424.

Kimura, N. and Tsuge, T. (1993) Gene cluster involved in melanin biosynthesis of the filamentous fungus *Alternaria alternata*, *Journal of Bacteriology* **175**, 4427-4435.

Kore-eda, S., Murayama, T. and Uno, I. (1991) Isolation and characterization of the adenylate cyclase gene of *Neurospora crassa*, *The Japanese Journal of Genetics* **66**, 317-334

Kubo, Y. and Furusawa, I. (1991) Melanin biosynthesis: prerequisite for successful invasion of the plant host by appressoria of *Colletotrichum* and *Pyricularia*, in G. T. Cole and H. C. Hoch (eds.), *The Fungal Spore and Disease Initiation in Plants and Animals*, Plenum Publishing Corp., New York, pp. 205-218.

Kuspa, A. and Loomis, W. F. (1992) Tagging developmental genes in *Dictyostelium* by restriction enzyme-mediated integration of plasmid DNA,. *Proceedings of the National Academy of Sciences USA* **89**, 8803-8807.

Lee, Y.-H. and Dean, R. A. (1993) cAMP regulates infection structure formation in the plant pathogenic fungus *Magnaporthe grisea*, *The Plant Cell* **5**, 693-700.

Lee, Y.-H. and Dean, R. A. (1994) Hydrophobicity of contact surface induces appressorium formation in *Magnaporthe grisea*, *The FEMS Microbiology Letters* **115**, 71-76.

Liu, S. and Dean, R. A. (1997) G protein □ subunit genes control growth, development , and pathogenicity of *Magnaporthe grisea*, *Molecular Plant-Microbe Interactions* **10**, 1075-1086.

Loubradou, G., Begueret, J. and Turcq, B. (1996) An additional copy of the adenylate cyclase-encoding gene relieves developmental defects produced by a mutation in a vegetative incompatibility-controlling gene in *Podospora anserina*, *Gene* **170**, 119-123.

Mitchell, T. K. and Dean, R. A. (1995) The cAMP-dependent protein kinase catalytic subunit is required for appressorium formation and pathogenesis by the rice blast pathogen *Magnaporthe grisea*, *The Plant Cell* **7**, 1869-1878.

Money, N. P. and Howard, R. J. (1996) Confirmation of a link between fungal pigmentation, turgor pressure, and pathogenicity using a new method of turgor measurement, *Fungal Genetics and Biology* **20**, 217-227.

Neer, E. J. (1995) Heterotrimeric G proteins: organizers of transmembrane signals, *Cell* **80**, 249-257.

Neiman, A. M. (1993) Conservation and reiteration of a kinase cascade, *Trends in Genetics* **9**, 390-394.

Nicholson, R. I. (1996) Adhesion of fungal propagules, in M. Nicole and V. Gianinazi-Pearson (eds.), *Histology, Ultrastructure and Molecular Cytology of Plant-Microorganism Interactions*, Kluwer Academic Publishers, Amsterdam, pp. 117-134.

Nishida, Y. M. and Gotoh, Y. (1993) The MAP kinase cascade is essential for diverse signal transduction pathways, *Trends in Biological Science* **18**, 128-131.

Ou, S. H. (1985) *Rice Diseases*, 2nd ed. Commonwealth Mycological Institute, Surrey, UK.

Pascholati, S. F., Deising, H., Leite, B., Anderson, D. and Nicholson, R. L. (1993) Cutinase and non-specific esterase activities in the conidial mucilage of *Colletotrichum graminicola*, *Physiological and Molecular Plant Pathology* **42**, 37-51.

Perpetua, N. S., Kubo, Y., Yasuda, N., Takano, Y. and Furusawa, I. (1996) Cloning and characterization of a melanin biosynthetic *THR1* reductase gene essential for appressorial penetration of *Colletotrichum lagenarium*, *Molecular Plant-Microbe Interactions* **9**, 323-329.

Rasmussen, J. B. and Hanau, R. M. (1989) Exogenous scytalone restores appressorial melanization and pathogenicity in albino mutants of *Colletotrichum graminicola*, *Canadian Journal of Plant Pathology* **11**, 349-352.

Regenfelder, E., Spellig, T., Hartmann, A., Lauenstein, S., Bolker, M. and Kahmann, R. (1997) G proteins in *Ustilago maydis*: transmission of multiple signals? *The EMBO Journal* **16**, 1934-1942.

Schiestl, R. H. and Petes, T. D. (1991) Integration of DNA fragment by illegitimate recombination in *Saccharomyces cerevisiae*, *Proceedings of the National Academy of Sciences USA* **88,** 7585-7589.

Simon, M. I., Strathmann, M. P. and Gautam, N. (1991) Diversity of G proteins in signal transduction, *Science* **252**, 802-808.

Talbot, N. J., Ebbole, D. J. and Hamer, J. E. (1993) Identification and characterization of *MPG1*, a gene involved in pathogenicity from the rice blast fungus *Magnaporthe grisea*, *The Plant Cell* **5**, 1575-1590.

Tanaka, M., Clouston, W. M. and Herr, W. (1994) The oct-2 glutamine-rich and proline-rich activation domains can synergize with each other or duplicates of themselves to activate transcription, *Molecular and Cellular Biology* **14,** 6046-6055.

Teng, P. S. (1994) The epidemiological basis for blast management, in R. S. Zeiger, S. A. Leong, and P. S. Teng (eds.), *Rice Blast Disease*, CABI, Wallingford, pp. 409-434.

Thines, E., Eilbert, F., Sterner, O. and Anke, H. (1997) Glisoprenin A, an inhibitor of the signal transduction pathway leading to appressorium formation in germinating conidia of *Magnaporthe grisea* on hydrophobic surfaces, *The FEMS Microbiology Letters* **151**, 219-224.

Uchiyama, T., Ogasawara, N., Nanba, Y. and Ito, H. (1979) Conidial germination and appressorial formation of the plant pathogenic fungi on the coverglass or cellophane coated with various lipid components of plant leaf waxes, *Agricultural and Biological Chemistry* **43**, 383-384.

Uno, I. and Ishikawa, T. (1973) Metabolism of adenosine 3', 5'-cyclic monophosphate and induction of fruiting bodies in *Coprinus macrorhizus*, *Journal of Bacteriology* **11**, 1249-1255.

Uno, I., Matsumoto, K., Adachi, K. and Ishikawa, T. (1983) Genetic and biochemical evidence that trehalase is a substrate of cAMP-dependent protein kinase in yeast, *Journal of Biological Chemistry* **258**, 10867-10872.

Walsh, D. A. and van Patten, S. M. (1994) Multiple pathway signal transduction by the cAMP-dependent protein kinase, *The FASEB Journal* **8**, 1227-1236.

Woloshuk, C. P., Sisler, H. D., Tokousbalides, M. C. and Dutky, S. R. (1980) Melanin biosynthesis in *Pyricularia oryzae*: site of tricyclazole inhibition and pathogenicity of melanin-deficient mutants. *Pestical and Biochemical Physiology* **14**, 256-264.

Xu, J.-R. and Hamer, J. E. (1996) MAP kinase and cAMP signaling regulate infection structure formation and pathogenic growth in the rice blast fungus *Magnaporthe grisea*, *Genes and Development* **10**, 2696-2706.

Xu, J.-R., Urban, M., Sweigerd, J. A. and Hamer, J. E. (1997) The *cpkA* gene of *Magnaporthe grisea* is essential for appressorial penetration, *Molecular Plant-Microbe Interactions* **10**, 187-194.

Zhu, H., Whitehead, D. S., Lee, Y.-H. and Dean, R. A. (1996) Genetic analysis of development mutants and rapid chromosome mapping of *APP1*, a gene required for appressorium formation in *Magnaporthe grisea*, *Molecular Plant-Microbe Interactions* **9**, 767-774.

Zhu, H., Choi, S., Johnston, A. K., Wing, R. A. and Dean, R. A. (1997) A large-insert (130 kb) bacterial artificial chromosome library of the rice blast fungus *Magnaporthe grisea*: genome analysis, contig assembly, and gene cloning, *Fungal Genetics and Biology* **21**, 337-347.

MOLECULAR ASPECTS OF HOST-PATHOGEN INTERACTIONS IN THE RICE-BLAST SYSTEM

T. BHARGAVA and J. E. HAMER
Department of Biological Sciences,
Purdue University
West Lafayette, IN-47907, USA.

1. Introduction

Rice blast disease, caused by the haploid filamentous fungus *Magnaporthe grisea* (Hebert) Barr (Barr, 1977), is the single most devastating disease of cultivated rice (*Oryza sativa*) worldwide. *M. grisea*, a heterothallic Ascomycete, is also phytopathogenic to more than fifty Graminaceous species. These include forage grasses and economically important cereal crops like wheat, barley, maize and fingermillet. Although the organism as a whole infects many different species of grasses, individual isolates have a more restricted host range. The rice blast fungus invades the above ground parts of the rice plant and in severe epidemics, large ellipsoid lesions can engulf the entire leaf surface. After the flowering stage of the plant, the fungus spreads into the panicle (the inflorescence that holds the rice grains), causing neck blast that can devastate an entire crop (Talbot, 1995). The fungus can also invade the nodes of the plant, turning them black and making the stem fragile enough to break (Ou, 1980).

According to the Commonwealth Mycological Institute, blast disease has been recorded in over eighty countries (Ou, 1980) but especially poses a serious threat to the tropical rice growing regions of the world, which feed more than one third of the global population. Rice is a staple food crop for nearly two billion people. It is predicted that an additional thirteen million tons of rice per year will be required to keep up with the rapid growth in rice consuming populations (Lampe, 1994). This translates to an annual increase of 1.7% in world rice production between 1990 and 2025 (Lampe, 1994). Genetic improvements by conventional methods and the use of agrochemicals, which totals tens of billions of dollars already, have been of limited scope in controlling the pathogen. One hundred and fifty seven million tons of rice have been lost to rice blast disease from 1975 to 1990, a figure that equals 11 to 30 % of global rice production (Baker *et al.*, 1997).

To attain increases in rice productivity, it is essential to acquire an understanding of the basic biology of the host and parasite alike. A host-pathogen system involves an interaction between two organisms. The molecular and biochemical changes triggered by an initial encounter between a potential host plant and a pathogen, likely involve many aggressive responses by the pathogen to overcome pre-existing defense barriers,

S. Sreenivasaprasad and R. Johnson (eds.), Major Fungal diseases of Rice Recent Advances, 61-86.
Prof. K. Manibhushan Rao Festschrift

as well as counter defense responses induced in the host. Thus, the pathogen's ability to colonise a host is closely intertwined with the latter's ability to recognise the foreign organism and activate its defense responses. The outcome of such host-pathogen interactions is unlikely to be determined by any single event or mechanism.

In the past decade, rapid progress has been made in understanding the molecular basis of pathogenicity and virulence in *M. grisea*, and this is expected to facilitate the development of novel disease control strategies. Our knowledge of the molecular basis of host resistance is just beginning to grow. A better handle on the complexities of the plant-pathogen interaction is expected from a molecular analysis of pathogenicity genes, avirulence genes and the host defense response. This chapter will examine recent findings and advances made in these areas in the rice-*M. grisea* pathosystem. Several reviews and papers describe the experimental tractability of *M. grisea* (Valent, 1990; Valent and Chumley, 1991; Howard, 1994). Molecular genetic manipulations in this fungus have been additionally facilitated by the availability of a large international strain collection, genetic markers, several high density restriction fragment length polymorphism (RFLP) maps, probes for population genetic studies, and numerous genomic and cDNA libraries.

2. Life cycle of *M. grisea*

M. grisea exhibits a complex level of morphological differentiation and pathogenic specialisation for successful pathogenesis. The life cycle of *M. grisea* is initiated when a conidium lands on the surface of a rice leaf, to which it attaches firmly by releasing a glue-like substance from the apex, called the spore tip mucilage (STM, Hamer *et al.*, 1988). The STM is stored in the periplasmic compartment of the tip cell of the conidium and consists of a complex of protein, carbohydrate and lipid (Howard, 1994). The carbohydrate component is most likely composed of a 1,2-mannose disaccharide which has a high affinity toward the plant lectin Concanavalin-A (Hamer *et al.*, 1988; Howard *et al.*, 1991a). These results were confirmed by the finding that spore adhesion is suppressed by prior treatment with protease, α-mannosidase and α-galactosidase (Xiao *et al.*, 1994).

One or more germ tubes emerge from any of the three cells of the conidium, although germination from the middle cell is rare (Bourett and Howard, 1990). Apical growth ceases within a few hours and the germ tube tip hooks and begins to form an appressorium. Hydrophobicity of the substratum is a primary determinant of appressorium formation (Lee and Dean, 1994). As it grows, the appressorium becomes melanised, except at a well-defined pore between the appressorium and the substratum. This melanisation is critical in allowing the build-up of high turgor pressure. The gray pigmented melanin layer acts as a solute barrier and is composed of polymerised dihydroxynaphthalene (DHN melanin), one of the end products of the polyketide biosynthetic pathway (Chumley and Valent, 1990). A granular substance, believed to be an adhesive, accumulates at the substratum interface (Howard and Ferrari, 1989). As hydrostatic pressure increases within the appressorium, a narrow hypha (penetration peg) emerges through the pore, and punctures the plant cuticle and cell walls of the epidermal cell underlying the appressorium (Howard and Valent, 1996). *M. grisea* also produces cell wall degrading enzymes such as xylanase, cutinase, cellulase and pectin lyase (Sumizu *et al.*, 1961; Hirayama *et al.*, 1976; Bucheli *et al.*, 1990; Valent and Chumley, 1991), which may play a role in invasion. In highly compatible interactions, approximately 80% of the penetrating appressoria form primary hyphae that develop into fat, bulbous and branching infectious hyphae. The infectious hyphae then ramify inside the plant tissue both inter-and intracellularly (Heath *et al.*, 1990a; Heath *et al.*, 1990b). Biomass estimates have shown that the invading fungus can constitute up to 10% of the total biomass in infected plant tissues (Talbot *et al.*, 1993). In compatible interactions, the spread of the fungus manifests itself by the formation of large coalescing lesions, which, under conditions of high humidity, will sporulate to produce more conidia and reinitiate the disease cycle. In less compatible interactions, penetration by the appressoria is rare (Heath *et al.*, 1992).

3. Molecular genetic approaches to the study of pathogenicity genes

In recent years, substantial breakthroughs have been achieved in the identification and characterisation of pathogenicity genes in *M. grisea*. This has greatly aided in furthering our knowledge of the infection process in this fungus (see also the chapters by Talbot *et al.* and Kershaw and Talbot in this volume). In this section we illustrate some of the molecular genetic techniques used to clone genes directly or indirectly involved in pathogenicity of *M. grisea* with selected examples.

3.1. CLONING GENES BY FUNCTION OR SEQUENCE HOMOLOGY

Many plant pathogenic fungi employ extracellular degradative enzymes, such as cutinases and hydrolases, to degrade components of the plant cuticle and cell wall in order to penetrate the host cells. Cutinase has been implicated in facilitating enzymatic degradation of the plant cuticle by foliar fungal pathogens which directly penetrate the leaf surface (Schafer, 1994). Plant cell wall degrading enzymes may be required for fungal nutrition, or to permit the invasive growth of the fungus throughout the host tissue. Breakdown products from certain plant cell wall degrading enzymes, such as xylanases, are also known to trigger plant defense responses (Walton, 1994). Heterologous probes and degenerate polymerase chain reaction (PCR) techniques were used to identify such putative pathogenicity genes in *M. grisea*. To study the role of cutinase in penetration in *M. grisea*, Sweigard and co-workers cloned a putative cutinase gene (*CUT1*) by homology to a cutinase cloned from *Colletotrichum gloeosporoides* (Sweigard *et al.*, 1992a). Since the gene is present in a single copy in the genome of *M. grisea*, it was possible to engineer a *cut1*⁻ strain by replacing the wildtype gene with a disrupted copy in a single-step gene replacement strategy. However no difference in pathogenicity was observed between the wildtype strain and the engineered *cut1*⁻ disruptant (Sweigard *et al.*, 1992b). Nevertheless, a role for this enzyme in the infection process cannot be ruled out with certainty, as the *cut1*⁻ strains +still exhibited significant activity in cutinase assays, suggesting that there may be other genes with this activity (Sweigard *et al.*, 1992b).

Two β 1,4-D-xylanases have been cloned from *M. grisea* by degenerate PCR (Wu *et al.*, 1995). These xylanases were purified from culture filtrates when the fungus was grown on rice cell walls as a carbon source. Degenerate oligonucleotide primers derived from the peptide sequences were used to amplify DNA fragments from *M. grisea*. The genes encoding these xylanases, designated *XYL1* and *XYL2*, were cloned and the deletion of either or both of these genes had no effect on pathogenicity. The deletion of *XYL2* revealed residual activity corresponding to additional isozymes (Wu *et al.*, 1997). These experiments illustrate the problems associated with testing the roles of extracellular enzymes by gene disruption. The existence of multiple enzyme forms encoded by different genes means that disruption experiments may only eliminate a component of the total enzyme activity under investigation.

3.2. MUTANT ISOLATION

The isolation of mutants, and the subsequent identification of the gene by complementation of the mutant phenotype with cloned DNA fragments, has been a reliable strategy in cloning genes from filamentous fungi. Cosmid clones carrying the wildtype allele are isolated from libraries either by their ability to complement the mutation and restore pathogenicity, or by hybridisation to sequences flanking the mutation. However, since this involves the screening of thousands of transformants, this technique can be problematic in *M. grisea*, where transformation frequencies are typically low despite the advent of improved transformation protocols. Mutant analysis has resulted in the identification of many genes involved in different aspects of pathogenicity. Some of these are discussed below.

3.2.1. *Pigmentation mutants*

Wildtype strains of *M. grisea* produce a dark polyketide-derived pigment, called melanin. Studies using antipenetrant blast fungicides such as tricyclazole, which inhibit melanin biosynthesis and block penetration of the host cell (Woloshuk *et al.*, 1983), have shown that the melanin layer surrounding the appressorium is essential for maintaining the high turgor pressure generated prior to and during penetration. The melanin layer acts by decreasing the porosity of the appressorium (Howard, 1994). This allows the generation and maintenance of enormous amounts of turgor pressure in the appressoria of *M. grisea*. The turgor pressure is believed to provide the mechanical force to pierce the host epidermis and allow successful penetration (Howard *et al.*, 1991b).

Chumley and Valent identified three phenotypic classes of pigmentation mutants in *M. grisea*, named albino (*ALB1*), rosy (*RSY1*) and buff (*BUF1*), for their altered appearance (Chumley and Valent, 1990). The mutations at these unlinked loci were shown to be due to single gene defects in the melanin biosynthetic pathway. All three mutants produce nonfunctional appressoria, and are therefore nonpathogenic on rice. The pigmentation mutants were identified as spontaneous mutants as well as from mutagenized strains. Among the spontaneous mutants, the preponderance of buff mutants in some strains was found to be due to deletion of the *BUF1* gene and adjacent genomic sequences, with some deletion events spanning as much as 40 Kb (Valent and Chumley, 1994).

3.2.2. *Conidiation mutants*

Several genes involved in asexual sporulation (conidiation) and implicated in pathogenicity have been isolated by visual screening of mutants. Among them are *SMO1*, a gene that affects spore and appressorial shapes (Hamer *et al.*, 1989b), and the *CON* genes involved in conidiogenesis and conidium shape determination (Shi and Leung, 1995). For all of these genes, the gene products remain unknown.

The *SMO* (spore morphology) locus, was first identified by Hamer and co-workers while screening UV-mutagenised cultures for mutants defective in appressorium formation (Hamer *et al.*, 1989b). Fourteen mutants were identified, that define a single genetic locus, and these were characterised by abnormally shaped spores and appressoria. smo mutants also produce misshapen asci when crossed to one another, and form appressoria on surfaces not normally conducive to the formation of infection structures. Although their pathogenicity toward weeping lovegrass appears unaltered, smo- strains appear to form fewer and smaller lesions on rice and this defect in rice pathogenicity cosegregates with the *smo* $^-$ mutation (Hamer *et al.*, 1989b; Hamer and Givan, 1990).

Seven genes (*CON1-CON7*) which control conidiogenesis and the determination of conidial shape, were identified by analysing both spontaneous and induced mutations affecting the sporulation pathway (Leung and Shi, 1994). In wildtype strains, conidia are borne in clusters of four to five on aerial conidiophores. Each conidium is a three-celled spore with each cell containing a single nucleus. *con1* mutants form a single elongated conidium that can germinate but cannot form appressoria. Consequently, *con1*$^-$ strains are nonpathogenic. Diepoxyoctane induced mutations at the *CON2* locus result in a suppression of sporulation. Both con1 and *con2* mutations are controlled by single genes and double mutant analysis has shown that *con2* is epistatic to *con1* (Leung and Shi, 1994). The two loci are closely linked, suggesting that genes controlling the sporulation pathway in *M. grisea* may be clustered (Leung and Shi, 1994).

Mutations in *CON5* and *CON6* completely abolish conidiation and pathogenicity. Mutations at *CON4* and *CON7* result in a reduced number of abnormally shaped conidia. Whereas the *con7* mutants fail to develop appressoria and are nonpathogenic, *con4* mutants have dramatically reduced pathogenicity. It is intriguing that all mutations in spore formation isolated thus far should also affect infection structure formation and pathogenicity. The gene products may act during the formation of both structures, and the molecular mechanisms controlling these processes may be similar.

3.2.3. *Appressorial mutants*

UV mutagenesis was used to isolate developmental mutants affected in appressorium formation (Zhu *et al.*, 1996). This screen identified two mutants with greatly reduced appressorium formation (*app1* and *app2*) and one mutant capable of forming appressoria on hydrophilic surfaces of Gelbond (*app3*). Genetic analysis showed that all three phenotypes were the result of single gene mutations. Allelism tests confirmed that the three mutations map to different loci, with *APP1* and *APP3* being closely linked. The presence of flanking markers at 0.5 cM on either side of *APP1* indicates that this gene can be rapidly cloned by chromosome walking.

3.2.4. *Mutants regulating nitrogen starvation*

It has been demonstrated in a number of phytopathogens, that nitrogen starvation, a type of metabolic stress, plays a role in the regulation of pathogenicity genes (Rahme *et al.*, 1992; Schulte and Bonas, 1992; Wei *et al.*, 1992; Van Den Ackervenken *et al.*, 1993, 1994). Although regulatory mechanisms controlling pathogenic growth are largely unknown, studies in these systems implicate a regulatory pathway involving starvation for preferred nitrogen or carbon sources. To study the role of nitrogen starvation in *M. grisea*, Lau and Hamer initiated a classical genetic screen for chlorate resistant mutants and identified two loci, *NPR1* and *NPR2* (Nitrogen and Pathogenicity Regulation, Lau and Hamer, 1996).

npr1 and *npr2* mutants grow poorly on a wide range of nitrogen sources. In addition to affecting nitrogen regulation, these mutants exhibit a dramatic loss of pathogenicity (Lau and Hamer, 1996). The authors showed that these loci were not allelic to *NUT1*, the wide-domain transcriptional regulator of nitrogen control in *M. grisea*, which is not required for pathogenicity (Froeliger and Carpenter, 1996). NPR1 and *NPR2* are required for the starvation-induced expression of *MPG1*, pathogenicity genes and presumably other genes under nitrogen control, and may code for trans-acting positive effectors of genes involved in pathogenicity (Lau and Hamer, 1996). These studies demonstrate a link between metabolic gene regulation and pathogenicity in the rice blast fungus.

3.3. INSERTIONAL MUTAGENESIS

With the advent of transformation vectors carrying dominant selectable markers, and the development of genetic transformation systems, the insertional tagging of genes important for pathogenicity in *M. grisea* has become a viable option. Transformation results in the integration of exogenous DNA by homologous and nonhomologous recombination between the vector and the genomic sequences (Fincham, 1989). Although transformation frequencies are generally low, in many cases, the addition of restriction enzymes to the transformation mixture (the technique of restriction enzyme-mediated integration or REMI) has been found to increase the frequency of transformation. A ten-fold increase in transformation frequency was achieved with the addition of the restriction endonuclease *Bam*H1 to the transformation mixture (Shi *et al.*, 1995). To verify that the mutant phenotype is due to the tagged marker, genetic segregation tests, molecular cloning and complementation are used. In those cases where the gene has been tagged, multiple insertion patterns can arise (see Figure 1), making cloning more difficult. In spite of these problems, large scale mutant hunts have identified a number of nonpathogenic mutants.

3.4. MAP-BASED CLONING

Map-based cloning by chromosome walking in *M. grisea* has been largely possible due to the availability of a well developed genetic map with a high density of markers (Sweigard *et al.*, 1993). The recent construction of a bacterial artificial chromosome

(BAC) library is expected to further facilitate positional cloning (Zhu *et al.*, 1997). The *SMO1* locus discussed earlier has been mapped by RFLP using a family of dispersed

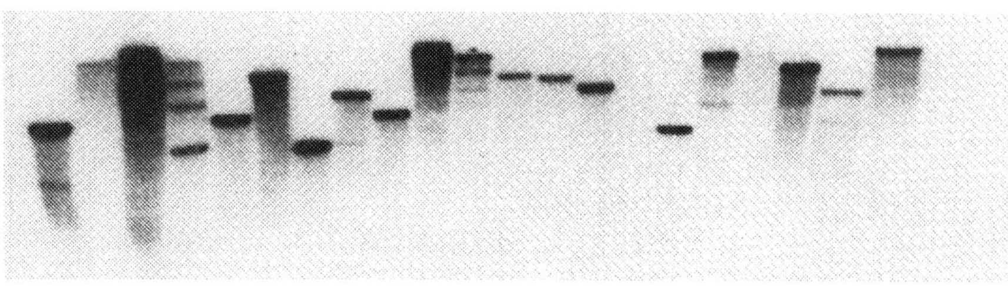

Figure 1. Insertional mutants generated with the bleomycin vector pAC905 in the *M. grisea* strain Guy11 showing multiple tandem integration patterns in some transformants.

repetitive sequences called MGR, which appear to be highly conserved in all rice pathogens (Hamer *et al.*, 1989a). However, efforts to clone the *SMO1* locus have been hampered, presumably by the presence of highly repetitive DNA sequences throughout the genome of *M. grisea*. Map-based cloning has been successful in the cloning of the avirulence genes *PWL2* and *AVR2-YAMO* where linked markers were found to be adjacent to the target genes. These are discussed in section 4.

3.5. DIFFERENTIAL cDNA SCREENING

Recently, differential cDNA screening was used to isolate *M. grisea* genes that are highly expressed in the plant (Talbot *et al.*, 1993). Fungal genes which are expressed exclusively or preferentially during growth in the plant, are likely to constitute important pathogenicity determinants and may provide clues to the metabolic requirements of the fungus during growth in planta.

The *M. grisea* gene, MPG1 (Magnaporthe pathogenicity gene 1) was cloned by differential cDNA cloning in a screen designed to identify genes expressed during infectious growth of *M. grisea* in rice (Talbot *et al.*, 1993). The *MPG1* gene is covered extensively in an accompanying chapter by Kershaw and Talbot in this volume. This screen also yielded a number of rice genes that were preferentially expressed during infection (see section 6).

3.6. CANDIDATE GENE KNOCK-OUTS

A database of DNA sequences, and biochemical studies in model fungi has made it possible to identify a panel of "candidate" genes that may be crucial components of pathways regulating steps in pathogenesis. This approach involves cloning candidate genes by homology, and analysing the phenotypes of knock-out mutants to assess their function in the process in question. One area in which this approach has been successfully applied is in the identification and characterisation of genes involved in appressorium formation. For a detailed discussion on signal transduction pathways controlling appressorium formation, refer to the chapter by Fang and Dean in this volume.

3.6.1 *cAMP-dependent protein kinase A*

cAMP has been shown to be part of the initial signal pathway controlling infection structure formation (Lee and Dean, 1993). The addition of cAMP or inhibitors of cAMP degradation promotes appressorium formation in germinating conidia or vegetative hyphae on a non-inducing hydrophilic surface (Lee and Dean, 1993). cAMP is also capable of rescuing the appressorium defect of *mpg1* mutants. The intracellular target of cAMP is the regulatory subunit of protein kinase A. The *CPKA* gene of *M. grisea*, encoding the catalytic subunit of cAMP-dependent protein kinase A was cloned by Mitchell and Dean, who reported that a deletion of the gene causes a dramatic reduction in appressorium formation (Mitchell and Dean, 1995). Photomicrographs of *cpkA⁻* strains show that germ tubes fail to undergo hooking and swelling, though other aspects of *M. grisea* growth and development are unaffected. Together, these results provide strong evidence that a cAMP-activated signal transduction pathway is necessary for infection-related morphogenesis.

The appressorial assays conducted by Mitchell and Dean were performed after eight hours of incubation. When these were repeated by Xu and co-workers for a longer period of time, using three different mutants in the *CPKA* gene, it was discovered that the *cpkA* mutants were not entirely defective in appressorium formation (Xu *et al.*, 1997). Instead, appressorium formation appeared delayed, with the appressoria being smaller in size than those in the wildtype strains. Pathogenicity remained affected. This suggested that the lack of pathogenicity in these mutants was due to an inability to penetrate plant cell surfaces. Since protein kinase A is known to play a role in the mobilisation of carbohydrate reserves in both animal and fungal cells, the failure of *cpkA* - appressoria to penetrate may be due to insufficient turgor generation (Xu *et al.*, 1997).

3.6.2. *MAP kinases*

In *Saccharomyces cerevisiae*, at least five mitogen activated protein (MAP) kinases are known to participate in various signaling pathways regulating responses to external signals ranging from increased osmolarity and stress to the presence of mating pheromones (Herskowitz, 1995). Xu and Hamer reasoned that infection structure formation and growth of the fungus in planta may be regulated by a MAP kinase mediated signaling pathway (Xu and Hamer, 1996). To test their hypothesis, they designed degenerate oligonucleotide primers to conserved regions of the yeast MAP kinases, and PCR-amplified homologous fragments from a *M. grisea* cDNA library. They identified three MAP kinases designated *PMK1* through *PMK3* (Pathogenicity MAP Kinase). *PMK1* was most similar to the *FUS3* and *KSS1* MAP kinases, involved in the yeast pheromone response (Herskowitz, 1995). The *pmk1* knock-out mutant was totally deficient in appressorium formation and hence pathogenicity, though growth rates and mating ability remained unaffected. *pmk1*⁻ strains are nonpathogenic in infection assays conducted either by spray inoculation or by conidial injection. This indicated that the defect is not just confined to appressorium formation, but extends to the ability of the fungus to grow in plant cells. Although pmk1- strains do not form appressoria, their germ tubes undergo extensive hooking and swelling on a hydrophobic surface, and on a hydrophilic surface in the presence of exogenous cAMP. These results support a model where cAMP signaling and MAP kinase signaling act cooperatively, with *PMK1* mediating events downstream of the cAMP signal, to stimulate appressorium formation.

3.6.3. *Adenylate cyclase*

In fungi, cAMP is synthesised by a membrane associated adenylate cyclase (Kronstad, 1997). The *M. grisea* adenylate cyclase (*MAC1*) gene has been cloned and sequenced (K. Adachi and J. E. Hamer, unpublished results). The *MAC1* gene is most homologous to the *Neurospora crassa* adenylate cyclase encoded by the *cr-1* gene. Functional conservation of these genes can be demonstrated by the observation that *MAC1* can complement *cr-1* mutants. *mac1*⁻ mutants are currently being assayed for appressorium defective phenotypes.

4. **Genetic and molecular characterisation of host specificity**

Host specificity in *M. grisea* is either cultivar-based or species-based. This is because in addition to infecting rice, different forms of the pathogen infect a wide range of cereals and forage grasses; however individual isolates have a limited host range. Depending upon the rice cultivars that they successfully infect, rice pathogens are subdivided into different races or pathotypes. Evidence to date suggests that the ability of any individual isolate of *M. grisea* to infect a specific rice host, is governed by a gene-for-gene interaction (Leung *et al.*, 1988; Ellingboe *et al.*, 1990; Valent *et al.*, 1991; Silue *et al.*, 1992a, b). Races of *M. grisea* are therefore inferred to have a unique combination of AVR genes. The durability of new blast resistant rice cultivars in the field is limited by the appearance of new pathotypes or races which can infect previously resistant rice varieties. Instability of resistance to *M. grisea* has been attributed to mutational changes in the pathogen population which give rise to new virulent strains from previously avirulent ones (Valent and Chumley, 1994). To obtain more durable resistance in rice against this pathogen, it is imperative that we study the molecular mechanisms underlying host specificity.

Early attempts to genetically study gene-for-gene interactions were hindered by the lack of fertility in wild isolates of *M. grisea* (Valent *et al.*, 1986). Subsequently, strain improvement efforts directed toward developing fertile pairs, showed single gene segregation (1:1) for host specificity (Leung *et al.*, 1988; Ellingboe *et al.*, 1990; Valent *et al.*, 1991; Silue *et al.*, 1992a, b; Lau *et al.*, 1993). Table 1 lists the AVR genes identified so far in *M. grisea*.

Not all examples of differences in host specificity segregate as single genes (Valent *et al.*, 1991). In genetic crosses between the weeping lovegrass pathogen 4091-5-8 (which elicits a hypersensitive response on rice) and the virulent rice pathogen O-135, many progeny formed intermediate-sized lesions on rice. This result indicates that lesion size in these interactions, is controlled by the presence of several "minor" genes segregating in the cross.

Molecular cloning of host specificity genes has provided insight into some of the mechanisms responsible for the pathogenic diversity in this fungus. In the following sections, we review the recent advances made in this area.

TABLE 1. Avirulence genes identified in *M. grisea*

AVIRULENCE GENE	CORRESPONDING HOST	REFERENCE
POS1	51583	Leung *et al.*, 1988
POS2	Sha-tiao-tsao	Leung *et al.*, 1988
POS3	K59	Leung *et al.*, 1988
POS4	Kinadong, Patong	Leung *et al.*, 1988
Avr1-CO39	CO-39	Valent *et al.*, 1991
Avr1-M201	M201	Valent *et al.*, 1991
Avr1-YAMO	Yashiro-mochi	Valent *et al.*, 1991
Avr1-C1C6	CICA6, CICA8, Med Noi	Silue *et al.*, 1992b
Avr1-KU86	Ku 86	Silue *et al.*, 1992b
Avr1-IrT7	IRAT 7, DJ 8-341	Silue *et al.*, 1992b
Avr1-Pin04	Pi - n0 4	Silue *et al.*, 1992a
P11, P12	Katy	Lau *et al.*, 1993
Avr1-KB	Kusabe	Hayashi and Naito, 1993
Avr1-TA	Tsuyuake	Hayashi and Naito, 1993
Avr1-XM	Yoshiro-mochi	Hayashi and Naito, 1993
Avr1-K59	K 59	Hayashi and Naito, 1993
Avr2-YAMO	Yashiro-mochi	Valent and Chumley, 1994
Avr1-MARA	Maratelli	Valent and Chumley, 1994
Avr1-TSUY	Tsuyuake	Valent and Chumley, 1994
Avr1-MINE	Minehikari	Valent and Chumley, 1994
avr-CO39	CO-39	Leong *et al.*, 1994
PWL1	Weeping Lovegrass	Valent and Chumley, 1994
PWL2	Weeping Lovegrass	Sweigard *et al.*, 1995

4.1. *AVR2-YAMO*

The *AVR2-YAMO* gene confers specificity toward the rice cultivar Yashiro-mochi. This gene was identified in a genetic cross between two rice pathogens and cosegregated with telomere-linked markers. Map-based cloning was used to clone the gene, which encodes a low molecular weight protein of 223 amino acids, with some homology to neutral zinc metalloproteases (Bryan *et al.*, 1997). When previously virulent strains of *M. grisea* were transformed with a functional copy of *AVR2-YAMO*, they became avirulent on rice cultivar Yashiro-mochi. This demonstrates that *AVR2-YAMO* functions as an AVR gene.

Sequence comparisons with nonfunctional alleles suggests that the putative zinc protease motif is essential for the function of *AVR2-YAMO* (Bryan *et al.*, 1997). In addition, mutation of specific amino acid residues in the putative zinc-binding region abolishes avirulence (Bryan *et al.*, 1997). The function of *Avr2-Yamo* is at present not known, but it is speculated that it may function by releasing an active elicitor from a plant or fungal precursor molecule (Knogge, 1996).

The *AVR2-YAMO* locus appears to be very unstable, as indicated by the recovery of frequent spontaneous mutants, virulent on rice cultivar Yashiro-mochi (Valent and Chumley, 1994). To determine the cause of instability of this locus, Valent and co-workers have carried out extensive studies on *avr2-yamo* mutant alleles (Valent and Chumley, 1994). Their findings on eight independent mutants from one strain, indicate, that although point mutations and insertions also contributed to the instability of the locus, the single most common event was deletions, ranging in size from 100 bp to over 12.5 Kb (Valent and Chumley, 1994). The frequent occurrence of deletions at this locus is consistent with its telomeric location (Valent and Chumley, 1994).

A second gene has been identified, that controls the ability to infect rice cultivar Yashiro-mochi (Valent and Chumley, 1994). This gene, called *AVR1-YAMO*, is unlinked to *AVR2-YAMO* and is derived from a grass pathogen. *AVR2-YAMO* and *AVR1-YAMO* may either represent two distinct genes or may be copies of the same gene present in separate locations in the genome of *M. grisea* (Valent and Chumley, 1994).

4.2. *PWL2*

Resistance at the plant species level can also be controlled by single fungal genes that function in a way very similar to cultivar-specific AVR genes. An example is the *PWL2* gene. This gene, identified in a cross between two rice pathogens, prevents infection of weeping lovegrass. *PWL2* behaves as an AVR gene because, when transformed into virulent mutants, it conditions avirulence toward weeping lovegrass. However, the transformed strain retains its pathogenicity toward rice and barley hosts, indicating that *PWL2* is highly specific toward weeping lovegrass (Valent and Chumley, 1994).

The cloning of *PWL2* was facilitated by virtue of its tight linkage to a RFLP marker (Sweigard *et al.*, 1995). The *PWL2* locus appears to be highly unstable in certain strains of the pathogen, with virulent mutants arising at a high frequency. The mechanisms of strain dependent genetic instability are unknown. In the *pwl2* mutants though, deletions occurring as a result of homologous recombination between flanking repetitive DNA sequences, are thought to have resulted in the loss of avirulence (Sweigard *et al.*, 1995). In another case, sequence analysis of a virulent allele of *PWL2* has detected a single base pair change in the coding region of the protein; an aspartic acid residue is converted to asparagine generating a putative glycosylation site (Sweigard *et al.*, 1995). It has not been verified whether the loss of avirulence is due to glycosylation of the protein.

PWL2 encodes a glycine-rich hydrophilic protein of 145 amino acids, with no homology to any protein in the database. It has a putative amino terminal signal sequence that suggests that the protein may be secreted, but this is at present unknown. Secretion of the *PWL2* gene product would appear to be consistent with the nature of AVR genes cloned from other fungal pathogens. Examples are *NIP1* from the barley pathogen *Rhynchosporium secalis* and *AVR9* from the tomato pathogen *Cladosporium fulvum* (Rohe *et al.*, 1995; Kooman-Gersmann *et al.*, 1996), where interaction with plant molecules is thought to occur at the plant cell plasma membrane.

PWL2-like sequences are present in most fungal strains isolated from different grass species. *PWL1* is a functional homologue of *PWL2* that was isolated from a fingermillet pathogen. But not all members of the PWL gene family can function as AVR genes. Many strains containing PWL sequences in their genomes are still able to infect weeping lovegrass. In addition, strains transformed with *PWL3* and *PWL4*, allelic homologues of *PWL2* isolated from grass pathogens, remained pathogenic on weeping lovegrass (Kang *et al.*, 1995). It would be interesting to know if other members of the PWL family condition specificity on other hosts.

5. Host defense responses

Research on plant defense responses has focused on the more tractable dicotyledonous species, with the development of *Arabidopsis thaliana* as a model system. Understanding the defense mechanisms in rice is essential, given its great economic importance and the devastating nature of rice blast disease. In this section we discuss progress on studying the defense response in rice and highlight relevant results from other monocotyledonous species.

Studies on dicot plants have revealed that they have evolved a multicomponent system of defense mechanisms to stave off pathogens. These mechanisms include the utilisation of pre-formed physical and chemical barriers as well as elaborate active mechanisms exemplified by the triggering of the hypersensitive response (HR). For recent reviews, refer to (Dangl *et al.*, 1996; Hammond-Kosack and Jones, 1996; Ryals *et al.*, 1996). Briefly, the occurrence of HR coincides with a barrage of physiological, molecular and biochemical events. These include: transient ion fluxes across the cell membrane, intracellular oxidative burst of active oxygen species, necrosis and localised cell death at the infection site, production of phytoalexins, increased expression of genes encoding enzymes in the phenylpropanoid pathway, biosynthesis and incorporation of cell wall and cuticle strengthening components, increased peroxidase activity, expression of genes encoding proteinase inhibitors and the induction of genes encoding the pathogenesis-related (PR) proteins, which include cell wall degrading enzymes such as chitinases and β-1,3 glucanases and proteins with antifungal properties (PR-1).

5.1. DEFENSE GENE EXPRESSION

In contrast to the extensive literature on the induction of defense genes in dicot plants, there is only a handful of studies in monocots, primarily in barley and wheat [for example, (Kragh *et al.*, 1990; Hertig *et al.*, 1991; Rasmussen *et al.*, 1991; Rebmann *et al.*, 1991; Thordal-Christensen *et al.*, 1992)].The thrust of studies on rice defense response has tended to be in elicitor-treated suspension cell cultures (Nelson *et al.*, 1994; Kuchitsu *et al.*, 1995) and in detached leaf assays (Rakwal *et al.*, 1996). In dicots, expression analysis of PR genes in true pathosystems has shown that they are very good physiological markers of the defense response. Transgenic plants containing various combinations of PR genes have also been shown to exhibit increased levels of resistance to many fungal pathogens (Mauch *et al.*, 1988; Roberts and Selitrennikoff, 1990; Alexander *et al.*, 1993; Sela-Buurlage *et al.*, 1993; Uknes *et al.*, 1996).

In an effort to characterise the rice defense response to *M. grisea*, rice cDNAs whose genes exhibited elevated expression during a susceptible infection were isolated by differential screening (Talbot *et al.*, 1993). Sequence analysis of these Blast Induced Transcripts (BITs) has shown that some of these genes encode PR proteins whereas others appear to be novel proteins, bearing little homology to genes in the database (Bhargava and Hamer, unpublished results). Gene-specific probes for these BITs have been used to study defense gene expression in rice following infection with compatible and incompatible races of *M. grisea*. Results indicate that, similar to dicots, PR genes are induced earlier in incompatible interactions than in compatible interactions (Bhargava and Hamer, unpublished results). A rice PR-10 gene identified in the differential screen was shown to be part of a family of three clustered genes that are differentially expressed during the rice-*M. grisea* interaction (McGee, 1995).

5.2. ACQUIRED RESISTANCE AND SIGNAL MOLECULES

In some dicots, the HR may lead to systemic acquired resistance (SAR) (reviewed in Lawton *et al.*, 1993 and Ryals *et al.*, 1996). SAR is characterised by the increased ability of an entire plant to resist subsequent pathogen attack following an initial infection by a necrotising pathogen. SAR develops in the uninfected parts of the plant after the initial infection, is long-lasting and is effective against a broad spectrum of pathogens. SAR can also be induced by certain chemical compounds and is characterized by the coordinate induction of PR proteins just before its onset. In the last few years, studies on SAR have accelerated considerably, given that it displays all the important characteristics for an efficient disease control strategy.

Little is known of the phenomenon of acquired resistance in monocots. Systemic protection in rice against *M. grisea* was documented when the first leaves of rice plants were preinoculated with the HR causing bacterial pathogen, *Pseudomonas syringae* pv. *syringae* (Smith and Metraux, 1991). The systemically protected leaves showed an 85% decrease in lesion number and a 50% reduction in lesion size. A local increase in the levels of phenylalanine ammonia lyase, peroxidase, β-1,3 glucanase and chitinase was detected in the *P. syringae* infection sites; the systemic sites which had been protected against *M. grisea* did not show an increase in these enzyme activities. Chemically induced resistance in rice has been correlated with elevated levels of momilactone phytoalexin biosynthesis, but it remains to be determined if this is also true for biologically induced resistance.

Induced resistance in the barley-powdery mildew interaction has demonstrated that wheat powdery mildew has a greater resistance inducing ability than either virulent or avirulent strains of barley powdery mildew [(Thordal-Christensen *et al.*, 1987) and references therein]. In contrast, preinoculation of rice with a weeping lovegrass pathogen does not protect it from a challenge inoculation with a rice pathogen (Bhargava and Hamer, unpublished results).

5.2.1. *Salicylic acid*

The phenomenon of SAR implies the existence of a signal that spreads systemically from the site of the HR to the rest of the plant. The observation that endogenous levels of salicylic acid (SA) increase transiently in the phloem after pathogen infection and before the onset of resistance in infected tobacco and cucumber (Malamy *et al.*, 1990; Metraux *et al.*, 1990), suggested that SA may be a signal molecule in the induction of resistance. To date, SA is the only plant-derived substance that has been demonstrated to be an inducer of SAR. Exogenously applied SA has been shown to induce PR genes and acquired resistance in both tobacco (Yalpani *et al.*, 1991) and *Arabidposis* (Uknes *et al.*, 1992). By far the most compelling evidence for the involvement of SA in SAR was the demonstration that transgenic tobacco plants containing the *NahG* gene from *Pseudomonas putida*, which encodes a SA degrading enzyme, were unable to accumulate SA or mount a SAR response against tobacco mosaic virus (Gaffney *et al.*, 1993). Based on research on the isolation and characterisation of SAR mutants, a SAR signal transduction pathway involving SA has been established [see (Lawton *et al.*, 1993; Bowling *et al.*, 1994; Cao *et al.*, 1994; Dietrich *et al.*, 1994; Weymann *et al.*, 1995) and (Cao *et al.*, 1997; Ryals *et al.*, 1997)]. However, it is at present still controversial whether SA is the translocated signal (Yalpani *et al.*, 1991; Vernooij *et al.*, 1994; Shulaev *et al.*, 1995).

Rice has one of the highest endogenous levels of SA (Raskin *et al.*, 1990; Silverman *et al.*, 1995). The biochemistry of SA biosynthesis and metabolism in rice is different from that of dicot plants (Silverman *et al.*, 1995), and no SA-binding activity has been detected in rice (Sanchez-Casas and Klessig, 1994). Although SA has been shown to induce the accumulation of extracellular chitinase, and a thickening of the cell wall in suspension cultured rice cells (Masuta *et al.*, 1991), SA levels did not increase upon rice inoculation with the pathogens, *M. grisea*, *Rhizoctonia solani* or with an avirulent strain of *P. syringae* (Silverman *et al.*, 1995). These observations suggest that in contrast to dicots, SA may not play a role in acquired resistance in rice.

5.2.2. *Other chemical inducers*

The compound probenazole has been used successfully for many years to control rice blast disease (Watanabe *et al.*, 1977). Probenazole is thought to act by inducing resistance in rice since no direct antifungal activity has been detected, and no probenazole-resistant strains have been reported (Midoh and Iwata, 1996).

A number of enzymes involved in plant defense, have been shown to have elevated activity in probenazole-treated rice plants infected with the blast fungus than in the nontreated infected controls (Iwata *et al.*, 1980). Probenazole also caused a significant accumulation of "anticonidial factors" (Kato *et al.*, 1984). These factors were identified as unsaturated fatty acids including linolenic acid [see section 6.2.3, (Shimura *et al.*, 1983)]. Probenazole has been shown to potentiate the oxidative burst and the accumulation of superoxide anion radicals, normally observed after infection with the blast fungus (Sekizawa *et al.*, 1985). This is an interesting observation since reactive oxygen species have been shown to be involved in the defense of plants against microbes (reviewed in Mehdy, 1994; Baker and Orlandi, 1995).

Recently, a number of probenazole induced transcripts were detected (Minami and Ando, 1994; Midoh and Iwata, 1996). One of the probenazole induced proteins (*PBZ1*) shows significant homology to intracellular PR proteins (Midoh and Iwata, 1996). *PBZ1* mRNA accumulation did not fully correlate with chemically induced anti-rice blast activity. For example, treatment with a probenazole-related compound that has strong anti-blast activity (Uchiyama *et al.*, 1973), did not induce *PBZ1*. This suggests that the gene may not play a direct role in the primary defense response of rice, but may be needed for other secondary functions, that ensue after the primary defense response has been mounted. Though the gene encoding this protein was also induced by *M. grisea*, it was not induced after treatment with sodium salicylate.

Two well characterised chemical inducers of SAR are 2,6 dichloroisonicotinic acid (INA) and benzo (1,2,3) thiadiazole-7-carbothioic acid S-methyl ester (BTH, Vernooij *et al.*, 1995; Friedrich *et al.*, 1996; Gorlach *et al.*, 1996; Lawton *et al.*, 1996). INA induces resistance in barley against powdery mildew concomitantly with the induction of PR-1 and other defense-related genes (Wasternack *et al.*, 1993; Kogel *et al.*, 1994). The isonicotinic acid derivative, N-cyanomethyl-2-chloroisonicotinamide, has been reported to be highly active against rice blast (Seguchi *et al.*, 1992). Rice plants treated with this chemical inducer exhibited high levels of lipoxygenase and peroxidase activities. This data suggests that activated lipoxygenase pathways are involved in rice plants exhibiting acquired resistance (see section 6.2.3). It also indicates that genes other than the typical SAR genes induced in tobacco (PR genes), are involved in a SAR-like response in monocots.

BTH induces acquired resistance in wheat against the pathogens *Erysiphe graminis*, *Puccinia recondita* and *Septoria* spp. (Gorlach *et al.*, 1996). According to these authors, BTH also provides season-long protection to rice seedlings from infection by *M. grisea*, but no details have been published on the effect of BTH in rice. In wheat, a novel set of genes was induced on treatment with BTH. These wheat chemically induced (WCI) genes encode a novel lipoxygenase, a cysteine proteinase and three other proteins whose functions are unknown (Ryals *et al.*, 1996). These were also induced by INA, SA and fungal pathogens although with different induction kinetics (Gorlach *et al.*, 1996). Induction of these genes with SA was transient in comparison to induction with INA and BTH. This pattern of induction also correlated with the effective protection of wheat plants by INA and BTH but not by SA (Gorlach *et al.*, 1996).

Figure 2. Lesions induced in the Sekiguchi lesion mimic mutant 72 h after inoculation with the avirulent *M. grisea* strain V04. These lesions are very similar to blast-induced lesions but show no evidence of fungal proliferation.

The availability of disease lesion mimic mutants in rice such as the Sekiguchi mutant (Marchetti *et al.*, 1983) could contribute towards unraveling the nature of the signal molecule involved in acquired resistance. The Sekiguchi mutant responds to infection by *M. grisea* with the formation of numerous HR lesions, which enlarge rapidly, mimicking actual disease lesions (see Figure 2). Histological and pathological analyses have revealed no detectable fungal biomass or proliferation of the pathogen, even from the original infection sites of the Sekiguchi lesions (Marchetti *et al.*, 1983). The Sekiguchi mutant may therefore represent a flaw in the biological mechanisms regulating the HR, leading to uncontrolled cell death, and may be linked to the SAR pathway in rice. PR gene expression in this mutant may be elevated after the formation of lesions, which can also be induced with abiotic agents (Marchetti *et al.*, 1983).

5.2.3. *Jasmonic acid*

Several lines of evidence indicate that jasmonates play an important role in defense responses [reviewed in (Reinbothe *et al.*, 1994; Creelman and Mullet, 1997)]. A number of defense-related genes are induced by jasmonic acid (JA) or methyl jasmonate in tomato, alfalfa, tobacco and barley. These genes include proteinase inhibitors (Farmer and Ryan, 1990; Farmer *et al.*, 1992), thionins (Andresen *et al.*, 1992), proline-rich proteins, enzymes involved in phenylpropanoid metabolism and ribosome-inactivating proteins (Reinbothe *et al.*, 1994). In addition, wounded plants show elevated levels of JA as do plant cell cultures treated with elicitors (Gundlach *et al.*, 1992). A model has been proposed, in which JA has a central role in an octadecanoid signal transduction cascade (Farmer and Ryan, 1992). According to the model, perception of an extracellular signal results in the activation of a plasma membrane-bound lipase that releases linolenic acid from membrane phospholipids. The linolenic acid then undergoes lipoxygenase (LOX)-dependent oxidation and is converted into the octadecanoid signal molecule, JA.

In transgenic rice plants, the promoter for a potato proteinase inhibitor gene was systemically induced by wounding and methyl jasmonate (Xu *et al.*, 1993). This finding suggests that there is a systemic signaling mechanism(s) in rice in response to wounding and that certain steps of this signal transduction pathway may be conserved between dicots and monocots. A number of gymnosperms as well as monocots and dicots, accumulate JA in cell cultures elicited with fungal cell wall fragments (Mueller *et al.*, 1993). The elevation in JA levels was found to be due to the rapid release of a-linolenic acid from the lipid pool of the plant cells.

There is considerable evidence for a role of fatty acid derivatives in constitutive as well as inducible resistance of rice against *M. grisea* (Kato *et al.*, 1983; Kato *et al.*, 1984). Infection of rice with *M. grisea* resulted in the accumulation of 13-hydroperoxylinolenic acid, a hydroxylated fatty acid which is a metabolic precursor of JA. The exogenous application of this precursor led to the accumulation of momilactone phytoalexins in rice (Xin *et al.*, 1991). This accumulation of hydroxylated fatty acids and phytoalexins was prevented when inhibitors of LOX were used. Accumulation of the phytoalexin momilactone A in suspension-cultured rice cells was also observed after treatment with an elicitor (Nojiri *et al.*, 1996). Prior to the accumulation of momilactone, a rapid and transient rise in JA levels was detected. The induction of momilactone A could be reproduced by treatment of the cells with exogenous JA, and inhibited in elicitor-treated cells by ibuprofen, a JA biosynthesis inhibitor (Nojiri *et al.*, 1996). JA has also been implicated as a signal molecule in copper chloride-elicited phytoalexin production in rice (Rakwal *et al.*, 1996). Copper chloride-elicited rice leaves showed a rapid increase in endogenous JA levels. Exogenously applied JA caused a large amount of phytoalexin production which decreased when rice leaves were treated with JA biosynthesis inhibitors. These results strongly suggest that JA functions as a signal transducer in the elicitation process leading to phytoalexin production.

Rice plants infected with *M. grisea* also build up high levels of antimicrobial oxygenated C18 fatty acids, which are products of linolenic acid (Namai *et al.*, 1993). Furthermore, a LOX pathway is activated during the infection of rice with an incompatible strain of *M. grisea* (Ohta *et al.*, 1991). Peng et al. (1994) have isolated a novel LOX from rice leaves inoculated with an incompatible race of *M. grisea*, which is induced at the same time that the pathogen begins to penetrate the rice leaf. The gene was only weakly induced in the compatible interaction. This LOX may represent a very early response of the plant against attack by *M. grisea*.

It has recently been reported that the exogenous application of JA to rice plants resulted in the induction of certain PR genes which were also induced by *M. grisea* (Schweizer *et al.*, 1997). The induction of the PR proteins was not preceded or accompanied by an elevation in endogenous JA levels in rice plants, nor did the exogenous application of JA protect rice plants from infection by *M. grisea*. Similarly, in barley plants it was found that resistance toward the powdery mildew fungus *E. graminis* is not associated with enhanced levels of endogenous jasmonates (Kogel *et al.*, 1995) and that exogenously applied JA did not induce acquired resistance against this pathogen (Schweizer *et al.*, 1993). PR-1 protein expression in response to inoculation by *M. grisea* was lowered when the plants were treated with a JA biosynthesis inhibitor, but this effect could be reversed by the application of exogenous JA. PR-2 protein expression in response to *M. grisea* was not affected. Taken together, these findings indicate that besides the octadecanoid pathway, there may be other defense pathways that coparticipate in the subtle regulation of defense responses in rice.

6. Conclusion

There has been considerable progress in our understanding of the mechanisms that underlie pathogenicity in *M. grisea*. Recombinant DNA techniques have enabled the identification of numerous components of the pathogenic process and many unexpected results have been obtained. Formerly less known genes such as the hydrophobins play a critical role in the pathogenic process whereas genes like the cutinase and xylanase, which were suspected to have roles in pathogenicity, appear to be dispensable. The task now is to elucidate the role(s) of different genes specifically involved in the pathogenic process per se and to uncover the other components of the signal transduction pathways regulating infection structure formation and recognition of the host surface.

Research on the mechanisms controlling resistance in the rice-blast system is rapidly advancing. The recent molecular cloning of AVR genes from *M. grisea* now makes it possible to pursue critical questions, regarding the nature of these genes, how they trigger resistance responses in rice, and the molecular basis of the recognition event. Since it appears unlikely that *M. grisea* evolved AVR genes to bypass host tissue colonisation, it is thought that they fulfill some other primary function in the pathogen. Determining what these functions are, may provide insight into the mechanisms of pathogen variability and host resistance stability. Other questions that need to be addressed are how and where the AVR gene products are being recognised. So far, the only demonstration of a direct interaction between an AVR gene product and its corresponding resistance gene product has been that of the *AvrPto* and *Pto* gene of tomato, in a yeast two-hybrid system (Tang *et al.*, 1996). If this was also the case for the rice-blast system, the yeast two-hybrid system could prove to be a powerful tool to look for new rice resistance genes for which the AVR genes have already been cloned.

Gene-for-gene interactions have been shown to operate in both dicots and monocots. Evolutionarily diverse plant species like tomato and rice have structurally similar disease resistance genes (Martin *et al.*, 1993; Song *et al.*, 1995). Studies on the defense mechanisms of rice and other monocot species will identify which, if any, components are shared in the signal transduction pathways mediating the defense response in dicots and monocots. The role of SA in SAR is less clear. From the current evidence, the jasmonate signaling pathway is not the only one involved in rice defense response. Thus there appear to be some differences in the defense response of monocots and dicots.

Recent evidence suggests that plants fine-tune many of their defenses depending on what resistance gene they possess and what pathogen they encounter (Reuber and Ausubel, 1996; Ritter and Dangl, 1996). Undoubtedly, these defense responses will also be influenced by the interplay of the environment with the genotypes of the interacting organisms. Although basic mechanisms of signal transduction may be highly conserved, it is likely that fine-tuning of the response by different plant species may manifest itself in the form of the activation of distinct and specific sets of responses, which may at times overlap. It would appear therefore, that certain intervening steps in the signal transduction pathways might be divergent between different plant species. These divergent steps might include the nature of the signal, its transmission from the attacked sites to systemic areas as well as the transcriptional activation of different subsets of defense genes (Gorlach *et al.*, 1996). Fine-tuning of the defense response could explain why lesion mimic mutants in *Arabidopsis* and maize display a wide array of spatial and temporal cell death patterns (the acd mutants of *Arabidopsis* and the les mutants of maize, Greenberg, 1997). Many of these resemble the cell death patterns triggered by various kinds of pathogens.

Subtle differences in the plant defense response are likely to be a reflection of the nature and spectrum of the interaction between various hosts and pathogens. Since each plant-pathogen interaction is unique, the experimental tractability of the *Oryza sativa-M. grisea* pathosystem should prompt more extensive investigations into controlling this economically important disease and increasing world rice productivity.

7. References

Alexander, D., Goodman, R. M., Gut-Rella, M., Glascock, C., Weymann, K., Friedrich, L., Maddox, D., Ahl-Goy, P., Luntz, T., Ward, E. and Ryals, J. (1993) Increased tolerance to two oomycete pathogens in transgenic tobacco expressing pathogenesis-related protein 1a, *Proceedings of the National Academy of Sciences USA* **90**, 7327-7331.

Andresen, I., Becker, W., Schluter, K., Burges, J., Parthier, B. and Apel, K. (1992) The identification of leaf thionin as one of the main jasmonate-induced proteins of barley (*Hordeum vulgare*), *Plant Molecular Biology* **19**, 193-204.

Baker, B., Zambryzski, P., Staskawicz, B. and Dinesh-Kumar, S. P. (1997) Signaling in Plant-Microbe Interactions, *Science* 276, 726-733.

Baker, C. J. and Orlandi, E. W. (1995) Active oxygen in plant pathogenesis. *Annual Review of Phytopathology* **33**, 299-321.

Barr, M. E. (1977) *Magnaporthe*, *Telimenella* and *Hyponectria* (Physosporellaceae), *Mycologia* **69**, 952-966.

Bourett, T. M. and Howard, R. J. (1990) *In vitro* development of penetration structures in the rice blast fungus *Magnaporthe grisea, Canadian Journal of Botany* **68**, 329-342.

Bowling, S. A., Guo, A., Cao, H., Gordon, A. S., Klessig, D. F. and Dong, X. (1994) A mutation in *Arabidopsis* that leads to constitutive expression of systemic acquired resistance, *The Plant Cell* **6**, 1845-1857.

Bryan, G. T., Farrall, L., Orbach, M., Kang, S., Sweigard, J. A. and Valent, B. (1997) Functional analysis of the *Magnaporthe grisea* avirulence gene *AVR2-YAMO, Nineteenth Fungal Genetics Conference*, Asilomar Conference Center, Pacific Grove, California,

Bucheli, P., Doares, S. H., Albersheim, P. and Darvill, A. (1990) Host-pathogen interactions xxxvi. Partial purification and characterization of heat-labile molecules secreted by the rice blast pathogen that solubilize plant cell wall fragments that kill plant cells, *Physiological and Molecular Plant Pathology* **4**, 489-496.

Cao, H., Bowling, S. A., Gordon, S. and Dong, X. (1994) Characterization of an *Arabidopsis* mutant that is non responsive to inducers of systemic acquired resistance, *The Plant Cell* **6**, 1583-1592.

Cao, H., Glazebrook, J., Clarke, J. D., Volko, S. and Dong, X. (1997) The *Arabidopsis NPR1* gene that controls systemic acquired resistance encodes a novel protein containing ankyrin repeats, *Cell* **88**, 57-63.

Chumley, F. G. and Valent, B. (1990) Genetic analysis of melanin deficient, nonpathogenic mutants of *Magnaporthe grisea*, *Molecular Plant Microbe Interactions* **3**, 135-143.

Creelman, R. A. and Mullet, J. E. (1997) Oligosaccharins, brassinolides, and jasmonates: nontraditional regulators of plant growth, development, and gene expression, *The Plant Cell* **9**, 1211-1223.

Dangl, J. L., Dietrich, R. A. and Richberg, M. H. (1996) Death don't have no mercy: cell death programs in plant-microbe interactions, *The Plant Cell* **8**, 1793-1807.

Dietrich, R. A., Delaney, T. P., Uknes, S. J., Ward, E. R., Ryals, J. A. and Dangl, J. L. (1994) *Arabidopsis* mutants simulating disease resistance response, *Cell* **77**, 565-577.

Ellingboe, A. H., Wu, B.-C. and Robertson, W. (1990) Inheritance of avirulence/virulence in a cross of two isolates of *Magnaporthe grisea* pathogenic to rice, *Phytopathology* **80**, 108-111.

Farmer, E. E., Johnson, R. J. and Ryan, C. A. (1992) Regulation of expression of proteinase inhibitor genes by methyl jasmonate and jasmonic acid, *Plant Physiology* **98**, 995-1002.

Farmer, E. E. and Ryan, C. A. (1990) Interplant communication: Airborne methyl jasmonate induces synthesis of proteinase inhibitors in plant leaves, *Proceedings of the National Academy of Sciences USA* **87**, 7713-7716.

Farmer, E. E. and Ryan, C. A. (1992) Octadecanoid precursors of jasmonic acid activate the synthesis of wound-inducible proteinase inhibitors, *The Plant Cell* **4**, 129-134.

Fincham, J. R. S. (1989) Transformation in fungi, *Microbiological Reviews* **53**, 148-170.

Friedrich, L., Lawton, K., Dincher, S., Winter, A., Staub, T., Uknes, S., Kessmann, H. and Ryals, J. (1996) Benzothiadiazole induces systemic acquired resistance in tobacco, *Plant Journal* **10**, 61-70.

Froeliger, E. H. and Carpenter, B. E. (1996) NUT1, a major nitrogen regulatory gene in *Magnaporthe grisea* is dispensible for pathogenicity, *Molecular and General Genetics* **251**, 647-656.

Gaffney, T., Friedrich, L., Vernooij, B., Negrotto, D., Nye, G., Uknes, S., Ward, E., Kessmann, H. and Ryals, J. (1993) Requirement of salicylic acid for the induction of systemic acquired resistance, *Science* **261**, 754-756.

Gorlach, J., Volrath, S., Knauf-Beiter, G., Hengy, G., Beckhove, U., Kogel, K. H., Oostendorp, M., Staub, T., Ward, E., Kessmann, H. and Ryals, J. (1996) Benzothiadiazole, a novel class of inducers of systemic acquired resistance, activates gene expression and disease resistance in wheat, *The Plant Cell* **8**, 629-643.

Greenberg, J. T. (1997) Programmed cell death in plant-pathogen interactions, *Annual Review of Plant Physiology and Plant Molecular Biology* **48**, 525-545.

Gundlach, H., Muller, M. J., Kutchan, T. M. and Zenk, M. H. (1992) Jasmonic acid is a signal transducer in elicitor-induced plant cell cultures, *Proceedings of the National Academic of Sciences USA* **89**, 2389-2393.

Hamer, J. E., Farrall, L., Orbach, M. J., Valent, B. and Chumley, F. G. (1989a) Host species-specific conservation of a family of repeated DNA sequences in the genome of a fungal plant pathogen, *Proceedings of the National Academy of Science USA* **86**, 9981-9985.

Hamer, J. E. and Givan, S. (1990) Genetic mapping with dispersed repeated sequences in the rice blast fungus: Mapping the *SMO* locus, *Molecular and General Genetics* **223**, 487-495.

Hamer, J. E., Howard, R. J., Chumley, F. G. and Valent, B. (1988) A mechanism for surface attachment in spores of a plant pathogenic fungus, *Science* **239**, 288-290.

Hamer, J. E., Valent, B. and Chumley, F. G. (1989b) Mutations at the *SMO* genetic locus affect the shape of diverse cell types in the rice blast fungus, *Genetics* **122**, 351-361.

Hammond-Kosack, K. E. and Jones, J. D. G. (1996) Resistance gene-dependent plant defense responses, *The Plant Cell* **8**, 1773-1791.

Hayashi, N. and Naito, H. (1993) Inheritance of avirulence in a cross of *Magnaporthe grisea* pathogenic to rice, *Abstracts of the 6th ICPP*, Montreal, Canada,

Heath, M. C., Valent, B., Howard, R. J. and Chumley, F. G. (1990a) Correlations between cytologically detected plant-fungal interactions and pathogenicity of *Magnaporthe grisea* toward weeping lovegrass, *Phytopathology* **80**, 1382-1386.

Heath, M. C., Valent, B., Howard, R. J. and Chumley, F. G. (1990b) Interactions of two strains of *Magnaporthe grisea* with rice, goosegrass and weeping lovegrass, *Canadian Journal of Botany* **68**, 1627-1637.

Heath, M. C., Howard, R. J., Valent, B. and Chumley, F. G. (1992) Ultrastructural studies of one strain of *Magnaporthe grisea* with goosegrass and weeping lovegrass, *Canadian Journal of Botany* **70**, 779-787.

Herskowitz, I. (1995) Map kinase pathways in yeast : for mating and more, *Cell* **80**, 187-197.

Hertig, C., Rebmann, G., Bull, J., Mauch, F. and Dudler, R. (1991) Sequence and tissue-specific expression of a putative peroxidase gene from wheat (*Triticum aestivum* L.), *Plant Molecular Biology* **16**, 171-174.

Hirayama, T., Sudo, T., Nagayama, H., Matsuda, K. and Tamari, K. (1976) Number and interrelation of components of carboxymethyl cellulase enzyme from *Pyricularia oryzae*, *Agricultural and Biological Chemistry* **40**, 2137-2142.

Howard, R. J. (1994). Cell biology of pathogenesis, in R. S. Zeigler, S. Leong and P. S. Teng (eds.), *Rice Blast Disease*, CABI, IRRI, Manila, The Philippines, pp. 3-22.

Howard, R. J., Bourett, T. M. and Ferrari, M. A. (1991a). Infection by *Magnaporthe grisea*: An *in vitro* analysis, in K. Mendgen and D. E. Lesemann (eds.), *Electron Microscopy of Plant Pathogens*, Springer-Verlag, Berlin, pp. 251-264.

Howard, R. J. and Ferrari, M. A. (1989) Role of melanin in appressorium function, *Experimental Mycology* **13**, 403-418.

Howard, R. J., Ferrari, M. A., Roach, D. H. and Money, N. P. (1991b) Penetration of hard substances by a fungus employing enormous turgor pressures, *Proceedings of the National Academy of Science USA* **88**, 11281-11284.

Howard, R. J. and Valent, B. (1996) Breaking and entering: Host penetration by the fungal rice blast pathogen *Magnaporthe grisea*, *Annual Review of Microbiology* **50**, 491-512.

Iwata, M., Suzuki, Y., Watanabe, T., Mase, S. and Sekizawa, Y. (1980) Effect of probenazole on the activities related to the resistant reaction in rice plants, *Annals of the Phytopathological Society of Japan* **46**, 297-306.

Kang, S., Sweigard, J. A. and Valent, B. (1995) The PWL host specificity gene family in the blast fungus *Magnaporthe grisea*, *Molecular Plant Microbe Interactions* **8**, 939-948.

Kato, T., Yamaguchi, Y., Harano, T., Yokohama, T., Uyehara, T., Namai, T., Yamanaka, S. and Harada, N. (1984) Unsaturated hydroxy fatty acids, the self defensive substance in rice plants against rice blast disease, *Chemistry letters* **3**, 409-412.

Kato, T., Yamaguchi, Y., Uyehara, T., Yokoyama, T., Namai, T. and Yamanaka, S. (1983) Defense mechanisms of the rice plant against rice blast disease, *Naturwissenschaften* **70**, 200-201.

Knogge, W. (1996) Fungal infection of plants, *The Plant Cell* **8**, 1711-1722.

Kogel, K., Beckhove, U., Dreschers, J., Munch, S. and Romme, Y. (1994) Acquired resistance in barley, *Plant Physiology* **106**, 1269-1277.

Kogel, K.-H., Ortel, B., Jarosch, B., Atzorn, R., Schiffer, R. and Wasternack, C. (1995) Resistance in barley against the powdery mildew fungus (*Erisyphe graminis* f. sp. *hordei*) is not associated with enhanced levels of endogenous jasmonates, *European Journal of Plant Pathology* **101**, 319-332.

Kooman-Gersmann, M., Honee, G., Bonnema, G. and DeWit, P. J. G. M. (1996) A high affinity binding site for the AVR9 peptide elicitor of *Cladosporium fulvum* is present on plasma membranes of tomato and other Solanaceous plants, *The Plant Cell* **8**, 929-938.
Kragh, K. M., Jacobsen, S. and Mikkelsen, J. D. (1990) Induction, purification and characterization of barley leaf chitinase, *Plant Science* **71**, 55-68.
Kronstad, J. W. (1997) Virulence and cAMP in smuts, blast and blights, *Trends in Plant Sciences* **2**, 193-199.
Kuchitsu, K., Kosaka, H., Shiga, T. and Shibuya, N. (1995) EPR Evidence for generation of hydroxyl radical triggered by N-acetylchitooligosaccharide elicitor and a protein phosphotase inhibitor in suspension-cultured rice cells, *Protoplasma* **188**, 138-142.
Lampe, K. J. (1994). Foreword, in R. S. Zeigler, S. A. Leong and P. S. Teng (eds.), *Rice Blast Disease*, CABI, IRRI, Manila, Philippines. Pp. ix-x.
Lau, G. and Hamer, J. E. (1996) Genetic identification of two regulatory loci controlling metabolism and pathogenicity gene expression in the rice blast fungus, *The Plant Cell* **8**, 771-781.
Lau, G. W., Chao, C. T. and Ellingboe, A. H. (1993) Interaction of genes controlling avirulence/virulence of *Magnaporthe grisea* on rice cultivar Katy, *Phytopathology* **83**, 375-382.
Lawton, K., Friedrich, L., Hunt, M., Weymann, K., Staub, T., Kessmann, H. and Ryals, J. (1996) Benzothiadiazole induces disease resistance in *Arabidopsis* by activation of the systemic acquired resistance signal transduction pathway, *Plant Journal* **10**, 71-82.
Lawton, K., Uknes, S., Friedrich, L., Gaffney, T., Alexander, D., Goodman, R., Metraux, J.-P., Kessmann, H., Ahl-Goy, P., Gut-Rella, M., Ward, E. and Ryals, J. (1993). The molecular biology of systemic acquired resistance, in B. Fritig and M. Legrand (eds.), *Mechanisms of Defence Responses in Plants*, Kluwer Academic Publishers, Dordrecht, pp. 410-420.
Lee, Y.-H. and Dean, R. A. (1993) cAMP regulates infection structure formation in the plant pathogenic fungus *Magnaporthe grisea, The Plant Cell* **5**, 693-700.
Lee, Y.-H. and Dean, R. A. (1994) Hydrophobicity of contact surface induces appressorium formation in *Magnaporthe grisea*, *FEMS Microbiology Letters* **115**, 71-75.
Leong, S. A., Farman, M., Smith, J., Budde, A., Tosa, Y. and Nitta, N. (1994). Molecular genetic approach to the study of cultivar specificity in the rice blast fungus, in R. S. Zeigler, S. Leong and P. S. Teng (eds.), *Rice Blast Disease*, CABI, IRRI, Manila, The Philippines, pp. 87-111.
Leung, H., E. S. Borromeo, Bernardo, M. A. and Notteghem, J. L. (1988) Genetic analysis of virulence in the rice blast fungus *Magnaporthe grisea*, *Phytopathology* **78**, 1227-1233.
Leung, H. and Shi, Z. (1994). Genetic regulation of sporulation in the rice blast fungus, in R. S. Zeigler, S. A. Leong and P. S. Teng (eds.), *Rice Blast Disease*, CABI, IRRI, Manila, Philippines, pp. 35-50.
Malamy, J., Carr, J. P., Klessig, D. F. and Raskin, I. (1990) Salicylic acid: a likely endogenous signal in the resistance response of tobacco to viral infection, *Science* **250**, 1002-1004.
Marchetti, M. A., Bollich, C. N. and Uecker, F. A. (1983) Spontaneous occurence of the Sekiguchi lesion in two american rice lines: its induction, inheritance and utilization, *Phytopathology* **73**, 603-606.
Martin, G. B., Brommonschenkel, S. H., Chunwongse, J., Frary, A., Ganal, M. W., Spivey, R., Wu, T., Earle, E. D. and Tanksley, S. D. (1993) Map-based cloning of a protein kinase gene conferring disease resistance in tomato, *Science* **262**, 1432-1436.
Masuta, C., Bulcke, M. V., Bauw, G., Montagu, M. V. and Caplan, A. B. (1991) Differential effects of elicitors on the viability of rice suspension cells, *Plant Physiology* **97**, 619-629.
Mauch, F., Mauch-Mani, B. and Boller, T. (1988) Antifungal hydrolases in pea tissue II. Inhibition of fungal growth by combinations of chitinase and β-1,3 glucanase, *Plant Physiology* **88**, 936-942.
McGee, J. D. (1995). Characterization of a pathogenesis-related gene family induced in rice during infection with *Magnaporthe grisea,* Ph. D thesis, Department of Botany and Plant Pathology. West Lafayette, IN, Purdue University.
Mehdy, M. C. (1994) Active oxygen species in plant defense against pathogens, *Plant Physiology* **105**, 467-472.
Metraux, J.-P., Signer, H., Ryals, J., Ward, E., Wyss-Benz, M., Gaudin, J., Raschdorf, K., Schmid, E., Blum, W. and Inverardi, B. (1990) Increase in salicylic acid at the onset of systemic acquired resistance in cucumber, *Science* **250**, 1004-1006.
Midoh, N. and Iwata, M. (1996) Cloning and characterization of a probenazole-inducible gene for an intracellular pathogenesis-related protein in rice, *Plant Cell Physiology* **37**, 9-18.
Minami, E. and Ando, I. (1994) Analysis of blast disease resistance induced by probenazole in rice, *Journal of Pesticide Science* **19**, 79-83.

Mitchell, T. K. and Dean, R. A. (1995) The cAMP-dependent protein kinase catalytic subunit is required for appressorium formation and pathogenesis by the rice blast fungus *Magnaporthe grisea*, *The Plant Cell* **7**, 1869-1878.

Mueller, M. J., Brodschelm, W., Spannagl, E. and Zenk, M. H. (1993) Signaling in the elicitation process is mediated through the octadecanoid pathway leading to jasmonic acid, *Proceedings of the National Academy of Sciences USA* **90**, 7490-7494.

Namai, T., Kato, T., Yamaguchi, Y. and Hirukawa, T. (1993) Anti-rice blast activity and resistance induction of C-18 oxygenated fatty acids, *Bioscience, Biotechnology and Biochemistry* **57**, 611-613.

Nelson, A. J., Doerner, P. W., Zhu, Q. and Lamb, C. J. (1994) Isolation of a monocot 3-hydroxy-3-methylglutaryl coenzyme A reductase gene that is elicitor-inducible, *Plant Molecular Biology* **25**, 401-412.

Nojiri, H., Sugimori, M., Yamane, H., Nishimura, Y., Yamada, A., Shibuya, N., Kodama, O., Murofushi, N. and Omori, T. (1996) Involvement of jasmonic acid in elicitor-induced phytoalexin production in suspension-cultured rice cells, *Plant Physiology* **110**, 387-392.

Ohta, H., Shida, K., Peng, Y.-L., Furusawa, I., Shishiyama, J., Aibara, S. and Morita, Y. (1991) A lipoxygenase pathway is activated in rice after infection with the rice blast fungus *Magnaporthe grisea*, *Plant Physiology* **97**, 94-98.

Ou, S. H. (1980) A look at world wide rice blast disease control, *Plant Disease* **64**, 439-445.

Peng, Y.-L., Shirano, Y., Ohta, H., Hibino, T., Tanaka, K. and Shibata, D. (1994) A novel lipoxygenase from rice, *Journal of Biological Chemistry* **269**, 3755-3761.

Rahme, G. L., Mindrinos, M. N. and Panopoulos, N. J. (1992) Plant and environmental sensory signals control the expression of *hrp* genes in *Pseudomonas syringae* pv *phaseolica*, *Journal of Bacteriology* **174**, 3499-3507.

Rakwal, R., Tamogami, S. and Kodama, O. (1996) Role of jasmonic acid as a signaling molecule in copper chloride-elicited rice phytoalexin production, *Bioscience, Biotechnology and Biochemistry* **60**, 1046-1048.

Raskin, I., Skubatz, H., Tang, W. and Meeuse, B. J. D. (1990) Salicylic acid levels in thermogenic and non-thermogenic plants, *Annals of Botany* **66**, 369-373.

Rasmussen, S. K., Welinder, K. G. and Hejgaard, J. (1991) cDNA cloning, characterization and expression of an endosperm-specific barley peroxidase, *Plant Molecular Biology* **16**, 317-327.

Rebmann, G., Hertig, C., Bull, J., Mauch, F. and Dudler, R. (1991) Cloning and sequencing of cDNAs encoding a pathogen-induced putative peroxidase of wheat (*Triticum aestivum* L.), *Plant Molecular Biology* **16**, 329-331.

Reinbothe, S., Mollenhauer, B. and Reinbothe, C. (1994) JIPS and RIPS: the regulation of plant gene expression by jasmonates in response to environmental cues and pathogens, *The Plant Cell* 6, 1197-1209.

Reuber, T. L. and Ausubel, F. M. (1996) Isolation of *Arabidopsis* genes that differentiate between resistance responses mediated by the *RPS2* and *RPM1* disease resistance genes, *The Plant Cell* **8**, 241-249.

Ritter, C. and Dangl, J. L. (1996) Interference between two specific pathogen recognition events mediated by distinct plant disease resistance genes, *The Plant Cell* **8**, 251-257.

Roberts, W. K. and Selitrennikoff, C. P. (1990) Zeamatin, an antifungal protein from maize with membrane-permeabilising activity, *Journal of General Microbiology* **136**, 1771-1778.

Rohe, M., Gierlich, A., Hermann, H., Hahn, M., Schmidt, B., Rosahl, S. and Knogge, W. (1995) The race-specific elicitor, *NIP1*, from the barley pathogen, *Rynchosporium secalis*, determines avirulence on host plants of the *Rrs1* resistance genotype, *The EMBO Journal* **14**, 4168-4177.

Ryals, J., Weymann, K., Lawton, K., Friedrich, L., Ellis, D., Steiner, H.-Y., Johnson, J., Delaney, T. P., Jesse, T., Vos, P. and Uknes, S. (1997) The *Arabidopsis thaliana NIM1* protein is homologous to the mammalian transcription factor inhibitor IKB, *The Plant Cell* 9, 425-439.

Ryals, J. A., Neuenschwander, U. H., Willits, M. G., Molina, A., Steiner, H.-Y. and Hunt, M. D. (1996) Systemic acquired resistance, *The Plant Cell* **8**, 1809-1819.

Sanchez-Casas, P. and Klessig, D. F. (1994) A salicylic acid-binding activity and a salicylic acid-inhibitable catalase activity are present in a variety of plant species, *Plant Physiology* 106, 1675-1679.

Schafer, W. (1994) Molecular mechanisms of fungal pathogenicity to plants, *Annual Review of Phytopathology* **32**, 461-477.

Schulte, R. and Bonas, U. (1992) A Xanthomonas pathogenicity locus is induced by sucrose and sulfur-containing amino acids, *The Plant Cell* **4**, 79-86.

Schweizer, P., Buchala, A., Silverman, P., Seskar, M., Raskin, I. and Metraux, J.-P. (1997) Jasmonate-inducible genes are activated in rice by pathogen attack without a concomitant increase in endogenous jasmonic acid levels, *Plant Physiology* **114**, 79-88.

Schweizer, P., Gees, R. and Mosinger, E. (1993) Effect of jasmonic acid on the interaction of barley (*Hordeum vulgare* L.) with the powdery mildew *Erisyphe graminis* f. sp. *Hordei*, *Plant Physiology* **102**, 503-511.

Seguchi, K., Kurotaki, M., Sekido, S. and Yamaguchi, I. (1992) Action mechanism of N-cyanomethyl-2-chloroisonicotinamide in controlling rice blast disease, *Journal of Pesticide Science* **17**, 107-113.

Sekizawa, Y., Haga, M., Iwata, M., Hamamoto, A., Chihara, C. and Takino, Y. (1985) Probenazole and burst of respiration in rice leaf tissue infected with blast fungus, *Journal of Pesticide Science* **10**, 225-231.

Sela-Buurlage, M. B., Ponstein, A. S., Bres-Vloesmans, S. A., Melchers, L. S., Van Den Elzen, P. J. M. and Cornelissen, B. J. C. (1993) Only specific tobacco (*Nicotiana tabacum*) chitinases and β-1,3 glucanases exhibit antifungal activity, *Plant Physiology* 101, 857-863.

Shi, Z., Christian, D. and Leung, H. (1995) Enhanced transformation in *Magnaporthe grisea* by restriction enzyme-mediated integration of plasmid DNA, *Phytopathology* **85**, 329-333.

Shi, Z. and Leung, H. (1995) Genetic analysis of sporulation in *Magnaporthe grisea* by chemical and insertional mutagenesis, *Molecular Plant Microbe Interactions* **8**, 949-959.

Shimura, M., Mase, S., Iwata, M., Suzuki, A., Watanabe, T., Sekizawa, Y., Sasaki, T., Furihata, K., Seto, H. and Otake, N. (1983) Anti-conidial germination factors induced in the presence of probenazole in infected host leaves III. Structural elucidation of substance A and C, *Agricultural and Biological Chemistry* **47**, 1983-1989.

Shulaev, V., Leon, J. and Raskin, I. (1995) Is salicylic acid a translocated signal of systemic acquired resistance in tobacco? *The Plant Cell* **7**, 1691-1701.

Silue, D., Notteghem, J. L. and Tharreau, D. (1992a) Evidence for a gene-for-gene relationship in the *Oryza sativa-Magnaporthe grisea* pathosystem, *Phytopathology* **82**, 577-580.

Silue, D., Tharreau, D. and Notteghem, J. L. (1992b) Identification of *Magnaporthe grisea* avirulence genes to seven rice cultivars, *Phytopathology* **82**, 1462-1467.

Silverman, P., Saskar, M., Kanter, D., Schweizer, P., Metraux, J.-P. and Raskin, I. (1995) Salicylic acid in rice- biosynthesis, conjugation and possible role, *Plant Physiology* **108**, 633-639.

Smith, J. A. and Metraux, J.-P. (1991) *Pseudomonas syringae* pv. *syringae* induces systemic resistance to Pyricularia oryzae in rice, *Physiological and Molecular Plant Pathology* **39**, 451-461.

Song, W.-Y., Wang, G.-L., Chen, L.-L., Kim, H.-S., Pi, L.-Y., Holsten, T., Gardner, J., Wang, B., Zhai, W.-X., Zhu, L.-H., Fauquet, C. and Ronald, P. (1995) A receptor kinase-like protein encoded by the rice disease resistance gene *Xa21*, *Science* **270**, 1804-1806.

Sumizu, K., Yoshikawa, M. and Tanaka, S. (1961) Studies of xylanase of *Pyricularia oryzae Journal of Biochemistry* **50**, 538-543.

Sweigard, J. A., Carroll, A. M., Kang, S., Chumley, F. G. and Valent, B. (1995) Identification, cloning and characterization of *PWL2*, a gene for host-species specificity in the rice blast fungus, *The Plant Cell* **7**, 1221-1233.

Sweigard, J. A., Chumley, F. G. and Valent, B. (1992a) Cloning and analysis of CUT1, a cutinase gene from *Magnaporthe grisea*, *Molecular and General Genetics* **232**, 174-182.

Sweigard, J. A., Chumley, F. G. and Valent, B. (1992b) Disruption of a *Magnaporthe grisea* cutinase gene, *Molecular and General Genetics* **232**, 183-190.

Sweigard, J. A., Valent, B., Orbach, M. J., Walter, A. M., Rafalski, A. and Chumley, F. C. (1993). A genetic map of the rice blast fungus *Magnaporthe grisea.*, in S. O'Brien (eds.), Genetic Maps:3.110, Cold Spring Harbor, NY.

Talbot, N. J. (1995) Having a blast: exploring the pathogenicity of *Magnaporthe grisea*, *Trends in Microbiology* **3**, 9-16.

Talbot, N. J., Ebbole, D. J. and Hamer, J. E. (1993) Identification and characterization of *MPG1*, a gene involved in pathogenicity from the rice blast fungus *Magnaporthe grisea*, *The Plant Cell* **5**, 1575-1590.

Tang, X., Frederick, R. D., Zhou, J., Halterman, D. A., Jia, Y. and Martin, G. B. (1996) Initiation of plant disease resistance by physical interaction of *AvrPto* and *Pto kinase*, *Science* **274**, 2060-2065.

Thordal-Christensen, H., Brandt, J., Cho, B. H., Rasmussen, S. K., Gregersen, P. L., Smedegaard-Petersen, V. and Collinge, D. B. (1992) cDNA cloning and characterization of two barley peroxidase transcripts induced differentially by the powdery mildew fungus *Erysiphe graminis*, Physiological and Molecular Plant Pathology **40**, 395-409.

Thordal-Christensen, H., Gregersen, P. L., Andersen, J. B. and Smedegaard-Petersen, V. (1987) Induction of defence reaction in plants, *Journal of Agricultural Science of Finland* **59**, 231-249.

Uchiyama, M., Abe, H., Sato, R., Shimura, M. and Watanake, T. (1973) Fate of 3-allyloxy-1,2-benzisothiazole 1,1-dioxide (Oryzemate) in rice plants, *Agricultural and Biological Chemistry* **37**, 737-745.

Uknes, S., Mauch-Mani, B., Moyer, M., Williams, S., Dincher, S., Chandler, D., Potter, S., Slusarenko, A., Ward, E. and Ryals, J. (1992) Acquired resistance in *Arabidposis, The Plant Cell* **4**, 645-656.

Uknes, S., Morris, S., Vernooij, B. and Ryals, J. (1996). The role of benzoic acid derivatives in systemic acquired resistance, in J. T. Romeo, J. A. Saunders and P. Barbosa (eds.), *Phytochemical diversity and redundancy in ecological interactions*, Plenum Press, New York, pp. 253-263.

Valent, B. (1990) Rice blast as a model system for plant pathology, *Phytopathology* **80**, 33-36.

Valent, B. and Chumley, F. G. (1991) Molecular genetic analysis of the rice blast fungus, *Magnaporthe grisea, Annual Review of Phytopathology* **29**, 443-467.

Valent, B. and Chumley, F. G. (1994). Avirulence genes and mechanisms of genetic instability in the rice blast fungus, in R. S. Zeigler, S. A. Leong and P. S. Teng (eds.), *Rice Blast Disease*, CABI, IRRI, Manila, Philippines. pp 111-134.

Valent, B., Crawford, M. S., Weaver, C. G. and Chumley, F. G. (1986) Genetic studies of fertility and pathogenicity in *Magnaporthe grisea* (*Pyricularia oryzae*). *Iowa State Journal of Research* **60**, 569-594.

Valent, B., Farrall, L. and Chumley, F. G. (1991) *Magnaporthe grisea* genes for pathogenicity and virulence identified through a series of backcrosses, *Genetics* 127, 87-101.

Van Den Ackervenken, G. F. J. M., Dunn, R. M., Cozzijnsen, T. J., Vossen, P., Denbroek, V. and DeWit, P. J. G. M. (1994) Nitrogen limitation induces expression of the avirulence gene *AVR9* in the tomato pathogen *Cladosporium fulvum, Molecular and General Genetics* **243**, 277-285.

Van Den Ackervenken, G. F. J. M., Kan, J. A. L. V., Joosten, M. H. A. L., Muisers, J. M., Verbakel, H. M. and DeWit, P. J. G. M. (1993) Characterization of two putative pathogenicity genes of the fungal tomato pathogen *Cladosporium fulvum, Molecular Plant Microbe Interactions* **6**, 210-215.

Vernooij, B., Friedrich, L., Ahl-Goy, P., Staub, T., Kessmann, H. and Ryals, J. (1995) 2,6-dichloroisonicotinic acid-induced resistance to pathogens without the accumulation of salicylic acid, *Molecular Plant Microbe Interactions* **8**, 228-234.

Vernooij, B., Friedrich, L., Morse, A., Reist, R., Kolditz-Jawhar, R., Ward, E., Uknes, S., Kessmann, H. and Ryals, J. (1994) Salicylic acid is not the translocate signal responsible for inducing systemic acquired resistance but is required in signal transduction, *The Plant Cell* **6**, 959-965.

Walton, J. D. (1994) Deconstructing the plant cell wall, *Plant Physiology* **104**, 1113-1118.

Wasternack, C., Atzora, R. and Kogel, K. H. (1993) Induction of a thionin, the jasmonate-induced 6 kDa protein of barley by 2,6-dichloroisonicotinic acid, *Journal of Phytopathology* **140**, 280-284.

Watanabe, T., Igarashi, H., Matsumoto, K., Seki, S., Mase, S. and Sekizawa, Y. (1977) The characteristics of probenazole (Oryzemate) for the control of rice blast, *Journal of Pesticide Science* **2**, 291-296.

Wei, Z.-M., Sneath, B. J. and Beer, S. V. (1992) Expression of *Erwinia amylovora hrp* genes in response to environmental stimuli, *Journal of Bacteriology* **174**, 1875-1882.

Weymann, K., Hunt, M., Uknes, S., Neuenschwander, U., Lawton, K., Steiner, H. and Ryals, J. (1995) Suppression and restoration of lesion formation in *Arabidopsis* mutants, *The Plant Cell* **7**, 2013-2022.

Woloshuk, C. P., Sisler, H. D. and Vigil, E. L. (1983) Action of the antipenetrant, tricyclazole, on appressoria of *Pyricularia oryzae, Physiological Plant Pathology* **22**, 245-249.

Wu, S. C., Ham, K. S., Darvill, A. G. and Albersheim, P. (1997) Deletion of two endo-b-1,4-xylanase genes reveals additional isozymes secreted by the rice blast fungus, *Molecular Plant Microbe Interactions* **10**, 700-708.

Wu, S. C., Kauffmann, S., Darvill, A. G. and Albersheim, P. (1995) Purification, cloning and characterisation of two xylanases from *Magnaporthe grisea* , the rice blast fungus, *Molecular Plant Microbe Interactions* **8**, 506-514.

Xiao, J.-Z., Ohshima, A., Kamakura, T., Ishiyama, T. and Yamaguchi, I. (1994) Extracellular glycoprotein(s) associated with cellular differentiation in *Magnaporthe grisea, Molecular Plant Microbe Interactions* 7, 639-644.

Xin, L. W., Kodama, O. and Akatsuka, T. (1991) Role of oxygenated fatty acids in rice phytoalexin production, *Agricultural and Biological Chemistry* **55**, 1041-1047.

Xu, D., McElroy, D., Thornburg, R. W. and Wu, R. (1993) Systemic induction of a potato pin2 promoter by wounding, methyl jasmonate, and abscisic acid in transgenic rice plants, *Plant Molecular Biology* **22**, 573-588.

Xu, J.-R. and Hamer, J. E. (1996) MAP-kinase and cAMP signaling regulate infection structure formation and pathogenic growth in the rice blast fungus *Magnaporthe grisea, Genes and Development* 10, 2696-2706.

Xu, J.-R., Urban, M., Sweigard, J. and Hamer, J. E. (1997) The *CPKA* gene of *Magnaporthe grisea* is required for appressorial function, *Molecular Plant Microbe Interactions* **10**, 187-194.

Yalpani, N., Silverman, P., Wilson, T. M. A., Kleier, D. A. and Raskin, I. (1991) Salicylic acid is a systemic signal and an inducer of pathogenesis-related proteins in virus-infected tobacco, *The Plant Cell* **3**, 809-818.

Zhu, H., Choi, S., Johnston, A. K., Wing, R. A. and Dean, R. A. (1997) A large-insert (130 Kbp) bacterial artificial chromosome library of the rice blast fungus *Magnaporthe grisea*: genomic analysis, contig assembly, and gene cloning, *Fungal Genetics and Biology* **21**, 337-347.

Zhu, H., Whitehead, D. S., Lee, Y.-H. and Dean, R. A. (1996) Genetic analysis of developmental mutants and rapid chromosome mapping of *APP1*, a gene required for appressorium formation in *Magnaporthe grisea*, *Molecular Plant Microbe Interactions* **9**, 767-774.

CYTOLOGICAL ASPECTS OF INFECTION BY THE RICE BLAST FUNGUS *PYRICULARIA ORYZAE*

HIRONORI KOGA
Research Institute of Agricultural Resources, Ishikawa Agricultural College, 1-308 Suematsu, Nonoichi, Ishikawa 921, Japan.

1. Introduction

Rice blast, caused by *Pyricularia oryzae* Cav. (telemorph: *Magnaporthe grisea* (Hebert) Barr) (Barr 1977) is a major factor limiting yields of rice (*Oryza sativa* L.) world wide. An understanding of the mechanisms underlying pathogenesis and host specificity is critical to future progress in both applied and basic aspects of rice blast pathology. Many studies on the penetration process of *P. oryzae* have been made on artificial substrate (Bourett and Howard, 1990; Howard and Ferrari, 1989; Howard *et al.*, 1991), on *Bryophyllum pinnatum* (Woloshuk *et al.*, 1983), on goosegrass and weeping lovegrass (Heath *et al.*, 1990,1992) and on rice (Hashioka *et al.*, 1968; Koga, 1994 a,b; Koga, 1995; Yoshii, 1937).

On the other hand, many cytological studies have been made on mechanisms of race-specific resistance of rice to *Pyricularia oryzae* Cav. Race-specific resistance was expressed as a hypersensitive response of cells to invasion by an incompatible race of *P. oryzae* which was characterised by granulation of the cytoplasm of attacked epidermal cells (Kawamura and Ono, 1948; Ohata *et al.*, 1963; Koga and Kobayashi, 1982a,b; Tomita and Yamanaka, 1983; Koga *et al.*, 1984; Peng and Shishiyama, 1988, 1989) and cell death (Ikata *et al.*, 1931; Sakamoto, 1968;Takahashi, 1956; Koga, 1994 b). Microscopic studies under blue light have revealed that these changes are usually accompanied by autofluorescence of the infected host cell (Heath *et al.*, 1990; Koga, 1994 a).

The importance of these interactions may be: recognition of parasite by host, and vice-versa; the secretion by the parasite of substances causing the modifications of cellular metabolism in the host which lead to its accommodation or rejection; the secretion by the host, in incompatible interactions, of toxins which contribute to the cessation of fungal growth; and the transfer of nutrients and other essential metabolites from host to parasite. Precise investigation of the histology and cytology of infection is an indispensable first step towards biochemical and molecular biological definition of these processes.

S. Sreenivasaprasad and R. Johnson (eds.) Major Fungal Diseases of Rice Recent Advances, 87-110.
Prof. K. Manibhushan Rao Festschrift

2. Infection process

2.1. GERMINATION AND APPRESSORIAL FORMATION

Conidia of *P. oryzae* germinated and formed appressoria at the tips of germ tubes if the surface was wet. The surface of the leaf blade was covered with waxes (Figure 1), but that of panicle and spikelet was not covered with waxes (Figure 2). The fact that panicles were more subjected to wetting condition than leaf blade could be the possible explanation for the lack of wax. The waxes around the appressorium disappeared on the surface of the leaf blade (Figure 1). It appears that the secretion from the appressorium dissolved the waxes around. Appressoria and a part of germ tube were held firmly against the cuticle by means of mucilage. The mucilage spread some distance (0.7 - 1.7 µm) from the appressorium (Figures 5 and 6). A melanised layer (Woloshuk *et al.*, 1983; Howard and Ferrari, 1989; Koga, 1994b) was formed around the entire appressorium except in the wall surrounding the penetration peg (Figure 5). It has been demonstrated that the melanin layer acts as the differentially permeable layer that is responsible for enormous appressorium turgor pressures (Howard, 1994).

2. 2. PENETRATION

Direct entry of the epidermal cell was achieved by a penetration peg, which emerged from the appressorium. The protoplast of the appressorium began protruding from the centre of the lower appressorial surface, causing indentation of the host cell cuticle (Figure 3). The tip of this protrusion lacked a fungal cell wall (Figure 4). Small electron dense vesicles, gathered at the tip of the protrusion (Figures 3 and 4). As the protrusion became an identifiable penetration peg, the host cuticle layer was perforated (Figure 4) and sometimes folded inward (Figure 5). Penetration pegs were 0.4-1.1 µm in diameter. A portion of the host epidermal cell wall subtending the penetration peg became more electron-dense than other cell wall regions (Figure 4). The microfibrils of the host cell wall around penetration pegs were disorganised, non-homogeneous, and electron dense in contrast to the regular pattern in non-penetrated cell walls (Figure 5). A small electron transparent zone was sometimes present in the area of disorganised microfibrils (Figure 6).

No ultrastructural changes were observed in the cytoplasm of host epidermal cells while penetration pegs were passing through the cuticle (Figures 3 and 4), but when penetration pegs were passing within host cell wall, collar-like wall appositions were formed between the host cell wall and host plasma membrane (Figure 5). It was clear that complete penetration of the wall and entry into the epidermal cell lumen was not necessary for apposition formation. Most appositions contained electron-dense globules within less electron dense material (Figures 5 and 6), however, some consisted of only electron-dense material.

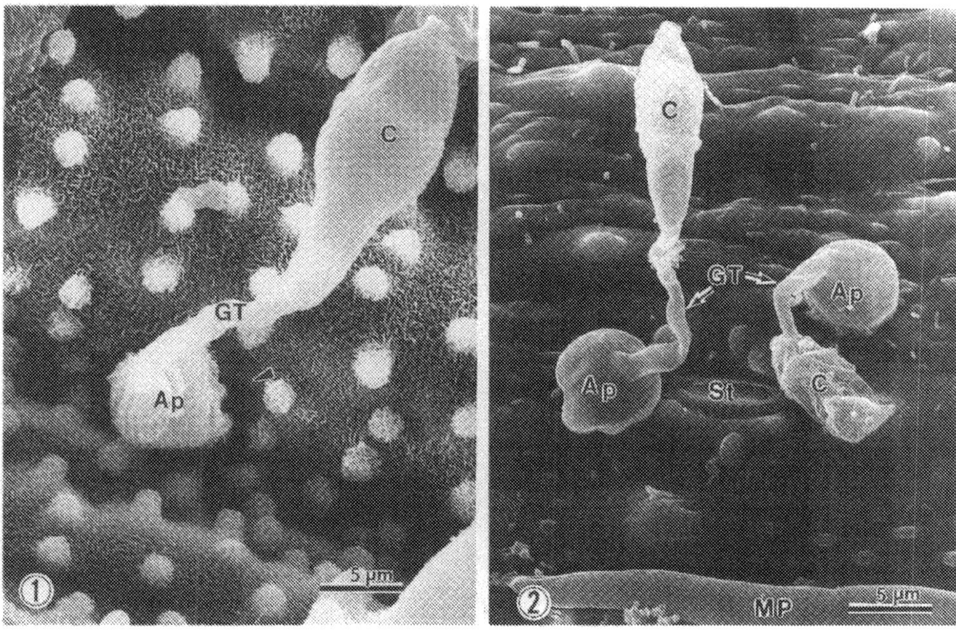

Figure 1. SEM image of spore, germ tube and appressorium on the epidermis of rice leaf blade. Note that the waxes around the appressorium have disappeared (arrow).

Figure 2. SEM image of spore, germ tube and appressorium on the epidermis of rice panicle neck. Note that the surface of the panicle was not covered with waxes.

Reprinted from H. Koga (1994). Copyright Phytopathological Society of Japan. Abbreviations used in figures: Ap, appressorium; ACW, appressorial cell wall; APM, appressorial plasma membrane; APR, appressorial protoplast; C, conidium; cg, granulation of the epidermal cellular contents; Cu, cuticle; dbc, epidermal cell with deep browning; EC, epidermal cell; GT, germ tube; HCW, host cell wall; HPM, host plasma membrane; IH, invading hypha; M, mitochondrion; ML, melanized layer; MP, minute protuberance; Mu, mucilage; N, nucleus; No, nucleolus; PC, parenchymatous cell; Pit, pit-pair; PP, penetration peg; SC, sclerencyma cell; WA, wall apposition.

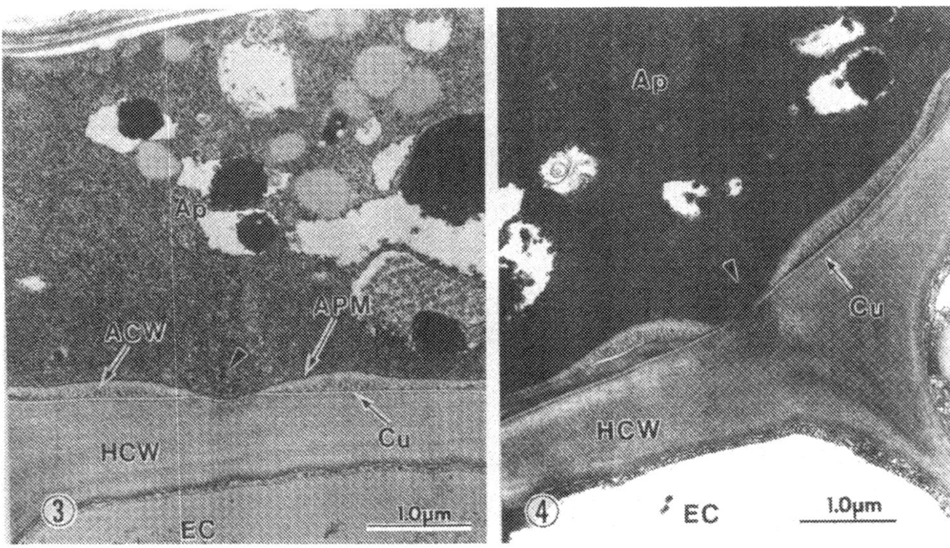

Figure 3-6. Penetration of epidermal cell wall of rice leaf-sheath of ZTS (susceptible) by *P. oryzae* at 25 h.

Figure 3. Protrusion of protoplast from the lower central portion of the appressorium. The cell wall of the appressorium at the protrusion site is almost absent, and the host cell cuticle layer and host cell wall are slightly depressed. Cytoplasmic vesicles of unknown origin (arrowhead). Reprinted from H. Koga (1994). Copyright National Research Council of Canada.

Figure 4. Penetration by the appressorium through the host cell cuticle layer and host cell wall. Note that the portion of host cell wall in advance of the penetration peg is electron dense. Cytoplasmic vesicles of unknown origin (arrow head). Reprinted from H. Koga (1994). Copyright National Research Council of Canada.

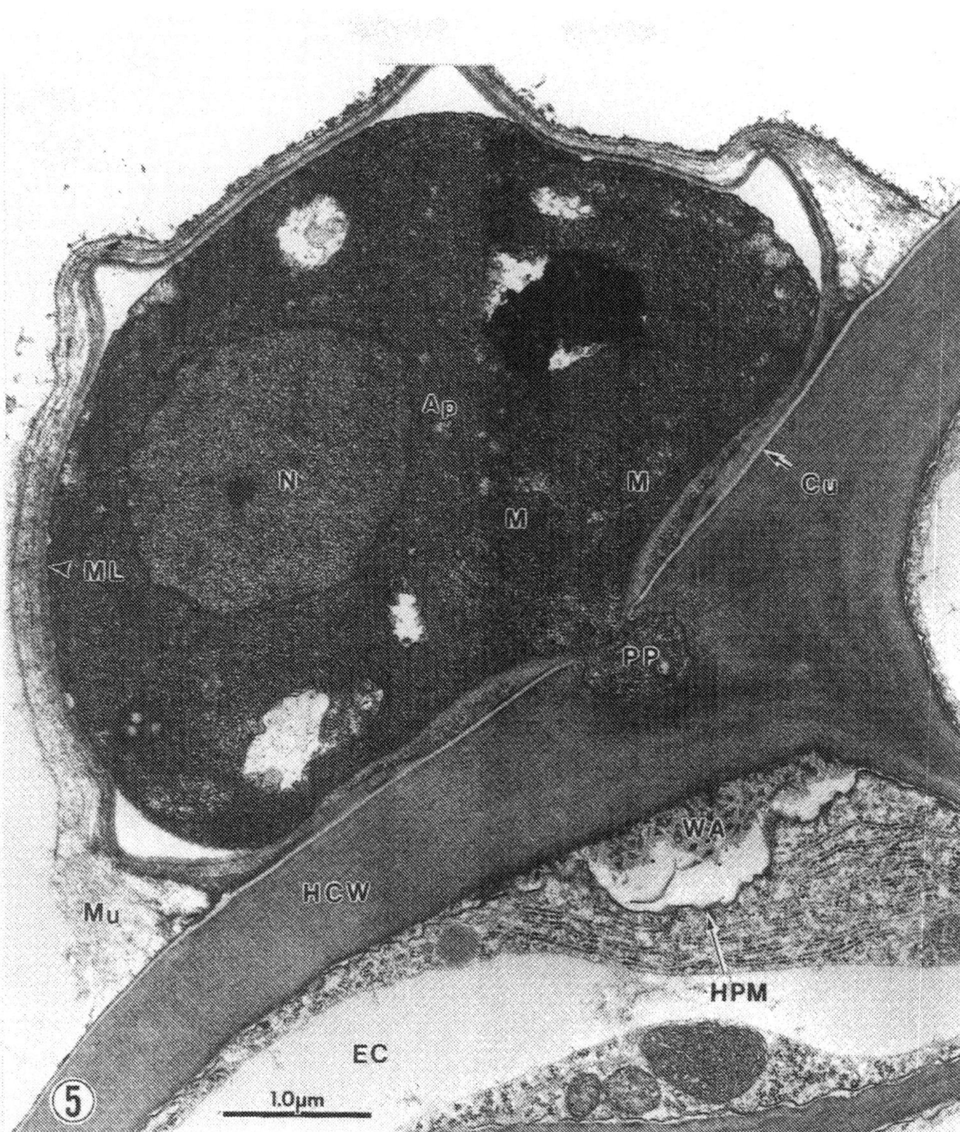

Figure 5. A median section through a developing penetration peg. The host cuticle and wall microfibrils appear digested. The penetration peg wall is ill-defined and the host microfibrils surrounding the penetration peg appear disorganized. The wall apposition is formed between the host cell wall and plasma membrane before the host cell wall is completely pierced by the penetration peg. Note that the host cuticle layer is perforated and folded inward. Reprinted from H. Koga (1994). Copyright National Research Council of Canada.

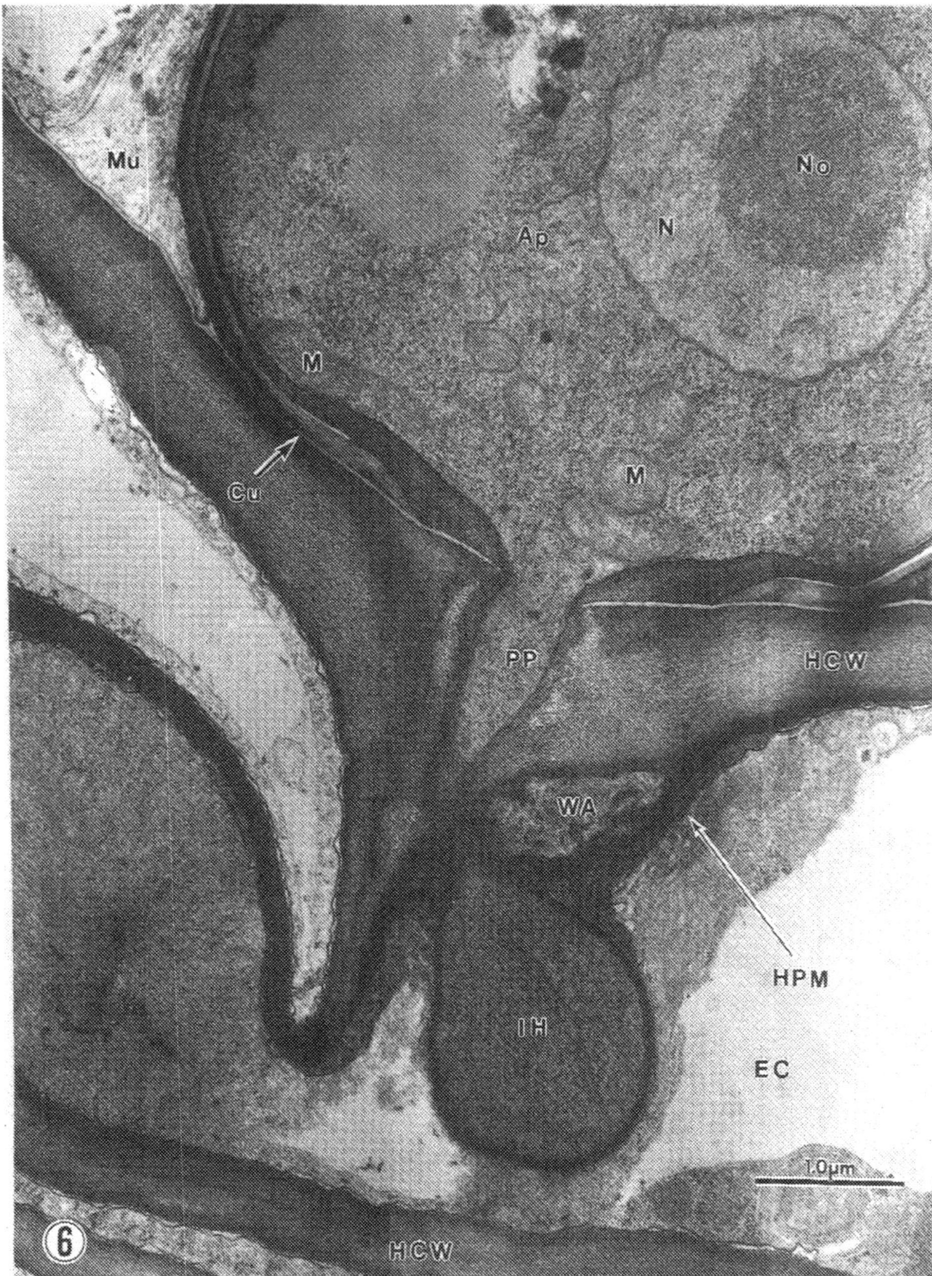

Figure 6. Appressorium with penetration peg and invasion hypha formed on host cell wall following piercing of the wall apposition. Reprinted from H. Koga (1994). Copyright National Research Council of Canada.

Wall apposition formation was observed only in living epidermal cells (Figures 5 and 6). Wall appositions were pierced by the penetration peg, which expanded into an invading hypha when it emerged inside the cell lumen (Figure 6). In living host cells invading fungal hyphae were surrounded by host cytoplasm (Figure 6). Other studies have also indicated that appositions can be formed before the host cell wall is pierced by the penetration peg (Edwards and Allen, 1970; Stanbridge *et al.*, 1971; Politis and Wheeler, 1973; Wehtje *et al.*, 1979). Most wall appositions were not effective as a barrier to fungal attack as has been reported by Hashioka *et al.* (1968) and Peng and Shishiyama (1989).

Electron micrographs of early stages of infection are consistent with the view that the penetration peg breached the pore wall overlay (Bourett and Howard, 1990; Heath *et al.*, 1992) and the cuticle is breached mechanically. They showed inward indentation of the host cell cuticle and cell wall. These indications of mechanical pressure by appressoria are similar to other reports relating to penetration of leaf cuticles (Heath *et al.*, 1992) and cellophane membranes (Bourett and Howard, 1990; Howard *et al.*, 1991). It is not clear, however, whether enzymatic dissolution is also involved in cuticle penetration, as has been reported for some other pathogens (McKeen, 1974; Ingram *et al.*, 1976; Valsangiacomo and Gessler, 1988). Penetration of cell walls showed clear evidence to support previous suggestions (Hashioka *et al.*, 1968) of enzyme action. Micrographs in Figures 5 and 6 showed disorganisation of microfibrils around infection pegs and a change in electron density. Gross swelling of the cell wall as reported by Mansfield and Richardson (1981) for broad bean leaves infected by *Botrytis cinerea* was not observed here.

The application of transmission electron microscopy (TEM) to the study of penetration into spikelet and panicle neck has been limited because the hardness of the tissue has resulted in poor infiltration of the embedding resins (Koga and Yoshino, 1990). However, application of a modified embedding technique for hard tissues (Park *et al.*, 1990) has enabled infection of *P. oryzae* in the spikelet (Koga, 1995) and panicle neck (Koga, 1994a) of rice to be studied. It was clarified that mode of penetration of blast fungus into the epidermal cells of the spikelets (Figure 7; Koga, 1995) and panicle necks (Figures 8 and 9; Koga, 1994a) was also similar to those of leaf blade (Hashioka *et al.*, 1968) and leaf sheath (Koga, 1994b). Although the outer cell walls of epidermis of spikelets were several times thicker than those of leaf blades and leaf sheaths (Figure 7), the penetration pegs, 0.3-0.8 μm in diameter, had pierced the outer cell walls of the epidermal cells 24 h after the spikelets were inoculated.

2. 3. DEVELOPMENT OF INVADING HYPHAE

Invading hyphae in the epidermal cells of panicle neck and spikelets developed into neighboring epidermal cells, sclerenchyma, parenchyma (Figure 8) and vascular bundles by 48 h after inoculation. Although sclerenchyma had a thick cell wall, most cells were invaded by the blast fungus 48 h after inoculation (Figure 8). It was observed that the invading hyphae often penetrated into the neighboring sclerenchyma cells through pit-pairs (Figure 8), where the cell wall is rather narrow. Invading hyphae often formed septa where they penetrated through the host cell wall. These septa were

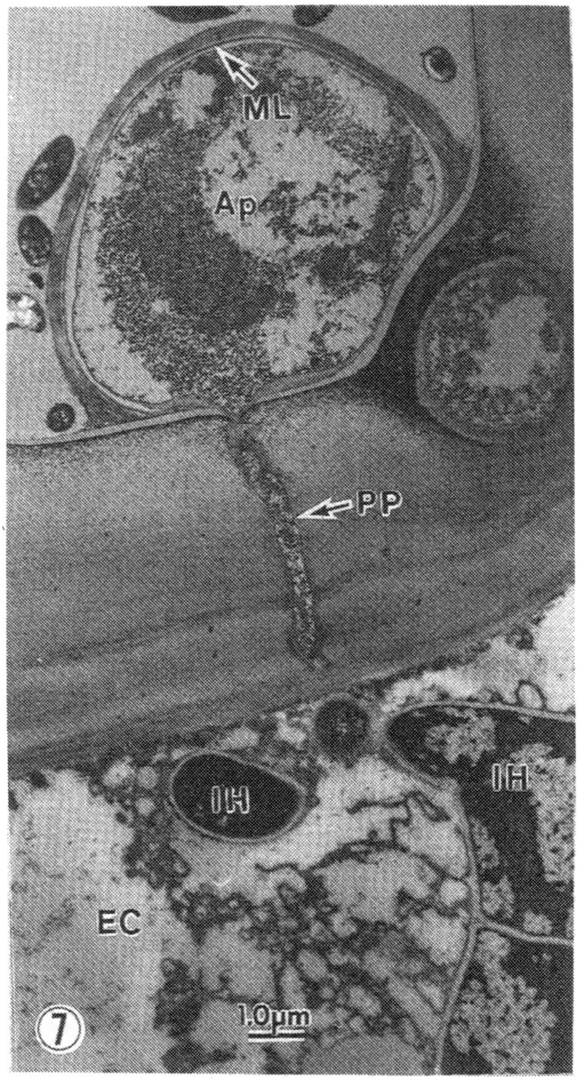

Figure 7. Section of an appressorium with a penetration peg in the epidermal outer wall of a spikelet 48 h after inoculation. Note the presence of a small electron transparent zone in the cell wall near the tip of the penetration peg. Reprinted from H. Koga (1995). Copyright Blackwell.

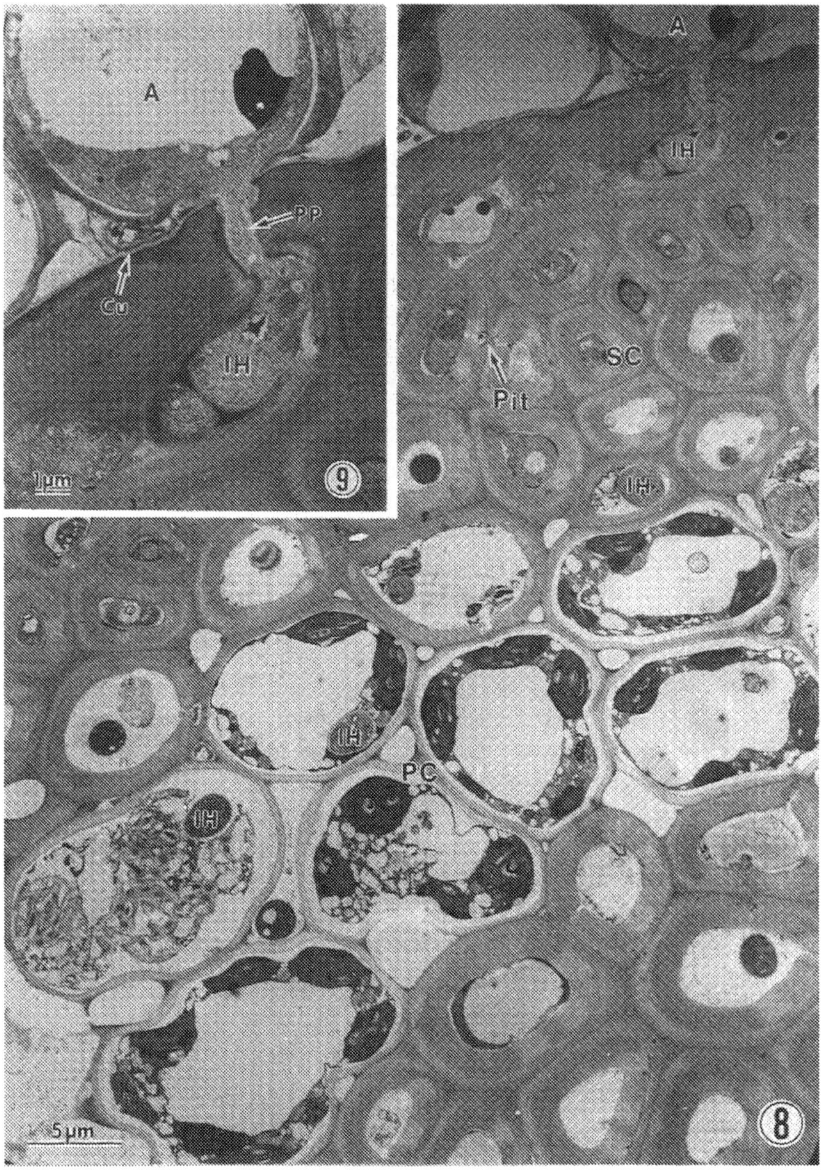

Figure 8. A penetration peg through an epidermal outer wall and invading hyphae in sclerenchyma and parenchyma 48 h after inoculation. Note that the chroloplasts in the parenchymatous cells invaded by blast fungi show the various degree of disorganization. Reprinted from H. Koga (1994). Copyright Phytopathological Society of Japan.

Figure 9. A higher magnification of part of Figure 8. A median section through a penetration site. Note that penetration is not followed by collar-like wall apposition. Reprinted from H. Koga (1994). Copyright Phytopathological Society of Japan.

sometimes accompanied by several Woronin bodies (Koga, 1994 a, 1995). Chroloplasts in parenchymatous cells invaded by the blast fungus showed various degrees of disorganisation, such as swelling, rupture of outer, inner, and thylakoid membranes ,and digestion by the blast fungus 48 h after inoculation (Figure 8), otherwise little ultrastructural change was observed in the cells of the sclerenchyma (Figure 8) and vascular bundle. Transections of parenchymatous cells 4 days after inoculation showed that parts of host cell wall might be digested enzymatically by the blast fungus. In contrast with parenchymatous cells, the cell wall of vessel and surrounding sclerenchyma cells remained comparatively intact 4 days after inoculation, but their cytoplasms were decompartmentalised by the fungal invasion (Koga, 1994a, 1995).

The ultrastructural study (Koga, 1994a) also provides information about the possible cause of the white head symptoms which occur when panicle necks become infected with the blast fungus. Yoshii (1937) suggested that these symptoms occur because of cessation of water supply caused by the collapse of vascular bundles and surrounding parenchymatous cells. In this study, it was elucidated that most vascular bundles and surrounding parenchymatous cells were invaded by the blast fungus and decompartmentalised by 4 days after inoculation. This may support Yoshii's hypothesis. On the other hand, Kozaka *et al.* (1985) reported that cultures of *P. oryzae* produced an extracellular phytotoxic glycopeptide capable of inciting white head symptoms quite similar to those of neck blast.

Anatomical examination of rice stems treated with the purified toxin revealed obvious blockage of the vascular system due to occlusion of the vessels, particularly in parts of the vascular system where invading hyphae had not yet developed. In the present study, no blockage of the vascular system was observed where invading hyphae were present. The results (Koga, 1994a) suggest that if invading hyphae of the blast fungus secrete toxin(s) they do not kill host cells of susceptible varieties immediately. Epidermal and parenchymatous cells of panicle neck invaded by the blast fungus appeared to be alive shortly after the fungal invasion (Figure 8), and thereafter began to degenerate slowly. It seems likely that such degeneration is caused not by the toxin(s), but hydrolytic enzymes secreted by the fungus.

2. 4. OUTTHRUST OF CONIDIOPHORE

The conidiophore cells of *P. oryzae* in rice leaf blades developed onto the intra-epidermal -cellular hypha, and outthrust outside the host plant by means of piercing through the outer wall of the host epidermal cell due to severe constriction of the conidiophore-mother cell and the first (basal) conidiophore cell. Thrust out of a conidiophore through stomatal slit without piercing of outer wall is also common in contradiction to limitation of invasion of the fungus into host through a stomal slit (Hashioka and Nakai, 1974). On lemma and palea of spikelets conidiophores outthrusted from large trichomes, minute protuberance and directly from epidermis. Also conidiophores erupted on empty glumes, rarely on rachilla ,on rudimentary glumes. On the panicle neck conidiophores outthrusted from stomata in addition to them (Koga and Yoshino, 1990; Koga, 1995). Conidiophores were often observed between empty glumes and rachilla, between neck and panicle branch, and between primary axis and branch , where wetting condition remains for long time (Koga and Yoshino, 1990). Conidiophores on panicles and spikelets erupted 1-2 days earlier than those on leaf blades (Koga, unpublished data).

3. Race -cultivar specificity and hypersensitive response

3. 1. QUANTATIVE ANALYSIS BY BRIGHT - FIELD AND FLUORESCENCE MICROSCOPY

3. 1. 1. *Leaf blade*

Spore germination and appressorial formation of *P. oryzae*, isolate Ken 60-19 (race 037) on rice leaves were compared among 4 resistant (Toride 1, Pi-No.4, Fukunishiki and Yashiromochi) and 6 susceptible varieties (Shin 2, Aichi asahi, Ishikari shiroke, Kanto 51, Tsuyuake and Koshihikari). Spore germination and appressorial formation were observed under the fluorescent microscope after samples were stained with Calcofluor White (Koga, 1983). The percentages of spore germination and appressorial formation of *P. oryzae* showed no statistical differences among 4 resistant and 6 susceptible varieties. From these results, it was suggested that specific resistance of rice varieties is not expressed before fungal penetration into leaf blades (Koga, 1983).

Early infection process of *P. oryzae* in leaves of susceptible and resistant varieties was observed under the light microscope with the aid of the whole-leaf clearing and staining technique (Figures 10 and 11, Koga and Kobayashi, 1982 ab). In the susceptible variety (Koshihikari), the penetration peg of *Magnaporthe grisea* pierced through host cell wall and developed invading hyphae. There were some appressoria with invading hyphae 36 h after inoculation and their proportion increased between 48 h and 72 h after inoculation. Invading hyphae developed and branched in the epidermal cell (Figure 10). The percentages of successful penetration on leaves of Toride 1 (highly resistant variety), and Fukunishiki (resistant variety), were less than on Koshihikari (susceptible) 48 h after inoculation. Growth of invading hyphae on leaves of Toride 1 and Fukunishiki were less than on Koshihikari 48 h after inoculation. In the highly resistant and resistant varieties, invading hyphae ceased growing within the primarily invaded epidermal cell even 144 h after inoculation (Figure 11, Koga and Kobayashi, 1982 ab).

The host-parasite encounter sites were grouped into 6 types, 72 h after inoculation, (Figure 12, N, A-E) according to fungal development and host responses. The rates of unsuccessful penetration without any host reactions at the penetration sites (Type N) in the 4 incompatible combinations were almost the same as those of 2 compatible ones,

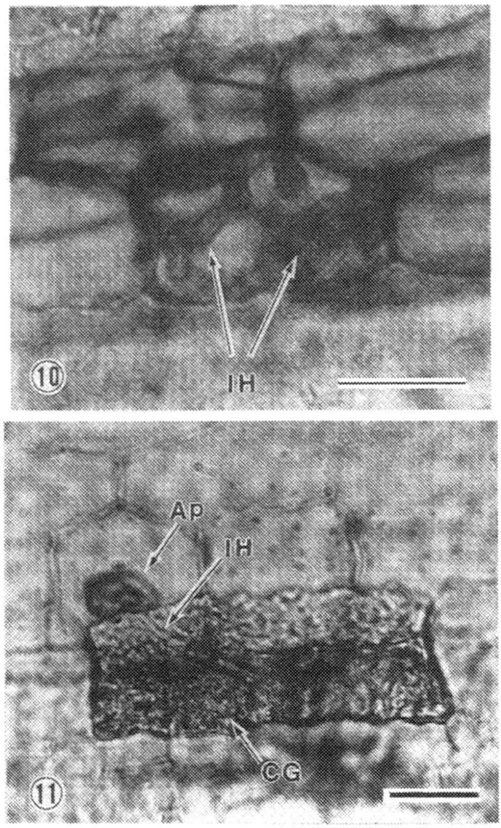

Figures 10-11. Microphotographs of epidermal cells of rice leaves invaded by *P. oryzae* isolate Ken 60-19. The scale in each picture represents 20 μm.

Figure 10. Invading hyphae that are branched are observed, but the epidermal cellular contents are not undergoing granulation 72 h after inoculation; susceptible cultivar Koshihikari.

Figure 11. A short and slender invading hypha which is not branched is observed and granulation of the epidermal cellular content is going on 144 h after inoculation; resistant cultivar Toride 1.

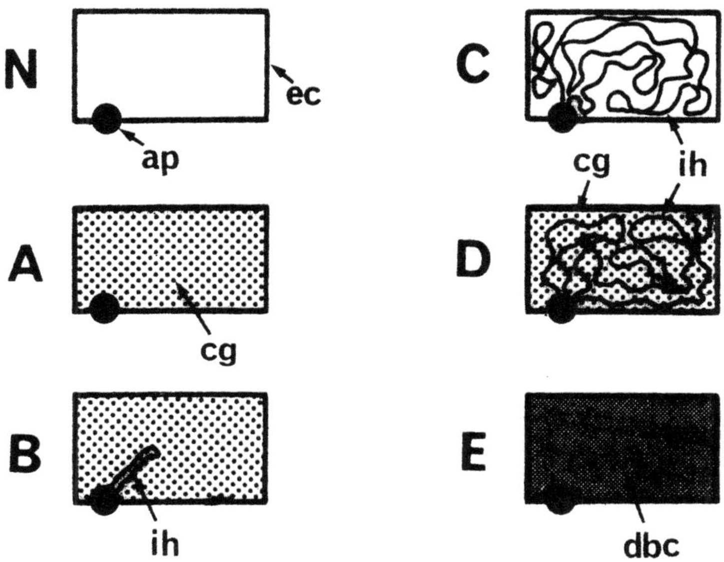

Figure 12. Classification of infection type according to the penetration of rice blast fungus, fungal growth and the reactions of the rice cells.

N: Unsuccessful penetration without any host reaction, A: Unsuccessful penetration with host cytoplasmic granulation in an epidermal cell, and with light browning late hours after inoculation, B: Poorly-developed invaded hyphae with host cytoplasmic granulation in an epidermal cell, and with light browning late hours after inoculation, C: Well-developed invaded hyphae without host cytoplasmic granulation in an epidermal cell, D: Well-developed invaded hyphae with host cytoplasmic granulation in an epidermal cell, E: Deep browning of the infected epidermal and parencymatous cells.

and were 55-65% in all the combinations. Therefore, the occurrence rate of each type except Type N is shown in Figure 13. The host-parasite encounter sites where fungal penetration or hyphal growth ceased within the primary attacked cells except reaction Type N were accompanied with the granulation of epidermal cellular contents (Type A and B). The occurrence rates of Type A and B were more than 80% in the 4 incompatible combinations, but about 20% in two compatible combinations. The epidermal cells with cytoplasmic granulations were often pale-brown 144 h after inoculation, where slender non-branched invading hyphae ceased within the primary attacked cells (Figure 11). Occurrence of well-developed invading hyphae (Figure

10,Type C) was frequently observed in the 2 compatible combinations, but not in the 4 incompatible ones (Figure 13). Some of them accompanied the cytoplasmic granules (Type D). It seemed that the granules in the type D appeared after the hyphae developed well. Therefore, the cytoplasmic granulation that occurred hypersensitively during and/or after penetration in Type A and B was qualitatively different from that in Type D.

Occurrence rates of the brown spot lesions (Type E) were 9.3 and 9.9% in Yashiromochi and Fukunishiki, respectively, rarely in Toride 1 and Pi No.4, about 15% in two compatible combinations (Figure 13). It was clarified from the electron-microscopic observation of the brown-spot lesion that degenerate hyphae developed sparsely in the necrotic parenchymatous cells. In the necrotic parenchymatous cells, the invading hyphae of *P. oryzae* were transformed at various grades. Some hyphae contained swollen mitochondria, extremely developed vacuoles, and the other hyphae appeared dead with granulation of their contents (Koga *et al.*, 1982). Therefore, it has been suggested that the necrotic lesions develop as a result of reactions of the parencymatous cells to the invading hyphae.

In the compatible combinations the epidermal cells where invading hyphae developed well and branched, were often surrounded by the browned neighbouring epidermal cells (Koga, unpublished data). These infection sites would become necrotic lesions. It seems that the sizes of necrotic lesions depend on the duration that the browning cells surrounded invading hyphae. As a result, many various sizes of necrotic lesions as well as acute type lesions would appear in the compatible combinations. It seems that the timing was closely affected by the environmental conditions such as temperature, the amounts of nitrogen in soil and so on.

From these results it appears that the granulation of the epidermal cellular contents closely correlated with the expression of the major genes for resistance. Peng and Shishiyama (1988) reported that cellular type showing light browning was observed at high rates of the infection sites in Aichiasahi (resistance gene, *Pi-a*) and Ishikarishiroke (*Pi-i*), and these browning as well as cytoplasmic granules closely correlated with the specific resistance.

3. 1. 2. *Leaf sheath*

Inoculation of the inner epidermis of excised rice leaf sheaths with the blast fungus following the method of Sakamoto (1949) has facilitated continuous observation of the host-parasite interaction in living tissue. The expression of resistance in the leaf-sheath during early stages of the infection process is similar to that in the leaf blade (Takahashi, 1956; Ohata *et al.*, 1963; Tomita and Yamanaka, 1983). Additionally fungal penetration is more frequent and better synchronised in the leaf sheath than in the leaf blade (Koga *et al.*, 1984) and thus this tissue is very suitable for use in studies on the relationship between cytological resistance responses, penetration and fungal growth within responding cells. Excised leaf sheaths are also amenable to experimental treatments (e.g. with heat and inhibitory chemicals) which can help elucidate the relative importance of the various distinct cytological events.

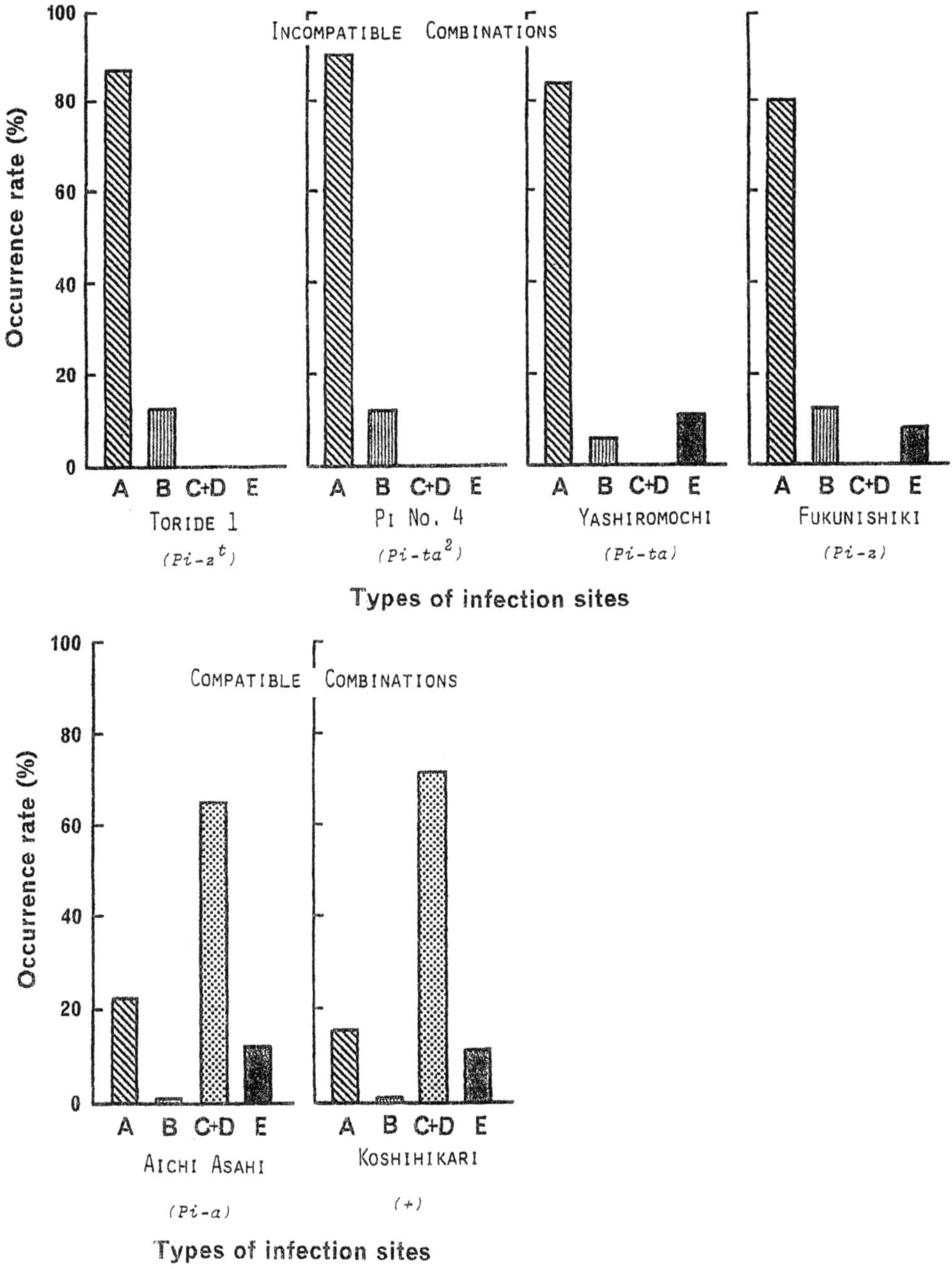

Figure 13. Occurrence rates of various types of infection sites in compatible and incompatible combinations of rice cultivars and *P. oryzae*, isolate Ken 60-19, race 037, 72 h after inoculation.

In both susceptible (ZTS) and resistant (ZTR) near-isogenic lines (Pi-z[t]), the first difference in microscopic events between susceptible and resistant lines was recognised as a loss of ability of some epidermal cells in the resistant line bearing *P. oryzae* appressoria to plasmolyse in hypertonic sucrose solution (Figure 14 a). Cell death in resistant ZTR as judged by the absence of plasmolysis was not observed 13 h after inoculation but was detected at low frequency at 16 h. It reached its maximum at 28 - 31 h when 77 % of attacked cells were dead.

The percentage of attacked epidermal cells which failed to plasmolyse was almost coincident with those which showed autofluorescence as reported in resistant barley leaves attacked by *Eryisiphe graminis* f. sp. *hordei* (Koga *et al.*, 1988). At 16 and 19 h most non-plasmolysed epidermal cells had weak autofluorescence, with only a few in which autofluorescence could not be detected. However, within 10-20 min after cell death the autofluorescence of such cells became sufficiently strong to be detected. After 22 h all epidermal cells which had lost the capability for plasmolysis autofluoresced strongly (Figure 15 a, b). Autofluorescence and loss of capability for plasmolysis were initially restricted to the first invaded epidermal cell and occasionally in a neighbouring cell, but by 72 h neighbouring cells frequently exhibited autofluorescence. In the susceptible ZTS line 1-7% of attacked cells were judged dead at each of the observation times after 13 h. Other invaded epidermal cells had the capability for plasmolysis, and showed autofluorescence only as faint halos surrounding the point of penetration (Figure 14a, b). Capability to plasmolyse was recorded in 70 - 80 % of the epidermal cells under appressoria, 28 h after inoculation in ZTS. However by 45 h after inoculation most of the first invaded epidermal cells were filled with hyphae, had lost their capability for plasmolysis, and showed weak overall autofluorescence. Neighbouring infected epidermal and parenchymatous cells were capable of plasmolysis and did not autofluoresce. The percentages of penetration in ZTS were similar to those in ZTR, when cells that were unable to plasmolyse were counted as penetrated. The length of hyphae in ZTR was less than in ZTS from 25 h after inoculation. There was no significant growth in ZTR from 19 h after inoculation.

The present microscopic evidence supports the view that the specific outcome of the interaction is determined during penetration into the epidermal cell (Koga and Kobayashi, 1982 ab; Koga, 1983). Degradation of the host cell nucleus and cytoplasm, and inability to plasmolyse in a hypertonic solution, are indicative of hypersensitive cell death. These reactions were first observed by light microscopy to occur between the time of appearance of a penetration pore in the appressorium and the first observation of an invading hypha in the epidermal cell. Kitazawa *et al.* (1973) clarified by cinephotomicrography that in an incompatible interaction between potato and a race of *Phytophthora infestans*, the protoplasmic streaming of the host cell stopped soon after

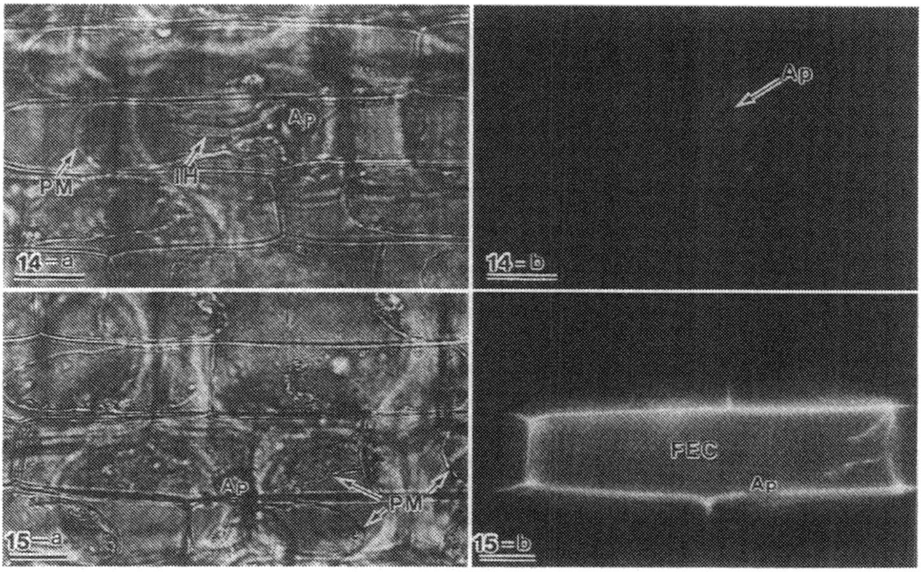

Figures 14-15. Light (a) and fluorescent (b) micrographs of the epidermal cells of leaf sheaths inoculated with *P. oryzae*. The leaf sheaths were treated with 0.8 M sucrose solution for 20 min. before observation. Scale bars = 20 µm. Reprinted from H. Koga (1994). Copyright National Research Council of Canada.

Figure 14. An infected epidermal cell of the susceptibel line (ZTS) 27 h after inoculation. The invaded epidermal cell plasmolysed (Fig. 14-a), and showed very weak local autofluorescence in a halo observed beneath the appressorium.

Figure 15. A dead epidermal cell of the resistant line (ZTR) showing a hypersensitive response to *P. oryzae* 24 h after inoculation. Note that the dead cell has lost the ability for plasmolysis and exhibited strong autofluorescence, whereas surrounding cells plasmolysed and showed no autofluorescence. Ap: appressorium, IH: invading hypha, FEC: autofluorescence epidermal cell, PM: plasma membrane.

penetration by the pathogen. Bushnell (1981) also reported that cytoplasmic streaming stopped in host epidermal cells of barley under attack by *Erysiphe graminis* f. sp. *hordei* when the Mla gene conditioned host-parasite incompatibility. In the present study cytoplasmic granulation appeared almost at the same time as hypersensitive cell death. Many cytoplasmic vesicles of unknown origin were observed around wall appositions. The relationship between cytoplasmic streaming or cytoplasmic aggregation and the cytoplasmic granules reported here are under detailed investigation.

3. 2. ULTRASTRUCTURAL CHANGES IN THE RESISTANT LINE

In ZTR, the ultrastructure of the host-parasite interaction was the same as that in ZTS until the penetration peg pierced the cuticle layer. Epidermal cells in which the penetration peg had entered the cell wall beneath the cuticle or passed into the lumen showed disorganisation typical of the hypersensitive reaction (Figures 16 and 17). The host plasma membrane was fragmented, and cell organelles were collapsed. As a result debris and vesicles derived from them were observed in the invaded epidermal cells. No wall appositions were found under penetration pegs. Some invading hyphae were moribund, but others were filled with cytoplasm, and cell organelles such as mitochondria (Figures 16 and 17). Electron microscopy results also indicated that the first signs of hypersensitivity can occur very early on in the interaction in the resistant line, certainly before the infection hypha had expanded into the lumen to become visible by light microscopy. Furthermore, initiation of the hypersensitive response may sometimes take place before the host cell wall is completely breached by the infection hypha.

In the susceptible ZTS line, most of the invaded host cells remained alive and wall appositions were seen in cells in which the penetration peg had incompletely penetrated the cell wall. In the resistant ZTR line, the invaded host cells died hypersensitively without accompanying wall appositions. These results suggested that the determination of resistance or susceptibility in some infection sites might occur in cells in which the penetration peg had incompletely penetrated the cell wall. In the resistant line it is presumed that hypersensitivity was triggered by the release of a 'cell death inducing substance(s)' as reported by Bucheli *et al.* (1990). Similarly, it is postulated that substances that stimulate the production of apposition material are released from encounter sites of the penetration peg and cell wall. In both cases the active materials would need to diffuse through part of the cell wall before initiating a response.

3. 3. EFFECT OF HEAT SHOCK AND INHIBITOR

Heat shock did not affect the viability of cells of non-inoculated leaf sheaths of ZTR. In inoculated leaf-sheaths, death of infected epidermal cells occurred a little earlier in sheaths that had received a heat shock than in those that had not, but by 25 h about 80% of infected cells had lost their ability to plasmolyse in both treated and untreated sheaths. However, heat shock did have a marked effect on the degree of autofluorescence of dead cells, and on fungal growth. Invaded epidermal cells of heat-treated sheaths showed little or no autofluorescence but contained long hyphae whereas those from untreated sheaths exhibited strong autofluorescence and very limited fungal

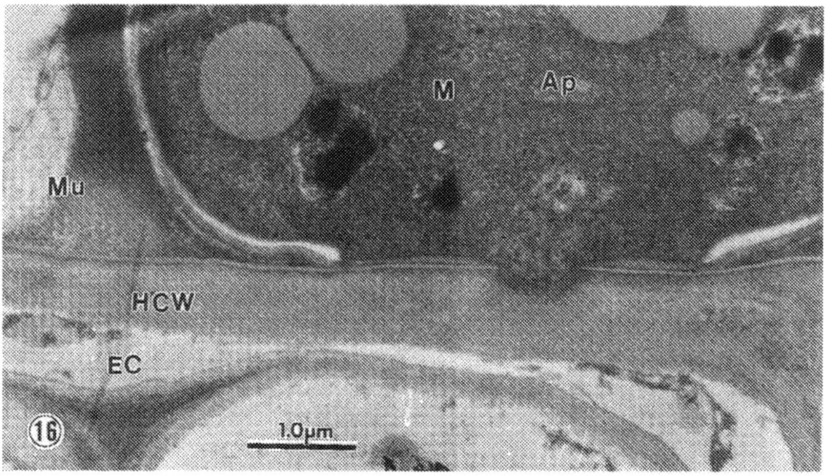

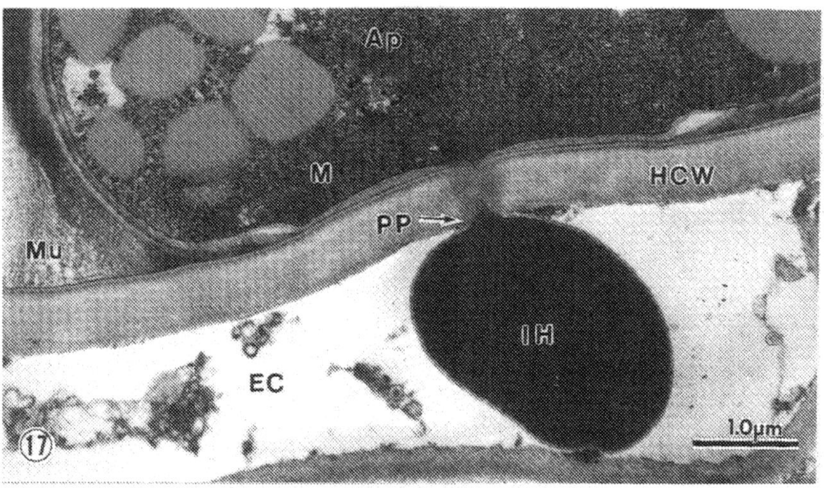

Figures 16-17. Penetration of epidermal cell wall rice leaf sheath of ZTR (resistant line) by *P. oryzae* 27 h after inoculation. Reprinted from H. Koga (1994). Copyright National Research Council of Canada.

Figure 16. A median section through a developing penetration peg. The host cuticle and wall microfibrils appear digested. The host cell wall in advance of the penetration peg is electron dense. The epidermal cell under attack has lost compartmentalisation. A wall apposition is not seen under the appressorium.

Figure 17. Development of an invading hypha of *P. oryzae* in the epidermal cell. Note that the epidermal cell has lost compartmentalisation and there is no wall apposition at the site of penetration.

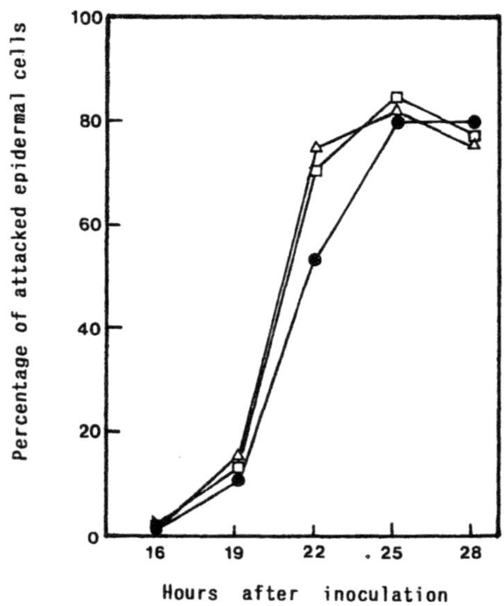

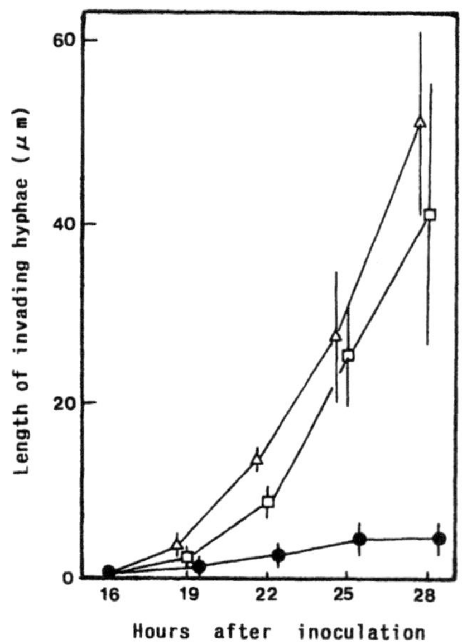

Figure 18. Effect of NaAsO2 on the percentage of epidermal cells of the resistant line (ZTR) infected with P. Oryzae that failed to plasmolyze in hypertonic sucrose solution. 0 μM $NaAsO_2$ (Control), 200 μM $NaAsO_2$, 1000 μM $NaAsO_2$.
Reprinted from H. Koga (1994). Copyright National Research Council of Canada.

Figure 19. Effect of $NaAsO_2$ on the growth of hyphae within epidermal cells of the resistant line (ZTR) infected with *P. oryzae.* Each point represents mean hyphal length from 30 infection sites. Vertical bars indicate standard deviation.
0 μM $NaAsO_2$ (Control), 200 μM $NaAsO_2$, 1000 μM $NaAsO_2$.
Reprinted from H. Koga (1994). Copyright National Research Council of Canada.

growth. Death of epidermal cells in inoculated leaf sheaths of ZTR treated with 200 μM or 1000 μM $NaAsO_2$ solution occurred a little earlier than in control (0 μM) leaf sheaths, but the percentage of cells which failed to plasmolyse reached about 80% in each treatment 25 h after inoculation (Figure 18). As described above for leaf sheaths receiving heat treatment, those treated with 200 μM and 1000 μM $NaAsO_2$ showed rapid fungal development which was associated with weak or no autofluorescence of the infected cells, while in control (0 μM) leaf sheaths, fungal growth was restricted and autofluorescence was strong (Figure 19).

The identity of antifungal compounds in autofluorescent cells is not known. However, the ability of $NaAsO_2$, an inhibitor of lipid synthesis in higher plants, to break resistance and facilitate fungal growth, is consistent with the view that production of an antifungal environment within hypersensitive cells is associated with the synthesis of unsaturated fatty acids such as those known to have strong antifungal activity towards *P. oryzae* (Kato *et al.*, 1986). Results from the present study (Koga, 1994b) support the previous findings of Tomita *et al.* (1984) that inhibitors of fatty acid synthesis affect the expression of resistance in a highly resistant rice variety. The $NaAsO_2$ treatment had no effect on hypersensitive cell death unlike some other resistance-breaking treatments in other host-parasite systems. For example, cell death was reported to be greatly reduced following removal of potato tissue beneath cells responding to *Phythophthora infestans* (Kitazawa *et al.*, 1973) and after raising the incubation temperature of rust-infected wheat lines containing the temperature-sensitive *Sr6* gene (Skipp *et al.*, 1974). Thus $NaAsO_2$ appears to prevent the accumulation of antifungal compounds rather than modify the hypersensitive response.

The results of this study (Koga, 1994b) support the view that the characteristic manifestation of race specific resistance to *P. oryzae* controlled by *Pi-z'* gene is the hypersensitive death of epidermal cells. Growth of invading hyphae in the resistant line was restricted within cells that had died as the result of a hypersensitive response during penetration of the host cell. However, it is clear from the results of the experiments using heat shock and $NaAsO_2$ treatments that cell death alone is not responsible for restriction of hyphal growth within the penetrated cell. Fungal hyphae grew unimpeded in treated tissue where cell death was not accompanied by autofluorescence. It seems likely that the compounds responsible for autofluorescence, or others which accumulate in cells capable of developing autofluorescence, are antifungal and arrest fungal growth in dead cells. This antifungal activity restricted hyphal development rather than causing a rapid killing of the invading hypha. Thus the ultrastructure of hyphae seen in hypersensitive cells often indicated the retention of cellular compartmentalisation, and some hyphal growth occurred after cell death, as has also been reported by Ohata *et al.* (1963).

4. References

Barr,M.E. (1977) *Magnaporthe*, *Telimenella* and , *Hyponectria* (Physosporellaceae), *Mycologia* **69,** 952-956.

Bourett, T. M. and Howard, R. J. (1990) *In vitro* development of penetration structures in the rice blast fungus *Magnaporthe grisea*, *Canadian Journal of Botany* **68**, 329-342.

Bucheli, P., Doares, S.H., Albersheim, P., and Darvill, A. (1990) Host-pathogen interactions XXXVI. Partial purification and characterization of heat-labile molecules secreted by the rice blast pathogen that solubilize plant cell wall fragments that kill plant cells, *Physiological and Molecular Plant Pathology* **36**, 159-173.

Bushnell, W.R. (1981) Incompatibility conditioned by the Mla gene in powdery mildew of barley: the halt in cytoplasmic streaming, *Phytopathology* **71**, 1062-1066.

Edwards, H.H. And Allen ,P.J. (1970) A fine structure of the primary infection process during infection of barley by *Erysiphe graminis* f. sp. *hordei*, *Phytopathology* **60**,1504-1509.

Hashioka, Y., Ikegami, H. and Murase, T. (l968) Fine structure of the rice blast. III. The mode of invasion of *Pyricularia oryzae* into rice epidermal cells, *Research Bulletin of the Faculty of Agriculture*, Gifu University **26**, 23-30.

Hashioka, Y. and Nakai (l974) Fine structure of the rice blast. XI. Outthrust of conidiophore cells of *Pyricularia oryzae*, etc., through an epidermal outer wall of a host leaf, *Research Bulletin of the Faculty of Agriculture*, Gifu University **36**, 9-18.

Heath, M. C., Valent, B., Howard, R. J., and Chumley, F. G. (1990) Interactions of two strains of *Magnaporthe grisea* with rice, goosegrass and weeping lovegrass, *Canadian Journal of Botany* **68**, 1627-1637.

Heath, M. C.,Howard, R. J. , Valent, B., and Chumley, F. G. (1992) Ultrastructural interactions of one strain of *Magnaporthe grisea* with goosegrass, *Canadian Journal of Botany* **70**, 779-787.

Howard,R.J. And Ferrari, M.A. (1989) Role of melanin in appressorium function, *Experimental Mycology* **13**, 403-418.

Howard, R. J., Bourett, T. M. and Ferrari, M. A. (1991) Infection by *Magnaporthe*: An *in vitro* analysis, in K. Mendgen and D.-E. Lesemann (eds.), *Electron microscopy of plant pathogens*, Thomson Press, Springer- Verlag, Berlin, Heidelberg, New York, pp. 251-264.

Howard, R.J. (1994) Cell biology of pathogenesis, in R.S. Zeigler, S. A. Leong and P. S. Teng (eds.*) Rice blast disease*, CAB International, UK, pp. 3-22.

Ikata, S., Matuura, K., and Taguchi, S. (1931) Results of experiments on the control 'of rice blast. 1. Studies on cultivar resistance to rice blast disease, *Noji Kairyo Shiryo* **20**, 1-140.

Ingram, D. C., Sargent, J. A. and Tommerup, I. C. (1976) Structural aspects of infection by biotrophic fungi, in J. Friend and D. R. Threfall (eds.) *Biochemical aspects of plant-parasite relationships*, Academic Press, London, pp. 43-78.

Kato, T., Yamaguchi, Uyehara, T. and Namai, T. (1986) Rice blast resistance and unsaturated hydroxy fatty acids, *Kagaku to Seibutsu* **24**,183-188.

Kawamura,E. and Ono, K.(1948) Resistance of foreign rice varieties to rice blast disease, *Journal of the National Agricultural Experiment Station* **4**, 13-22.

Kitazawa, K., Inagaki, H.and Tomiyama, K. (1973) Cinephotomicrographic observation on the dynamic responses of protoplasm of potato cell to infection by *Phytophthora infestans*, *Journal of Phytopathology* **76**, 80-86.

Koga, H. and Kobayashi, T. (1982 a) Comparison of the leaf infection process of *Pyricularia oryzae* Cav. in rice leaves of compatible and incompatible combinations, *Annals of the Phytopathology Society of Japan* **48**, 506-513.

Koga, H. and Kobayashi, T. (1982 b) Cytological reactions in leaves of the rice variety, Toride 1 to the incompatible race of *Pyricularia oryzae* Cav. *Proceedings of the Association of Plant Protection* Hokuriku **30**,12-18.

Koga, H., Kobayashi, T. and Horino, O. (L982 c) Electron microscopical observation of rice leaves infected with *Pyricularia oryzae* Cav. in compatible and incompatible combinations. I. Fine structure of invaded hyphae in host cells, *Annals of the Phytopathology Society of Japan* **48**, 281-289.

Koga, H. (1983) Comparison of spore germination and appressorial formation of *Pyricularia oryzae* Cav. on rice leaves in compatible and incompatible combinations, *Proceedings of the Association of Plant Protection* Hokuriku **31**, 7-12.

Koga, H., Horino, O. and Kobayashi, T. (1984) Ultrastructural studies on the resistance mechanisms of rice plants infected with *Pyricularia oryzae* Cav., Bulletin of the Hokuriku *National Agricultural Experiment Station* **28**,1-56.

Koga, H., Zeyen, R.J. Bushnell, W.R. and Ahlstrand, G.G. (1988) Hypersensitive cell death, and insoluble silicon accumulation in barley leaf epidermal cells under attack by *Erysiphe graminis* f. sp. *hordei*, Physiological and Molecular Plant Pathology 32, 395-409.

Koga, H. and Yoshino, R. (1990) A scanning electron microscopic study of conidiophores of *Pyricularia oryzae* Cav. on rice panicle, *Proceedings of the Association of Plant Protection* Hokuriku **38**, 3-8.

Koga, H. (1994 a). An electron microscopy of early infection processes in the panicle neck of rice inoculated with *Pyricularia oryzae* Cav., *Annals of the Phytopathology Society of Japan* **60**, 89-98.

Koga, H. (1994 b) Hypersensitive death, autofluorescence and ultrastructural changes in cells of leaf sheaths of susceptible and resistant near-isogenic line of rice (Pi-zt) in relation to penetration and growth of of *Pyricularia oryzae*, *Canadian Journal of Botany* **72**, 1463-1477.

Koga, H. (1995) An electron microscopic study of the infection of spikelets of rice by *Pyricularia oryzae*, *Journal of Phytopathology* **143**, 439-445.

Kozaka, T., Tsuchizawa, M.,Hanaue, M. and Watanabe, M. (1985) Phytotoxic glycopeptide inducing white head of rice plant produced by *Pyricularia oryzae* Cav., *Annals of the Phytopathology Society of Japan* 51, 199-205.

Mansfield, J. W. and Richardson, A. (1981). The structure of interactions between *Botrytis* species and broad bean leaves, *Physiological Plant Pathology* 19, 41-48.

McKeen, W.E. (1974). Mode of penetration of epidermal cell wall of *Vicia faba* by *Botrytis cinerea*, *Phytopathology* **64**, 461-467.

Ohata, K., Goto, K. and Kozaka, T. (1963) Observations on the reaction of rice cells to the infection of different races of *Pyricularia oryzae*, *Annals of the Phytopathology Society of Japan* **28**, 24-30.

Park, P., Ohno, T., Nishimura, S., Tanabe, K., Kohmoto, K. and Otani, H. (1990) Improved fixation and embedding methods for electron microscopy of *Alternaria alternata* spores, *Annals of the Phytopathology Society of Japan* **56**, 16-25.

Peng, Y. -L., and Shishiyama, J. (1988) Temporal sequence of cytological events in rice leaves infected with *Pyricularia oryzae*, *Canadian Journal of Botany* **66**, 730-735.

Peng, Y. -L., and Shishiyama, J. (1989) Timing of a cellular reaction in rice cultivars associated with differing degrees of resistance to *Pyricularia oryzae*, *Canadian Journal of Botany* **67**, 2704-2710.

Politis, D. J. and Wheeler, H. (1973) Ultrastructural study of penetration of maize leaves by *Colletotricum graminicola*, *Physiological Plant Pathology* **3**, 465-471.

Sakamoto, M. (1949) On the new method of sheath- inoculation of rice plants with blast fungus, *Pyricularia oryzae* Cav., for the study of the disease-resistant nature of the plant, *Bulletin of the Institute of Agricultural Research* Tohoku University **1**, 120-129.

Sakamoto, M. (1968) Study on blast resistance in rice, in *The sixtieth anniversary issue of Professor Sakamoto's birth*, Turuoka Printed Co., Turuoka, Yamagata Prf., pp.1-139.

Skipp, R.A., Harder, D.E. and Samborski, D.J. (1974) Electron microscopy studies on infection of resistant (Sr6 gene) and susceptible near-isogenic wheat lines by *Puccinia graminis* f. sp. *tritici*. *Canadian Journal of Botany* **52**, 2615-2620.

Stanbridge, B. , Gay, J. L. and Wood, R. K. S. (1971) Gross and fine structural changes in *Erysiphe graminis* and barley before and during infection, in T. F. Preece and C. H. Dickinson (eds.) *Ecology of leaf surface micro-organisms*, Academic Press , London, pp.367-379.

Takahashi, Y. (1956) Studies on the mechanism of rice plants to *Pyricularia oryzae*. 2. Pathological changes microscopically observed in host cells in which fungus hyphae do not grow well, *Bulletin of the Yamagata University of Agricultural Science* **2**, 37-51.

Tomita, H. and Yamanaka, S. (1983) Studies on the resistance reaction in the rice blast disease caused by Pyricularia oryzae Cav. 1. The pathological changes in the early infection stage of the inner epidermal cells of leaf sheath, *Annals of the Phytopathology Society of Japan* **49**, 514-521.

Tomita, H., Yamanaka,S., Namai, T. and Kato, T. (1984) Effect of the inhibitors of fatty acid synthesis on the resistance expression of rice plants to blast fungus, *Annals of the Phytopathology Society of Japan* 50, 385 (Abstract).

Valsangiacomo, C. and Gessler, C. (1988) Role of the cuticular membrane in ontogenic and Vf-resistance of apple leaves against *Venturia inequalis*, *Phytopathology* 78, 1066-1069.

Wehtje, G., Littlefield, L. J. and Zimmer, D. E. (1979) Ultrastructure of compatible and incompatible reactions of sunflower to *Plasmopara halstedii*, *Canadian Journal of Botany* **57**, 315-323.
Woloshuk, C. P., Sisler, H. D., and Vigil, E. L. (1983) Action of the antipenetrant, tricyclazole, on appressoria of *Pyricularia oryzae*, *Physiological Plant Pathology* **22**, 245-259.
Yoshii. H. (1937) Pathological studies on the rice blast caused by *Piricularia oryzae*. II. On the mode of infection of the pathogen, *Annals of the Phytopathology Society of Japan* **6**, 205-218.

CONDITIONS FOR SPORULATION AND PRESERVATION OF CONIDIA OF RICE BLAST FUNGUS *PYRICULARIA GRISEA*

SUN GUOCHANG and SUN SHUYUAN
Plant Protection Institute, Zhejiang Academy of Agricultural Sciences, Hangzhou 310021, P.R.China.

1. Introduction

Rice blast caused by *Pyricularia grisea* (Rossman *et al.*, 1990), is a serious and destructive disease around the world (Ou, 1985; Latterell, 1986; Guochang, 1994). In experiments involving the screening of rice varieties to blast resistance (Shuyuan *et al.*, 1986), conidia were required in large quantity for use as inoculum. Sporulation in *P. grisea* is a process distinct from mycelial growth in many respects. A certain amont of vegetative growth is a pre-requisite to the formation of conidia but did not always lead to sporulation. The present paper describes the conditions required for sporulation and the effects of some factors upon sporulation namely, different light sources and various media. Information on the conditions required for sporulation is likely to be useful in understanding the disease epidemics and control and also to produce sufficient spores for large scale inoculation experiments. Unfortunately, not all isolates of *M. grisea* sporulate well. In order to determine the best methods and optimal conditions for increasing sporulation of *P. grisea*, 17 media and BLB irradiation (Ahn, 1974) have been tested.

Many methods have been used to preserve fungal propagules and these have been reviewed by Heckly (1978) and Smith *et al.* (1983). One of the most useful methods is freeze drying and the straw preservation technique (Staplers *et al.*, 1987). Freeze drying provides a very stable storage technique as cells remain dormant and do not undergo metabolic activity. Although the straw preservation technique is simple, a good sporulating culture is required and a number of *P. grisea* isolates do not sporulate well (Guochang *et al.*, 1989). Thus, we have a problem in synchronising good sporulating cultures of *P. grisea* and the growth of rice seedling for artificial inoculation experiments. In order to overcome this problem, we undertook this study to optimise the conditions for the sporulation of *P. grisea* and to study the effects of preservation conditions on the viablity and pathogenicity of *P. grisea* conidia.

2. Materials and methods

In the present study, 41 *P. grisea* isolates and 11 rice varieties were used. For routine maintenance, all *P. grisea* isolates were grown on a yeast starch agar medium and

S. Sreenivasaprasad and R. Johnson (eds.), Major Fungal diseases of Rice Recent Advances, 111-117.
Prof. K. Manibhushan Rao Festschrift

stored at 28° C. 17 media and 3 irradiation levels were used to test the sporulation of blast fungal isolates. Cultures were irradiated continuously at a distance of30 cm from a 20 w black light florescent bulb (BLB, wave length 310-420nm). Test tubes containing 5 ml each of different media were inoculated with a 3 mm^2 agar plug (mycelium). Cultures were incubated at 28° C for 12 days. Five ml distilled water was added to each tube and the contents were shaken for 5 min to collect the conidia. Conidia were counted using a haemocytometer. Each experiment was replicated.

To develop a rapid sporulation method, cultures were grown in liquid medium for2 days at 28° C and the mycelium from these was used to inoculate cornmeal and rice straw agar medium. The conidia were counted after 4 days (Guochang *et al.*, 1990a,b). As a comparison, cornmeal and rice straw agar medium was also inoculated with a 3 mm2 agar plug (mycelial). Similar experiment was performed to test barley grains as medium for the sporulation of *P. grisea.*

Freshly produced conidia were washed from barley grains with distilled water. The spore suspension was passed through a filer paper in a funnel, dried at 30°-35° C and stored in an aluminium box at different temperatures (-5° C to 35° C.) and in a dessicator at room temperature, 4° C and –20° C. Conidial viability was evaluated every 3-4 months in a drop of distilled water on a slide inculated at 28° C for 24 h. Pathogenicity was tested by inoculating 11 rice varieties which included both *indica* and *japonica* varieties and seven Chinese rice differentials (Guochang *et al.*, 1989).

3. Results

3.1. EFFECT OF VARIOUS MEDIA ON SPORULATION

The results indicated that cornmeal and rice straw agar medium supported maximum sporulation of the rice blast fungus. Rice leaf and potato agar medium, rice straw agar medium and oatmeal agar medium were also effective (Table 1).

3.2. EFFECT OF CONTINUOUS IRRADIATION ON SPORULATION

Conidial production by seven *P. grisea* isolates growing on five media and under continuous light or continuous darkness was studies. Conidia were produced both in light and darkness. Generally, light enhanced the sporulation of each isolate. The effect of continuous exposure to BLB on the amount of conidia produced was different. Sporulation increased when the cultures were exposed to continuous irradiation with BLB, 310-420 nm (Table 2).

3.3. TECHNIQUE FOR RAPID SPORULATION OF THE RICE BLAST FUNGUS

To improve the level of sporulation of *P. grisea*, mycelium from liquid culture was used as inoculum on cornmeal and rice straw agar medium. The time required for sporulation was reduced from 11 days to 6 days, and the spore yields were higher than the standard method (Table 3). Similar results were obtained on barley grains as medium for sporulation of *P. grisea*, where the time required for sporulation decreased from 20 days to 9 days with increased spore yields (Table 4).

TABLE 1. Effect of various media on sporulation of the rice blast fungus

Medium	Mean sporulation (10^4/ml)[1]
Cornmeal rice straw agar (40, 50,20)[2]	27.89^{a}
Rice leaf potato agar (100, 200, 20)[3]	11.93^{b}
Rice straw agar (100, 20)[3]	11.91^{b}
Oatmeal agar (40, 20)	10.68^{c}
Cornmeal agar (50, 20)	9.46^{d}
Corn leaf potato agar (100, 200, 20)[3]	9.23^{d}
Rice bran rice leaf agar (20, 50, 20)	8.85^{d}
Rice leaf agar (100, 20)[4]	7.83^{e}
Rice bran agar (20, 20)	5.94^{f}
Rice polish agar (20, 20)	5.61^{fg}
Cornmeal beef extract agar (40, 40, 20)[4]	5.13^{gh}
Prune agar (1, 20)5	4.48^{hi}
Yeast starch agar (2, 10, 20)	4.26^{i}
Barley agar (50, 20)[5]	3.51^{j}
Potato agar (200, 20)	3.13^{j}
V-8 juice rice straw agar (250, 50, 20)[6]	2.89^{j}
Protein peptone yeast agar (10, 5, 20)	1.45^{k}

[1] Average of two replicates of 41 isolates.
Means followed by a common letter are not significantly different at the 1% level by DMRT.
[2] contains 40 g cornmeal, 50 g rice straw, 20 g agar and 1000 ml distilled water.
[3] plus 20 g sucrose; 4 plus 10 g glucose; [5] plus 1 g lactose and 1 g yeast extract; [6] plus 3 g CaCO3.

TABLE 2. Effect of continuous light treatment on sporulation of the rice blast fungus

Light treatment	Sporulation mean (104/ml)[1]
Black light blue fluorescent lamp	13.46^{a}
White fluorescent lamp	5.94^{b}
Continuous darkness (control)	4.77^{c}

1 Average of 2 replications of 41 isolates.
Means followed by a common letter are not significantly different at the 1% level by DMRT.

3.4. EFFECT OF PRESERVATION CONDITIONS ON CONIDIAL VIABILITY OF *P. GRISEA*

The results indicated that conidia stored for four months in an aluminium box at ambient temperature lost their viability. Conidia stored for four months in a dessicator retained their original level of viability, which did not differ significantly with temperature. Viability of conidia stored in a dessicator for more than 9 months varied significantly with temperature and fungal race (Figure 1).

TABLE 3. Effect of various cultural conditions on sporulation of *P. grisea* on cornmeal rice straw agar medium

Race	Culture method	Mycelial culture (days)	Sporulation culture	Total no. of days	Mean sporulation (105/ml)
ZB1 (926)	improved	2	4	6	10.57[a]
ZB1 (926)	standard	7	4	11	5.17[b]
ZF1 (80-38)	improved	2	4	6	5.92[b]
ZF1 (80-38)	standard	7	4	11	3.33[c]

Means followed by common letters are not significantly different at the 5^ level by DMRT.

TABLE 4. Effect of various cultural conditions on sporulation of *P. grisea* on barley grain medium

Race	Culture method	Mycelial culture (days)	Barley grain culture (days)	Sporulation culture	Total no. of days	Mean sporulation[1]
ZB1 (926)	improved	2	6	1	9	+++
ZB1 (926)	standard	7	12	1	20	++
ZF1 (80-38)	improved	2	6	1	9	+
ZF1 (80-38)	standard	7	12	1	20	+

[1] Mean sporulation: +, low; ++, moderate; +++, high.

3.5. PATHOGENICITY OF *P. GRISEA* CONIDIA FROM VARIOUS PRESERVATION CONDITIONS

Conidia of all *P. grisea* isolates tested retained their original level of pathogenicity in general, except ZA63 which yielded pathogenicity reactions equivalent to ZA61. Preservation of *P. grisea* conidia dessicated at –20° C enabled ready revival of the culture and sporulation. Consequently, no problems were encountered in synchronising good sporulation of the *P. grisea* isolates with rice seedling growth for pathogenicity tests (Table 5).

4. Discussion

Sporulation of *P. grisea* is a process influenced by several factors. Some of the factors basically necessary for the production of vegetative mycelium might be listed as necessary for sporulation, e.g. pH of the medium and temperature. Other factors such as specific substrates etc. may be requirements for vegetative growth but possibly unsuitable for sporulation. In most previous studies, sporulation has been simply examined after a certain period of incubation without considering the time pattern of each isolate under given conditions. Thus the important characteristic of dynamic role of light has been overlooked. BLB was effective for the sporulation of almost all isolates.

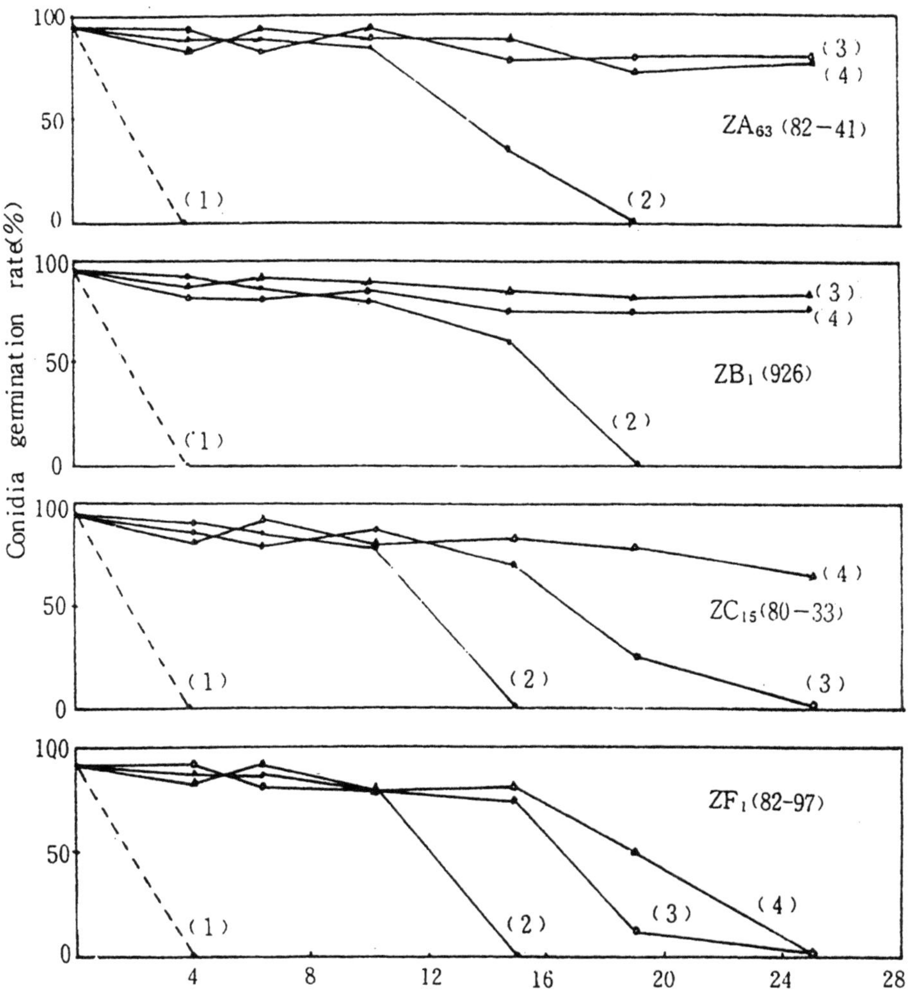

Preservation stage (months)

Figure 1. Effect of preservation conditions on conidial viability of *P. grisea*.stored in an aluminium box at ambient temperature, (2) stored in a dessicator at ambient temperature, (3) stored in a dessicator at 4° C and (4) stored in a dessicator at –20° C.

TABLE 5. Pathogenicity of *P. grisea* conidia from various preservation conditions

Race	Preservation condition	0	4	6.5	9.5	15	19	25
ZA63 (82-41)	Aluminum box ambient	ZA63	-*	-	-	-	-	-
	Dessicator ambient	ZA63	ZA63	ZA63	ZA61	ZA61	-	-
	Dessicator 4° C	ZA63	ZA63	ZA63	ZA61	ZA61	ZA61	ZA61
	Dessicator –20° C	ZA63	ZA63	ZA63	ZA61	ZA61	ZA61	ZA61
ZB1 (926)	Aluminum box ambient	ZB1	-*	-	-	-	-	-
	Dessicator, ambient	ZB1	ZB1	ZB1	ZB1	ZB1	-	-
	Dessicator 4° C	ZB1	ZB1	ZB1	ZB1	ZB1	ZB1	ZB1
	Dessicator –20° C	ZB1	ZB1	ZB1	ZB1	ZB1	ZB1	ZB1
ZC15 (80-33)	Aluminum box ambient	ZC15	-	-	-	-	-	-
	Dessicator, ambient	ZC15	ZC15	ZC15	ZC15	-	-	-
	Dessicator 4° C	ZC15	ZC15	ZC15	ZC15	ZC15	-	-
	Dessicator –20° C	ZC15	ZC15	ZC15	ZC15	ZC15	ZC15	ZC15
ZF1 (82-97)	Aluminum box ambient	ZF1	-	-	-	-	-	-
	Dessicator, ambient	ZF1	ZF1	ZF1	ZF1	-	-	-
	Dessicator 4° C	ZF1	ZF1	ZF1	ZF1	ZF1	ZF1	ZF1
	Dessicator –20° C	ZF1	ZF1	ZF1	ZF1	ZF1	ZF1	ZF1

*Conidia non-viable

In general, all the isolates produced abundant conidia on cornmeal rice straw agar medium under BLB irradiation. Cultures grown on protein peptone yeast agar medium showed poor or no sporulation even under BLB irradiation. This is indicative of possible involvement of stimulating substances present in cornmeal or rice straw, although the differences in the nutritional value of the two media could also play a role. For general purposes, the cornmeal, rice straw agar medium appears best suited for sporulation of the rice blast fungus and is simple to prepare.

We have also shown that conidia of *P. grisea* isolates preserved dessicated at –20° C survived up to 25 months. This technique appears suitable for preserving the conidia of *P. grisea* without losing their viability and pathogenicty.

5. References

Ahn, S.W. (1974) Effects of near ultraviolet irradiation on sporulation by *Pyricularia oryzae* Cavara on cultural media, *Annals of the Phytopathological Society of Japan* **40**, 337-343.

Heckly, R.J. (1978) Preservation of microorganisms, *Advances in Applied Microbiology* **24**, 1-53.

Latterell, F.M (1986) Longevity and pathogenic stability of *Pyricularia oryzae*, *Phytopathology* **76**, 231-235.

Ou, S.H. (1985) *Rice diseases* (2nd edition), Commonwealth Mycological Institute, Surrey, England, pp.109-201.

Rossman, A.Y., Howard, r.J. and Valent, B. (1990) *Pyricularia grisea*, the correct name for the rice blast fungus, *Mycologia* **82**, 509-512.

Smith, D. and Agnes, H.S.O. (1983) A comparison of some preservation techniques for fungi, *Transactions of the British Mycological Society* **81**, 535-540.

Staplers, J.A., De Hoog, A. and Vlug, I.J. (1987) Improvement of the straw technique for the preservation of fungi in liquid nitrogen, *Mycologia* **79**, 82-89.

Shuyuan, S., Minzhong, J., Zhimin, Z., Xinlan, T., Rongxiang, T. and Fang, D. (1986) Rice blast disease and its control, Shanghai Sciences and Technology publishing, Shanghai, P.R. China, pp. 1-182.

Guochang, S., Zongtan, S., Shuyuan, S., Yongqiang, T.and De, S. (1989) Preservation of *Pyricularia oryzae* conidia, *Chinese Journal of Rice science* **3**, 40-43.

Guochang, S., Zongtan, S. and Shuyuan, S (1990a) Studies on the condition of sporulation of rice blast fungus, *Pyricularia oryzae*, *Acta Agriculturae Univeristatis Zhejiangensis* **16**, 51-54.

Guochang, S., De, S., Shuyuan, S. and Zongtan, S. (1990b) Studies on the techniques for rapid sporulation of *Pyricularia oryzae*, *Acta Agriculturae Zhejiangensis* **2**, 93-94.

Guochang, S. (1994) Notes on the correct name of pathogen of rice blast, *Acta Mycologica Sinica* **13**, 158-159.

EPIDEMIOLOGY OF RICE BLAST DISEASE IN KOREA

CHANG KYU KIM
Plant Pathology Division,
National Institute of Agricultural Science & Technology
Rural Development Administration,
Suweon 441-707, Korea.

1. Introduction

Blast disease caused by *Pyricularia grisea* is one of the most destructive diseases in the world where rice is grown. In Korea, the occurrence of the disease was first recorded in the early 1900s. To date acreage of rice cultivation is more or less one million hectares, and leaf blast occurs in one hundred and fifty thousand hectares annually, and panicle blast in eleven thousand hectares on average (RDA, 1995).

Numerous programmes and continuous effort have been directed towards the control of this disease, at both national and international levels, and considerable progress has been made. Among various methods available to control the blast disease, use of resistant varieties is considered to be most effective and economical (Ou, 1972, 1977; Crill and Khush, 1979). Coons (1953) has stated that the control of rice blast disease by using disease resistant varieties is a painless method that does not levy on the farmer's pocket except to pay for the care and harvest of a larger crop. Experience in Korea with rice blast epidemics of 1978, 1980 and so on has shown the fallacy of this statement. Resistant cultivars do not necessarily remain resistant indefinitely and for scientists, rice researchers and rice producers to avoid devastating blast epidemics, it is necessary to have a complete and thorough understanding of the disease.

Thus, despite of a long history of research over one hundred years, blast is still a problem in rice production. One of the characteristics of the rice blast fungus is that it is very sensitive to minute changes of temperature and other environmental parameters. The life cycle of the fungus in nature is very complicated. In Korea, a series of rice blast epidemiology investigations has been conducted since 1974.

2. Materials and methods

The main experimental details are as follows: experiment site, Icheon; experiment duration, 1974-1995; variety tested, Jinheung (Japonica type, transplanting - May 24, heading - Aug

S. Sreenivasaprasad and R. Johnson (eds.), Major Fungal diseases of Rice Recent Advances, 119-129.
Prof. K. Manibhushan Rao Festschrift

14, maturity - Sep 30); fertiliser used, 1N ($N:P_2O_5:K_2$)=110:70:80kg/ha) and 2N (220kg/ha).

A rotation sampler (Suzuki 1969a) was installed between 1N and 2N plots. Spore trapping was done for one hour from 0100 to 0200 h using two glass slides, and spores were counted within a 864 mm^2 area, KY-type spore traps (Kim and Yoshino, 1987) were used to measure the amount of sporulation and conidial release from undetached lesions in the field. Leaf blast severity was checked from June 28 to August 2 at one week interval. Panicle blast was checked two weeks before maturity. For early detection of leaf blast lesion, I-KI (Iodine-potassium iodide) method was used.

3. Results

3.1. SPORE TRAPPING

Conidia were first collected on July 11 in 1992, June 11 in 1993 and July 11 in 1995. In mean year the date was June 20. The difference in collection dates was due to weather conditions during early stage of plant growth. The amount of conidia collected was also variable depending upon growth stages (Table 1), and the peak was at the end of July exhibiting ten thousand spores per day.

TABLE 1. Mean number of blast fungus conidia collected by a rotation sampler in Icheon, Korea during 1974-1995

Duration	No. of conidia	Duration	No. of conidia
June 3	4	July 6	51995
4	8	Aug. 1	10944
5	286	2	16419
6	1513	3	14181
July 1	10521	4	10462
2	25991	5	7748
3	48744	6	9702
4	55923	Sep. 1	11937
5	46912	2	7343

3. 2. EARLY DETECTION OF LEAF BLAST LESION

In the beginning of leaf blast season, it is very difficult and time consuming to find the first symptom. Since the I-KI method was known to be very effective for detection of infection sites by the blast fungus (Yoshino, 1971), we applied this technique into the field condition. Before the start of the leaf blast season, healthy-looking leaf samples were collected, chlorophyll removed with methanol and treated with 0.3% I-KI solution. By I-KI treatment, healthy parts changed to dark purple colour by the reaction of iodine and starch, but infection sites remained as greyish circles. As a result, it was possible to detect blast infection four days prior to lesion formation (Kim, 1982).

3. 3. LEAF AND PANICLE BLAST OUTBREAK

Leaf blast severity was surveyed in terms of mean number of lesions per hill from June 28 to August 2 at 7-day intervals (Table 2). In mean year, first day of leaf blast outbreak was June 25 but it varied year to year. The earliest date was June 18 in 1979, 1983 and 1993, and the latest date was July 10 in 1992. The peak of outbreak was July 19 until early 1980s but it became one week earlier thereafter. Leaf blast was most severe in 1984, and number of lesions per hill on July 5 was 176 in 1N plot and 257 in 2N plot. Percent panicle blast per hill in mean year was 22.4 in 1N plot and 61.1 in 2N plot. In general, blast severity was affected by temperature during rice crop season, and the disease was especially severe when low temperature prevailed. In 1980, low temperature passed at heading stage thus panicle blast was very severe, and in 1993 it passed at the effective tillering stage so leaf blast was very severe.

TABLE 2. Mean number of leaf blast* lesions per hill and percent panicle blast in Icheon, Korea during 1974-1995

1N Plot					
June 28	July 5	July 12	July 19	July 26	Aug. 2
4.0	20.6	37.1	32.2	21.1	13.0
Panicle blast 22.4					
2N Plot					
June 28	July 5	July 12	July 19	July 26	Aug. 2
3.5	60.7	98.7	98.8	74.9	52.2
Panicle blast 61.1					

*First day of leaf blast outbreak June 25

3. 4. SPORULATION AND CONIDIAL RELEASE

A series of experiments was conducted to measure the amount of sporulation and conidial release from naturally infected leaf blast lesions since 1985. As a result, a KY-type spore trap (Figure 1, Kim and Yoshino, 1987) was devised in 1985, and using the trap large amount of data was collected under different weather conditions. A discrete lesion-bearing leaf was fixed on the acrylic rain protector (using double-sided adhesive tape) without detaching the leaf from the plant. A basin-shaped acrylic capsule 40 mm long was placed just beneath the lesion. Spore collection was at 1 h or 24 h intervals. Spores were either counted immediately or sealed in the capsule with masking tape and counted a few weeks or months later. Before counting with a heamocytometer, 0.1 ml of 0.02% Tween 20 solution was added to the spores in the capsule. Spores collected in one capsule can be counted in 10 min.

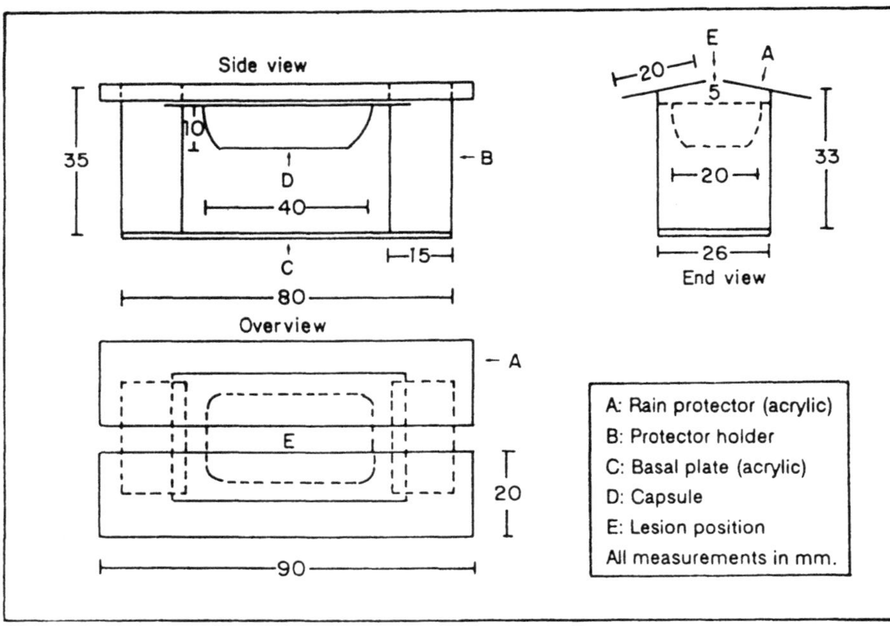

Figure 1. The KY – type spore trap (Korea, 1985).

The mean number of conidia produced on detached lesions in 1985 leaf blast season ranged from 1,050 to 7,500, and the maximum number of conidia produced was 16,200 from a single lesion sampled on July 22 (Table 3). There was a big difference in sporulation capacity between old and new lesions, with the newly developed lesions exhibiting high sporulation capacity. A discrete lesion-bearing leaf was fixed on the KY-type spore trap and sporulation amount was measured irregularly from different type and sized lesions, the amount varied greatly from a few hundred to few hundred thousand (Table 4). The highest count was 220,500 from 8.8 mm-long chronic-type lesion. Sporulation amount was also different between natural lesions (pre-existing conidiophores and conidia were not removed) and from removed lesions. More conidia were formed on natural lesions. In most cases, more spores were formed on the abaxial surface of leaves.

Conidial release patterns vary with weather conditions, type and age of lesions, as we have observed measured since 1987. Figures 2 and 3 illustrate spore catches in different devices: the KY-type spore trap (Kim and Yoshino, 1987) is used for measurement of conidia release; the rotation sampler (Suzuki, 1969a) for measurement of dynamic conidial dispersal and a horizontal spore trap for measurement of static conidial dispersal. All the measurements were done at 1-h intervals from 1800 h on day 1 to 1700 h on day 2. Under clear day conditions during 9-10 July, 1987 (Figure 2), conidial release measured by the KY-type spore trap and conidial dispersal measured by the other two traps exhibited one typical peak at 0600-0700 h. However, under continuous cloudy conditions during 14-15 July, 1987 there were 2-3 conidial release and dispersal peaks between 0100 and 0800 h (Figure 3). Conidial release and dispersal were also observed under rainy conditions (Kim *et al.*, 1989). The data presented in Figure 2 and 3 are representative of the spore release patterns observed on many days over many years, and are presented for illustrative purposes. Successive behavior of conidial release during 48 hours from 1800 h of July 19 to 1700 h of July 21, 1989, exhibited two distinct periods from midnight to early morning, revealing that the conidial release cycle of blast fungus repeats at 24 h interval. The relationship between rainfall and leaf blast severity at Icheon, Korea during June – August, 1988 is shown in Figure 4.

TABLE 3. Lesion length and number of *Pyricularia grisea* conidia produced from naturally infected lesions on a cultivar Jinheung at different stages in Icheon, Korea, 1985

	Lesions appeared on July 8-10		Lesions appeared on July 15-16	
Date	Lesion (mm)	No. of spores*	Lesion (mm)	No. of spores
July 18	14.7	1,300	5.8	7,500
20	16.4	1,800	10.7	4,900
22	19.6	1,700	12.0	7,500
24	21.2	4,100	17.1	1,050
26	21.4	1,300	18.7	4,200

* Average of 3 lesions and the number indicates the amount of conidia produced at 28C for 15 hours at >95% RH.

4. Discussion

Despite the conidial release and dispersal peak at 0600-0700 h (Kim and Kim, 1991), spore catch by rotation sampler in the present study was done from 0100 h to 0200 h. In the middle part of Korea, dew starts to form at 2000 h and disappears at 1000 h under normal weather conditions. For the blast fungus to successfully penetrate into the host, at least 6 to 7 h are necessary from adhesion to appressorium formation. Considering the role of effective inoculum source, it is desirable to catch spores early in the morning to forecast leaf blast outbreak.

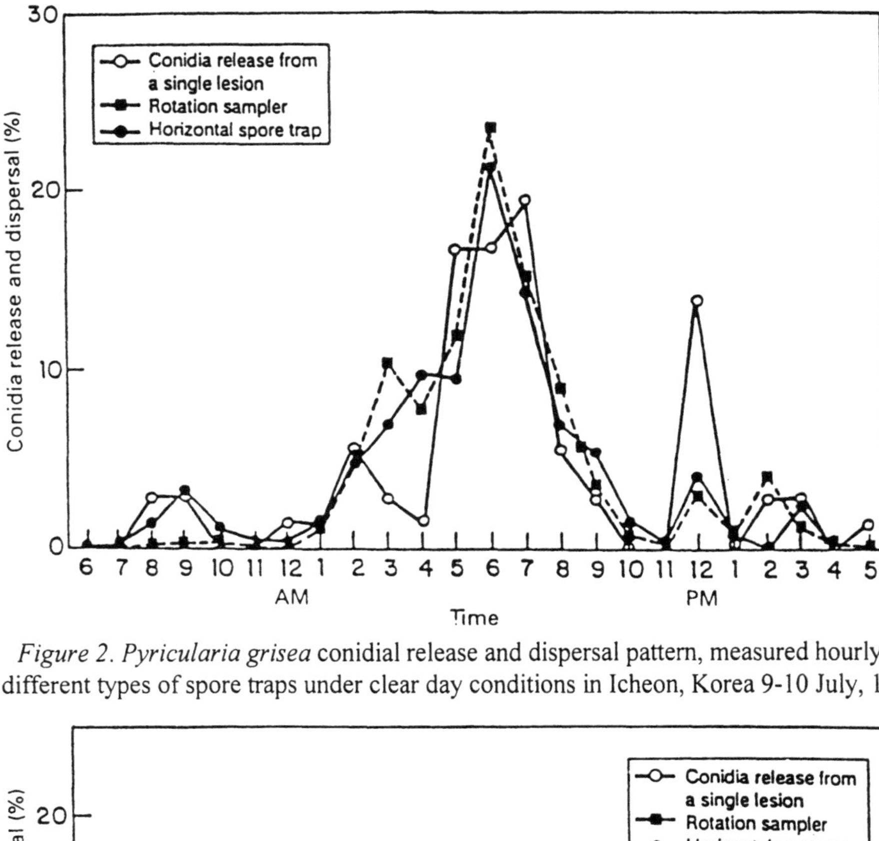

Figure 2. Pyricularia grisea conidial release and dispersal pattern, measured hourly by different types of spore traps under clear day conditions in Icheon, Korea 9-10 July, 1987.

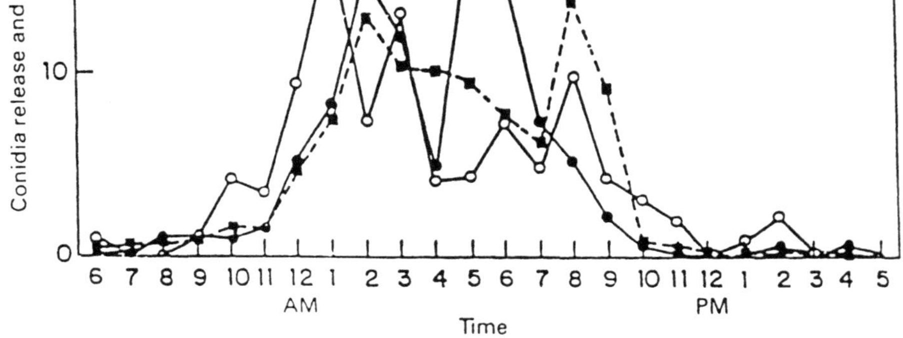

Figure 3. Pyricularia grisea conidial release and dispersal pattern, measured hourly by different types of spore traps under cloudy conditions in Icheon, Korea 14-15 July, 1987.

TABLE 4. Sporulation of *Pyricularia grisea* from natural leaf blast lesions in Icheon, Korea, 1991

Date	Lesion size (mm)	Lesion type[a]	No. released[b]	No. remained[c]	Total sporulation[d]
July 5	6.2	Acute	360	8,720	9,080
9	8.9 6.7	Chronic Acute	3,400 4,880	24,400 48,080	27,800 52,960
10	8.8 8.3	Chronic Chronic	14,800 2,080	205,700 48,760	220,500 50,840
11	15.9 4.4	Chronic Acute	2,620 300	58,860 1,620	61,480 1,920
12	14.2 6.4	Chronic Acute	5,700 1,940	19,520 13,520	25,520 15,460
24	28.0	Chronic	200	140	340

[a] Acute type : new greyish lesion without brown margin; chronic type : old lesion with yellow surroundings and distinct brown margin. Lesion type is much affected by weather conditions.

[b] Total number of conidia collected from 1800h on day 1 until 1700h on day 2 by the KY-type spore trap (Kim and Yoshino, 1987).

[c] Total number of conidia remaining on both sides of lesion, Remaining conidia were thoroughly scraped with a 1 ml hypodermic syringe needle and counted using a hemocytometer.

[d] Total sporulation amount derived from released and remained conidia during 1700-0900h. Crop growth stages during the experiment were maximum tillering to panicle initiation and the period was the rainy season. Rainfall pattern - 5, 9 and 12 Jul: 0 mm; 10 Jul: 0.5 mm; 11 Jul: 1.4 mm and 24 Jul:15.5 mm.

I-KI technique was good for early detection of infection site by the fungus at the beginning of leaf blast season, but direct quantification of infection site with lesion number during leaf blast season was difficult. It is very important to realise the complexity of the nature of the inoculum population. The average number of hills per hectare is 240,000; if there is only one lesion per hill and the daily number of conidia release is 2000, then the total number of conidia per day becomes 4.8×10^8. Sporulation from one lesion continues for more than 20 days (Kato, 1974; Kim, 1987) and there are different types of lesions in one field: fresh, old, short and long. This complexity makes it difficult to quantify number of leaf blast lesions by I-KI detected infection sites.

Weather conditions that cannot be controlled are the most important factor governing blast epidemic. Temperature exceeding 30°C suppresses blast, but below 19° C for a few days during the reproductive stage causes cold damage to the rice plants, and also results in severe blast epidemics in temperate countries. The dew period should be continued for more than 10 h per day. Rainfall intensity over 3.5 mm/h is known to be a suppressive factor against blast fungus attack (Suzuki, 1969b). In recent years, we frequently experienced strong showers during the leaf blast season, and this became one critical factor in reducing leaf blast. The data presented in Figure 4 are representative of the relationship between rainfall and leaf blast severity in Icheon, Korea, during late June-late July, 1988. Total amount of rainfall for every 10 days in July was more or less 100 mm in a normal year, but for the first 10 days in July 1988 it was 150 mm and for the middle ten days in July it was 420 mm. On the contrary, the number of leaf blast lesions per plant in the 2N plot on July 19 was 35.7 in 1988 and 107.1 in a normal year. This phenomenon demonstrates clearly that leaf blast development is greatly affected by rainfall intensity and amount during a limited period of time.

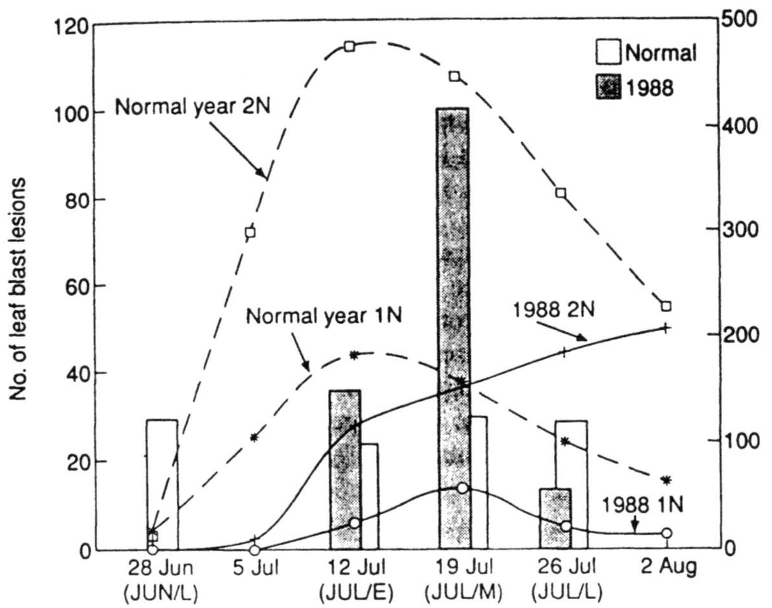

*Figure 4.*The relationship between rainfall and leaf blast severity at Icheon, Korea during June – August, 1988.

Sporulation pattern of the rice blast fungus was studied using naturally infected lesions. Total number of spores ranged from 340 on an old lesion to 220,500 spores on a relatively fresh lesion. Kato (1974) defined sporulation potential as the capacity of a fungus to produce conidia under optimum conditions per unit time. The maximum number per lesion recorded was 4×10^4 (Kato, 1974), 14.8×10^3 for variety Fujiminori an 20.12×10^3 for variety Ootori (Kato and Sasaki, 1974). We measured sporulation capacity of naturally infected lesions exposed under field conditions during 1700 h on day 1 to 1700 h on day 2, and we recorded a maximum of 12,540 spores, during successive measurements from a single lesion. On the other hand, many detached lesions formed more than 5×10^4 spores at 0900 h on day 2. The followings points can be considered as possible reasons for the difference in sporulation capacity between the two cases stated above. It is practically impossible to remove only conidia in a successive measurement using the same lesion, and conidiophores were also removed by scraping with syringe needle during the collection of spores. Toyoda and Suzuki (1952) reported that conidiophores were produced five hours after the placement of lesion in a moist chamber, and sporulation started two to three hours later resulting in seven to nine conidia per a single conidiophore during one night. It is expected that conidiophore and conidia production process is stopped during daytime under clear day conditions. In case of a lesion from which pre-existing conidiophores and conidia were removed, sporulation was much lesser than that of intact lesions (Kim *et al.*, 1990). This might be the case with the scraped lesion in this experiment. Intact lesions have numerous conidiophores which may produce conidia two to three hours after leaf wetness, and new conidiophores are also produced. Once the lesion starts to wet, the conidiophores and conidia react differently: i.e. a conidiophore without conidia sometimes remains as it is or starts to sporulate, mature conidia on the conidiophore start to release or retrogression of immature conidia occurs (Yoshino *et al.*, 1988). Considering the above reports, low sporulation capacity of scraped lesions might be possible, and this fact will make a big difference in sporulation capacity between the two cases of measurement.

Through this study, we found that less than 10% spores out of the total produced per day was released depending upon weather condition during night time. There is no evidence as to what per cent of the released conidia serves as inoculum source for next cycle in the nature. By inoculation test at 25-27°C, it was revealed that percent penetration of *P. grisea* into leaf epidermal cells of rice variety Nihonkai was 15.2% under standard nitrogen fertiliser level (Yoshino, 1972). It indicates that only a minor proportion of the released conidia is involved in lesion production under natural conditions.

In the present study, abaxial surface produced several fold to more than tenfold conidia than adaxial surface of the same lesion. Iwano (1984) reported the release of 10,000 conidia from adaxial surface and 26,000 conidia from abaxial surface of the same lesion during 6 to 20 h after treatment.

Sporulation capacity was higher in the order of intact lesion, conidia and conidiophore-removed lesion and successively measured lesion. The main objective of this experiment was to measure the amount of sporulation under natural conditions as we did for conidial release (Kim *et al.*, 1990). Considering the duration of sporulation for more than 20 days (Kato, 1974) and conidia release for 30 days (Kim and Yoshino, 1987), it is desirable to check sporulation capacity for two to three weeks. However, the method used for estimating sporulation in this study was laborious. It is very important to know the successive sporulation pattern of a single lesion to understand the nature of sporulation and leaf blast cycle. For this purpose, development of new methods for assessing sporulation is required. Successive collection of spores with a syringe needle from the same lesion tends to destroy the lesion within 7 to 10 days. Once the lesion was broken, sporulation capacity suddenly decreased or stopped.

In the study of conidial release patterns in the field under different weather conditions, we found some new trends. Conidial release started from 0100 h and the peak was 0600-0700 h under the clear day conditions. Under cloudy conditions, conidial release started at 2300 h and revealed 2-3 peaks. Conidial release was continuously occurring in the rain with differences in number to some extent, and the amount was greatly increased immediately after heavy rainfall probably due to the physical shock. Based on the successive conidial release pattern for 48 h, it was clear that there was a regular rhythm at 24 h interval with some differences in peaks depending upon microclimate as well as experimental conditions, such as leaf age of rice plants and type of blast lesions. The success or failure of released conidia as inoculum in the next cycle of blast epidemic seems to depend totally up on microclimate and growth stage of the host plants.

5. References

Coons, G. H. (1953) *Breeding for resistance to disease*, Yearbook of Agriculture 1953, U.S. D. A., pp. 174-192.

Crill, J. P. and Khush, G. S. (1979) Techniques and procedures for effective and stable control of rice blast with monogenic resistance, in *Meeting on Rice Blast Disease,* pp. 351-368.

Iwano, M (1984) Studies on the spore release of *Pyricularia oryzae* Cavara from lesions on rice leaves, *Bulletin of the Hokuriku National Agricultural Experimental Station* **26**, 67-95.

Kato, H. (1974) Epidemiology of rice blast disease, *Review of Plant Protection Research* **7**, 1-20.

Kato, H. and Sasaki, T. (1974) Epidemiological studies of rice blast disease, with special reference to reproductive process in lesions on rice plants and disease forecasting, *Bulletin of the National Institute of Agricultural Sciences Series C* **28**, 1-61.

Kim, C. K. (1982) Improved methods for rice blast forecasting, *Korean Journal of Plant Protection* **21**, 19-22.

Kim, C. K. (1987) Disease forecasting: Past and future in special reference to rice leaf blast, *Korean Journal of Plant Pathology* **3**, 304-312.

Kim, C. K. and Yoshino, R. (1987) Epidemiological studies of rice blast disease caused by *Pyricularia oryzae* Cavara. I. Measurement of the amount of spores released from a single lesion, *Korean Journal of Plant Pathology* **3**, 120-123.

Kim, C. K. and Kim, C. H. (1991) Predicting rice blast outbreaks in Korea, in *Rice blast modeling and forecasting*, International Rice Research Institute, P.O.Box 933, Manila, Philippines, pp. 53-67.

Kim, C. K., Min, H. S. and Yoshino, R. (1989) Epidemiological studies of rice blast disease caused by *Pyricularia oryzae* Cavara. IV. Conidia release under rainy conditions, *Korean Journal of Plant Pathology* **5**, 60-64.

Kim, C. K., Min, H. S. and Yoshino, R. (1990) Epidemiological studies of rice blast disease caused by *Pyricularia oryzae* Cavara. V. Successive conidia release from different aged lesions in the nature, *Korean Journal of Plant Pathology* **6**, 214-218.

Ou, S. H. (1972) *Rice Diseases*, Commonwealth Mycological Institute, Kew, Surrey, U. K. p. 368.

Ou, S. H. (1977) Genetic defense of rice against disease, in *The genetic basis of epidemics in agriculture,* The New York Academy of Science, pp. 275-286.

Rural Development Administration (RDA, 1995) *Crop protection report for 1995*. Suweon, Korea, pp. 102-103.

Suzuki, H. (1969a) Studies on the behavior of rice blast fungus spore and application to outbreak forecast of rice blast disease, *Bulletin of the Hokuriku Agricultural Experiment Station* **10**, 1-118.

Suzuki, H. (1969b) Interrelationship between the occurrence of the rice blast disease and the meteorological conditions, *Journal of Agricultural Meteorology* Japan **24,** 211-218.

Toyoda, S. and Suzuki, N. (1952) Histochemical studies on the lesions of rice blast caused by *Piricularia oryzae* Cav. I. Some observations on the sporulation on lesions of different types occurring on leaves of the same variety, *Annals of the Phytopathological Society of Japan* **17**, 1-5.

Yoshino, R. (1971) Influence of temperature on the incubation period of *Pyricularia oryzae*, and early detection of lesions by staining with iodine-potassium iodide, *Proceedings of the Association of Plant Protection* Hokuriku **19**, 11-14.

Yoshino, R. (1972) Ecological studies on the infection in rice blast epidemics. I. Infection rates and hyphal growth in epidermal cells, *Proceedings of the Association of Plant Protection* Hokuriku **20**, 4-9.

Yoshino, R., Arai, M. and Kim, C. K. (1988) Sporulation and release of *Pyricularia oryzae* under different environmental conditions, in *Proceedings of the 5th International Congress of Plant Pathology*, abstract, Kyoto, Japan. **Sec. VIII (2-6)**, 280.

EVALUATION OF GENETIC RESISTANCE TO RICE BLAST: SEARCHING FOR VARIETAL RESISTANCE STABLE OVER SPACE AND DURABLE OVER TIME

SANG-WON AHN
International Network for Genetic Evaluation of Rice, Genetic Resources Center, International Rice Research Institute, PO Box 933, 1099 Manila, Philippines.

1. Introduction

Demand of farmers and consumers for rice cultivars with specific traits such as disease and pest resistance is increasing because they are the most economically viable and, environmentally sound option, and the preference of a health-conscious society. Blast is one of the major constraints to stable rice production in most blast-prone rice areas; evaluation of genetic resistance to rice blast is one essential and critical phase in a systematic and effective varietal improvement program.

In blast-prone areas, various methods have been utilized to assess the efficacy and stability of genetic resistance to blast. Although several rice cultivars with durable blast resistance have been successfully developed in different countries, our ability to select promising lines and to predict their performance over time is still limited. This is mainly due to the presence of diverse races of the blast pathogen *Pyricularia grisea* , its inherent ability to generate genetic variability (Ou, 1985; Zeigler *et al* 1997) and its complex interaction with the rice genotype over space and time (Ahn , 1994). To improve the current evaluation system, it is essential to have a better understanding of the dynamics of the rice blast pathogen population, particularly knowledge on spatial and temporal interactions of the rice genotype and the blast pathogen under different rice production environments. Recent progress in molecular techniques both in the study of genetic structure of blast population (Hamer et al, 1989; Chen *et al*, 1995) and in identifying specific blast resistance genes provides useful tools for a detailed study on this interaction. An improved scheme for blast resistance evaluation based on knowledge about spatial and temporal interaction of rice blast with rice genotype pathogen, will allow 1) reliable prediction of efficacy and durability of target gene(s), (2) screening for desirable breeding lines, and (3) prediction of adaptability and durability of given rice genotypes. It will also be helpful in designing a better strategy on deploying host genotypes in specific rice production areas.

S. Sreenivasaprasad and R. Johnson (eds.), Major Fungal diseases of Rice Recent Advances, 131-143.
Prof. K. Manibhushan Rao Festschrift

This paper reviews principles behind some of the widely used schemes for evaluation of genetic resistance to rice blast. It points to dynamic research on long-term interactions between rice genotype and blast pathogen under diverse edapho-climatic conditions as one key parameter in the evaluation of genetic resistance to rice blast.

2. Varietal reaction to rice blast

Phenotypic expressions of varietal reaction to rice blast infection and subsequent phases of pathogen cycle under diverse environmental conditions vary widely. It is not easy to group them into a simple classification. The terms such as true resistance (Ezuka *et al.*, 1969), qualitative resistance (Ou *et al.*, 1971; Ahn and Ou, 1982), single gene resistance (Crill *et al.*, 1981), and complete resistance (Wang *et al.*,1994) have been used to describe hypersensitive reactions or incompatible interactions. To describe varietal differences in disease severity of compatible interaction between pathogen and rice plant, terms such as field resistance (Ezuka *et al.*, 1969), slow blasting (Chung *et al.*, 1980; Ahn, 1982), dilatory resistance (Marchetti, 1983), quantitative resistance (Ahn and Ou, 1982) and partial resistance (Rodriguez and Galves, 1975; Yeh and Bonman, 1986; Ahn *et al.*, 1989) have been used. To emphasise useful characters of varietal resistance, terms such as stable resistance (Ou *et al.*, 1971), broad spectrum resistance (Ahn and Ou, 1982), and durable resistance (Johnson, 1979; Lee *et al.*, 1989) are also frequently used.

Use of different terms often brought some confusion. Nevertheless, various concepts, ideas, and approaches to 1) understand the nature of diverse mode of interaction between host and pathogen population under a wide range of environments, and 2) characterise host resistance, have gradually broadened our knowledge base of disease resistance evaluation and disease management (Fry, 1982).

3. Variation of rice blast fungus over space and time

If the blast fungus, *Magnaporthe grisea* Barr (anamorph *Pyricularia grisea* (Cooke) Sacc. syn. *P. oryzae* Cav.) had narrow genetic variation and were stable in its virulence character under natural conditions, probably breeding for blast resistance could have been much easier and straightforward in the past. Once resistant rice cultivars have been developed, their performance would have been stable over space and time. Contrary to the hopes of rice farmers and scientists, however, resistance of many rice cultivars in the past often did not last long. Due to the dynamic nature of occurrence of new pathogenic variant of blast fungus and race composition in field pathogen population, the average life span of host resistance in rice cultivars released in Colombia was estimated only at 2-3 years until the early 1980s (Ahn, 1982).

Pathogenic, as well as, other cultural variation of the fungus has been known for decades and has been well documented (Ou 1980, 1985). Although the sexual stage of *P. grisea* was observed from the mating between some selected blast isolates in artificial conditions (Yaegashi and Nishihara, 1976), natural occurrence of a sexual stage has not been reported. A recent study on a blast pathogen population using DNA fingerprint analysis with the middle repetitive element MGR586 (Hamer *et al.*, 1989) showed that several to many distinct groups of pathogen populations, presumably representing clonally propagated lineages, exist in different rice-growing countries (Chen et al., 1995; Hamer et al., 1989; Han et al. 1993; Shen et al., 1996). It is believed that all isolates belonging to each lineage were derived from a common ancestor. High haplotypic diversities are commonly observed, indicating a high rate of mutation or parasexual exchange of DNA (Zeigler *et al.*, 1997). In some cases, genetic instability in the fungus appears to be a consequence of the chromosomal location of the gene (Valent and Chumley, 1994). These mechanisms may be useful for generating genetic variation or eliminating disadvantageous genetic factors accumulated in clonal lineages.

Different races of the rice blast pathogen from different geographical areas were reported (Ou, 1985). Due to this spatial variation in race composition, different rice cultivars show a remarkable degree of differential reaction to rice blast. Recent analysis of genotype (G) x environment (E) interaction in multilocation trials of rice blast has indicated that a significant proportion of variation due to G x E interaction, which is often higher than that due to genotype or environmental effect alone, can be attributed to spatial variation of virulence frequency in test sites (Ahn and McLaren, 1995). This spatial variation is closely associated with varietal type (*indica* or *japonica*) of commonly cultivated rice in the area of the test site (Ahn *et al.*, 1997). It is most unlikely that these spatial differences would remain constant. Population structure of the pathogen would respond sensitively over time to any changes in biotic as well as abiotic factors of the environment.

With the application of molecular techniques, knowledge of the pathogen population structure of rice blast has been broadened recently (Leung *et al.*, 1993). Yet our understanding of the dynamic evolution of blast pathogen population in relation to rice host genotype under different rice ecosystems, is quite limited. Only a few studies have attempted to deal with the temporal change of population structure. This is generally true also with other microorganisms. Mundt (1995) concluded in a recent review that fitness of a microorganism in nature is difficult to measure or predict with much accuracy, and is unlikely to be constant with respect to either environmental conditions or time. It is possible that one rice cultivar could have different durability against different blast populations. Also, the same combination of resistance genes may have different effects in reducing disease severity and durability because of varying genic interactions from one rice genotype to another.

Early studies have indicated that an increase of a certain pathotype as well as highly adapted isolates of blast pathogen in farmers' fields was highly correlated with the planting area of rice cultivars of particular genotypes (Kiyosawa and Shiiyomi, 1976). Extensive and intensive cultivation of one rice cultivar possessing certain resistance gene over a large area for a prolonged period would exert a high selection pressure favoring a rapid increase of particular pathotype with corresponding virulence gene. This was clearly observed in several countries.

In Korea, semidwarf Tong-il type of rice cultivars derived from the cross between japonica and indica parents were released to farmers since 1972. They soon became a dominant varietal type in more than 80 percent of 1.2 million hectares planted in 1978. Tong-il type rice cultivars exhibited a remarkably high yield potential responding well to an increased nitrogen fertilizer application. They possess resistance to stripe virus disease, and rice blast but they were susceptible to cold, and grain quality was inferior to that of japonicas. Through active government promotion of these high-yielding rice cultivars, Korea was able to achieve self-sufficiency in rice in 1977. Accordingly, industrialisation of Korean economy was able to develop in the mid-1970s without the problem of food shortage, mainly due to a stable rice production.

Compatible isolates virulent to sister lines of Tong-il cultivars (IR667-98) were first found at the blast nursery located in Suweon, Korea in 1971. Occurrence of blast on Tong-il type cultivars in farmers' fields was first observed in 1976 in one small area, then again reported from 20 locations in 1977. Actually, the author spent two weeks in 1972 to collect blast lesions from Tong-il type cultivars in major planting areas, but failed. It was only possible to collect lesions from Jinheung, an improved japonica. With prevailing favorable weather conditions highly conducive to blast, a severe nationwide blast epidemic on Milyang 29 (Raekyung) and Iri 327 (Nopung), both Tong-il types occurred in 1978, resulting in substantial yield losses (7 %). It was quite impressive to observe that Tong-il type cultivars were more severely affected than well known susceptible cultivars such as Jinheung. The same group of varieties showed entirely opposite reaction to blast over the period 1972 to 1978. Fortunately, no blast epidemic of such magnitude occurred again in Korea since 1978, except another epidemic associated with low temperature in 1993. Coupled with susceptibility of Tong-il types to cold damage in 1980 and consumers preference for japonica type cultivars, total planting area of Tong-il type was gradually decreased. A long-term survey of blast race composition during this period clearly indicated that the frequency of KJ and KI race groups, pathogenic to *japonica* and *indica* cultivars, respectively, has been closely following the change of the proportion of planting area of these two varietal groups (Ryu *et al.*, 1987).

Kiyosawa and Shiyomi (1976) calculated the mean longevity of new cultivars with vertical resistance as 5 years in Japan. Epidemics whose increase in intensity takes many years, termed as "polyetic epidemics" (Zadoks and Schein, 1979) is a challenging area of study to develop durably resistant rice cultivars. The damage level of rice cultivars could change over time due to a dynamic process of selecting particular virulence as well as increase in aggressiveness of isolates even if they belong to the same lineage group or race. A significant gradual increase in aggressiveness of blast isolates was observed in the seventh crop cycle of C101A51, which was sequentially planted at 15-day intervals (Ahn *et al.*, 1996). On the other hand, aggressiveness of isolates from IR64 and IR36 , known for high partial resistance, was much lower than C101A51. This indicates that durably resistant cultivars have a genetic mechanism which does not readily allow the building up of highly aggressive strain of virulent race. It is not yet known whether this low aggressiveness of isolates to IR64 an IR36 is due to the fitness cost related to presence of virulence gene(s) necessary to infect these cultivars. The interaction of genetically well defined rice genotypes and blast populations under diverse production situations should be studied to develop proper strategies to prolong efficacy and durability of host resistance.

4. Commonly used blast evaluation procedure

Blast evaluation is a continuous step-wise procedure, which should be well integrated into every phase of varietal improvement. Usually, leaf blast evaluation is carried out in early generations , and panicle blast evaluation in advanced generations. In general, a good correlation between leaf and panicle blast resistance has been observed. Yet under field conditions, it is common to observe different levels of leaf and panicle blast infection in the same rice genotype in a crop season. Through carefully designed staggered planting of diverse rice cultivars, variation due to differences in climatic and racial composition at two distinct growth stages could be minimized, allowing comparison of varietal difference under field conditions (Ahn and Rubiano, 1984). In some cases, however, panicle blast resistance was found to be independent from leaf blast resistance (Higashi, 1995). Often, heavy leaf blast incidence affects crop growth and productive tillers, resulting in small panicles with few grains and low panicle blast incidence under field conditions (Filippi and Prabhu, 1997). Genetic composition of blast populations fluctuates during the rice plant's crop growth (Chen *et al.*, 1995) and this change may also result in differential reactions between two stages.

The most widely used method at the initial phase of blast evaluation in many rice programs may be a uniform blast nursery test (Ou, 1965). It aims to evaluate the leaf blast reaction of test entries at the early vegetative stage of the rice plant, a vulnerable period for blast infection. This method can accommodate a large number of test entries requiring small seed quantity. Relatively uniform infection can be obtained under proper management. The key assumptions in this evaluation scheme are:

- Optimal conditions favorable for blast development are provided. This includes dense planting in the upland nursery, high rate of nitrogen application, prolonged dew period by covering the plots with plastic film at night and /or supplementary overhead sprinkler irrigation or misting 2-3 times each rainless day.
- All test entries are uniformly exposed to the inoculum consisting of diverse races of the blast pathogen with a wide range of aggressiveness. This is achieved by planting a mixture of several broadly susceptible cultivars including local susceptible check cultivars. All locally available isolates and infected tissues of rice plants of interest are continuously added to the blast nursery to represent the blast population in farmers' fields. If particular host cultivars for specific blast lineages are known, these can be also included as spreader rows.
- Compatible reaction scores of 4-9 of the Standard Evaluation System (SES) based on 0-9 scale (International Rice Research Institute, 1996) indicate the presence of virulent isolates of the blast pathogen. Blast severity of test cultivars, often used as a parameter for quantitative resistance, may vary depending upon the level of partial

resistance as well as the spectrum of qualitative resistance controlled by major genes (Ahn and Koch, 1988). On the other hand, reaction scores of 0-3 indicate the absence or extremely low frequency of virulent isolates. Genetic resistance of the test entries exhibiting scores of 0-3 would be effective as long as race composition of the blast population remains the same. However, this method can not determine damage severity level against compatible isolates which are absent or an extremely low frequency at the time of evaluation.

In advanced generations, efficacy, and stability of partial resistance or the spectrum of qualitative resistance must be intensively assessed against diverse races in national or international multilocation trials. Because of the global importance and complexity of this disease, international collaboration on breeding for resistance to rice blast has been carried out since the 1960s. The International Network for Genetic Evaluation of Rice (INGER, formerly the International Rice Testing Program) has been coordinating the International Rice Blast Nursery (IRBN) since 1975. This trial was specifically designed to provide rice scientists working on rice blast easy access to diverse resistance donors and improved breeding lines as well as information on their location specific and overall reactions to rice blast under a wide range of test environments (Ahn, 1994).

Intensive evaluation of durably resistant rice cultivars in different ecosystems could provide an insight into general response patterns and efficacy against existing blast populations in the target area of breeding. Also common characters of durably resistant cultivars and possible evaluation parameters for identifying potentially durable cultivars could be formulated.

5. Relationship between resistance over space and time

One of the common features of the durably resistant cultivars in the IRBN test was low disease severity(DSI) values, mostly near 5.0, whereas the DSIs of all the known susceptible cultivars are higher than 6.0. The frequency of incompatible reactions of durably resistant cultivars is higher than 60%, whereas the frequency of extremely high reaction scores, 8-9, is much lower than those of nondurable resistant cultivars (Ahn, 1994). Accordingly, these parameters have been utilized in identifying promising entries from IRBN trials. However, some cultivars exhibiting these evaluation parameters in IRBN were not necessarily durably resistant over time under intensive cultivation over wide planting area or sequential planting trial at IRRI in the past.

CICA 8 was released in 1978 as a blast resistant cultivar for the Eastern Plains of Colombia in 1978, but became highly susceptible under favorable upland conditions in 1980. To develop CICA8, Tetep, one of the broadly resistant cultivars in IRBN, was used as one of the donor parents for blast resistance. It has also showed compatible reactions since 1980, and often was severely damaged in experimental plots. CICA 8 , however, showed incompatible reactions in over 75% of IRBN test sites (Table 1). Thus, a broad spectrum of blast resistance observed in a multilocation test could be one good parameter, but may not be a sufficient parameter to identify durably resistant cultivars.

Some of the rice cultivars exhibiting resistance in many test sites were evaluated at IRRI using the sequential planting technique at 15-day intervals to simulate an intensive cultivation(Ahn and Ou, 1982, Kim and Ahn, 1991). Cultivars such as IR64 and Samgangbyeo known as durably resistant cultivars in South and South Asia and Korea, respectively, consistently showed high partial resistance over 15 consecutive sequential plantings (Ahn *et al.*, 1997). However, Tetep, Carreon, and C101A51 exhibiting a broad spectrum of resistance over many test sites often showed severe blast infection in sequential plantings (Table 1). This indicates that broad spectrum qualitative resistance is a necessary condition, but may not be a sufficient condition for durably resistant cultivars. Probably high partial resistance which is broadly effective against many compatible isolates of blast pathogen is required for durably resistant cultivars (Ahn and Koch, 1988; Wang *et al.*, 1994).

TABLE 1. Blast reactions of selected rice cultivars in the International Rice Blast Nursery (IRBN) from 1990 to 1995, the sequential evaluation plots and farmers' fields

Cultivar	Origin	IRBN				Evaluation plot[a]	Farmers' field[b]
		Frequency score (%)			Disease severity index		
		0-3	4-6	7-9			
Tetep	Vietnam	81	16	3	5.3	7	-[c]
Oryzica Llanos 5	Colombia	78	22	0	4.4	-	1
Carreon	Philippines	76	22	2	4,7	6	-
CICA8	Colombia	76	21	3	4.7	-	7
IR64	Philippines	70	28	2	4.7	3	1
C101A51	IRRI	69	21	10	5.6	8	-
Samgangbyeo	Korea	64	28	8	5.4	3	3
ITA 212	Nigeria	64	32	4	4.9	-	7
Todorokiwase	Japan	44	40	16	5.5	-	3
IR50	Philippines	42	27	31	6.6	8	8
Yushin	Korea	25	38	87	6.4	8	9

[a]Standard evaluation scores in sequential evaluation plots, Philippines.
[b]Standard evaluation scores observed in farmers' fields in respective countries of origin or areas where cultivated.
[c]No data.

Oryzica Llanos 5, a blast resistant cultivar developed at a hot spot site, the Santa Rosa experimental and breeding farm in eastern Colombia, was released as a commercial variety in 1989 for the blast-prone region in Colombia. This variety is being planted on more than 50,000 ha, and no typical compatible blast lesions were observed until 1995. It showed a broad spectrum of resistance over many test sites in IRBN (Table 1). Controlled inoculations under greenhouse conditions with more than

500 isolates representing all the known blast pathogen diversity at Santa Rosa have supported the observation of a wide spectrum of resistance of Oryzica Llanos 5 (Correa-Victoria *et al.*, 1994). In 1996, some susceptible lesions were observed in evaluation plots in the Santa Rosa and Matazul sites located in Altillanura. The number of lesions increased both in the experiment station as well as in farmers' fields in 1997, but the severity was quite low (Correa-Victoria, pers. comm.). This gradual increase of lesions was probably due to a severe infection of Oryzica Caribe 8, another Colombian rice cultivar released in 1995 in the region. Blast isolates virulent to both cultivars were classified as blast lineage group SRL-4, which have been in a low frequency in the past. However, due to the large planting area of Oryzica Caribe 8 and its high susceptibility to isolates belonging SRL-4, inoculum of isolates virulent to Oryzica Llanos 5 was apparently increased in farmers' fields.

In the past, average span of durability of blast resistance in rice cultivars in Meta region of Colombia was 1-2 years (Ahn, 1982). Durable resistance is resistance that remains in a cultivar that is widely grown for a long period of time in an environment favorable to the disease (Johnson, 1983). Accordingly, Oryzica Llanos 5 can be called as durably resistant. It is also clear that Oryzica Llanos 5 apparently exhibited high level of partial resistance compared to Oryzica Caribe 8 in the presence of enough compatible inoculum. The combined characters of a broad spectrum of qualitative resistance against predominant isolates and high partial resistance to compatible isolates might be also responsible for durability of blast resistance in Oryzica Llanos 5. It will take some time to determine whether partial resistance in this cultivar could be sustained or not.

6. Evaluation scheme for blast resistance in future

In the past, only the static and immediate responses of test entries to existing sets of blast isolates, major race groups, or prevailing inoculum present in field population in major test sites were mainly considered in identifying resistant rice cultivars. This was based on the assumption that stability over space or diverse isolates is highly correlated with durability over time. Although in most cases, this is true and it could be a necessary condition for durable resistance, it may not be a sufficient parameter to predict the durability of a given rice genotype. Accordingly, this conventional method is a primary parameter in identifying effective resistant cultivars against existing population structure at the time of varietal development and release. Therefore, the currently utilized method of blast nursery test conducted in multilocation tests or qualitative test against existing diverse groups of isolates, races, or lineages should be considered only as a first step in identifying promising resistant cultivars. In practical breeding, emphasis should be given to identifying those entries with extremely susceptible reactions over space or isolates. These should be eliminated as early as possible unless they have other good traits to consider.

After the static phase of evaluation and selection using existing key basic blast isolates of different races or lineages, evaluation should proceed to the next step of assessing partial resistance level using micro-plot evaluation against virulent isolates present in the region (Marchetti, 1983). This technique would allow 2-3 cycles of blast infection (oligocyclic in a strict sense) during an evaluation period of one test. Test entries with extremely low partial resistance can be readily detected and eliminated using this technique. If possible, each rice genotype should be challenged to all available virulent isolates, since some partial resistance may show race specific reaction (Bonman *et al.*, 1989; Ikehashi and Kiyosawa, 1981). In the past, only one broadly virulent isolate was commonly utilised to differentiate genotypic differences of test entries by eliminating possible source of variation associated with different isolates such as differences in aggressiveness. This may be a suitable approach for some specific research purpose, but can not be a practical method of predicting long-term consequence of host and pathogen interaction in a specific production system. This is mainly because of our inability to predict the most prevalent compatible isolate on a given rice genotype in the future. More research is needed to predict the evolution of the blast population structure with a specific rice genotype and its surrounding environments.

The final stage of evaluation in a given test site will be the process of determining the efficacy and stability of partial resistance over a prolonged period of time. Unlike the existing evaluation schemes, this new evaluation strategy aims to monitor the dynamic response of rice genotype to blast infection over time. Although no standard method for this important phase of evaluation has been established, our experience has indicated that evaluation of partial resistance using the sequential planting technique will serve this purpose (Ahn and Ou, 1982; Kim and Ahn, 1991). This technique allows monitoring of qualitative and quantitative change of blast population structure as a long term consequence of interaction between one rice genotype and compatible blast isolates in a given environment (Ahn *et al.* 1996). One important assumption of this evaluation method is that a relatively low rate of change in aggressiveness and/or population increase of compatible isolates over time is a good indication of genetic effect of durable partial resistance. It is not clear whether this reduced aggressiveness is related to the fitness cost associated with virulence to matching major gene(s) in a rice genotype. Aggressiveness can be measured by various parameters of blast isolates such as infection efficiency, latency, lesion size, and sporulation. Any outcome of fitness change in compatible isolates over time would certainly result in different damage levels in rice genotypes.

The sequential planting evaluation technique can be integrated best at the final stage of selection when only several promising lines targeted for blast prone areas are evaluated. This method can be readily utilised in tropical conditions without major problems. However, it may need some modifications for use in temperate regions. Actually, only 4 sequential plantings were possible under field conditions in Korea (Y.T. Lee, pers. comm.). In temperate regions, sequential planting can be done at weekly intervals, and plants can be raised in suitable plastic containers under greenhouse conditions. Initial inoculum is prepared using a mixture of all available compatible isolates. Visible lesions are collected before sporulation, and incubated in moisture chamber for spore production. Blast spores collected from incubated blast lesions can be utilized as an inoculum for the following batch of sequential planting in the greenhouse (H. Leung, pers. comm.). Disease development in different sequential plantings is monitored, and aggressiveness of isolates from different plantings can be simultaneously compared.

All phases of blast evaluation should be well integrated into the whole process of varietal development for greater efficiency. The scheme shown in Table 2 emphasizes the identification of blast resistance which is stable over space and durable over time. It also emphasises mutual benefit through international collaboration in blast evaluation such as IRBN or inter-country trials. Our knowledge on host and pathogen interactions and ability to predict the outcome of such interaction are quite limited. With the progress in molecular biology and related technology, it will be possible to develop various potentially durable rice cultivars. Accordingly, an improved evaluation technique is necessary to identify rice genotypes with durable blast resistance in a given rice ecosystem. This will be possible only when we gain enough understanding on evolution of the blast pathogen as a consequence of long-term interaction of rice genotype and blast populations under diverse rice environments, using genetically well defined test materials.

TABLE 2. An integrated evaluation scheme to develop durable blast resistance

Step	Activity
National level:	
1. Blast Nursery (BN) or greenhouse evaluation	• Eliminate highly susceptible entries (negative selection) • Identify low infection type entries (0-3) which are agronomically excellent • Select potential candidates for partial resistance evaluation (those that show compatible reaction at a test site)
2-1 Microplot/sequential/farmer's field trials	• Assess partial resistance using matching virulent isolates only • Monitor stability and efficacy of partial resistance
2-2 Multilocation test	• Assess the spectrum of qualitative resistance of low infection types identified in Step 1 • Obtain virulent isolates from promising entries for microplot evaluation.
International level:	
1. International Rice Blast Nursery (IRBN) evaluation	• Estimate the efficiency of qualitative resistance and levels of quantitative resistance • Obtain diverse sources of resistance
2. Intercountry evaluation of partial resistance	• Evaluate partial resistance of entries from participating country where no virulent isolates is available • Exchange of information and visit for joint evaluation

7. Summary

Qualitative (complete) and partial resistance to rice blast are important components of durable blast resistance. Qualitative resistance is known to be race-specific. Different spectra of qualitative resistance can be assessed using a diverse set of races or lineages, Stability of such resistance over space is assessed using multilocation tests such as IRBN. Some partial resistance is unstable over time. Sequential planting has demonstrated that isolates belonging to a particular lineage could gain significantly enhanced aggressiveness over time without changing lineage or race group. Stability over space apparently is not always associated with durability over time indicating the need for another evaluation strategy to determine durability. Sequential planting appears promising for that purpose. An extensive research in long-term interaction of rice genotype and blast population should be made to develop proactive breeding strategy for blast resistance as well as appropriate gene deployment scheme.

8. References

Ahn, S. W. (1982) The slow blasting resistance to rice blast, in IRAT/GERDAT (ed.), *Proceedings of the Symposium on Rice Resistance to Blast*. 18-12 March 1981, IRAT/GERDAT, Montpellier, France, pp.343-370.

Ahn, S. W. (1994) International collaboration on breeding for resistance to rice blast, in R. S. Zeigler, S.A. Leong, and P. S. Teng (eds.) , *Rice Blast Disease*, CABI and International Rice Research Institute, Wallingford, UK, pp. 137-153.

Ahn, S. W. and Ou, S. H. (1982) Epidemiological implications of the spectrum of resistance to rice blast, *Phytopathology* **72**, 282-284.

Ahn, S. W. and Koch, M. F. (1988) A conceptual model of disease resistance in rice pathosystems, and its implications for evaluating resistance, *International Rice Research Newsletter* **13**, 12-13.

Ahn, S. W. and Rubiano, M. (1984) Relationship between susceptibility to leaf blast and panicle blast severity, *International Rice Research Newsletter* **9**, 13.

Ahn, S. W., Makihara, D., Imbe, T., Chen D. H. and Barrios, H. C. (1996) Population dynamics of *Pyricularia grisea* in rice cultivars with partial resistance. *Phytopathology* **86**, 646 (abstract).

Ahn, S. W. and McLaren, C. G. (1995) Multilocation evaluation of quantitative resistance to rice leaf blast, *European Journal of Plant Pathology*, in Abstract of 13th International Plant Protection Congress, Kluwer Academic Publishers, p.469.

Ahn, S. W., Seshu, D. V. and Bonman, J. M. (1989) Prediction of partial resistance to rice blast using multilocation trials, *Phytopathology* **79**, 1183 (abstract).

Ahn, S.W., Ynalvez, A.H., McLaren, C.G., Lopez, V.C. and Barrios, H.A. (1997) Rice genotype by environment (G x E) interaction in multilocation evaluation of rice blast, *Phytopathology* 87 (S3), (abstract).

Bonman, J. M.. Bandong, J. M., Lee, E. J. and Valent, B. (1989) Race specific partial resistance to blast in temperate japonica rice cultivars. *Plant Disease* **73**, 496-499

Chen, D. H., Zeigler, R. S., Leung, H., and Nelson, R. J. (1995) Population structure of *Pyricularia grisea* at two screening sites in the Philippines, *Phytopathology* **85**, 1011-1020.

Chung, H. S., Choi, G. H. and Shakya, D. D. (1980) Slow blasting rice cultivars at leaf and heading stages in the paddy field, *ORD, AIC 80-15, Office of Rural Development*, Suweon, Korea.

Correa-Victoria, F. J., Zeigler, R. S. and Levy, M. (1994) Virulence characteristics of genetic families of *Pyricularia grisea* in Colombia, in R. S. Zeigler, S.A. Leong , and P. S. Teng (eds.), *Rice Blast Disease*, CABI and International Rice Research Institute, Wallingford, UK, pp. 211-229.

Crill, J. P., Ham, Y. S. and Beachell, H. M. (1981) The blast disease in Korea and its control with race prediction and gene rotation, *Korean Journal of Breeding* **13**, 106-114.

Ezuka, A., Yunoki, T., Sakurai, Y. Shinoda, H. and Toriyama, K. (1969) Studies on the varietal resistance to rice blast. 2. Test for field resistance in paddy fields and upland nursery beds (in Japanese, English summary). *Bulletin of the Chugoku Agricultural Experiment Station Series* **E4**, 33-53.

Filippi, M. C. and Prabhu, a. S. (1997) Integrated effect of host plant resistance and fungicidal seed treatment on rice blast control in Brazil, *Plant Disease* **81**, 351-355.

Fry, W. E. (1982) *Principles of plant disease management*, Academic Press, New York.

Hamer, J., Farrall, L., Orbach, M. J ., Valent, B. and Chumley, F, G. (1989) Host species-specific conservation of a family of repeated DNA sequences in the genome of a fungal plant pathogen, *Proceedings of the National Academy of Sciences USA* 86, 9981-9985.

Han, S. S., Ra, D. S. and Nelson, R. J. (1993) Comparison of RFLP-based phylogentic trees and pathotypes of *Pyricularia oryzae* in Korea, *RDA Journal of Agricultural Science* **35**, 315-323.

Higashi, T, (1995) Genetic studies on field resistance of rice to blast disease, *Bulletin of the Tohoku National Agricultural Experimental Station* No. **90**.

Ikehashi, H. and Kiyosawa, S. (1981) Strain specific reaction of field resistance of Japanese rice varieties revealed with Philippine strains of rice blast fungus, *Pyricularia oryzae* Cav., *Japan Journal of Breeding* **31**, 293-301.

Johnson, R. (1979) The concept of durable resistance, *Phytopathology* **69**, 198-199.

Johnson, R. (1983) Genetic background of durable resistance, in F. Lamberti, J. Waller and N. A. Van der Graaff (eds.) *Durable resistance in crop*, Plenum Press, New York, pp.5-26.

Kim, C. H. and Ahn, S. W. (1991) Sequential evaluation for field assessment of partial resistance to rice blast, *Korean Journal of Breeding* **23**, 103-110.

Kiyosawa, S. and Shiyomi, M. (1976) Simulation of the process of breakdown of disease resistant varieties, *Japan Journal of Breeding* **26**, 339-352.

Lee, E. J., Zhang, Q. and Mew, T. W. (1989) Durable resistance to rice disease in irrigated environments, in *Progress in Irrigated Rice Research*, International Rice Research Institute, Manila, Philippines, pp. 93-110.

Leung, H., Nelson, R. J. and Leach J. E. (1993) Population structure of plant pathogenic fungi and bacteria, *Advances in Plant Pathology* **10**,157-205.

Marchetti, M. A. (1983) Dilatory resistance to rice blast in USA rice, *Phytopathology* **73**, 645-649.

Mundt, C. C. (1995) Models from plant pathology on the movement and fate of new genotype of microorganisms in the environment, *Annual Review of Phytopathology* **33**, 467-488.

Ou, S. H. (1965) A proposal for an international program of research on the rice blast disease, in *The Rice Blast Disease*, The Johns Hopkins Press, Baltimore, Maryland, pp. 441-446.

Ou, S. H. (1980) Pathogenic variability and host resistance in rice blast disease, *Annual Review of Phytopathology* **18**, 167-187.

Ou, S. H. (1985) *Rice Diseases*, 2nd edn. Commonwealth Mycological Institute, Kew, Surrey, U.K.

Ou, S.H., Nuque, F.L. and Awoderu, V. A. (1971) A type of stable resistance to blast disease of rice, *Phytopathology* **61**, 703-706.

Rodriquez, M. and Galves, G. C. (1975) Indications of partial resistance of rice to the fungus *Pyricularia oryzae* Cav., in *Proceedings of Seminar on Horizontal Resistance to the Blast Disease of Rice*, 1971, Centro Internacional de Agricultura Tropical, pp.137-154.

Ryu, J.D., Yeh, W.H., Han, S.S., Lee, Y.H. and Lee, E.J. (1987) Regional and annual fluctuation of races of *Pyricularia oryzae* during 1977-1985 in Korea, *Korean Journal of Plant Pathology* **3**, 174-179.

Shen Y., Zhu, P., Yuan, X., Zhao, X., Manry, J., Rojas, C.,Shahjahan, A.K. M.and Levy, M. (1996) The genetic diversity and geographic distribution of *Pyricularia grisea* in China, *Scientia Agricultura Sinica* **29**, 39-46.

Valent, B. and Chumley, F. G. (1994) Avirulence genes and mechanisms of genetic instability in the rice blast fungus, in R. S. Zeigler, S.A. Leong, and P. S. Teng (eds.) *Rice Blast Disease*, CABI, Wallingford, U.K., pp. 111-134.

Wang, G. I., Mackill, D. J., Bonman, J. M. , McCouch, S. R., Champoux, M. C. and Nelson, R. J. (1994) RFLP mapping of genes conferring complete and partial resistance to blast in a durably resistanct rice cultivar, *Genetics* **136**, 1421-1434.

Yaegashi, H. and Nishihara, N. (1976) Production of the perfect stage in *Pyricularia* from cereals and grasses. *Annals of the Phytopathological Society of Japan* **42**, 511-515.

Yeh, W. H. and Bonman, J. M. (1986) Assessment of partial resistance to *Pyricularia oryzae* in six rice cultivars, *Plant Pathology* **35**, 319-323.

Zadoks, J. C. and Schein, R. D. (1979) *Epidemiology and Plant Disease Management*, Oxford University Press. Inc.

Zeigler, R. S., Cuoc, L. X., Scott, R., P., Bernardo, M. A., Chen, D. H., Valent, B. and Nelson, R. J. (1995) The relationship between lineage and virulence in *Pyricularia grisea* in the Philippines, *Phytopathology* **85**, 443-451.

Zeigler, R. S., Scott, R. P., Leung H., Bordeo, A. A., Kumar, J and Nelson, R.J. (1997) Evidence of parasexual exchange of DNA in the rice blast fungus challenges its exclusive clonality, *Phytopathology* **87**, 284-294.

GENETIC STUDIES ON RICE BLAST RELATIONSHIPS

S. KIYOSAWA[1] and Z. LING[2]
[1]*Sakuragaoka 31-8, Kukizaki, Inashiki, Ibaraki 300-12, Japan.*
[2]*Breeding and Cultivation, Chinese Academy of Agricultural Sciences, Beijing, China.*

1. Introduction

Genetics of disease resistance is different from genetics of other characters because it is genetics of interaction of two organisms, the host and the pathogen. Genetics of blast resistance in rice advanced as follows: 1) the first studies on genetics and discovery of races of blast fungus (Sasaki, 1922a,b); 2) breeding of highly resistant varieties using introduced rice varieties (Kitamura, 1962); 3) resistance tests with many varieties and many fungus isolates and establishment of the first differential varieties (Goto, 1961, 1964; Kozaka, 1972); 4) the first genetic studies using single fungus strains (Niizeki, 1960; Iwata and Narita, 1961); 5) beginning of systematic studies on genetics of resistance (Yamasaki and Kiyosawa, 1966); 6) finding of gene-for-gene interaction in rice blast relationship; 7) breakdown (high infection) due to failure of varieties with initially high resistance introduced from foreign varieties (Yamada, 1965); 8) identification of different fungal strains and classification of varieties based on their resistance pattern to fungal strains; 9) establishment of nine Japanese differential varieties with single resistance genes (Yamada *et al.*, 1976) and twelve Kiyosawa's differential varieties (Kiyosawa, 1984b); 10) finding of 13 blast resistance genes (Kiyosawa, 1974, 1986; Imbe and Matsumoto, 1985); 11) beginning of international investigations at IRRI and 12) establishment of near-isogenic lines by several investigators.

Steps 1 - 10 mentioned above were reviewed by Kiyosawa and Ando (1997). Three proposals for future investigations are mentioned below: a) gene-for-gene theory is to be applied to the host-pathogen relationship of rice/blast exploiting the variations in the avirulence genes of the pathogen (Kiyosawa, 1965a); b) dynamic investigations on population genetics are required (Kiyosawa, 1965b) and c) biotechnological approaches including molecular breeding (Kiyosawa 1965c). Studies on resistance to blast disease based on the consideration that genetic analyses are essential for these three proposals, and also population genetic studies were carried out by Kiyosawa and his colleagues. The results of these studies have been reviewed (Kiyosawa 1974, 1976b). Further, Niizeki and Oono (1968), and Tanaka and Nakata (1969) published work on anther culture of rice and tobacco, respectively laying the basis for

S. Sreenivasaprasad and R. Johnson (eds.), Major Fungal diseases of Rice Recent Advances, 145-162.
Professor K. Manibhushan Rao Festschrift

biotechnological approaches to breeding, but it took nearly two decades to begin systemic studies in Japan.

In developing breeding strategies to counter breakdown of resistant varieties four methods were proposed: a) accumulation of several resistance genes in one variety, b) replacement of varieties with those carrying different resistance genes, c) usage of varieties with field resistance (Kiyosawa, 1965a), and d) multilines (Kiyosawa, 1973b). These studies were carried out theoretically and the results reviewed by Kiyosawa (1977, 1982, 1983, 1986). Later, investigations clarified the following: 1) regional variation in relative fitness of Av-a+ (virulence gene) was also recognised in a district (Niigata Prefecture) (Kiyosawa *et al.*, 1993a,b); 2) replacement of varieties with different genotypes for blast disease in farmers' fields led to non-random associations between two virulence genes, and 3) if relative fitness of virulence genes are moderately large, relative fitness of virulence genes could be estimated from the pattern of the non-random association of virulence genes before and after replacement of varieties with different resistance genes (Kiyosawa *et al.*, 1996a,b,c).

2. Specific resistance or major-gene resistance

In crop - pathogen relationships, gene-for-gene interaction was first found between flax and rust. In rice blast, the first study was by Nakamori (1936). Many varieties were tested in fields in Gifu Agricultural Experimental Station in Shichigo, Motosu and its branch in Furukawa, Hida. Two varieties showing different responses between two stations were selected and F2 progeny of the two varieties were tested in both stations. In both stations, 3:1 segregations were obtained with different responses of two varieties (RS/SR). Yamasaki and Kiyosawa (1966) confirmed the gene-for-gene relationship in the rice blast system using many varieties with seven selected fungus strains. To date, thirteen resistance genes were found as resistance to act in Japanese varieties including ones, which were bred by introduction of genes from foreign varieties (Table 1). One gene (*Pi-sh*) was added later (Imbe and Matsumoto, 1985). Multiple alleles were found in *Pi-k* (5 alleles), *Pi-ta* (2) and *Pi-z* (2) loci. Two are RS/RR (Table 2) and one is RS/SR (Table 3).

TABLE 1. Reaction of *Pi-k* locus

Rice variety	Gene	Fungus strain (genotype)			
		A	B	C	D
Aichi Asahi	$Pi\text{-}k^-$	S	S	S	S
Shin 2	$Pi\text{-}k^s$	MR	S	S	S
K 60	$Pi\text{-}k^p$	R	R	S	S
Kusabue	*Pi-k*	R	R^h	R	S
Tsuyuake	$Pi\text{-}k^m$	R^h	R^h	R^h	R^h

TABLE 2. Reaction of *Pi-ta* locus

Rice variety	Gene	Fungus strain (genotype)		
		A	B	C
Aichi Asahi	+	S	S	S
K 1	*Pi-ta*	R	S	S
Pi No.4	$Pi\text{-}ta^2$	R	R	S

TABLE 3. Reaction of *Pi-z* locus

Rice variety	Gene	Fungus strain (genotype)		
		A	B	C
Aichi Asahi	+	S	S	S
Fukunishiki	*Pi-z*	R	S	S
Toride 1	$Pi\text{-}z^1$	S	R	S

Different genes at different loci generally show RS/SR reaction for two different avirulence genes. This is a characteristic of gene-for-gene relationship. Additionally, gene-for-gene relationships have following characteristics. 1) The number of resistance genes found is dependent upon the number of avirulence genes present in fungus strains used. 2) Varieties with single genes have the highest ability to differentiate pathogen isolates into races and vice versa. 3) Fungal mutants can be efficiently used for identification of resistance genes. 4) Fungal mutants can be efficiently used for gene analysis of resistance. 5) Fungal mutants can be efficiently used for gene analysis using cumulative frequency distribution curve method (Kiyosawa, 1970, 1976b). These principles make it possible to analyse the genes for host resistance by using pathogen strains carrying different genes for avirulence.

3. Differential varieties

As shown in Tables 4 and 5, differential varieties were developed in various countries. Development of a set of differential varieties having single genes is very important especially for investigation of population genetics and brief estimation of genotypes of varieties. Kiyosawa differential varieties are used in Japan (Kiyosawa *et al.*, 1993a). Use of Japanese fungus strains permitted division of about 2600 varieties and lines into

12 genotypes up to 1979 in Japan (Kiyosawa *et al.*, 1979a,b), China (Pan *et al.*, 1991), Korea (Kan *et al.*, 1990; Kim *et al.*, 1995; Shin *et al.*, 1996), and Russia (Kolmiets and Kovalenko, 1995a,b). In Korea, a set of eight differential varieties was used for genetic analysis and classification of races of blast fungus (Kan *et al.*, 1990; Kim *et al.*, 1995; Shin *et al.*, 1996) (Table 4). In China, Chinese differential varieties (Du *et al.*, 1996) and Kiyosawa differential varieties (Pan *et al.*, 1991) have been used.

TABLE 4. Differential varieties in some countries I

JAPAN			IRRI (Inukai *et al.*, 1994b)		
Japanese dif.	Kiyosawa's dif.	Gene		Dif. var.	Gene
Shin 2	Shin 2	*Pi-sh*	I	C101LAC	*Pi-1 (t)*
Aichi Asahi	Aichi Asahi	*Pi-a*		C103TTP	*Pi-1 (t)*
Ishikari Shiroke	Fujisaka 5	*Pi-I*		C104LAC	*Pi-1 (t)*
Kanto 51	Kusabue	*Pi-k*	II	C101A51	*Pi-2 (t)*
Tsuyuake	Tsuyuake	*Pi-k*^{m}		C102A51	*Pi-2 (t)*
Fukunishiki	Fukunishili	*Pi-z*		C103A51	*Pi-2 (t)*
Yashiro Mochi	K 1	*Pi-ta*		C104A51	*Pi-2 (t)*
Pi No.4	Pi No.4	*Pi-ta*^{2}		C105A51	*Pi-2 (t)*
Toride 1	Toride 1	*Pi-z*^{t}	III	C101PKT	*Pi-4 (t)*
	K 60	*Pi-k*^{p}		C101TTP-1	*Pi-4 (t)*
	BL 1	*Pi-b*		C101TTP-2	*Pi-4 (t)*
	K 59	*Pi-t*		C101TTP-3	*Pi-4 (t)*
				C101TTP-4	*Pi-4 (t)*
Yamada	Kiyosawa,			C102TTP	*Pi-4 (t)*
et al., 1976	1984b			C105TTP-1	*Pi-4 (t)*
				C105TTP-2L23	*Pi-4 (t)*
				C105TTP-4L6	*Pi-4 (t)*
			IV	C102PKT	
				C105TTP-2L9	
			V	C104PKT	*Pi-3 (t)*
			VI	C105TTP-4L23	*Pi-4 (t)*
				C039	

TABLE 5. Differential varieties in some countries II

China (Kiyowawa and Ling, 1984)			Taiwan (Chien *et al.*, 1989)
Tetep	*P-1*	*P-6* [1]	Kun-shan-wu-shen-ken
Zhenlong 10			Taichung 65
Sifeng 43	*Pi-a*	*Pi-b* [2]	Pai-kan-tao
Dongnong 363	*Pi-a*	*Pi-k* [2]	Taichung 171
Kanto 51	*Pi-k* [2]		Chianung 280
Hejiang 18	*Pi-a*	*Pi-1* [2]	Taichung line 33
Korea (Naqvi *et al.*, 1995)			Kanto 51
Tetep	*P-1*	*P-6* [1]	Norin 21
Tabaegbeo			Sensho
Tongil	*Pi-b*	*Pi-10 (t)* [3]	Cutsugulcul
Yushin			Kao-chio-liu-chou
Kanto 51	*Pi-k*		Kaohsiumg-ta-li-chen-yu
Nongbaeg			Taichung-ti-chio-wu-chien
Jinheung			
Nagdongbyeo			

[1] He and Shen, 1990; [2] Kiyosawa and Ling, 1984; [3] Naqvi *et al.*, 1995.

Inukai *et al.* (1994b) compared a set of 5 near-isogenic lines (NILs), aset of 8 international differential varieties and a set of Kiyosawa's 12 differential varieties on their ability to classify Philippine fungal isolates. The integration of differential varieties with the NILs permitted differentiation of 46 Philippine isolates into 7 races. The combination of NILs and the differentials appeared to be tentatively useful to describe blast pathogen isolates in the Philippines. Kiyosawa's differential varieties developed in Japan were not adequate in the Philippines, because some of Kiyosawa's differential varieties showed intermediate reaction to Philippine isolates and some carried more than one resistance gene effective against Philippine isolates. Kiyosawa *et al.* (1994) divided 1549 isolates into 69 races with Kiyosawa's differential varieties in Japan. Japanese (J) or Kiyosawa's (K) sets of differential varieties are more effective than IRRI's (P) or international (I) sets of differential varieties to distinguish between isolates collected from japonica varieties, and the P and I sets are better for discrimination of isolates from *indica* varieties.

As a first step, it is better that differential varieties and differential fungus strains are developed in each country where import of pathogens and seeds are prohibited. Japanese and Kiyosawa's differential varieties were selected from varieties, which were confirmed to include single resistance genes. However, some varieties in them were found to contain the same gene (*Pi-k*^{s}), for example in Shin 2 and Fujisaka 5 (Kiyosawa, 1969). IRRI's differential varieties include some genes heterogeneously (Kiyosawa, unpublished). It is very important to produce an international set of differential varieties. For this purpose, a search for a universally susceptible variety is important. It has not been found among Japanese varieties and also a very susceptible variety CO39 is not suitable, because it and all others have resistance genes some strains collected from *japonica* regions (Hayashi *et al.*, 1996) and against those collected in *indica* regions (Yamamaoto and Ogawa, 1991).

4. Recent investigations for identification of major genes for resistance

During the last decade, rice blast genetic analyses advanced greatly and a large number of papers have been published. Registered genes are shown in Table 6, with chromosomes on which these genes are located. There are two new genes [*Pi-16*(t) and *Pi-17*(t)] which were regtistered but not published (Inukai, pers. comm.). Additionally, there are some genes, *Pi-11*(t)* in Hongjiaozhan (Lu *et al.*, 1994), *Pi-13*(t)* in Kasalath (Hayasaka *et al.*, 1995) and *Pi-14*(t)* in Aichi Asahi (Hayashi *et al.*, 1996), although these are not yet registered for allocation of the number.

In addition to these genes, many other genes have been found. Important ones are as follows: In Vietnamese *indica* variety Tetep, Kiyosawa (1973a) estimated the presense of *Pi-k*^{h} using mutational analysis. Mackill and Bonman (1992) and Inukai *et al.* (1994a,b) recently showed that Tetep has *Pi-1*, *Pi-ta* and an unnamed resistance gene. Inukai et al. (1995) found a fourth gene in Tetep. In USA, Marchetti *et al.* (1987) identified several genes that confer blast resistance (*Pi-k*^{s}, *Pi-z*, *Pi-k*^{h}, *pi-n*, *pi-g*, and *pi-d*)in Brazos, Gulfrose, Lebonnet, and Vista. They found that rice varieties often contain modifier genes combined with a major gene, or more than one gene that conferred resistance to some pathotypes. Moldenhauer *et al.* (1992) identified dominant and monogenic genes in a variety Katy and two complementary genes in NP 125.

A total of 596 *japonica* rice varieties, inoculated by spraying with seven Japanese differential fungus strains, were divided into 35 groups. Genotypes of fifty varieties were determined and the following genes identified *Pi-a*, *Pi-i*, *Pi-k*, *Pi-ta*, *Pi-ta*^{2}, *Pi-a Pi-i* and *Pi-i Pi-k*. Three hundred and seventy one varieties showed unknown resistance patterns. Three varieties were considered to have genotype *Pi-ta Pi-k*^{m} *Pi-b* (Ling *et al.*, 1990a).

TABLE 6. Resistane genes found in rice varieties and chromosome on which the genes are Located, with references

Japan			IRRI		
Gene	Variety	Chromosome	Gene	Variety	Chromosome
Pi-sh	Shin 2 [1]		*Pi-1*	LAC23 [4]	11 [4,11]
Pi-a	Aichi Asahi [2]		*Pi-2*(t)	5173 [4]	6 [2,4]
Pi-i	Fujisaka 5 [2]	6 ?	*Pi-3*(t)	Pai-kan-tao [4]	6?
Pi-k	Kusabue [2]	11 [4]	*Pi-4*(t)	Tetep [4]	12 [4]
$Pi\text{-}k^{p}$	K 60 [2]	11 [4]	*Pi-5*(t)	Moroberekan [5]	4 [5]
$Pi\text{-}k^{m}$	Tsuyuake [2]	11 [4]	*Pi-6*(t)	Apura1 [3]	12 [12,13]
Pi-ta	K 1 [2]	12 [4]	*Pi-7*(t)	Moroberekan [14]	11 [5]
$Pi\text{-}ta^{2}$	Pi No. 4 [2]	12 [4,15]	*Pi-8*(t)	Kasalath [7]	6 [13]
Pi-z	Fukunishiki [2]	6 [4]	*Pi-9*(t)	*O.minuta* [13]	6 [13]
$Pi\text{-}z^{t}$	Toride 1 [2]	6 [4,20]	*Pi-10*(t)	Tongil [6]	5 [6]
Pi-b	BL 1 [2]	2 [11,19]	*Pi-11*(t)	Zhaiyeqing1 [3]	8 [13]
Pi-t	K 59 [2]		*Pi-12*(t)	Moroberekan [14]	
			Pi-13(t)	Mauwangu [8]	6 [8,21]
$Pi\text{-}k^{s}$	Shin 2 [2]	11 [4]	*Pi-14*(t)	Mauwangu [8]	2 [8,21]
$Pi\text{-}k^{h}$	HR 22, K3 [2]	11 [4]	*Pi-15*(t)	GA25 [7,22]	
			**Pi-13(t)*	Kasalath [10]	6 [10]
			**Pi-14(t)*	Aichi Asahi [9]	12 [9]

Investigators responsible for identification of the gene in the indicated variety and its localisation

1) Imbe and Matsumoto, 1985
2) Kiyosawa, 1972,1974
3) Inukai *et al.*, 1992a,b
4) Inukai *et al.*, 1994a
5) Wang *et al.*, 1994
6) Naqvi *et al.*, 1995*
7) Pan *et al.*, 1996c
8) Pan *et al.*, 1993
9) Hayashi *et al.*, 1996
10) Hayasaka *et al.*, 1995*
11) Miyamoto *et al.*, 1996*
12) Yu *et al.*, 1996*
13) Causse *et al.*, 1994
14) Inukai *et al.*, 1996
15) Ise, 1993
16) Mew *et al.*, 1994*
17) Naqvi and Chattoo, 1996*
18) Zheng *et al.*, 1995*
19) Rybka *et al.*, 1997*
20) Fukuchi *et al.*, 1992
21) Pan *et al.*, 1996a
22) Pan *et al.*, 1996b

* : Physical markers were used to know linkage.

Ling *et al.* (1990b) found new unknown genes in Chinese varieties, Mengwanggu-1 and Hongliandaogu by crossing to Kiyosawa's differentials. Pan *et al.* (1991) crossed two Yunnan (China) japonica varieties with Kiyosawa's differential varieties, and found that one of two genes in Mengwanggu-1 was an allele of *Pi-ta* locus and another was an unknown gene. Another variety Dabainuo had two genes both of which are unknown genes at new loci. The Korean variety Tongil has at least *Pi-a* and *Pi-b* genes for blast resistance in Japan (Kiyosawa and Cho, 1980). This variety showed high levels of resistance for about six years when widely planted under highly disease conductive conditions, before becoming susceptible. Tongil was found to carry a single dominant gene, designated *Pi-10t*, conferring resistance to isolates 106 of the blast pathogen from the Philippines (Naqvi *et al.*, 1995). He and Shen (1990) analysed 10 *indica* rice varieties with Chinese fungus strains, and found 11 genes (*P-1* to *P-11*). Among them, some resistance genes are linked e.g. *P-1* and *P-6* in Tetep (Table 7).

TABLE 7. Genes for resistance to two pathogen B1 races in 10 varieties, Hangzhou, China

Race	Resistance genes in each variety						
	Tetep	Chai-tang	Ai-mei-zao 3	IR36	IR58	Suweon 300	Hong-yang-ai 4
ZC3	*P-1*	*P-1*	*P-1*	*P-1*	*P-1*	*P-1, P-2*	*P-1*
ZG1	*P-6*	*P-6*	*P-6*	*P-8*	*P-9*	*P-6, P-7*	*P-6, P-7*

(Continued)

Race	Resistance genes in each variety		
	Ai-jiao-bai- mi-21	IR9782	Guo-ji-you-zhan-1
ZC3	*P-1, P-2*	*P-3*	*P-4, P-5*
ZG1	*P-7*	*P-8*	*P-10, P-11*

(He and Shen, 1990)

An *indica* variety Kasalath was crossed with 9 of Kiyosawa's 12 differential varieties, each carrying a single resistance gene at one of seven known loci. Allelism tests were performed in the F2 populations with rice blast races. Two dominant genes at different loci controlled the resistance of Kasalath. These two resistance genes were non-allelic to the nine known resistance genes at six loci, *Pi-a*, *Pi-i*, *Pi-k*, *Pi-ta*, *Pi-b* and *Pi-t*. One gene was determined to be *Pi-z*^{t}. Another gene is a new gene and designated *Pi-8* (Pan *et al.*, 1996c). Many of the investigations on mapping of chromosomes shown in Table 6 are dependent upon molecular genetic studies, indicated by *. In them, multiple alleles *Pi-ta* and *Pi-ta*^{2} were compared by molecular genetic analysis, indicating that both have a common origin (Rybka *et al.*, 1997).

5. Mapping of blast resistance genes and molecular genetics

Accumulation of resistance genes in a variety is not always easy. When specific resistance genes are used, selection of genes giving resistance to a wide range of races is essential. It is best to use resistance genes for which no virulence genes are present in each country. When two dominant resistance genes, A and B, are used for breeding, it is not possible to distinguish between three genotypes AABB, AA++, and ++BB. Molecular markers would make it possible to select AABB genotype. Accumulation of two or more true resistance genes, true and field resistance genes, and of several genes for field resistance are three challenges. Most of the resistance genes located by molecular methods are shown in Table 6. Additionally, tagging of genes for resistance to blast has been accomplished (Yu *et al.*, 1991, Sharma and Mohapatra, 1996).

As a first step in cloning, the rice blast resistance gene *Pi-b* was mapped near the terminal region of chromosome 2 using restriction fragment length polymorphism (RFLP) and random amplified polymorphic DNA (RAPD) markers. Pi-b cosegregated with RAPD (b-1) and RFLP (RZ123) markers and it was bracketed between flanking RFLP markers at 0.5 cM on the centromeric side and 1.9 cM on the telomeric side (Miyamoto *et al.*, 1996). McCouch *et al.* (1994) reviewed the mapping of blast resistance genes and mapped 20 resistance loci on chromosomes. Half these genes were present in Japanese varieties and the other half were found in varieties from IRRI and elsewhere. Causse *et al.* (1994) mapped 12 genes including *IPi-*(t) and *IPi-3*(t) found in varieties from IRRI and elsewhere on chromosomes. Varying results on linkage relationships between *Pi-z* and *Pi-i* loci have been obtained: some investigators suggested linkage and others denied it. Kinoshita and Kiyosawa (1997) considered this problem as follows. Ishikari Shiroke resistance is controlled by two complementary genes (*Pi-i* and *Pi-j*), as all parents of four of five varieties which are origins of the Ishikari Shiroke resistance in Japan are Shin 2 type (susceptible), except for one, Sekiyama, of which parents are unknown. *Pi-i* is linked to *Pi-z* but *Pi-j* is not linked.

5.1. EFFICIENCY IN USE OF MOLECULAR MARKERS

The accuracy of identification of homozygous resistance gene *Pi-2*(t) was 96% when RG64 marker (SAP and RFLP) alone was scored. The accuracy of selection increased to 100% when both the markers flanking the *Pi-2*(t) were scored simultaneously (Hittalmani *et al.*, 1984). Inukai *et al.* (1996) used RFLP for selection of lines likely to carry individual blast resistance genes and a minimum number of chromosomal segments from Moroberekan.

5.2. ISOLATION OF SINGLE RESISTANCE GENES

After systematic gene analyses, the second set of differential varieties with single genes for blast resistance were selected (Kiyosawa, 1984b), and about 3600 varieties and lines were divided into 14 groups based on genotypes including *Pi-a*, *Pi-i*, *Pi-k*, $Pi\text{-}k^{m}$, *Pi-kp*, *Pi-ta*, *Pi-ta2*, *Pi-z*, *Pi-zt*, *Pi-b* and *Pi-t* for blast resistance. Later, new varieties registered by the Ministry of Agriculture, Forestry and Fisheries had to include a description of their genotype for blast resistance in Japan. When a new resistance gene is found in a variety that possesses two or more genes, it is necessary to isolate single gene for making a set of differential varieties. Toriyama *et al.* (1983) and Purba *et al.* (1996) attempted two methods. Purba *et al.* (1996) used F3 lines to select lines having single genes from crosses with several genes. Toriyama *et al.* (1983) recommended use of resistant plants in lines showing 3 : 1 or 1 : 3 segregation in B_1F_2 to isolate single genes.

5.3. DEVELOPMENT OF ISOGENIC LINES

It is very important to make isogenic lines for disease resistance. It is also important to make multilines with uniformity of flowering date, quality and yielding ability and to

TABLE 8. Near-isogenic lines and origin of the genes

IRRI			Japan		
Isogenic line	Gene	Donor	Isogenic line	Gene	Donor
I C101LAC	*Pi-1*(t)	LAC23	Kanto IL 10	$\textit{Pi-ta}^2$	Etsunan 109
C101TTP		Tetep	Kanto IL 11	$\textit{Pi-ta}^2$	Etsunan 109
C101LAC		LAC23	Kanto IL 1	*Pi-b*	BL 1
II C101A51	*Pi-2*(t)	5173	Kanto IL 14	*Pi-b*	BL 1
C102A51	*Pi-2*(t)	"	Kanto IL 2	*Pi-k*	Kusabue
C103A51	*Pi-2*(t)	"	Kanto IL 12	*Pi-k*	Kusabue
C104A51	*Pi-2*(t)	"	Kanto IL 4	*Pi-z*	Ou 287
C105A51	*Pi-2*(t)	"	Kanto IL 5	*Pi-z*	Ou 287
III C101PKT	*Pi-4*(t)	Pai-kan-tao	Kanto IL 7	*Pi-i*	Todorokiwase
C101TTP-1	(=*Pi-ta*)	Tetep	Kanto IL 13	*Pi-i*	Todorokiwase
C101TTP-2	"	"	Kanto IL 8	$\textit{Pi-z}^t$	73-8205
C101TTP-3	"	"	Kanto IL 9	$\textit{Pi-z}^t$	73-8205
C101TTP-4	"	"	IL 78	$\textit{Pi-k}^m$	Tsuyuake
C102TTP	"	"	IL 77	$\textit{Pi-k}^m$	Tsuyiake
C105TTP-1	*Pi-4*(t)	"	IL 81	$\textit{Pi-k}^m$	Tsuyuake
C105TTP-2L23	"	"	IL 87	$\textit{Pi-k}^m$	Tsuyuake
C105TTP-4L6	"	"	IL 88A	$\textit{Pi-t}^a$	Yashiromochi
IV C102PKT	"	Pai-kan-tao	IL 88B	$\textit{Pi-t}^a$	Yashiromochi
C105TTP-2L9	"	Tetep	IL 92	$\textit{Pi-t}^a$	Yashiromochi
V C104PKT	*Pi-3*(t)	Pai-kan-tao			
VI C104PKY	*Pi-4*(t)	Tetep			
R*	CO39			Nipponbare	
	Inukai *et al.*, 1994			Ise and Horisue, 1988	

Others (China)

Isogenic lines (Donor of gene or genes), [References]
K80R, K79S (Hong Jiao Zhan); H7R, H7S (Tetep), [Zheng *et al.*, 1993]
F-80-1 (*Pi-k*), F-98-7 (*Pi-km*), F-124-1 (*Pi-ta*), F-128-1 (*Pi-ta2*),
F-129-1 (Pi-kp), F-145-2, (Pi-b), [Ling, 1995a]

Others (Japan)
ZTR, ZTS (*Pi-zt*), [Yokoo, 1983]

* : R means recurrent parent.

investigate the mechanism of resistance. Isogenic lines were produced as shown in Table 8. The *Pi-4b*(t) designated by Mackill and Bonman (1992) was demonstrated to be the same as *Pi-4a*(t) and was retracted later (Inukai, personal communication). Some of these near-isogenic lines are heterogeneous for Japanese fungus strains (Kiyosawa unpublished). CO39 used as a recurrent parent of these near-isogenic lines showed high resistance to five of seven Japanese differential fungus strains (Kiyosawa *et al.*, 1986). It is very difficult to remove these resistance gene(s) detected using Japanese fungus strains, as they could not be detected using Philippines isolates. Similarly, Japanese and Kiyosawa's differential varieties include some resistance gene(s) which cannot be recognised by the Japanese differential fungus strains (Kiyosawa, 1969). In China *indica* partial near-isogenic lines K80R and K79S,and H7R and H7S were used (Zheng *et al.*, 1993, 1995). Near-isogenic lines for a blast resistance gene were used for biochemical studies (Huang *et al.*, 1992). In Japan Koga (1994) investigated the penetration of blast fungus to resistant rice using an isogenic line pair on $Pi\text{-}z^t$ (resistant and susceptible), which was developed by Yokoo (1983).

5.4. ANALYSES WITH CUMULATIVE FREQUENCY DISTRIBUTION CURVE METHOD

Field resistance was defined to be resistance that can be tested only in a field, but not in a greenhouse, for late blight of potato (Gallegly and Niederhauser, 1959). Methods that can be carried out in a greenhouse were developed. These are based on (1) the use of aged seedlings (Niizeki, 1967) and (2) inoculation with low aggressive fungus strains (Kiyosawa, 1966a,b).

Recently, the word, durable resistance, is frequently used (see chapter by Ahn). Breeding of varieties with durable resistance is a hope of breeders. However, the determination of durability of resistance is very critical. In Japan, durable resistance was not found in the process of breeding of varieties with high resistance for ca. 10 years. Breakdown of almost all varieties with high resistance was noticed one or more years after their release to farmers. Ikehashi and Kiyosawa (1981) tested 24 Japanese varieties with high field resistance with some Japanese and Philippine blast fungus strains. These Japanese varieties showed higher resistance to Philippine fungus strains than to Japanese strains. Correlation was not found between their responses to fungus strains in both countries. Ou (1979) obtained similar data.

Field resistance was investigated by statistical methods in many cases. Field resistance in rice is not unusual. The topic is discussed by Higashi in this volume. Roumen (1994) investigated the field resistance in the cross IR 36 (recognised as possessing durable resistance in lowland cultivation, Bonman, 1992) x CO39. In F_2 and F_3 tests, narrow sense heritabilities were 0.06-0.16 and 0.41-0.68, respectively. This means that selection of resistant lines in F_3 generation is necessary to use the field resistance in the breeding programme. The cumulative frequency distribution curve method was devised first to analyse field resistance in F_3 generation (Kiyosawa, 1970, 1976b). In this method, 17 plants in F_3 line are used. Seventeen is enough to differentiate 3 : 1 ratio from 1 : 0 ratio with probability of 99.9%, but not to differentiate 15 : 1 ratio from 16 : 0. This method was devised as a method to analyse segregation in F_3 generation without such differentiations (Kiyosawa, 1970, 1976b, Ali *et al.*, this volume)

This method made it possible effectively to estimate the functional values of genes for field resistance. As narrow heritability of field resistance in F2 is in general low, resistance of 17 individual plants of each F3 line was evaluated by the size of lesions and their number. Ali *et al.* (this volume) showed that resistance in *japonica* rice variety Nipponbare reduced the lesion number and that of *indica* variety Juma reduced lesion size. The former resistance is largely affected by environmental conditions and active in summer and inactive in winter in greenhouse. This indicates that a change in epistasis was caused by a change in environmental conditions. It is thought that resistance active at penetration stage reduces the lesion number, and resistance active after penetration reduces lesion size.

Some varieties show true resistance controlled by complementary or additive action of minor genes. Norin 22 [Japan (J)] and Quan-wei-ai [China (C)] have high field resistance [functional values (rate of resistant plants in F3 line) are 0.12 and 0.00, respectively], and lines showing the functional value of 1 appeared in F3 lines. Similarly, in Nan Jing 11 (C) x Reimei (J), 0.00 and 0.00 respectively (functional values of individual genes in parents) make 1 (functional value in combination of both genes), in Raekeong [Korea (K)] x Reimei, 0.06 and 0.18 make 1.00, in Reimei x IR 24 (IRRI) 0.42 and 0.12 make 1, in RP9-3 [India (I)] x Reimei 0.76 and 0.24 make 1 and in CR 44 (I) x Reimei 0.24 and 0.12 make 1 (Purba *et al.*, 1994). In China, the cumulative frequency distribution curve method was used for gene analysis of resistant *japonica* variety, Dongnong 415. This variety was crossed with a Japanese susceptible variety Norin 22, and the F3 lines were inoculated with Japanese differential fungus strains. It was demonstrated that the resistance of Dongnong 415 was explained by two dominant genes for fungus strains Ken 53-33, Ina 72, and Ken 54-20, by two dominant and one recessive genes for Ken 54-04 and by three major genes for P-2b, Hoku 1 and Ina 168 (Li and Zou, 1996). This method was used in greenhouse tests (Duan and Li, 1990; Chen *et al.*, 1991), and in a blast nursery (He *et al.*, 1989a,b).

5.5 EFFECT OF RESISTANCE GENES

In Japan, many resistance genes were used in cultivated varieties. However, all of these varieties were broken down within a few years after the release to farmers. A variety with the largest number of resistance genes is Hama Asahi. This variety has four resistance genes, *Pi-a*, *Pi-i*, *Pi-k* and *Pi-b* (Kiyosawa 1994a). Even this resistance of Hama Asahi was broken down in the year when it was registered by the Ministry of Agriculture, Forestry and Fisheries. Most fungal isolates showed virulence to five resistance genes, *Pi-a*, *Pi-i*, *Pi-k*, *Pi-b*, *Pi-ta* or often $Pi\text{-}ta^2$. *Pi-ta* is included in a variety Shimokita which has *Pi-ta* and *Pi-a* genes. This indicates that the decrease of field resistance by accumulation of four genes in one variety is larger than the decrease of aggressiveness by accumulation of five virulence genes in blast fungus. Furthermore, it was shown that replacement of varieties by those having different resistance genes soon resulted in accumulation of virulence genes in the pathogen. Hama Asahi was also broken down in North-Eastern China (Kiyosawa *et al.*, 1988).

The breakdown of Tongil in Korea occurred on the 7th year after the release of this variety (Naqvi *et al.*, 1995). The number of resistance genes is two or three (Kiyosawa and Cho, 1980; Naqvi and Chattoo, 1996). Kiyosawa (1976a) predicted breakdown of Tongil and suggested three reasons for its greater than usual longevity: 1) Tongil has 2 - 3 genes, 2) it was bred outside the country, and 3) fungus strains in Korea, Japan and north-eastern area of China were not adapted to *indica* varieties (see chapter by Ahn for further discussion on durable resistance).

6. Summary

Many rice growers in developing countries find the cost of fungicides prohibitive, hence varietal resistance has been the preferred means of controlling the rice blast disease. Developing durably resistant rice lines, therefore, is a high priority for most rice breeding programmes where rice blast occurs. The rice variety Oryzica Llanos 5 was selected from a line with complete blast resistance and released for commercial cultivation in blast prone eastern Colombia. It is grown on approximately 5 x 104 ha/year and has remained resistant to blast over 10 consecutive growing seasons, both in experimental blast nurseries and in farmers' fields where blast normally is a severe problem. This variety also showed uniformly high levels of blast resistance when evaluated at seven high blast level sites in four Asian countries in blast nurseries with 72 other varieties with differing levels and origins of blast resistance. The results suggest that selecting high levels of resistance when diverse sources are combined can be used to develop varieties with stable resistance to rice blast (Correa-Victoria and Zeigler, 1995).

7. References

Ali, M. S., Purba, D., Kawase, M., Okuno, K. and Kiyosawa, S. (1999) Genetic analysis of two components of field resistances, lesion size and number, to rice blast in rice varieties, Nipponbare and Juma, (this volume).

Bhatt, J. C., Singh, R. A., Mani, S. C. and Shoran, J. (1994) Inheritance of blast resistance in rice, *Indian Genetics* **54**, 142-148.

Causse, M. A., Fulton, T. M., Cho, Y. G., Ahn, S. N., Chunwongse, J., Wu, K., Xiao, J., Yu, Z., Ronald, P. C., Harrington, S. E., Second, G., McCouch, S. R. and Tanksley, S. D. (1994) Saturated molecular map of the rice genome based on an inter specific backcross population, *Genetics* **138**, 1251-1274.

Chauhan, J. S., Variar, M., Maiti, D. and Shukla, V. D. (1995) Inheritance of leaf blast resistance in two crosses of upland rice (*Oryza sativa*), *Indian Phytopathology* **48**, 436-438.

Chien, C. C., Hsieh, C. and Chang, Y. C. (1989) Studies on the break-down of resistance in rice cultivar Taichung 70 to rice blast, *Journal of Agricultural Research China* **38**, 72-79.

Chen, B., Pen, Z. and Xu, Y. (1991) Inheritance of blast resistance of a rice variety, Ai-mei-zao No.3, *Acta Genetica Sinica* **20**, 354-361.

Correa-Victoria, F. J. and Zeigler, R. S. (1995) Stability of partial and complete resistance in rice to *Pyricularia grisea* under rainfed upland conditions in Eastern Colombia, *Phytopathology* **85**, 977-982.

Duan, Y. and Li, H. (1990) Analysis on the resistant genes of rice varieties, *Scientia Agriculture Sinica* **23**, 57-59.

Fukuchi, A., Suzuki, Y., Hirochika, H. and Kikuchi, F. (1992) Linkage relations between the photoperiod sensitivity gene *Se-1* and two other genes, *Pi-zt* and *Pgi-2*,. *Rice Genetics Newsletter* **9**, 47-50.

Gallegly, M. E. and Niederhauser, J. S. (1959) Genetic controls of host-parasite interactions in the *Phytophthora* late blight disease, in O. S. Holton, G. W. Fischer, R. W. Fulton, H. Hart and S. E. A. McCallan (eds.), *Plant Pathology, Problems and Progress*, The University of Wisconsin Press, Madison, U.S.A., pp. 168-182.

Goto, K. (1961) Joint work on the race of blast fungus, Piricularia oryzae (Fascicle 1), *Special Report on the Forecasting of Occurrence of the Disease and Insect Pest*, **No. 5**, pp. 89.

Goto, K. (1964) Joint work on the race of blast fungus, Piricularia oryzae (Fascicle 2), *Special Report on the Forecasting of Occurrence of the Disease and Insect Pest*, **No 18**, pp. 132.

Hayasaka, H., Takamatsu, M., Kuboki, Y., Yano, M., Matsunaga, K. and Sasaki, T. (1995) Mapping genes conferring rice blast resistance in rice variety Kasalath using RFLP markers. II. Linkage analysis of the resistance gene on chromosome 6, *Breeding Science* **45(Suppl. 2)**, 168.

Hayashi, N., Ando, I. and Naito, H. (1996) Gene analysis of a new blast resistance in the paddy rice variety, Aichi Asahi, *Breeding Science* **46(Suppl. 2)**, 168.

He, Y., Kiyosawa, S., Wang, Y., Li, J., Li, C. and Higashi, T. (1989a) Inheritance of blast resistance in Chinese upland rice varieties, Zhalulong, Mowanggu and Mongwangu, *Oryza* **36**, 288-298.

He, Y., Wang, Y., Li, J., Li, C., Kiyosawa, S., Higashi, T. and Horisue, N. (1989b) Gene analysis of blast resistance of a Yunnan upland rice variety, Haonaihuan, *Oryza* **26**, 173-182.

He, Z. and Shen, Z. (1990) Studies on blast (Bl) resistance genes in *indica* rice, *International Rice Research Newsletter* **15**, 4.

Higashi, T. (1999) Genetic analysis of field resistance to leaf blast in rice, (this volume)

Hittalmani, S., Foolad, M., Mew, T., Rodoriguez, R. and Huang, N. (1994) Identification of blast resistance gene, Pi-2(t) in rice plants by flanking markers, *Rice Genetics Newsletter* **11**, 144-146.

Huang, D., Yang, W., Wang, J., Fan, Z. and Zou, Q. (1992) The products of genes resistant to rice blast by using some pairs of near-isogenic lines, *Chinese Journal of Rice Science* **6**, 49-52.

Hayasaka, H., Takamatsu, M., Kuboki, Y., Yano, M., Matsunaga, K. and Sasaki, T. (1995) Mapping genes conferring rice blast resistance in rice variety Kasalath using RFLP markers. II. Linkage analysis of the resistane gene on chromosome 6, *Breeding Science* **45(Suppl. 2)**, 168.

Hayashi, N., Ando, I. and Naito, H. (1996) Gene analysis of a new blast resistance in the paddy rice variety, Aichi Asahi, *Breeding Science* **46(Suppl. 2)**, 168.

Ikehashi, H. and Kiyosawa, S. (1981) Strain-specific reaction of field resis tance of Japanese rice varieties revealed with Philippine strains of rice blast fungus, *Pyricularia oryzae* Cav., *Japanese Journal of Breeding* **31**, 293-301.

Imbe, T. and Matsumoto, S. (1985) Inheritance of resistance of rice varieties to the blast fungus strains virulent to the variety "Reiho", *Japanese Journal of Breeding* **35**, 332-339.

Inukai, T., Mackill, D. J., Bonman, J. M., Sarkarung, S., Zeiglar, R. S., Nelson, R., Takamure, I. and Kinoshita, T. (1992a) Blast resistance genes *Pi-2(t)* and *Pi-z* may be allelic, *Rice Genetics Newsletter* **9**, 90-92.

Inukai, T., Mackill, D. J., Bonman, J. M., Sarkarung, S., Zeigler, R. S., Nelson, R., Takamure, I. and Kinoshita, T. (1992b) A blast resistance gene *Pi-3(t)* in near-isogenic line C104PKT, *Rice Genetics Newsletter* **9**, 94-95.

Inukai, T., Nelson, R. J., Zeigler, R. S., Sarkarung, S., Mackill, D. J., Bonman, J. M., Takamure, I. and Kinoshita, T. (1994a) Allelism of blast resistance genes in near-isogenic lines of rice, *Phytopathology* **84**, 1278-1283.

Inukai, T., Nelson, R. J., Zeigler, R. S., Sarkarung, S., Takamure, I. and Kinoshita T (1994b) Identification of pathogenic races of rice blast fungus by using near-isogenic lines with *indica* genetic background, *Journal of the Faculty of Agriculture Hokkaido Univrsity* **66 (Pt.1)**, 27-35.

Inukai, T., Viet, D. L., Imbe, T., Zeigler, R. S., Kinoshita, T. and Nelson, R. J. (1995) Identification of a fourth blast resistance gene in the Vietnamese *indica* cultivar Tetep, *Rice Genetics Newsletter* **12**, 237-238.

Inukai, T., Zeigler, R. S., Sarkarung, S., Bronson, M., Dung, L. V., Kinoshita, T., Nelson, R. J. (1996) Development of pre-isogenic lines for rice blast-resistance by marker-aided selection from a recombinant inbred population, *Theoretical and Applied Genetics* **93**, 560-567.

Ise, K. (1993) A close linkage between the blast (Bl) resistance gene *Pi-ta2* and a marker on chromosome 12 in *japonica* rice, *International Rice Research Newsletter* **18**, 14.

Ise, K. and Horisue, N. (1988) Characteristics of several near-isogenic lines of rice for blast resistance genes, *Japanese Journal of Breeding* **Suppl.2**, 404-405.

Iwata, T. and Narita, T. (1961) Analyses of resistance genes in varieties, Bozu and Asahi, to an isolate Naga 89 (C-II race) [in Goto (1961)].

Kan, S. W., Cho, D. J., Shin, W. K. and Lee, Y. S. (1990) Occurrence of physiologic races of *Pyricularia oryzae* in Gyeongnam Province during 1984-1989, *Research Report RDA* **32**, 1-7.

Kim, Y. -G., Kim, Y. -J., Choi, H. -C. and Han, S. -S. (1995) Inheritance of leaf blast reactions to four races of blast fungus in segregating populations of eight rice crosses. *RDA Journal of Agricultural Science* **37**, 283-298.

Kinoshita, T. and Kiyosawa, S. (1997) Some problems on linkage relationship between *Pi-i* and *Pi-z* loci of blast resistance genes in rice, Rice Genetics Newsletter 13.

Kitamura, E. (1962) Tests and selection in breeding of blast resistance: especially on the introduction of resistance genes from foreign varieties, *Recent Advances of Breeding* **3**, 18-27.

Kiyosawa, S. (1965a) Mechanism of specific resistance in plant to pathogen, imagined from its inherent bebaviour, *Plant Protection* **16**, 353-360.

Kiyosawa, S. (1965b) Ecological analysis on breakdown of resistance in resistant varieties and breeding-counterplan against it, *Agr. Tech.* **20**, 465-470, 510-512.

Kiyosawa, S. (1965c) A dream of an investigator for breeding: Cell Breeding, *The Heredity* **19**, 33-34.

Kiyosawa, S. (1966a) Resistance of some rice varieties to a blast fungus strain, Ken 54-04 - An analysis of field resistance, *Agr. Hort.* **41**, 1223-1230.

Kiyosawa, S. (1966b) A comparison of resistances to a blast fungus strain Ken 54-04 among sister varieties of Norin 22, *Agr. Tech.* **12**, 580-582.

Kiyosawa, S. (1969) Inheritance of resistance of rice varieties to a Philippine fungus strain of *Pyricularia oryzae*, *Japanese Journal of Breeding* **16**, 61-73.

Kiyosawa, S. (1970) Inheritance of blast resistance of the rice varieties Homare Nishiki and Ginga. 1. Resistance of Homare Nishiki and Ginga to the fungus strain Ken 54-04, *Bulletin of the National Institute of Agricultural Science* **D21**, 73-105.

Kiyosawa, S. (1972) Genetics of blast resistance, *Rice Breeding*, IRRI, Philippines, pp. 203-225.

Kiyosawa, S. (1973a) Gene analysis of *indica* type varieties. Basic studies on Breeding Measures of Common Explosive Outbreaks of Blast Epidemics, in *Resistant Rice Varieties, Ministry of Agriculture, Forestry and Fisheries Council*, Tokyo, Japan, pp. 47-61.

Kiyosawa, S. (1973b) Some problems on disease resistance, *Recent Advances in Breeding for Disease Resistance* **13**, 45-50.

Kiyosawa, S. (1974) Studies on genetics and breeding of blast resistance in rice, *National Institute of Agricultural Sciences Miscellaneous Publication* **D1**, 1-58.

Kiyosawa, S. (1976a) Breakdown of Korean variety Tongil by blast disease: its possibility and counter measure, *Agr. Tech.* **31**, 391-395, 442-445.

Kiyosawa, S. (1976b) Methods for tests and gene analysis of blast resistance of rice varieties, *Oryza* **13**, 1-32.

Kiyosawa, S. (1977) Development of methods for the comparison of utility values of varieties carrying various types of resistance, *Annals of the New York Academy of Sciences* **287**, 107-123.

Kiyosawa, S. (1980) The possible application of gene-for-gene concept in blast resistance, *JARQ* **14**, 9-14.

Kiyosawa, S. (1982) Genetics and epidemiological modeling of breakown of plant disease resistance, *Annual Review of Phytopathology* **20**, 93-117.

Kiyosawa, S. (1983) Population genetics of rice-blast relationships, in *Genetics: New Frontiers*, Oxford & IBH Publishing Co., India, pp. 119-130.

Kiyosawa, S. (1984a) Tests for pathogenicity of blast fungus isolates collected from the rice variety, Hama Asahi, in Aomori Prefecture, *Japanese Journal of Breeding* **34**, 485-489.

Kiyosawa, S. (1984b) Establishment of differential varieties for pathogenicity test of rice blast fungus, *Rice Genetics Newsletter* **1**, 95-96.

Kiyosawa, S. (1986) Breakdown of blast resistance in rice in relation to general strategies of resistance gene deployment to prolong effectiveness of disease resistance in plants, in K. J. Leonard and W. E. Fry (eds.), *Plant Disease Epidemiology. Genetics, Resistance and Management*, Vol. 2, McGraw-Hill Publishing Company, New York, U.s.A., pp. 251-283.

Kiyosawa, S. and Cho, C. I. (1980) Identification of blast-resistance genes in Korean variety, Tongil, *Japanese Journal of Breeding* **30**, 73-82.

Kiyosawa, S. and Ling, Z. Z. (1984) Identification of genes for rice blast resistance in Chinese differential varieties with Japanese fungus strains, *Scientia Sinica (Series B)* **27**, 273-283.

Kiyosawa, S. and Ando, I. (1997) Inheritance of resistance. 1. Blast disease resistance. *Science of the Rice Plant: III (Genetics)*, Food and Agriculture Policy Research Center, Tokyo, pp. 470-500.

Kiyosawa, S., Fujimaki, Y. and Iwano, M. (1993a) Collections of blast fungus from resistant rice varieties and their frequency of races from resistant rice varieties and virulence genes, *Annals of the Phytopathological Society of Japan* **59**, 387-395.

Kiyosawa, S., Fujimaki, Y. and Iwano, M. (1993b) Estimation of fitness values for virulence genes in collections of rice blast fungus from Niigata Prefecture, *Annals of the Phytopathological Society of Japan* **59**, 396-406.

Kiyosawa, S., Aihara, J., Inoue, M. and Matsumoto N (1979a,b). Classification of rice varieties based on true-resistance to blast, I and II, *Japanese Journal of Breeding* **29**, 77-83, 166-170.

Kiyosawa, S., Ling, Z. Z., Plao, C. S. and Xue, Q. (1988) Breeding for blast resistance in rice in Japan and China, *Agr. Tech.* **63**, 695-706.

Kiyosawa, S., Mackill, D. J., Bonman, J. M., Tanaka, Y. and Ling, Z. Z. (1986) An attempt of classification of world's rice varieties based on reaction pattern to blast fungus strains, *Bulletin of the National Institute of Agrobiological Resources* **2**, 41-59.

Kiyosawa, S., Okinaka, Y., Purba, D. and Furutani, T. (1994) Collection of blast fungus from resistant rice varieties and their virulence analysis, *Breeding Science* **44**, 41-415.

Kiyosawa, S., Purba, D., Ali, M. S., Okinaka, Y., Shimizu, T. and Saito, A. (1996a) Influence of frequencies of host genotypes and fitness of virulence genes on nonrandom association type and pattern between avirulence loci, *Breeding Science* **45**, 287-293.

Kiyosawa, S., Purba, D., Ali, M. S., Okinaka, Y., Shimizu, T. and Saito, A. (1996b) Nonrandom association of avirulence loci in four disease resistance gene system. I. Principles of determination of nonrandom association within host groups having the same number of susceptibility genes, *Annals of the Phytopathological Society of Japan* **62**, 95-100.

Kiyosawa, S., Purba, D., Ali, M. S., Okinaka, Y., Shimizu, T. and Saito, A. (1996c) Nonrandom association of avirulence loci in four disease resistance gene system. II. Replacement between host groups having different number of susceptibility genes, *Annals of the Phytopathological Society of Japan* **62**, 101-107.

Kobayashi, D. Y., Tamaki, S. J. and Keen, N. T. (1989) Cloned avirulence genes from the tomato pathogen *Pseudomonas syringae* pv. *tomato* confer cultivar specificity on soybean, *Proceedings of the National Academy of Sciences USA* **86**, 157-161.

Koga, H. (1994) Hypersensitive death, autofluorescence, and ultrastructural changes in cells of leaf sheaths of susceptible and resistant near-isogenic lines of rice (*Pi-zt*) in relation to penetration and growth of *Pyricularia oryzae*, *Canadian Journal of Botany* **72**, 1463-1477.

Kolmiets, T. M. and Kovalenko, E. D. (1995a) Characteristics of blast resistance genotype in rice cultivars from different regions of the world, *Genetika* **31**, 382-384.

Kolmiets, T. M. and Kovalenko, E. D. (1995b) Identification of blast resistance genes in domestic rice cultivars, *Genetika* **31**, 219-221.

Kozaka, T. (1972) Joint work on the race of blast fungus, Piricularia oryzae (Fascicle 3), *Special Report on the Forecasting of Occurrence of the Disease and Insect Pest* **No 24**, pp. 296.

Li, N. and Zou, D. (1996) Resistant gene analysis of japonica rice variety, Dongnong 415, to *Pyricularia grisea* Saccardo, *Chinese Journal of Rice Science* **10**, 85-90.

Ling, Z. (1995) Development of near-isogenic lines as international differentials of the blast pathogen, *International Rice Research Newsletter* **20**, 13-14.

Ling, Z. Z., Pan, Q., Wang, J. and Li, M. (1990a) Gene analysis for blast resistance of Yunnan *japonica* type variety Mengwanggu-1, in L. Zhu, W. Lu and Y. Xie (eds), *Advances in Genetic Researches on Resistance to Diseases in Major Crops,* Jiangsu Science & Technology Publishing House, Nagjing, China, pp.105-119.

Ling, Z. Z., Pan, Q., Wang, J. and Li, M. (1990b) Analysis of Yunnan japonica rice Hongliandaogu for resistance to blast diesase, in L. Zhu, W. Lu and Y. Xie (eds), *Advances in Genetic Researches on Resistance to Diseases in Major Crops,* Jiangsu Science & Technology Publishing House, Nagjing, China, pp. 111-116.

Ling, Z. Z., Wang, J. L., Pan, Q. H., Li, M. F., Huang, Q. G. and Qiu, Z. E. (1990c) Classification for blast resistance of some japonica rice varieties in Yunnan Povince, *Scientia Agricultura Sinica* **23**, 5-11.

Lu, J., Qianm H. R., Zhuang, J. T., Lin, H. X. and Zheng, K. L. (1993) Identification of a RAPD marker linked ot a blast resistance gene in Oryza sativa, *Chinese Rice Research Newsletter* **2**, 2-3.

Mackill, D. J. and Bonman, J. M. (1992) Inheritance of blast resistance in near-isogenic lines of rice, *Phytopathology* **82**, 746-749.

Mackill, D. J., Bonman, J. M., Tenorio, P. D. Jr. andVergel De Dios, T. I. (1988) Near- isogenic *indica* rice lines with blast resistance genes, *Rice Genetics Newsletter* **5**, 98-101.

Marchetti, M. A., Lai, X. and Bollich, C. N. (1987) Inheritance of resistance to *Pyricularia oryzae* in rice cultivars grown in the United States, *Phytopathology* **77**, 799-804.

McCouch, S. K., Nelson, R .J., Tohme, J. and Zeigler, R. S. (1994) Mapping of blast resistance genes in rice, in R. S. Zeigler, S. A. Leong and P. S. Teng (eds.), *Rice Blast Disease*, IRRI, Philippines, pp. 167-187.

Mew, T. W., Parco, A. S., Hittalmani, S., Inukai, T., Nelson, R., Zeigler, R. S. and Huang, N. (1994) Fine-mapping of major genes for blast resistance in rice, *Rice Genetics Newsletter* **11**, 126-128.

Miyamoto, M., Ando, I., Rybka, K., Kodama, O., Kawasaki, S. (1996) High-resolution mapping of the *indica*-derived rice blast resistance genes, 1. *Pi-b*, *Molecular Plant Microbe Interaction* **9**, 6-13.

Moldenhauer, K. A. K., Bastawisi, A. O. and Lee, F. N. (1992) Inheritance of resistance in rice to races IB-49 and IC-17 of *Pyricularia grisea* rice blast, *Crop Science* **32**, 584-588.

Nakamori, E. (1936) Regional variation on blast resistance of lowland rice varieties, *Agr. Hort.* **11**, 823-834.

Naqvi, N. I., Bonman, J. M., Mackill, D. J., Nelson, R. J. and Chattoo, B. B. (1995) Identification of RAPD markers linked to a major blast resistance, Molecular Breeding 1, 341-348.

Naqvi, N, I. and Chattoo, B. B. (1996) Development of a sequence characterized amplified region (SCAR) based indirect selection method for a dominant blast resistance gene in rice, *Genome* **39**, 26-30.

Niizeki, H. (1960) On a gene for resistance to *Piricularia oryzae* in a Japanese rice variety, Aichi-asahi, *Agr. Hort.* **35**, 1321-1322.

Niizeki, H. (1967) On some problems in rice breeding for blast resistance, with special reference to variation of blast fungus, *Recent Advances in Breeding* **8**, 71-78.

Niizeki, H. and Oono, K. (1968) Induction of haploid rice plant from anther culture, *Proceedings of the Japanese Academy* **44**, 554-557.

Ou, S. H. (1979) Breeding rice for resistance to blast - a critical review, in *Proceeding of the Rice Blast Workshop*, IRRI, Philippines, pp. 81-137.

Pan, Q., Ling. Z. and Wang, J. (1991) Gene analysis of blast resistance of two *japonica* varieties from Yunnan province, *Chinese Journal of Rice Science* **5**, 61-66.

Pan, Q. H., Tanisaka, T. and Ikehashi, H. (1996a) Studies on the genetics and breeding of blast resistance in rice VI. Gene analysis for the blast resistance of two Yunnan native cultivars GA20 and GA25, *Breeding Science* **(Suppl. 2)**, 70.

Pan, Q. H., Tanisaka, T., Ikehashi, H. and Yamagata, H. (1996b) Studies on the genetics and breeding of blast resistance in rice. V. Identiication of two blast resistance genes in a Yunnan native variety, Maowangu, *Breeding Science* **46 (Suppl. 1)**, 79.

Pan, Q. H., Wang, L., Ikehashi, H. and Tanisaki, T. (1996c) Identification of a new blast resistance gene in the *indica* rice cultivar Kasalath using Japanese differential cultivars and isozyme markers, *Phytopathology* **86**, 1071-1075.

Pan, Q. H., Tanisaka, T., Okumoto, Y. and Yamagata, H. (1993) Gene analysis for the blast resistance of a Yunnan local variety, Maowanggu, *Japanese Journal of Breeding* **43 (Suppl. 2)**, 101.

Purba, D., Kiyosawa, S., Ando, I. and Furutani, T. (1994) Estimation of functional values of field-resistance genes to blast disease in some rice varieties, *Breeding Science* **44**, 285-293.

Purba, D., Ali, M. S., Furutani, T.and Kiyosawa, S. (1996) Identification and isolation of blast resistance genes in three *indica*-type rice varieties, *Breeding Science* **46**, 107-115.

Roumen, E. C. (1994) The inheritance of host plant resistance and its effect on the relative infection efficiency of *Magnaporthe grisea* in rice cultivars, *Theoretical and Applied Genetics* **89**, 498-503.

Rybka, K., Miyamoto, M., Ando, I., Saito, A. and Kawasaki, S. (1997) High resolution mapping of the *indica* derived rice blast resistance genes, Molecular Plant-Microbe Interaction.

Sasaki, R. (1922a) Inheritance of resistance to *Piricularia oryzae* in different varieties of rice.

Sasaki R (1922b). Existence of strains in rice blast fungus, *Journal of Plant Protection* **9**, 631-644.

Sharma, R. P. and Mohapatra, T. (1996) Molecular mapping and tagging of genes in crop plants, *Genetica* **97**, 313-320.

Shin, Y. –S., Parl, M. K., Kim, K. –W. and Cho, S.-Y. (1996) Inheritance of blast resistance in the Tongil-type rice cultivars, *RDA Journal of Agricultural Science* **38**, 462-465.

Tanaka, M. and Nakata, K. (1969) Tobacco plants obtained by anther culture and the experiment to get diploid seeds from haploids, *Japanene Journal of Genetics* **44**, 47-54.

Toriyama, K., Ezuka, A., Asaga, K. and Yokoo, M. (1989) A method of estimating true resistance genes to blast in rice varieties by testing their back-crossed progenies for race-specific reactions, *Japanese Journal of Breeding* **33**, 448-456.

Wang, G.-L., Mackill, D. J., Bonman, J. M., McCouch, S. R., Champoux, M. C. and Nelson, R. J. (1994) RFLP mapping of genes conferring complete and partial resistance to blast in a durably resistant rice cultivar, *Genetics* **136**, 1421-1434.

Yamada, M. (1965) Outbreak of blast disease on some highly resistant rice varieties derived from foreign varieties, *Shokubutsu Boeki* **19**, 231-234.

Yamada, M., Kiyosawa, S.,Yamaguchi, T., Hirano, T., Kobayashi, T., Kushibuchi, K. and Watanabe, S. (1976) Proposal of a new method for differentiating races of *Pyricularia oryzae* Cavara in Japan, *Annals of the Phytopathological Society of Japan* **42**, 216-219.

Yamamoto, Y. and Ogawa, T. (1991) Resistance of a rice *indica* cultivar CO 39 to blast fungus in Japan, *Japanese Journal of Breeding* **41(Suppl. 2)**, 416-417.

Yamasaki, Y., Kiyosawa, S. (1966) Studies on inheritance of resistance of rice varieties to blast 1. Inheritance of Japanese varieties to several strains of the fungus, *Bulletin of the National Institute of Agricultural Science* **D14**, 39-69.

Yokoo, M. (1983) Near-isogenic lines of rice with respect to a *Pi-zt* gene for resistance to blast disease, *Japanese Journal of Breeding* **33**, 341-345.

Yu, Z. H., Mackill, D. J., Bonman, J. M. and Tanksley, S. D. (1991) Tagging genes for blast resistance in rice via linkage of RFLP markers, *Theoretical and Applied Genetics* **81**, 471-476.

Yu, Z. H., Mackill, D. J., Bonman, J. M., Mccouch, S. R., Guiderdoni, E., Notteghem, K. and Tanksley, S. D. (1996) Molecular mapping of genes for resistance to rice blast (*Pyricularia grisea* Sacc.), *Theoretical and Applied Genetics* **93**, 859-863.

Zheng, K. L., Shen, B., Qian, H. R. (1993) Tagging rice blast resistance genes to RFLP markers. Biotechnology in Agriculture, *Current Plant Science and Biotechnology in Agriculture* **15**, 258-260.

Zheng, K. L., Zhung, J. Y., Lu, J., Qian, H. R. and Lin, H. X. (1995) Identification of DNA markers linked to a blast resistance gene in rice, in *Induced Mutations and Molecular Techniques for Crop Improvement*, Conference Proceedings, Vienna, Austria, 19-23 June 1995, pp. 253-261.

GENETIC ANALYSIS OF FIELD RESISTANCE TO LEAF BLAST IN RICE

T. HIGASHI
Department of Lowland Farming, Tohoku National Agricultural Experiment Station, Yotsuya, Omagari, Akita, 014-01 Japan.

1. Introduction

Resistance to rice blast can be classified into true resistance (qualitative resistance) and field resistance (quantitative resistance). True resistance is unstable and controlled by major genes that react specifically to different races of blast fungus. The breakdown of true resistance has often occurred and shocked rice breeders. On the other hand, field resistance is a quantitative difference between varieties in a field where true resistance is ineffective. A variety with high field resistance shows a slow disease progress even if it is infected by blast and susceptible lesions are formed. Field resistance is usually stable to changes in pathogenicity of blast fungus. The breakdown of field resistance had not been observed in actual rice cultivation and rice breeders are trying to improve field resistance of varieties in Japan.

When exposed to blast, Japanese lowland rice cultivars suffer considerable damage because the level of the field resistance is low. Therefore, chemical control is indispensable in the area where blast is prevalent.

The field resistance of upland rice is high and there are some foreign varieties with high field resistance. However, upland varieties and foreign varieties have many undesirable traits for Japanese conditions. Moreover, field resistance seems to be controlled by many genes, and their genetic analysis is insufficient. Therefore, it is difficult to introduce the resistance into Japanese lowland varieties in breeding. The elucidation of the mode of inheritance of field resistance is needed for efficient breeding of rice resistant to blast.

In this research, using Japanese upland varieties, lowland varieties, and a foreign variety, the mode of inheritance of the field resistance to leaf blast was asserted by various methods, and systematic elucidation was attempted. The field resistance was a quantitative trait and the mode of inheritance was analyzed by a diallel analysis. Next, the mode of segregation of progeny lines was analyzed in two crosses, and the number of effective factors affecting the field resistance was estimated from the ratio of segregating families in several crosses. Moreover, since some parts of the resistance of upland rice Sensho were introduced into many Japanese lowland varieties, and played important roles in the blast resistance breeding, the chromosomes on which the field resistance genes located were clarified using linkage testers with marker genes. The

S. Sreenivasaprasad and R. Johnson (eds.), Major Fungal diseases of Rice Recent Advances, 163-180.
Professor K. Manibhushan Rao Festschrift

results of these investigations have already been published (Higashi and Kushibuchi, 1978; Higashi and Saito, 1985; Higashi, 1995).

2. Diallel analysis

2.1. MATERIALS AND METHODS

A half diallel set of crosses was made using seven varieties and lines; Imochi 314, IR 179, Upland Rice Norin 12, Norin 22, Nipponbare, Koshihikai and Norin 29. These varieties and lines differed in their level of field resistance and had no true resistance genes effective to the race 003 predominant in the blast nursery. Imochi 314 derived from Upland Rice Norin-mochi 4, and IR 279 was selected at IRRI. The field resistance of both varieties is extremely high. Resistance of Upland Rice Norin 12 is also high and that of Norin 22 and Nipponbare is moderate, and the resistance is low in Koshihikari and Norin 29. Leaf blast was spread in the blast nursery by natural infection at the Central Agricultural Experiment Station, Konosu in 1976 (F_1) and 1977 (F_2). The leaf blast scores of each plot were assessed using eleven classes of 0-10 according to the investigation standard (Table 1). The diallel analysis was based on the method of Mather and Jinks (1971). For the analysis of variance of the diallel tables the model of Jones (1965) modified according to the model of Hayman (1954) was used because no reciprocal crosses were made.

TABLE 1. Scores of leaf blast resistance in a blast nursery (Asaga, 1976)

Score	General description of leaf blast	% of infected leaf area
0	No susceptible lesions observed	0
1	Few susceptible lesions	1
2	A few susceptible lesions	2
3	Several susceptible lesions	5
4	Many susceptible lesions	10
5	A great many susceptible lesions with few dead leaves	20
6	A few dead leaves	40
7	Several dead leaves	60
8	Many dead leaves	80
9	Almost all dead leaves	90
10	All leaves and stems dead	100

2.2. RESULTS

The leaf blast scores (field resistance) of the F_1 and F_2 were almost the mean of those of parents. On the analysis of variance of the F_1 diallel table the additive effects of genes (source of variation a) were significant (Table 2), whereas the dominance effects (b) were not significant. In the F_2 analysis, however, both the a and b sources were significant (Table 2). Under b, the source b_1 (mean deviation of the F_2's from their mid-parents) and the source b_3 (dominance deviation unique to each F_2: specific combining ability) were significant. The b_2 source (mean dominance deviation of the F_2 from their mid-parents within each array differs over arrays: general combining ability), however, was not significant. The average field resistance of the F_2 was partially dominant as shown in Figure 1.

TABLE 2. Variance analysis of blast resistance in the diallel cross of seven rice varieties

Source of variation		F1			F2		
		df	MS	VR	df	MS	VR
a		6	41.94	94.87**	6	95.36	499.28**
b		21	0.61	1.72	21	1.28	3.80**
	b_1	1	0.05	0.37	1	6.35	83.54*
	b_2	6	0.41	1.17	6	0.40	1.16
	b_3	14	0.73	1.99	14	1.30	3.71**
Blocks (B)		1	0.83		2	0.62	
B×a		6	0.44		12	0.19	
B×b		21	0.35		42	0.34	
	B×b_1	1	0.15		2	0.08	
	B×b_2	6	0.35		12	0.34	
	B×b_3	14	0.37		28	0.35	
Block interactions		27	0.37		57	0.31	

The relationships between the variance (*Vr*) and parent-offspring covariance (*Wr*) of members of the same array in the F_1 and F_2 diallel tables were tested. The graph of regression of *Wr* on *Vr* and the analyses of variances of *Wr*+*Vr* and *Wr*-*Vr* provided some genetic information. First, the simple additive-dominance model with the absence of non-allelic interaction (epistasis) is satisfactory for the field resistance. Second, the additive effects are larger than the dominance effects, that is, the dominance is incomplete; partial dominance in the direction of resistance. In addition, Imochi 314, IR 279 and Upland Rice Norin 12, which have very high field resistance, are presumed to have a larger number of dominant genes.

The estimates of genetic components are given in Table 3. The components of dominance effects (H_1, H_2) are smaller than that of the additive effects (D), and the narrow sense and broad sense heritabilities were high.

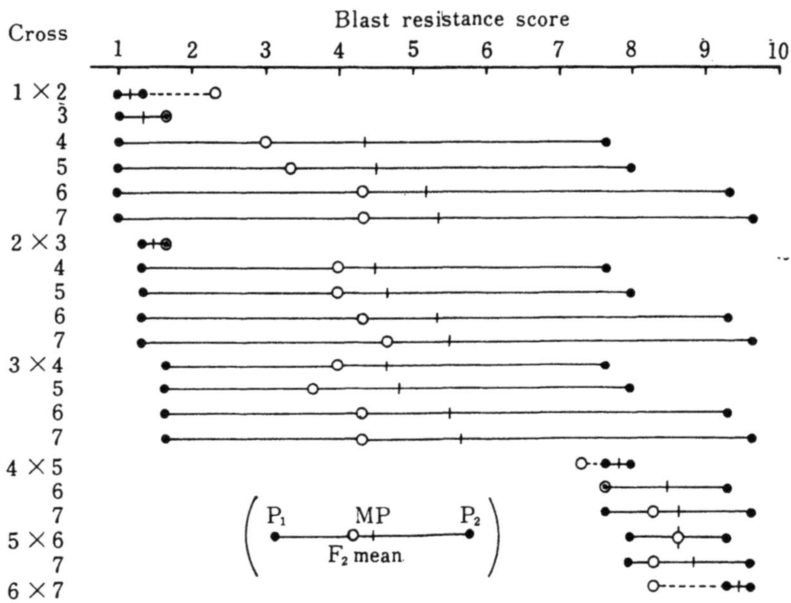

Figure 1. Blast resistance of parents and F2 means in the diallel cross of seven rice varieties. 1: Imochi 314, 2: IR 279, 3: Up. Nprin 12, 4: Norin 22, 5: Nipponbare, 6: Koshihikari, 7: Norin 29.

2.3. DISCUSSION

Since a simple additive-dominance model is satisfactory, the field resistance to leaf blast can be explained by additive effects and dominance effects only. The additive effects are larger and much more important than the dominance effects. The dominance is incomplete and partially dominant in the direction of resistance. However, the magnitude of the dominance effects is influenced by the time of recording and the rating method of the score (Higashi and Kushibuchi, 1978).

Abe *et al.* (1974, 1976) suggested partial dominance in some cases though the blast resistance of upland rice was mainly additive. Goto (1970) assumed that three incompletely dominant genes acted additively in Sensho. Asaga (1976, 1981) reported recessive genes with additive effects in some cases, though the field resistance was controlled by polygenes with additive effects and incomplete dominance effects as a whole. Roumen (1994) also observed the additive and dominance effects for partial resistance to leaf blast. These reports give general support to the present results.

TABLE 3. Estimates of genetic variance components in the diallel cross

Component	F1			F2			
	Block		Mean	Block			Mean
	I	II		I	II	III	
D	6.53	10.55	8.54	16.60	14.60	15.98	15.73
H_1	0.11	1.14	0.63	2.19	5.16	5.79	4.38
H_2	0.02	0.66	0.34	2.92	5.28	4.56	4.26
F	-2.87	1.43	-0.72	4.41	-0.44	8.19	4.06
E	0.37	0.37	0.37	0.31	0.31	0.31	0.31
$\sqrt{(H_1/D)}$	0.13	0.33	0.27	0.36	0.60	0.60	0.53
u v	0.04	0.14	0.13	0.33	0.26	0.20	0.24
h_B^2	0.93	0.93	0.93	0.96	0.96	0.95	0.96
h_N^2	0.93	0.90	0.91	0.94	0.92	0.91	0.92

In the present study, epistatic effects of the genes affecting the field resistance were not observed. However, Shinoda *et al.* (1971) presumed that more than three genes including supplemental genes controlled the high field resistance of upland rice Kuroka and suggested the occurrence of epistasis. It is thought that a result difference from the present findings was obtained because the quantitative resistance of Kuroka was not estimated in their research but was assessed qualitatively using translocation analysis. If quantitative moderate resistance is classified qualitatively with susceptibility some additive gene effects cannot be detected.

Abe *et al.* (1976) and Maruyama *et al.* (1983) found no difference between F_1s of reciprocal crosses by some combinations of lowland and upland rice varieties for field resistance to leaf blast; however, no reciprocal crosses were test in the present study.

Both the broad sense and narrow sense heritability estimates of the field resistance were high (0.9 or more). Abe *et al* (1974) and Yokoo *et al* (1974) obtained also high broad sense heritabilities (0.75-0.98 and 0.81 respectively) using F_3 lines derived from crosses between the Upland Rice Norin-mochi 4 and lowland rice varieties. Asaga (1981) calculated heritabilities of field resistance in the progenies of crosses of lowland varieties by various methods, and most of the values were low. This may be because the genetic variation of the field resistance of lowland varieties is not large. Wang *et al.* (1989) evaluated partial resistance of progenies derived from the crosses between partially resistant varieties (IR 36 and Milyang 42) and susceptible varieties (IR 50 and Milyang 57) by the inoculation method in a greenhouse. In their study, the heritabilities were low, though it was suggested that the partial resistance was controlled by polygenes, from the distribution of F_2. It is thought that the heritability estimates may have been low because the accuracy of the individual test in a greenhouse was low. Roumen (1994) reported that the heritability in the F_2 was low, while that of F_3 lines was much higher in the crosses of IR 36/CO 39 and IR 36/IR 64.

3. Analysis of progeny lines

3.1. MATERIALS AND METHODS

Leaf blast resistance of the F_6 lines (130 lines) derived from IR 279/Kanto 100, and the F_7 lines (135 lines) derived from Imochi 314/Kankei 623, which had been maintained randomly, was tested in the blast nursery at Konosu in 1976. In addition, the leaf blast resistance of the F_7 and F_8 families (one family consisted of two lines randomly selected from the F_6 and F_7 lines) was tested in 1977. Kanto 100 and Kankei 623 have no true resistance gene effective to race 003, and have medium field resistance. The scoring system was similar to that used in Experiment 1. The broad sense heritability of the field resistance was estimated from the variance of progeny lines and that of parents.

3.2. RESULTS

Scores for leaf blast resistance of the progeny lines in both crosses were normally distributed between those of the parents. The estimated broad sense heritabilities for field resistance to leaf blast in both crosses were extremely high (h^2_B = 0.915 in IR 279/Kanto 100, 0.862 in Imochi 314/Kankei 623). In addition, parent-offspring correlations were also high (r = 0.742**, 0.792**). Figures 2 and 3 show the distribution of progeny lines and the parent-offspring correlation in IR 279/Kanto 100. Similar figures were obtained in Imochi 314/Kankei 623. The number of effective factors affecting the field resistance was estimated using the expression Jinks and Towey (1976), shown in the next experiment. The minimum number of effective factors was seven in IR 279/Kanto 100 and eleven in Imochi 314/Kankei 623.

3.3. DISCUSSION

The progeny lines of the crosses showed normal distribution approximately between parents. Judging from the parents' differences and the sizes of each variation of the progeny lines, the field resistance to leaf blast is controlled by several multiple genes and/or polygenes with additive effects. The results were similar to the findings of Asaga (1981) where field resistance of Japanese lowland rice was controlled by polygenes with minute additive effects. However, Maruyama *et al.* (1983) reported an example that progeny lines of a cross were not normally distributed, though they show continuous distribution. This could be because some genes with comparatively large effects were present.

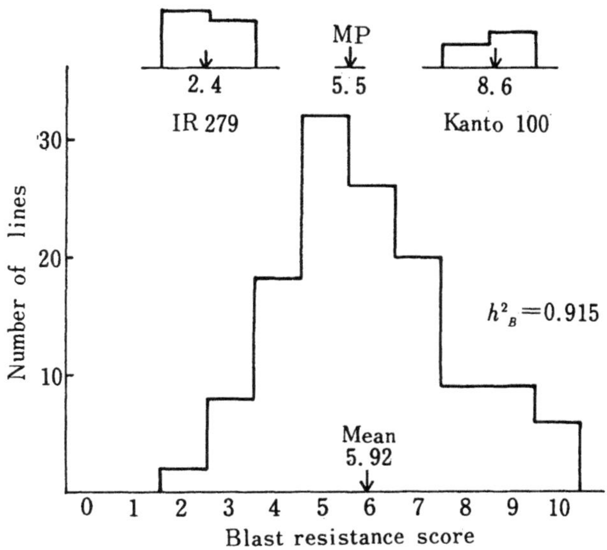

Figure 2. Segregations of blast resistance in F6 lines randomly selected from the cross of IR 279/Kanto 100.

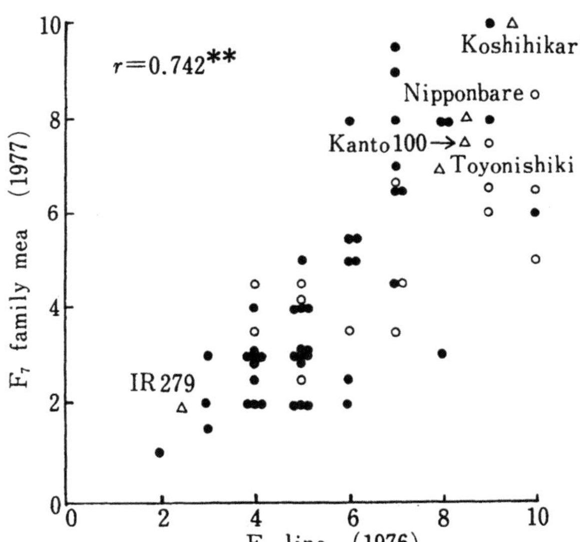

Figure 3. Parent-offspring correlation of blast resistance observed between F6 lies derived from IR 279/Kanto 100 and their F7 families. Open circle: Segregating families, Closed Circle: Non-segregating families, Triangle: Parents or standard varieties.

The minimum number of effective factors was estimated to be seven in IR 279/Kanto 100 and eleven in Imochi 314/Kankei 623. The effective factors in this case are not genes but are units of structural change (recombination) of a chromosome (Mather, 1949). These estimates are close to the eight genes estimated by Asaga (1981) from the variance of the F_3 lines of Toride 1/Yamabiko for field resistance. Roumen (1994) estimated the number of genes that reduce lesion number, a component of partial resistance to be five or more in IR 36. In other reports, fewer genes (3 to several) were presumed for field resistance including upland rice (Kiyosawa *et al.*, 1967; Kiyosawa, 1970; Goto, 1970; Shinoda *et al.*, 1971; Abe *et al.*, 1974, 1976). The discrepancy may be explained by the difference in detection power of the estimation method and the difference of the materials used.

4. Estimating the number of effective factors affecting field resistance

4.1. MATERIALS AND METHODS

All possible cross combinations except reciprocal crosses were made among five varieties and lines: upland rice Sensho [true resistance: +, field resistance: very high], upland irrigation rice Ishioka-mochi 18 [*Pi-a*, very high], lowland variety Ouu 247 [+, high], Reimei [*Pi-a*, high] and Moukoto [+, very low]. These materials except Moukoto are useful for the blast resistance breeding.

The materials were grown at random according to the procedure shown in Figure 4, at the Tohoku National Agricultural Experiment Station, Omagari. Each cross consisted of 120-150 families (2 lines each). The leaf blast was tested in the blast nursery where race 003 and 007 were predominant. These materials do not have any true resistance genes effective to these races. The investigation standard was similar to that of Experiment 1 (Table 1). When testing for leaf blast, two lines of a family were made into a pair. Segregating families were detected by the difference of blast scores of 0.5 or more between the lines in a family.

The number of effective factors affecting the field resistance to leaf blast was estimated by the following expression of Jinks and Towey (1976).

$$P=1-[1-5/(2^{n+2})]^{k}$$

Where n=3 and P is the ratio of the segregating families at the F_5 generation. k is the minimum number of effective factors. However, as mentioned above, the effective factors in this case are not genes but a unit of structural change (recombination) of a chromosome (Mather, 1949).

4.2. RESULTS

The leaf blast scores of F_5 families (means of paired lines) segregated continuously between parents or transgressed those of parents (Table 4). The distributions were not normal though the averages were close to the mid point between. When Sensho or

Ishioka-mochi 18 were used as parents, the modes were not clear. Transgressive segregations in both directions were observed in the crosses of resistant varieties. However, when Moukoto was used as a parent, no transgressive segregation was observed.

TABLE 4. Segregation of leaf blast scores in F5 families

Cross and parents	Number of F5 families with leaf blast score											Mean score
	0	1	2	3	4	5	6	7	8	9	10	
Sensho/Moukoto				3	15	23	27	27	23	22	7	6.71
Sensho			2	4	2							3.00
Moukoto										4	4	9.50
Mid-parent												6.10
Ouu 247/Sensho	1	8	20	23	21	12	12	12	10	1		4.22
Ouu 247						1	3	2				6.16
Sensho			2	5	1							2.88
Mid-parent												4.52

Leaf blast score of a family is a mean of two lines.

Eight and 7-8 effective factors were estimated in Ishioka-mochi 18/Moukoto and Sensho/Moukoto, respectively. In both crosses of Ouu 247/Moukoto and Reimei/Moukoto the effective factors were estimated to be 5-6. Moreover, 5, 6-7, 5-6, 4-5, 5-6 and 4 effective factors were estimated in Ishioka-mochi 18 and Ouu 247/Reimei, respectively.

4.3. DISCUSSION

No transgressive segregations were observed in the crosses with Moukoto as one parent. Because the most susceptible variety Moukoto did not have any resistance genes effective in the nursery test, or the resistant parents also have the same genes even with genes giving small effects. Moukoto had no resistance genes affecting the segregations, therefore, the number of effective factors estimated in these crosses are effective factors from the resistant parents. From this result, it can be concluded that Ishioka-mochi 18, Sensho, Ouu 247 and Reimei have at least 8, 7-8, 5-6, 5-6 effective factors, respectively. On the other hand, transgressive segregations in both directions were observed in the crosses between resistant parents. The additive genes on different loci of both parents may be accumulated, resulting in transgressions.

Ten effective factors were estimated in total from the evidence that Ishioka-mochi 18 had eight, Sensho had seven, and five factors were estimated from the cross between them. The symbols a- j are temporarily assigned to these effective factors for the explanation (Table 5). The effective factors existing together in both varieties are five (a, c, d, e, f). These five factors therefore do not affect the segregation of resistance in the cross between them. The factors present only in Ishioka-mochi 18 are three (b, i, j) and those in Sensho only are two (g, h), and it is these five effective factors (b, g, h, i, j) that will affect the segregation in Ishioka-mochi 18/Sensho. The effective factors can be accumulated up to 10 in this cross including the effective factors which both varieties have together. In addition, relationships of the effective factors in all the varieties used for this experiment can be explained by adding another effective factor (k).

As the major effects of the field resistance genes are additive as shown in Experiment 1, the differences of the leaf blast scores of resistant varieties and Moukoto are the total effects of the genes of resistant varieties. As shown in Table 5, assume effect of 3.0 for the effective factor (a) when the factor is homozygous, genotype (aa) and 2.0 for b, 1.0 for c and d, 0.6 for e, 0.2 for f, g, h, i, j and k (for example c+d=2.0 and g+h=0.4). Then, total effect value (expected value) of the factors affecting the resistance of Ishioka-mochi 18 is 8.2 which approximates the observed value of 8.7; the difference between Ishioka-mochi 18 and Moukoto. The expected and observed values of the total effects are calculated in the same say for the other resistant varieties.

TABLE 5. Mutual relation of effective factors among resistant varieties

Variety or cross	Estimated number of effective factors	Effective factors and their effects											Total effect [2]	
		a	b	c	d	e	f	g	h	i	j	k		
		3.0	2.0	1.0	1.0	0.6	0.2	0.2	0.2	0.2	0.2	0.2	Exp.	Obs.
		Effective factors of resistant varieties [1]												
Is. 18	8	O	O	O	O	O	O			O	O		8.2	8.7
Sensho	7-8	O		O	O	O	O	O	O				6.2	6.5
Ouu 247	5-6			O	O	O				O		O	3.0	3.3
Reimei	5-6			O	O		O				O	O	2.6	2.1
		Differences of effective factors between parents												
Is. 18/Sen.	5		*					*	*	*	*		2.8	2.2
0.247/Is.18	4-5	*	*				*				*	*	5.6	5.7
Reim./Is	5-6	*	*			*				*		*	6.0	6.0
0.247/Sen.	6-7	*					*	*	*	*		*	4.0	3.3
Reim./Sen.	5-6					*		*	*		*	*	4.4	4.3
0.247/Reim.	4					*	*			*	*		1.2	0.8

[1] Moukoto is supposed to have no effective factor affecting segregation.

[2] Expected effect : Total effect value of effective factors concerned.
Observed effect : Difference of blast scores between Moukoto and resistant variety, or between parents.

In the crosses of other resistant varieties, the effects of the expected and the observed values reached the approximated values. Assuming eleven effective factors and their effects of 3.0- 0.2, the number of effective factors and the levels of field resistance of Ishioka-mochi 18, Sensho, Ouu 247, Reimei and Moukoto can all be explained without contradiction.

5. Chromosomes on which field resistance genes of upland rice Sensho located.

5.1. MATERIALS AND METHODS

Sensho was crossed with twelve linkage testers having morphological marker genes provided by Hokkaido University. Field resistance to leaf blast and segregation of the markers was investigated using F_3 lines randomly selected (300-400 lines per cross). This experiment was also conducted at Omagari, from 1979 to 1982.

The field resistance of the F_3 lines of eleven crosses having no true resistance gene (+) or *Pi-a* gene were tested in a blast nursery. The field resistance of the F_3 lines of H-232/Sensho were tested in the greenhouse by inoculating race 007 (Naga 69-150) of blast fungus to the seedlings at the six-leaf stage, as the tester H-232 had the *Pi-i* gene. The leaf blast score was recorded in 11 classes of 0-10 according to the standard (Table 1), both in the nursery test and greenhouse test. Average scores of the F_3 lines were judged by observation, though many individual segregations of resistance in the lines were observed. The same materials were grown in a paddy field by direct sowing, and the presence and segregation of the marker genes were scored at maturing time. For the phenol reaction (*Ph*) of seed glumes, one of the marker traits, seeds of the F_3 lines were soaked in 2% phenol liquid and the staining of the glumes was examined. Brown pericarp (*Rc*) was investigated using hulled grains in a laboratory. The segregation of the glutinous rice character (*wx*) was investigated using the iodine reaction of grains of the F_2 individuals. All other traits were investigated in the field. Because Sensho has the *Ph* gene, the author investigated the phenol reaction by the cross with H-337 one of the tester without *Ph* gene.

The F_3 lines were classified into three groups, tester type fixed lines, Sensho type fixed lines, and segregating lines, for the marker genes. The difference of the average value of the leaf blast score of each group was determined using a *t*-test. There is a linkage relation between the marker gene locus and the field resistance gene (or genes) when the difference between the tester type group and the Sensho type group was significant. The gene symbols and chromosomes (linkage groups) followed the arrangement of Kinoshita (1993) and Kinoshita *et al.*, 1991.

5.2. RESULTS

Linkage relations were found between the field resistance to leaf blast and the marker loci of the testers on *bl-1* (brown leaf spot-1, chromosome 2) of H-68 (Table 6), *d-2* (ebisudwarf, chrom. 4), *lg* (liguleless, chrom. 4), *la* (lazy growth habit, chrom.11) and *bc-1* (brittle culm-1, chrom. 3) of H-79, *d-6* (ebisumochi, chrom. 7) of H-84, *d-14* (kamikawa tillering dwarf, chrom. 3) of H-147, *Ph* (phenol staining, chrom.4) of H-337, *sp* (short panicle, chrom. 11) of H-452 and *lax* (lax panicle, chrom.1) of H-453.

The linkage relations with resistance were observed with 10 marker loci which belonged to six linkage groups (chromosomes) as shown in Table 7. Among them, though the resistance of the tester type group was higher than that of the Sensho type group in *lax* (chrom. 1), the resistance of the tester type groups. The average values of the blast scores of the segregating groups were either midway between the tester type groups and the Sensho type groups, or shifted toward one side.

In contrast, no linkage relations were observed between the field resistance to leaf blast and the marker loci *ri* (verticillate rachis, chrom. 5) of H-68 (Table 6), *Rc* (brown pericarp, chrom. 7) and *I-Bf* (inhibitor for brown furrows, chrom. 9) of A-5, *wx* (glutinous endosperm, chrom. 6), *fs-1* (fine stripe-1, chrom. 6) and *bc-1* (brittle culm-1, chrom. 3) of H-84, *nl-l* (neck leaf-1, chrom.5) and *gl-1* (glabrous leaf and hull, chrom. 5) of H-103, *d-6* (ebisu-mochi, chrom.7) and *Hg* (hairy glume, chrom.3) of H-126, *Dn-1*

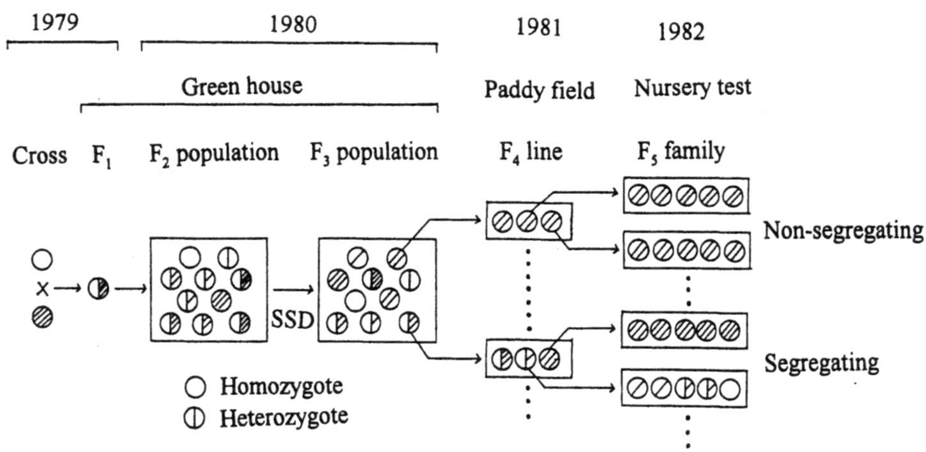

Figure 4. Method of growing materials for estimating number of effective factors.

(dense panicle-1, chrom.9) of H-147, *Ur-1* (undulated rachis-1, chrom. 6) and *Dn-1* of H-232, *Cl* (clustered spikelets, chrom. 6) of H-337 and *st-2* (stripe-2, chrom. 5) and *gh-1* (gold hull and internode-1, chrom. 5) of H-143. That is, no linkage relation with resistance was observed for 15 marker gene loci of five chromosomes (Table 7). However, the linkage is as mentioned above observed for other crosses, for *d-6* of the chromosome 7, and *Rc* also belongs to the same chromosome.

Moreover, the linkage of *bc-1* for the chromosome 3 was observed in another cross, and the linkage was observed for *d-14* of the same chromosome. *Hg* also belongs to the same chromosome. Therefore, chromosomes for which resistance genes are not observed are three (chrom. 5, 6 and 9).

TABLE 6. Relations between leaf blast resistance and marker genes in the F3 lines of the cross of H-68/Sensho

Marker gene (Chromo-some)	Pheno-type of F3 line	Number of F3 lines with leaf blast score										Mean score	Difference of Means[1]	
		1	2	3	4	5	6	7	8	9	10			
ri	*ri*		3	10	16	11	13	10	11	5	5	5.79	0.07	0.22
(5)	*ri*/+		6	23	46	27	34	29	23	14	9	5.72	0.15	
	+		3	10	14	16	15	4	9	6	3	5.57		
bl-1	*bl-1*			2	8	5	14	16	16	14	14	7.34	2.01**	2.39**
(2)	*bl-1*/+		5	24	46	30	31	21	20	8	1	5.33	0.38	
	+		7	17	22	19	17	6	7	3	2	4.95		
h-68	(*ri*, *bl-1*)								2	4	14	9.60	6.85	
Sensho	(+, +)		6	13	1							2.75		

[1] The values of 0.07, 0.15, 0.22 are differences of mean scores between *ri* and *ri*/+, *ri*/+ and +, *ri* and + respectively. So is *bl-1*.

5.3. DISCUSSION

Nine chromosomes among the 12 chromosomes of rice were examined here. Linkage relations between the field resistance to leaf blast and the marker gene loci was observed in six chromosomes: chromosome 1 (*lax*), 2 (*bl-1*), 3 (*bc-1, d-14*), 4 (*d-2, lg, Ph*), 7 (*d-6*) and 11 (*la, sp*). All resistance genes on these chromosomes are on the Sensho side except for chromosome 1 (*lax*). On the other hand, no linkage was observed in three chromosomes: chromosome 5 (*gh-1, st-2, ri, nl-1, gl-1*), 6 (*wx, fs-1, Cl, Ur-1*), and 9 (*Dn-1, I-B*). These marker gene loci are shown on the linkage map (Kinoshita, 1993) in Figure 5.

Goto (1970) assumed three incompletely dominant genes in Sensho by sheath inoculation method, and found that one of the loci was linked with *la* locus (chrom. 11) and later named it as *Pi-se* (Goto and Buluch, 1983), which agrees with the result of this research. However, linkage relations between blast resistance and the *lg* locus (chrom. 4) or *bc-1* locus (chrom. 3) were not observed in his research in contrast to the result of this research. As for the chromosome 4, two genes controlling the high field resistance of Juroka were located on it (Shinoda *et al.*, 1971). Moreover, a part of resistance of

TABLE 7. Linkage relations between field resistance of Sensho to leaf blast and marker loci, and effects of the resistance genes

Chromosone	Marker gene	Linkage between marker locus and resistance[1]	Effect of resistance gene linked with marker locus		
			Additive[2]	Dominance[2]	Tester
1	*lax*	**	-0.86	-	H-453
2	*bl-1*	***	2.39	+	H-68
3	*bc-1*	*	0.62	±	H-79
	bc-1	ns	(0.55)		H-84
	d-14	***	1.79	+	H-147
	Hg	ns	(0.23)		H-126
4	*d-2*	*	0.62	-	H-79
	lg	***	2.03	±	H-79
	Ph	***	1.73	+	H-337
5	*gh-1*	ns	(0.05)		H-143
	st-2	ns	(-0.01)		H-143
	ri	ns	(0.22)		H-68
	nl-1	ns	(-0.08)		H-103
	gl-1	ns	(-0.53)		H-103
6	*wx*	ns	(0.35)		H-84
	fs-1	ns	(0.13)		H-84
	Cl	ns	(-0.17)		H-337
	Ur-1	ns	(0.20)		H-232
7	*d-6*	*	0.69	-	H-84
	d-6	ns	(-0.02)		H-126
	Rc	ns	(0.08)		A-5
9	*Dn-1*	ns	(0.04)		H-232
	Dn-1	ns	(0.01)		H-147
	I-Bf	ns	(-0.16)		A-5
11	*la*	**	1.09	+	H-79
	sp	***	0.89	+	H-452

[1] ***, **, * : Significant at 0.1%, 1% and 5% levels, respectively

[2] Additive effect: Difference of the means of blast scores between tester type group and Sensho type group on the marker gene.

[3] + : Resistance is dominant, ± : No dominance, - : Resistance is recessive.

upland rice Norin-mochi 4 is reported to be linked with *Ph* gene located on the chromosome 4 (Kikuchi *et al.*, 1976; Maruyama *et al.*, 1983). As Kuroka and Upland Rice Norin-mochi 4 derived both from Sensho, the resistance gene(s) that located on chromosome 4 of three varieties might be the same gene(s). For the *bc-1* locus, a linkage relation with resistance was observed in H-79/Sensho but not in H-84/Sensho. The reason for this may be that the linkage is not tight. However, because the linkage relation with resistance is observed for the *d-14* locus of the same chromosome as the *bc-1* locus, the existence of a resistance gene in chromosome 3 is certain. Fukuoka *et al.* (1994) clarified by RFLP analysis that the field resistance genes with large effects of upland rice Owarihatamochi occur in chromosomes 4 and 11 in which the author also observed linkage relations.

Wang *et al.* (1994) conducted an RFLP analysis on partial resistance to blast using progenies of the cross between Moroberekan; highly resistant japonica type upland variety in west Africa, and CO 39; susceptible indica variety. They reported that 10 chromosome fragments in eight chromosomes affected partial resistance (almost synonymous with field resistance) of Moroberekan. That is, the RFLP markers on chromosome 1, 3, 5, 6, 7, 8, 11 and 12 were presumed to affect the partial resistance, and two chromosome fragments were presumed in each chromosome 1 and 6 among them. For four chromosomes of the remainder (chromosome 2, 4, 9, 10), no linkage to partial resistance was detected. In Sensho and Moroberekan, chromosome 3, 7 and 11 are common for the field resistance. Chromosomes 9 and 10 only have not yet been implicated in field resistance (partial resistance) of rice.

6. General Discussion

The mode of inheritance of field resistance to leaf blast was examined in various ways. Field resistance is controlled by multiple genes and/or polygenes with additive effects and some partial dominance effects. Estimated heritabilities are high. The number of effective factors (genes) of each resistant variety or line for leaf blast resistance are from several to ten, based on the score of 0.5 (the minimum varietal difference) that can be detected by observation in a nursery test. Moreover, the factors (genes) have various levels of effects from an average effect 0.2 calculated in Ouu 247/Reimei to 3.0, the maximum value of factors that Ishioka-mochi 18 and Sensho share. The number of effective factors (genes) and their effects calculated here is based on the segregation and linkage that can be detected in a nursery test. Therefore they are not necessarily the ultimate genetic units and there are also possibilities that they are gene groups of linked loci. However, it was analyzed under the same condition as the testing method used in blast resistance breeding and the information provided here could be applied directly in actual rice breeding. The gene effects were calculated to range from comparatively large to small. Though it is not appropriate to classify them as major or minor genes, they belong to the category of minor gene compared with the large effect of a true resistance gene.

Though there is no assurance that field resistance will not break down, it is certain that field resistance is more stable than true resistance and it will play an important role in blast resistance breeding in the future. The field resistance controlled by a major gene also was reported (Toriyama *et al.*, 1968; Inoue *et al.*, 1987; Mikami *et al.*, 1990).

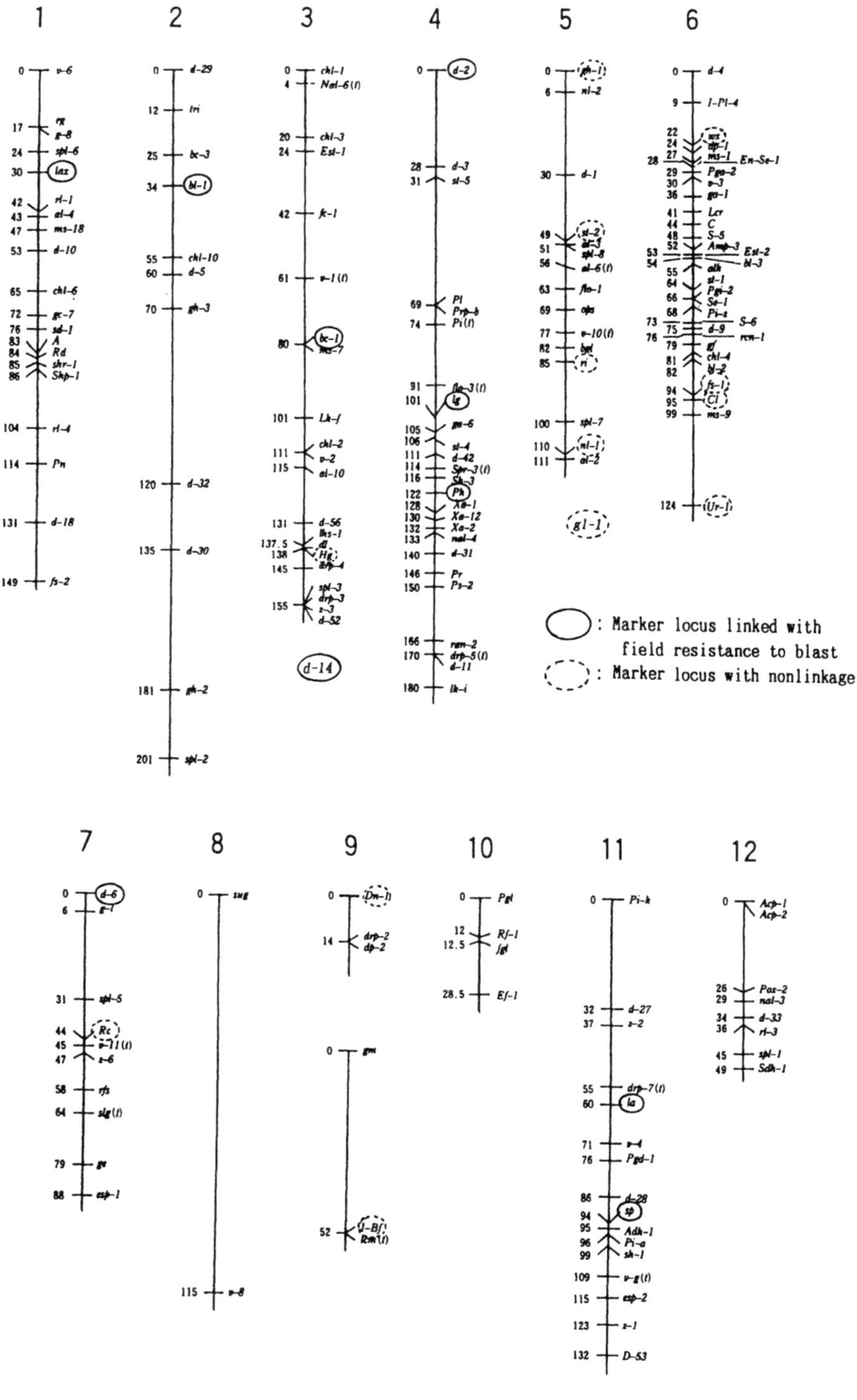

Figure 5. Linkage map of marker gene loci linked with field resistance to leaf blast (after Kinoshita, 1993).

They are special examples, which are genetically near the true resistance controlled by major gene, and there is a possibility of breakdown of the resistance though they can be called field resistance from symptoms. Stable resistance will be expected as a whole even if there is race specificity in an individual gene of the field resistance controlled by polygenes and multiple genes with the additive effects as clarified by the present study.

7. Summary

In a diallel analysis of leaf blast, the additive effects of genes were stronger than dominance effects and no epistasis was observed. Heritability was very high. Progeny lines in two crosses were normally distributed between parents for leaf blast. Field resistance to leaf blast was found to be controlled by several genes and/or polygenes with additive effects. Effective factors of 7-8, 8, 5-6 and 5-6 for leaf blast resistance were estimated in Sensho, Ishioka-michi 18, Ouu 247 and Reimei respectively. Linkage relations between blast resistance of upland rise Sensho and marker loci were observed in chromosome 1, 2, 3, 4, 7, and 11. Five chromosomes, excluding chromosome 1, affected the resistance of Sensho.

8. References

Abe, S., Kiyosawa, S. and Ono, S. (1974) Inheritance of blast resistance of upland rice varieties. 1. Inheritance of an upland Norin-mochi 4 (in Japanese), *Bulletin of Ibaraki Agricultural Experiment Station* **15**, 47-64.

Abe, S., Suga, T. and Ono, S. (1976) Inheritance of blast resistance of upland rice varieties. 2. Inheritance of an upland variety Norin-mochi 26 (in Japanese), *Bulletin of the Ibaraki Agricultural Experiment Station* **17**, 67-76.

Asaga, K. (1976) Investigation standard of leaf blast in blast nursery (in Japanese), *Nogyogijyutu Journal of Advanced Agriculture Science* **31**, 156-159.

Asaga, K. (1981) A procedure for evaluating field resistance to blast in rice varieties (in Japanese, English summary), *Journal of the Central Agricultural Experiment Station* **35**, 51-138.

Fukuoka, S., Schlater, A. R., Nakayama, H., Okuno, K., Kawase, M., Miura, K., Ohta, H., Kiyosawa, S. and Furutani, T. (1994) Genetic analysis for rice blast field resistance using RFLP markers (in Japanese), *Japan Journal of Breeding* **44** (Suppl.1), 233.

Goto, I. (1970) Genetic studies on the resistance of rice plant to the blast fungus. 1. Inheritance of resistance in crosses Sensho × H-79 and Imochi-shirzau × H-79, *Annals of the Phytopathological Society of Japan* **36**, 304-312.

Goto, I. and Buluch, A. A. (1983) Genetic studies on resistance of rice plant to blast fungus. VI. Additive effect of blast resistance genes of Sensho under natural infection (in Japanese, English summary), *Bulletin of the Yamagata University, Agricultural Science* **9**, 273-283.

Hayman, B. I. (1954) The analysis of variance of diallel crosses, *Biometrics* **10**, 235-244.

Higashi, T. and Kushibuchi, K. (1978) Genetic analysis of field resistance to leaf blast (*Pyricularia oryzae*) in rice (in Japanese, English summary), *Japan Journal of Breeding* **28**, 277-286.

Higashi, T. and Saito, S. (1985) Linkage groups of field resistance genes of upland rice variety Sensho to leaf blast caused by *Pyricularia oryzae* Cav. (in Japanese, English summary), *Japan Journal of Breeding* **35**, 438-448.

Higashi, T. (1995) Genetic studies on field resistance of rice to blast disease (in Japanese, English summary), *Bulletin of the Tohoku National Agriculture Experimental Station* **90**, 19-75.

Inoue, M., Saito, S. and Ikeda, R. (1987) Genetic analysis of field resistance of Etsunan 108 to rice blast (in Japanese), *Japan Journal of Breeding* **37** (Suppl. 1), 300-301.

Jinks, J. L. and Towey, P. (1976) Estimating the number of genes in a polygenic system by genotype assay, *Heredity* **37**, 69-81.

Jones, R. M. (1965) Analysis of variance of the half diallel table, *Heredity* **20**, 117-121.

Kikuchi, F., Yokoo, M. and Niitsuma, Y. (1976) Inheritance of blast resistance and its relationship with other characters in a cross between lowland- and upland-rice (in Japanese), *Japan Journal of Breeding* **26** (Suppl.2), 205-206.

Kinoshita, T. (1993) Report of the committee on gene symbolization, nomenclature and linkage groups, *Rice Genetics Newsletter* **10**, 7-39.

Kinoshita, T., Futsuhara, Y., Khush, G. S., Iwata, N., Oka, H. I., Takeda, K., Wu, H. K., Yano, M. and Yoshimura, A. (1991) Report of the meeting to discuss the chromosome numbering system in rice, 15 May 1990. *Rice Genetics* **II,** IRRI p.816-818.

Kiyosawa, S. (1970) Inheritance of blast resistance of the two rice varieties Homare Nishiki and Ginga. 1. Resistance of Homare Nishiki and Ginga to the fungus strain Ken 54-04, *Bulletin of the National Institute of Agricultural Science* **D21**, 73-105.

Maruyama, K., Kikuchi, F. and Yokoo, M. (1983) Gene analysis of field resistance to rice blast (*Pyricularia oryzae*) in Rikuto Norin Mochi 4 and its use for breeding (in Japanese, English summary), *Bulletin of the National Institute of Agricultural Science* **D35**, 1-31.

Mather, K. (1949) *Biometrical Genetics*, Methuen, London p.102-125.

Mather, K. and Jinks, J. L. (1971) *Biometrical Genetics*, 2nd Edition, Chapman and Hall, London p. 249-284.

Mikami T, Kawamura Y, Horisue N (1990). Estimation of resistance genes and field resistance to leaf blast of a rice cultivar "Fukei 138" (in Japanese), *Report of the Tohoku Branch, Crop Science Society of Japan* **33**, 87-88.

Roumen, E. C. (1994) The inheritance of host plant resistance and its effect on the relative infection efficiency of *Magnaporte grisea* in rice cultivars, *Theoretical and Applied Genetics* **89**, 498-503.

Shinoda, H., Toriyama, K., Yunoki, T., Ezuka, A. and Sakurai, Y. (1971) Studies on the varietal resistance of rice to blast. 6. Linkage relationship of blast resistance genes (in Japanese, English summary), *Bulletin of the Chugoku National Agriculture Experimental Station* **A20**, 1-25.

Toriyama, K., Yunoki, T. and Shinoda, H. (1968) Breeding rice varieties for resistance to blast. II. Inheritance of high field resistance of Chuugoku No.31 (in Japanese), *Japan Journal of Breeding* **18** (Suppl.1), 145-146.

Wang, G., Mackill, D. J., Bonman, J. M., McCouch, S. R., Champoux, M. C. and Nelson, R. J. (1994) RFLP mapping of genes conferring complete and partial resistance to blast in a durably resistant rice cultivar, *Genetics* **136**, 1421-1434.

Wang, Z., Mackill, D. J. and Bonman, J. M. (1989) Inheritance of partial resistance to blast in indica rice cultivars, *Crop Science* **29**, 848-853.

Yokoo, M., Kikuchi, F. and Kawai, M. (1974) Preliminary test for blast resistance in a cross "Koshihikari ^{x} Glutinous Upland Norin 4" of rice (in Japanese), *Japan Journal of Breeding* **24**, 146-147.

GENETIC ANALYSIS OF TWO COMPONENTS OF FIELD RESISTANCE, LESION SIZE AND NUMBER, TO RICE BLAST IN RICE VARIETIES, NIPPONBARE AND JUMA

M. S. ALI[1,a], D. PURBA[1,b], M. KAWASE[2,c], K. OKUNO[2] and S. KIYOSAWA[1,d].
[1]*Tsukuba International Agricultural Training Centre, Tsukuba 305, Japan.*
[2]*National Institute of Agrobiological Resources, Tsukuba 305, Japan.*

Present address
[a]*Plant Breeding Division, Bangladesh Rice Research Institute, Gazipur 1701, Bangladesh.*
[b]*Directorate of Food Crops Production Development, Jl. AUP 3, Pasar Minggu, Jakarta Selatan, Indonesia.*
[c]*Shikoku National Agricultural Experiment Station, Zentsuji 765, Japan.*
[d]*Sakuragaoka 31-8, Kukizaki, Inashiki, Ibaraki 300-12, Japan.*

1. Introduction

Disease resistance was divided into true resistance and field resistance (Muller and Haigh, 1953). Field resistance has been used for resistance to potato late blight and rice blast diseases. In rust of wheat and barley, resistance has been divided into seedling resistance and adult-plant resistance which are similar to true and field resistance, respectively. Partial resistance and quantitative resistance also have been used for low levels of resistance.

Genetic studies of true resistance to rice blast in Japanese varieties are nearly completed. To date, 14 genes for true resistance were found. Among them *Pi-a* and *Pi-i* were found in Japanese *japonica* varieties and in some *indica* varieties, *Pi-k* and $Pi\text{-}k^{m}$ were found in Chinese varieties, *Pi-ta* and $Pi\text{-}ta^{2}$ were in Philippine (*indica*) variety, *Pi-z* in a United States variety, and $Pi\text{-}k^{p}$, $Pi\text{-}z^{t}$, *Pi-b* and *Pi-t* in *indica* varieties (Kiyosawa, 1974, 1976). *Pi-sh* was found in Japanese varieties (Imbe and Matsumoto, 1985). $Pi\text{-}k^{s}$ was found in Japanese varieties only with a Philippine fungus strain (Kiyosawa, 1969). All genes controlling true resistance found were of a race specific nature and were major genes (genes of large effect). Genetic studies on field resistance have been conducted since 1967, and are increasing (Purba *et al.*, 1994). Genes controlling field resistance are minor genes. In some varieties, two or more minor genes were found by the Mendelian method of gene analysis, and many minor genes or polygenes were found by a method of statistical genetics (Purba *et al.*, 1994). Kiyosawa (1970), Abe *et al.* (1974) and Purba *et al.* (1994) estimated the functional values of genes controlling field resistance by the cumulative frequency

S. Sreenivasaprasad and R. Johnson (eds.), Major Fungal diseases of Rice Recent Advances, 181-199.
Professor K. Manibhushan Rao Festschrift

distribution curve method. This method was applied to a field test for gene analysis (He *et al.*, 1989a, b). In almost all cases, accumulation of minor genes resulted in additive effects. In a few cases, complementary gene action was observed in which resistance was not expressed by single genes but was expressed in combination. Particularly, some *japonica* and *indica* have complementary effects relating to minor genes for blast resistance (Purba *et al.*, 1994).

Field resistance has been divided into three components: infection frequency, latent period and sporulation capacity (Hwang and Heitefuss, 1982; Asher and Thomas, 1983, 1984, 1987); latent period, disease severity and sporulation (Cunfer *et al.*, 1988), or infection frequency (uredinium density); latent period and urediosorus (uredinium) size (Broers, 1989 a, b; Pretorius *et al.*, 1994). Shew *et al.* (1989) used four components: lesion number, the percentages of sporulating lesions, lesion diameters and conidial production per sporulating lesion. This indicates that there are at least four components of resistance: infection frequency, latent period, lesion size and sporulating capacity. In general, field resistance of Japanese varieties is low, but there are many *indica* varieties that have high field resistance (Kiyosawa *et al.*, 1986). In the present paper, the results of genetic analysis in the F_3 generation of a hybrid between Nipponbare (*japonica*) having field resistance that inhibits the increase of lesion number and Juma (*indica*) having field resistance that inhibits the increase of lesion size are reported.

2. Materials and Methods

Parents were tested with *Magnaporthe grisea* strains, P-2b, Ken 53-33, Ina 72, Hoku 1, Ken 54-20, 231-1, Kaminoho 33, 24-11-1-1, Niigatawase 2-18, Myojin D27 and Myojin D31 (Table 1). All tests were carried out in the greenhouse maintained between 22° C and 35° C, except for daily and seasonal changes of natural environmental conditions.

F_3 lines of Nipponbare x Juma were used for genetic analysis. Tests were replicated twice. Inoculation was performed on 15 June and 27 October, respectively. In Test 1, one hundred and sixteen F_3 lines (17 plants in each line) were sown in 12 wooden boxes (45 x 15 x 10cm). Two rows for each parent were also accommodated in two separate boxes. In Test 2, 98 F_3 lines that were different from lines in Test 1, F_1 plants and parents were sown. At stage 6.0 for test 1 and 4.0 for test 2, test plants were inoculated with blast fungus strain Ken 54-20 (race number 003.0) (Kiyosawa, 1984). Inoculated plants were kept in an incubation chamber at 28° C and 100 per cent RH for 24 hours. After that, the boxes were placed in a growth chamber maintained at 28° C and about 82 per cent RH. Disease evaluation was conducted 7 days after inoculation on an individual plant basis following Kiyosawa's method (Kiyosawa, 1976).

The distribution of parental reactions overlapped as shown in Table 2 (for number of lesions) and Table 3 (for type of lesions which were divided into four: b (brown spot), bg (small lesion having gray center and brown margin), bG (large lesion having gray center and brown margin) and pG (large lesion having gray center and no or purple margin). There

TABLE 1. Resistance of parents to 11 fungus strains

Variety	Fungus strain (race number)					
	P-2b (303.1)	Ken 53-33 (137.1)	Ina 72 (031.0)	Hoku 1 (007.0)	Ken 54-20 (003.0)	231-1 (033.3)
Nipponbare	S	S	S	S	S	S
Juma	R	R-	R	R-	R	MR

(continued)

Variety	Fungus strain (race number)				
	Kaminoho 33 (301.0)	24-11-1-1 (173.3)	Niigatawase 2-18 (047.0)	Myojin D27 (073.1)	Myojin D31 (077.1)
Nipponbare	S	S	S	S	MS
Juma	M	MR	MR	MS	MS

were significant differences in the number of lesions between parents. Significant differences were also recognized for predominant lesion type between parents. Therefore, individual plants in the F_3 lines were first divided into two groups based on the number of lesions: plants having a) less than five lesions and b) five or more lesions. The former group was again divided into two classes based on the number of lesions: R (no lesions) and N (Nipponbare type, with one to four lesions). The latter was divided into two classes based on predominant lesions: J (Juma type, having bg as predominant lesions) and S (susceptible type, having bG as predominant lesions). Individual plants of parents and the F_3 lines were classified into four groups: R, N, J and S (Table 4). R, N, J and S mean the number of plants showing four types in each line. Then, the ratio of R : N : J : S was obtained for each F_3 line including parents. F_3 analyses were, therefore, carried out on three ratios, R : (N + J + S), (R + N) : (J + S) and (R + N + J) : S. In each analysis, these segregations were analysed as r : s ratio. Here r means resistant and s means susceptible. Accordingly, the number and functional values of genes controlling R, R + N and R + N + J responses were estimated, respectively, in one inoculation test.

TABLE 2. The number of lesions formed on two parents

Variety		Number of plants with the number of lesions described											
		Number of lesions											
		0	5	10	15	20	25	30	35	40	45	Mean	Resistance
Nipponbare	a	17										1.18	M[1]
	b	15	1									1.82	M[1]
Juma	a	3	2	6	2	3			1			15.76	M
	b	1	8	1	4	2						10.88	M
Nipponbare	c	4	3	4	3			1				10.00	S
Juma	c	2	2	3	7				1			16.71	MR

Note: a and b are the result in this experiment, and c in other experiment.
[1] The resistance was evaluated R^h by Kiyosawa *et al.* (1981), and is evaluated M based on the predominant lesions in this test.

TABLE 3. The number of lesions belonging to each type on Nipponbare and Juma in two blocks in Test 1

		Number of lesions in type					Response
		b	bg	bG	pG	Total	
Nipponbase	a	.66	.12	.89		1.67	M
	b	.19	.81	.75	.06	1.81	M
Juma	a	5.59	4.47	5.71		15.76	M
	b	.81	6.57	3.50		10.88	M
Susceptible line in F_3	a	.20	3.12	5.47	3.65	12.44	MS
	b	.76	3.94	13.12	1.12	18.94	MS

a and b mean average of varieties in different boxes for Nipponbare and Juma, and average of different lines in different boxes for susceptible line in F_3

Generally, hybrids between *indica* and *japonica* show a high degree of hybrid sterility in F_1 and F_2 plants. In 1993 when F_2 plants were grown weather was very cool, which tended to increase sterility. The percentages of sterility of F_1 and F_2 plants were 67% and between 2 and 95%, respectively. Color of basal leaf sheath was examined to determine the influence of sterility on segregation ratios for resistance and plant color. Obtained relationships of these characters are shown in Table 5. No relationship was found between resistance and sterility, although statistically significant deviation ($\chi^2 = 5.38^*$) from random association was found between sheath color and resistance (R + N resistance). Segregation for blast resistance was not skewed by seed sterility.

TABLE 4. Ratio of four response types of two parents in Test 1

Variety	Experiment	Response types [1)] R	N	J	S	Total
Nipponbare	a	5	12			17
Nipponbare	b	4	11	1	1	17
Juma	a		2	15		17
Juma	b		3	11	2	16
Nipponbare	c	1	1	1	12	15
Juma	c	2	1	14		17

1): Classification into R, N, J and S is not always suitable for Expt. c. For comparison with F_3 analyses, response of plants was classified by the same method. Concentrations of spores of suspension inoculated are 100,000/ml for a and b and 380,000/ml for c.

The number of plants in disease resistance tests was limited by greenhouse space and inoculation chamber size. However, when evaluating the same F_2 lines in field tests, it was very difficult to obtain similar results. The number of resistance genes that can be detected frequently changes from test to test (Kiyosawa, 1970) due to variation in the environment. In this method, 116 lines with 17 plants per line were grown generally for analysis. Because only 17 plants per line were used, it was impossible to differentiate 1 : 0 segregation ratio from 15 : 1 and even to differentiate 3 : 1 from 15 : 1, if there were only major genes. Therefore, the cumulative frequency distribution curve method (Kiyosawa, 1974, 1976), which can be used without differentiating these segregation ratios, was used for genetic analysis.

TABLE 5. Relationship among blast resistance, color of plant base and seed sterility in Test 1

Character	^{2} in uniformity test: D. Purple base	E. Seed sterility
A. R type resistance	1.60	0.59
B. R + N type resistance	5.38*	1.48
C. R + N + J type resistance	0.21	1.28
D. Purple base		1.11

* means significance at the 5% level.

At first, observed frequencies (Table 6) of resistant plants per 17 plants were plotted against the number of resistant plants per line to obtain the frequency distribution curve shown in Figure 1. These frequencies were added from left to right to obtain a cumulative frequency distribution curve. To obtain expected frequency and cumulative frequency distribution curves, one gene was at first assumed for each of the parents: AA in Nipponbare and BB in Juma.

TABLE 6. Distribution of the number of resistant plants per line in Test 1

Number of resistant plants in line	The number of lines in segregation ratio of		
	R : (N + J + S)	(R + N) : (J + S)	(R + N + J) : S
0	44 J 0	9	0
1	15	7	1
2	20	9	0
3	12	11 J .1515	1
4	5	9	1
5	3 N .2647	8	2
6	1	4	2
7	2	8	3
8	3	3	3
9	1	8	2
10	2	11	5
11	0	8	4
12	0	3	8
13	0	4	11
14	0	2	9
15	0	2	17
16	0	1 N .9411	24 J .9394
17	0	1	15 N .9706
Total	108	108	108

N and J are the frequencies of resistant plants in two parents, Nipponbare and Juma.

These two genes produce the following genotypes and phenotypes at respective frequencies in the F_2 generation:

Genotypes	AABB,	AAB+,	AA++,	A+BB,	A+B+,	A+++,	++BB,	++B+,	++++
Phenotypes	r_A	r_A	r_A	r_A	r_A	r_A	r_B	r_B	s
Frequencies	1/16	2/16	1/16	2/16	4/16	2/16	1/16	2/16	1/16.

These F_2 plants show F_3 segregations of

genotypes	1:0	3:1	1:0	3:1	F_2seg	3:1	1:0	3:1	0:1,
phenotypes	r_A	r_A	r_A	r_A:r_B	r_A:r_B:s	r_A:s	r_B	r_B:s	s ,

if gene A is epistatic over gene B, and both genes are dominant. Here, r_A and r_B are phenotypes of genes A and B, respectively. F_2seg means the same segregation as in the F_2. Frequencies of resistant plants (functional values) in parents (9/34 = 0.2647 and 0/33 = 0 in Table 6) were given to genotypes, AA++ and ++BB, respectively, for R (r) in the R : (N + J + S) segregation. For other genotypes, various functional values were given in a range of the following restrictions on dominance and epistasis.

For A gene, AA $\geq$ A+ $\geq$ ++
For B gene, BB $\geq$ B+ $\geq$ ++
For A and B genes, AABB $\geq$ AA++ or $\geq$ ++BB
(here, AA means functional value of genotype AA).

These functional values were given to all genotypes in F_3 generations and frequencies of r plants in F_3 generation for individual F_2 plants were calculated. A binomial distribution of resistant plants in 17 F_3 plants sown for each F_2 genotype was calculated based on the following equation:
$f\,{}_{17}C_r\, p^r (1\text{-}p)^{(17\text{-}r)}$ (f is frequency in F_2, C is combination, r is the number of resistant plants in 17 plants and p is frequency of resistant plants in F_3 generation per F_2 genotype).

By adding frequencies in the distribution curves, an expected frequency distribution curve was drawn as shown in Figure 1. An expected cumulative frequency distribution curve was obtained by adding frequencies in the frequency distribution curve from left to right. Among cumulative frequency distribution curves obtained by giving various functional values to various genotypes, a curve with minimum difference between observed and expected curves was finally selected. Significance of the difference was tested by Kolmogorov-Smirnov method (Kiyosawa, 1974, 1976). When the difference was statistically significant under the assumption of two genes, the possibility of three genes was tested. In this case, two possibilities were considered: AACC in Nipponbare and BB in Juma, and AA in Nipponbare and AACC in Juma (hereafter, AACC/BB and AA/BBCC were used, respectively). In the case of AACC/BB, functional values (frequency of resistant plants) of parents are given to AA++CC and ++BB++, and one of F_1 is given to A+B+C+, if tested (see Purba *et al.*, 1994, 1996). In these cases, the following restrictions were added:

For C gene, CC ≥ C+ ≥ ++
For A and C genes, AACC ≥ AA++ or ≥ ++CC
For B and C genes, BBCC ≥ BB++ or ≥ ++CC
For A, B and C genes, AABBCC ≥ AABBC+≥ AABB++,
AABBCC ≥ AAB+CC≥ AA++CC,
AABBCC ≥ A+BBCC≥ ++BBCC.

Finally, a set of functional values that can commonly explain all three segregations was selected for each F_2 plant.

In Test 2, we could not differentiate Nipponbare from Juma by the number of lesions, due to overlapping of large lesions on Nipponbare. Therefore, responses were divided into three: r (more resistant than Juma), m (Juma type) and s (Nipponbare type), and r : (m + s) and (r + m) : s segregations were analysed by cumulative frequency distribution curve method.

3. Results

3. 1. TEST 1

The observed number of resistant plants per line is shown in Table 6 for R + (N + J + S), (R + N) : (J + S) and (R + N + J) : S segregations. When the two-gene model, AA in Nipponbare and BB in Juma, was tested, maximum differences between observed and expected cumulative frequency distribution curves were statistically significant for (R + N) : (J + S) (Figure 1), although not significant for the other two segregations (Table 7).

TABLE 7. Maximum difference between expected and observed curves when each segregation ratio was explained by two or three genes in Test 1

Segregation	Genotype assumed for Nipponbare and Juma		
	AA/BB	AACC/BB	AA/BBCC
R : (N + J + S)	0.0470	0.0599	0.0575
(R + N) : (J + S)	0.1943**	0.0641	0.2037**
(R + N + J) : S	0.0856	0.0679	0.0980

** means that the difference is significantly larger from 0 at the 1% level.

A three gene model was tested. For the AACC/BB, the maximum of differences between expected curve selected and observed curve was obtained: 0.0599, 0.0641 and 0.0679 for the three segregations, respectively. For the model, AA/BBCC, 0.0575, 0.2037 and 0.0980 was obtained. The latter does not show agreement between observed and expected cumulative frequency distribution curves. Therefore, the former showing good agreement was accepted. The frequency and cumulative frequency distribution curves are shown in Figures 2, 3 and 4. In these cases, functional values for three segregation ratios are shown in Table 8. The results suggest that resistance of hybrid of Nipponbare x Juma is controlled by three minor genes: two in Nipponbare and one in Juma. These three genes have a complementary effect, which is defined as the effect producing resistance [with functional value, 0.0588 (1/17) to 1] in accumulation of three genes (functional value of individual genes is near 0) in R : (N + J + S) segregation, and have additive effect in other segregations (Table 8). Legends for figures 1 – 6 are shown in figure 2.

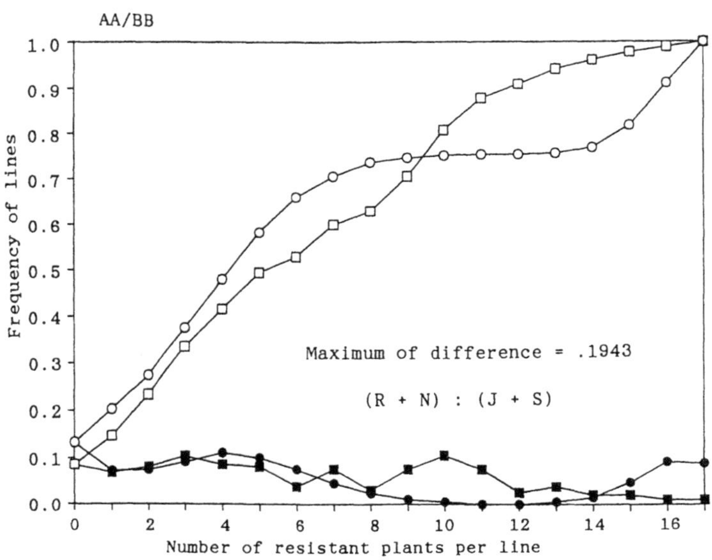

Figure 1. Frequency and cumulative frequency distribution curves of resistant segregants in F_3 of Nipponbare x Juma (Test 1)

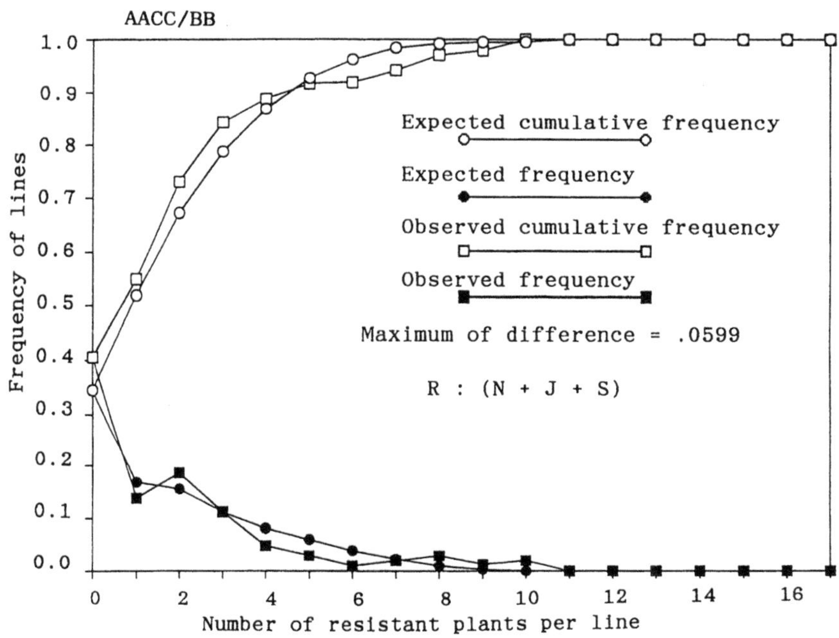

Figure 2. Frequency and cumulative frequency distribution curves of resistant segregants in F_3 of Nipponbare x Juma (Test 1)

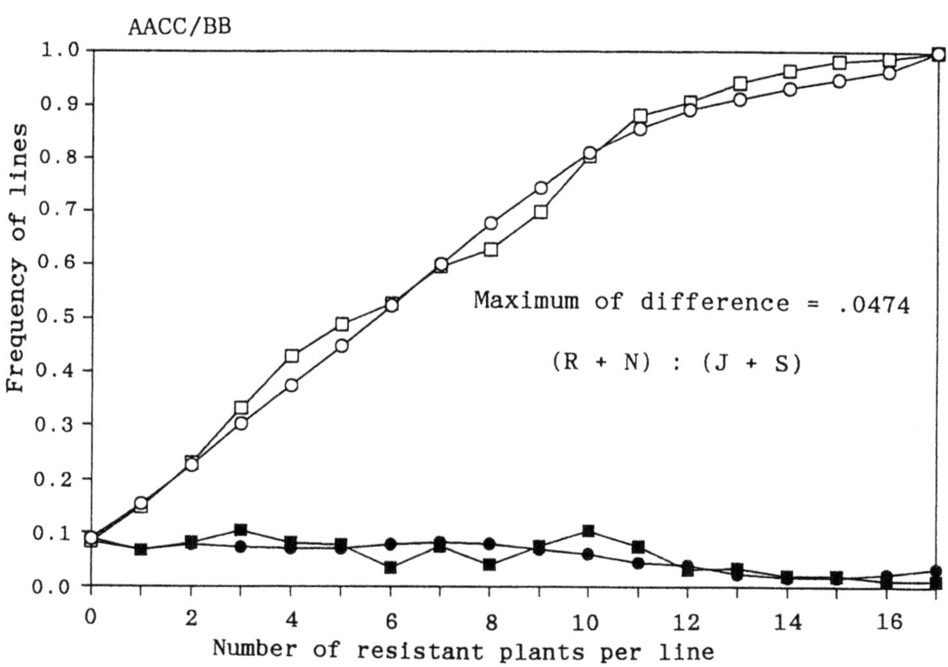

Figure 3. Frequency and cumulative frequency distribution curves of resistant segregants in F_3 of Nipponbare x Juma (Test 1)

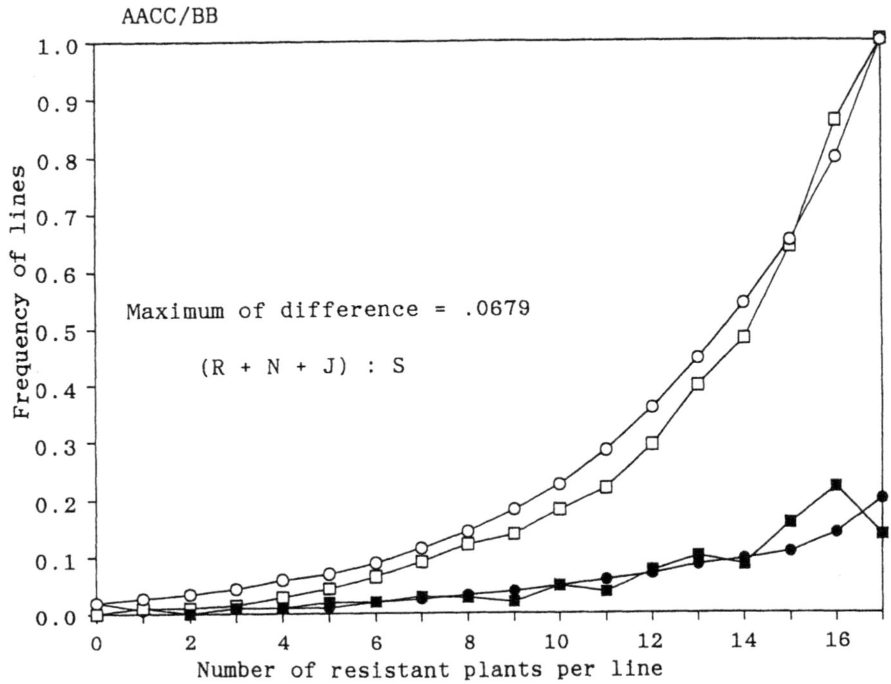

Figure 4. Frequency and cumulative frequency distribution curves of resistant segregants in F_3 of Nipponbare x Juma (Test 1)

TABLE 8. Functional values given to genotypes in hybrid of Nipponbare and Juma in Test 1

Genotype	Segregation ratio		
Parent[1]	R : (N + J + S)	(R + N) : (J + S)	(R + N + J) : S
AACCBB	0.3529	0.9411	1.0000
AACCB+	0.2647	0.9411	1.0000
AACC++	0.2647 N	0.9411 N	0.9706 N
AAC+BB	0.2647	0.8235	0.9412
AAC+B+	0.2353	0.7647	0.8235
AAC+++	0.1765	0.6471	0.7647
AA++BB	0.1765	0.7059	0.9412
AA++B+	0.1765	0.5882	0.8235
AA++++	0.0588	0.4706	0.7647
A+CCBB	0.2647	0.5294	1.0000
A+CCB+	0.1765	0.4706	0.9412
A+CC++	0.1765	0.2941	0.9394
A+C+BB	0.1765	0.4706	0.9394
A+C+B+	0.0588	0.2353	0.7647
A+C+++	0.0000	0.1176	0.7059
A+++BB	0.0000	0.3529	0.9394
A+++B+	0.0000	0.0000	0.7647
A+++++	0.0000	0.0000	0.7059

++CCBB	0.0000	0.1515	1.0000
++CCB+	0.0000	0.0000	0.7394
++CC++	0.0000	0.0000	0.2941
++C+BB	0.0000	0.1515	0.9394
++C+B+	0.0000	0.0000	0.2941
++C+++	0.0000	0.0000	0.1176
++++BB	0.0000 J	0.1515 J	0.9394 J
++++B+	0.0000	0.0000	0.2941
++++++	0.0000	0.0000	0.0000
Maximum difference	0.0599	0.0641	0.0679

N and J are functional values given to Nipponbare and Juma, respectively

[1] AACCBB means that AACC is in Nipponbare and BB is in Juma.

3. 2. TEST 2

In Test 2, 96 F_3 lines, excluding two lines showing poor germination, and F_1 plants were used for analysis. Reactions of parents, particularly Nipponbare, and segregation ratio in the F_3 generation were largely different from those of Test 1. The number of lesions on Nipponbare was smaller than on Juma, but the difference was not significant (Table 9). It could not be differentiated from most of the susceptible lines in the F_3 generation. There was no association between seed sterility (6 to 99% in F_2 plants) and blast resistance, although there was an association between color of the basal leaf sheath and seed sterility (Table 10). The frequencies of lines with various numbers of resistant plants in 17 plants of F_3 lines are shown in Table 11. Of the 96 lines tested, only two were homogeneously resistant. Therefore, the two-gene model could not explain the results.

TABLE 9. The number of lesions on Nipponbare and Juma in Test 2

Variety or line		Mean number of lesions in type					Freq. of immune plants	Response
		b	bg	bG	pG	Total		
Nipponbare	a				7.67	7.67	0	S
	b		0.78	1.11	9.30	11.19	0	S
Juma	a	1.33	11.67	9.77	22.77		0	M
	b	5.00	11.30	3.33	19.63		0	M
Susceptible F_3 line	a		1.24	2.53	9.88	13.65	0	S
	b	0.35	0.41	1.65	8.29	10.67	0	S

TABLE 10. Uniformity tests among characters evaluated in Test 2

	Purple base	Seed sterility
A. r : (m + s)	1.41	0.37
B. (r + m) : s	0.69	0.69
C. Purple base		3.75*

* means significance at the 5% level.

TABLE 11. Distribution of the number of resistant plants per line in Test 2

Number of resistant plants in line	The number of lines in segregation ratio of r : (m + s)	(r + m) : s
0	45 N J F1 0	4 N 0
1	18	9
2	13	9
3	4	9F1 .17
4	5	6
5	4	5
6	4	3
7	1	5
8	0	6
9	0	5
10	1	6
11	0	7
12	1	9
13	0	3
14	0	3
15	0	3
16	0	2
17	0	2 J
Total	96	96

N, J and F1 are the numbers of resistant plants in two parents, Nipponbare and Juma, and F_1 with their functional values.

Three genes in resistant parent, Juma, were required to explain the segregation. Functional values shown in Table 12 are given to the genotypes in r : (m + s) and (r + m) : s ratios to minimise the maximum difference between expected and observed cumulative frequency distribution curves (Figs. 5 and 6). Three genes in Juma and no gene in Nipponbare were considered to account for the segregations. This is very different from the conclusion in Test 1 where two genes in Nipponbare and one in Juma accounted for the data.

TABLE 12. Functional values given to genotypes in hybrid of Nipponbare and Juma in Test 2

	Segregation rate		
Genotype	r : (m + s)	Genotype	(r + m) : s
AADDEE	0.7059	AADDFF	1.0000 J
AADDE+	0.6041	AADDF+	1.0000
AADD++	0.0000 J	AADD++	0.9412
AAD+EE	0.5882	AAD+FF	0.9412
AAD+E+	0.2353	AAD+F+	0.9412
AAD+++	0.0000	AAD+++	0.7647
AA++EE	0.0588	AA++FF	0.7647
AA++E+	0.0000	AA++F+	0.7647
AA++++	0.0000	AA++++	0.7049
A+DDEE	0.2353	A+DDFF	0.5294
A+DDE+	0.1765	A+DDF+	0.5294
A+DD++	0.0000	A+DD++	0.4118
A+D+EE	0.0588	A+D+FF	0.4706
A+D+E+	0.0000F1	A+D+F+	0.1700F1
A+D+++	0.0000	A+D+++	0.1176
A+++EE	0.0000	A+++FF	0.3529
A+++E+	0.0000	A+++F+	0.1176
A+++++	0.0000	A+++++	0.0588
++DDEE	0.1176	++DDFF	0.3529
++DDE+	0.0588	++DDF+	0.1176
++DD++	0.0000	++DD++	0.0588
++D+EE	0.0000	++D+FF	0.2353
++D+E+	0.0000	++D+F+	0.0588
++D+++	0.0000	++D+++	0.0000
++++EE	0.0000 N	++++FF	0.0000
++++E+	0.0000	++++F+	0.0000
++++++	0.0000	++++++	0.0000 N
Maximum difference	0.0667		0.0844

N, J and F1 are the number of resistant plants and functional values of genotypes of Nipponbare, Juma and F_1, respectively

1) : In AAD+E+, AAD+ and E+ show segregation of genes from Nipponbare and Juma, respectively.

In Test 2, Nipponbare did not show any resistance. Two genes (AADD) in Juma and one gene (EE) in Nipponbare were considered to account for susceptibility of both parents and the small number of plants more resistant than the parents for r : (m + s) segregation. For (r + m) resistance, the assumption of three genes (AADDFF) in Juma was required to account for only two homogeneously resistant F_3 lines out of 96 F_3 lines and resistance of Juma. Functional values given to all genotypes for each segregation are shown in Table 12. Additive or complementary effects appeared in some lines of F_3 in (r + m) : s segregation. Namely, AA (0.7049) + DD (0.0588) _ AADD (0.9412) or DD (0.0588) + FF (0.0000) _ DDFF (0.3529).

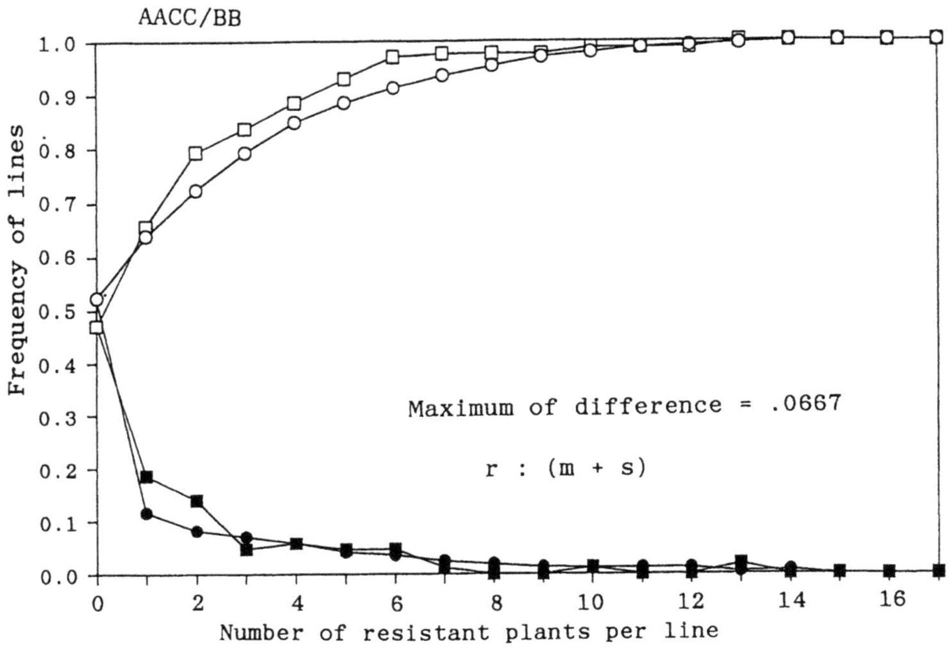

Figure 5. Frequency and cumulative frequency distribution curves of resistant segregants in F_3 of Nipponbare x Juma (Test 2)

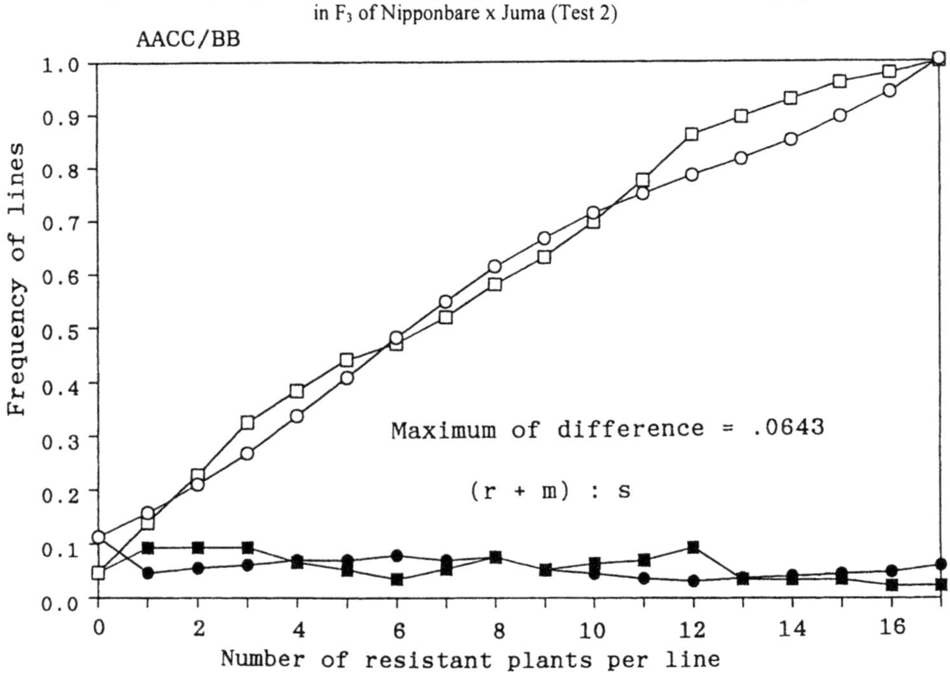

Figure 6. Frequency and cumulative frequency distribution curves of resistant segregants in F_3

The apparently contradictory results of Tests 1 and 2 are explained as follows. Two genes in Nipponbare assumed from result in Test 1 were not expressed under the conditions of Test 2, and two additional genes which were found in Juma in Test 2 were non-functional or not detected under the conditions in Test 1. Conditions were different in the greenhouse between June and October and in plant stages in leaf number, 6.0 and 4.0 in the two tests. Conditions in Test 2 were considered to be more favorable for disease development than those for Test 1. The difference between two tests indicates that the two genes which were functional in Test 1 were not functional in Test 2 and two minor genes in Juma which were not found under epistatic function of Nipponbare's genes were detected in Test 2.

4. Discussion

4. 1. COMPONENTS IN FIELD RESISTANCE

Field resistance is divided into three components by variable methods by different investigators as mentioned above (Hwang and Heitefuss, 1982; Asher and Thomas, 1983 1984, 1987; Denissen, 1993; Broers, 1989 a,b; Pretorius *et al.*, 1994). Shew *et al.* (1989) divided it into four: lesion number, the percentages of sporulating lesions, lesion diameter and conidial production per sporulating lesion. Bruno and Nelson (1990) divided partial resistance of wheat to septoria glume blotch into six components: incubation period, latent period, percentage of diseased leaf tissue, initial spore production at the end of latent period, total spore production at 100% necrosis, and maturation period.

In the present paper, it is suggested that different minor genes controlled lesion size and lesion number. Field resistance includes lesion number (infection frequency) and lesion size among other components. Among them, lesion size and number are controlled by different genes. The present study also showed that one or two genes controlling lesion number in Nipponbare were very sensitive to environmental conditions and/or plant stage. At least some partial resistance to rice blast is expressed as fewer and smaller lesions in the leaf blade but latent period does not appear to be an important component (Bonman, 1992; Roumen and de Boef, 1993).

4. 2. GENETIC RELATIONSHIPS AMONG RESISTANCE COMPONENTS

The present paper describes that two minor genes controlling field resistance (lesion size and lesion number) which are present in different varieties and these are independent of each other. Transgressive segregation was found in the F_3 generation. Roumen (1993) showed that transgressive segregations were found in rice blast resistance in *indica* hybrids: in this case at least one gene in each parent controlled the reduction of the number of lesions in the hybrid. According to Roumen (1993), the number of genes reducing the lesion number was five or more in *indica* varieties. In *indica* rice varieties, Roumen (1992) found that lesion size and lesion density were positively correlated.

Wang *et al.* (1989) examined lesion number and lesion size in F_2 generation of IR36 x Milyang 42 and IR50 x Milyang 57. Heritability obtained was low. Wang et al. (1994)

showed that different genes control decrease of lesion size and of lesion number and these genes are different from each other. It has been observed that, infection is inhibited by a reduction of spore attachment, penetration and growth of the infection hyphae in *O. minuta.* This means that host resistance should be operative in these processes. Wang *et al.* (1994) found significant associations between markers and the three individual parameters of partial resistance (lesion number, lesion size and diseased leaf area) on chromosomes 1, 3, 5, 6, 7, 8, 11, and 12 in a *japonica* variety, Moroberekan. Higashi (1995) studied field resistance of *japonica* upland rice variety, Sensho, and estimated the presence of 5 to 8 effective factors using a cross with a highly susceptible variety, Mokoto. Furthermore, 4 to 7 effective factors were estimated in 6 other combinations of varieties. These were located on chromosomes, 2, 3, 4, 7 and 11. Inukai *et al.* (1996) mapped genes for partial resistance on chromosomes 1 and 6 in 2 recombinant inbred lines obtained from resistant variety Moroberekan and susceptible variety CO39. In F_2 and F_3 generations of crosses of upland rice, Annada x Prasanna and BG 367-4 x RP 149-177, it was shown that additive and dominance effects were both significant but dominance effects were larger and predominant. The presence of large dominance gene effects limited the effectiveness of early generation selection in both crosses.

Genetic analysis showed that RIE [relative infection efficiency (the number of sporulating lesions that developed)] is controlled by several minor genes. RIE-reducing genes were shown to operate on a gene-for-gene basis with genes in the pathogen. Genotypes developing fewer lesions per plant in the greenhouse demonstrated a higher level of PR (partial resistance) in field studies. PR can be achieved in early generations of a breeding programme. A higher PR to leaf blast was usually linked to a higher PR to neck blast (Roumen 1993). It is generally known that major genes are functional in reducing RIE to 0, and shows a very small environmental variance. Some minor genes are also functional to reduce IRE. However, gene-for-gene relationships are generally not found for minor genes. There are some discrepancies in the conclusion by Roumen (1993).Other reports indicate that there are complicated genetic relationships among genes controlling these components. For example, Cunfer *et al.* (1988) found significant correlations among components of resistance to *Leptosphaeria nodorum* in wheat [latent period, the rate of disease progress up the plant (disease severity) and area under the disease progress curve], suggesting that a single or interrelated mechanisms control expression of resistance. Denissen (1993) found that the correlation coefficient between infection frequency and latency period was low in wheat leaf rust. Broers (1989 a,b) indicated the possibility that three components (infection frequency, latency period and urediosorus size) were controlled by different genes in wheat leaf rust, although he also found associated variation of the components. The latency period was the most important component among wheat varieties used. Latency period was controlled by one to three genes in wheat (Jacobs and Broers, 1989).

From the epidemiological standpoint, sporulation capacity is a very important component in field resistance to the rice blast pathogen. Sporulation of fungus strains varies markedly with artificial medium and host tissue. Some strains sporulate abundantly on oatmeal medium but not on host tissue. Other strains show the opposite tendency. There is not always a positive correlation between sporulation on the medium and on the plant.

For example, one commonly used strain of blast fungus, Ken 54-20, produces few spores on a medium and large lesions on the host leaf, and another strain Hoku 1 produces a large number of spores on artificial media, but small lesions on the host. A negative correlation was reported between sporulation on the host and on the medium (Kiyosawa and Cho, 1973). Shew *et al.* (1989) reported significantly higher disease in specific isolate and genotype combinations, among moderately and highly resistant genotypes. These facts indicate that there are genes acting commonly and also genes acting separately to control these components.

In tests including many *indica* varieties worldwide, many intermediate resistances were found to Japanese and Philippine fungus strains. This suggested that these varieties have many field resistance genes (Kiyosawa *et al.*, 1981). Purba *et al.* (1994) investigated the resistance of some hybrids of *indica* and *japonica* for blast, and found additive or complementary effects in a few hybrids. In the present test, Nipponbare and Juma have different field resistance genes. This indicates that there are many different field resistance genes in *indica* and *japonica* varieties.

Our results suggest that lesion type and lesion number, among others, are components of field resistance. Some of these components are thought to be expressed at different stages in penetration of fungus. In the present paper, a gene (A) in Nipponbare was epistatic over a gene (B) in Juma in Test 1

Changes in dominance of resistance genes has been found in the resistance of some plants to some pathogens (Kiyosawa, 1970; Kolmer and Dyck,1994). For rice blast, a change from complete dominance to complete recessiveness was reported for the *Pi-i* gene under conditions that included plant stage, method of inoculations and fungus strains (Kiyosawa 1970; Kiyosawa unpublished). In the present study, a change of epistasis was observed. In Test 1, the additive effect of AA and CC assumed in Nipponbare was epistatic over BB in Juma, but in Test 2, BB was epistatic over AA and CC, because AA and CC were not expressed in Test 2.

5. Summary

One hundred and twelve F_3 lines of rice varieties, Nipponbare (*japonica*) and Juma (*indica*), were used for genetic analysis of blast resistance. Nipponbare showed a small number of large lesions and Juma showed a large number of small lesions in one of two tests. F_3 plants with higher levels of resistance than their parents were observed in some lines. Resistance on an individual plant basis was divided into four reaction types, R (highly resistant), N (Nipponbare type, small number), J (Juma type, small size) and S (susceptible). Resistances in three classes, R, R + N and R + N + J, were analysed by the cumulative frequency distribution curve method. To explain these three segregations, three genes with minor effect were assumed: AACC in Nipponbare and BB in Juma. Additive effects of these three genes were considered for explaining the resistance, AACC for N type field resistance and BB for J type field resistance. In another test (Test 2), the F_3 segregation of different F_3 lines from the same F_1 was explained by BBDD genes in Juma and EE gene in

Nipponbare. This suggests that at least one gene in Nipponbare was not expressed in Test 2, indicating a change in epistasis between the two tests.

6. References

Abe, S., Kiyosawa, S. and Ono, S. (1974) Inheritance of blast-resistance of upland rice varieties.I. Inheritance of resistance of an upland rice variety Norin No.4, *Bulletin of Tsukuba Agricultural Experiment Station* **15**, 47-64.

Asher, M. J. C. and Thomas, C. E. (1983) The expression of partial resistance to *Erysiphe graminis* in spring barley, *Plant Pathology* **32**, 79-89.

Asher, M. J. C. and Thomas, C. E. (1984) Components of partial resistance to *Erysiphe graminis* in spring barely, *Plant Pathology* **33**, 123-130.

Asher, M. J. C. and Thomas, C. E. (1987) The inheritance of mechanisms of partial resistance to *Erysiphe graminis* in spring barley, *Plant Pathology* **36**, 66-72.

Bonman, J. M. (1992) Durable resistance to rice blast disease - environmental influences, *Euphytica* **63**, 115-123.

Broers, J. H. M. (1989a) Influence of development stage and host genotypes on three components of partial resistance to leaf rust in spring wheat, *Euphytica* **44**, 187-195.

Broers, J. H. M. (1989b) Partial resistance to wheat leaf rust in 18 spring wheat cultivars, *Euphytica* **44**, 247-258.

Bruno, H. H. and Nelson, L. R. (1980) Partial resistance to *Septoria* blotch analyzed in winter wheat seedlings, *Crop Science* **30**, 54-59.

Cunfer, B. M., Stooksbury, D. E. and Johnson, J. W. (1988) Components of partial resistance to *Leptosphaeria nodorum* among seven soft red winter wheats, *Euphytica* **37**, 129-140.

Denissen, C. J. M. (1993) Components of adult plant resistance to leaf rust in wheat, *Euphytica* **70**, 131-140.

He, Y., Wang, Y., Li, J., Kiyosawa, S., Higashi, T. and Horisue, N. (1989a) Gene analysis of a Yaonaihuan upland rice variety, Honaihuan, *Oryza* **26**, 173-182.

He, Y., Kiyosawa, S., Wang, Y., Li, J., Li, C. and Higashi, T. (1989b) Inheritance of blast resistance in Chinese upland rice varieties, Zhalulong, Mowanggu and Mongwangu, *Oryza* **26**, 288-298.

Higashi, T. (1995) Genetic studies on field resistance of rice to blast disease, *Bulletin of Tohoku National Agricultural Experiment Station* **90**, 19-75.

Hwang, B. K. and Heitefuss, R. (1982). Characterization of adult plant resis tance of spring barley to powdery mildew (*Erysiphe graminis* f. sp. *hordei*). 1. Race specificity and expression of resistance, *Journal of Phytopathology* **104**, 168-178.

Imbe, T. and Matsumoto, S. (1985) Inheritance of resistance of rice varieties to the blast fungus strains virulent to the variety "Reiho", *Japanese Journal of Breeding* **35** 332-339.

Inukai, T., Zeigler, R. S., Sarkarung, S., Bronson, M., Dung, L. V., Kinoshita, T. and Nelson, R. J. (1996) Development of pre-isogenic lines for rice blast-resistance by marker-aided selection from a recombinant inbred population, *Theoretical and Applied Genetics* **93**, 560-567.

Jacobs, Th. and Broers, L. H. M. (1989) The inheritance of host plant effect on latency period of wheat period of wheat leaf rust in spring wheat. I. Estimation of gene action and number of effective factors in F_1, F_2 and backcross generations, *Euphytica* **44**, 197-206.

Kiyosawa, S. (1969) Inheritance of resistance of rice varieties to a Philippine fungus strain of *Pyricularia oryzae*, *Japanese Journal of Breeding* **19**, 61-73.

Kiyosawa, S. (1970) Inheritance of blast resistance of the rice varieties Homare Nishiki and Ginga. I. Resistance of Homare Nishiki and Ginga to the fungus strain Ken 54-04, *Bulletin of the National Institute of Agricultural Science* **D21**, 73-105.

Kiyosawa, S. (1974) Studies on genetics and breeding of blast resistance in rice, *Miscellaneous Publication, National Institute of Agricultural Science* **D1**, 1-58.

Kiyosawa, S. (1976) Methods for tests and gene analysis of blast resistance of rice varieties, *Oryza* **13**, 1-32.

Kiyosawa, S. (1984) Establishment of differential varieties for pathogenicity test of rice blast fungus, *Rice Genetics Newsletter* **1**, 95-97.

Kiyosawa, S. and Cho, C. I. (1973) Relation of the sporulating ability on the rice plant to some other characters in blast fungus strains, *Japanese Journal of Breeding* **23**, 239-244.

Kiyosawa, S., Ikehashi, H., Kato, H. and Ling, Z. Z. (1981) Pathogenicity tests of Philippine isolates of blast fungus using two sets of rice varieties, *Japanese Journal of Breeding* **31**, 367-376.

Kiyosawa, S., Mackill, D. J., Bonman, J. M., Tanaka T. and Ling, Z. Z. (1986) An attempt of classification of world's rice varieties based on reaction pattern to blast fungus strains, *Bulletin of the National Institute of*

Agrobiological Resources **2**, 41-59.

Kolmer, J. A. and Dyck, P. L. (1994) Gene expression in the *Triticum aestivum-Puccinia recondita* f.sp.*tritici* gene-for-gene system, *Phytopathology* **84**, 437-440.

Muller, K. O. and Haigh, J. C. (1953). Nature of 'field resistance' of the potato to *Phytophthora infestans* de Bary, *Nature* **171**, 781-783.

Pretorius, Z. A. Kloppers, F. J. and Drijepondt, S. C. (1994) Effects of inoculum density and temperature on three components of leaf rust resistance controlled by *Lr34* in wheat, *Euphytica* **74**, 91-96.

Purba, D., Kiyosawa, S., Ando, I. and Furutani, T. (1994) Estimation of functional values of field-resistance genes to blast disease in some rice varieties, *Breeding Science* **44**, 285-293.

Purba, D., Ali, M. S., Furutani, T. and Kiyosawa, S. (1996) Identification and isolation of blast resistance genes in three *indica*-type rice varieties, *Breeding Science* **46**, 107-115.

Roumen, E. C. (1992) Effect of leaf age on components of partial resistance in rice to leaf blast, *Euphytica* **63**, 271-279.

Roumen, E. C. (1993) Inheritance of host plant effect oin the relative infection efficiency of *Magnaporthe grisea* in rice cultivars, in E.C. Roumen (ed.), *Partial Resistance in Rice to Blast and How to Select for it*, Wageningen Agricultural University, Wageningen, Netherlands, pp. 61-72.

Roumen, E. C. and de Boef, W. S. (1993) Latent period to leaf blast in rice and its importance as a component of partial resistance, *Euphytica* **69**, 185-190.

Roumen, E. C. (1993) *Partial reisistance in rice to blast and how to select for it*, Wageningen Agricultural University, Wageningen, Netherlands, pp. 108.

Shew, B. B. Sommartya, T. and Beute, M. K. (1989). Components of partial resistance in peanut genotypes to isolates of *Cercosporidium personatum* from the United States and Thailand, *Phytopathology* **79**, 136-142.

Wang, G. –L., Mackill, D. J., Bonman, J. M., McCouch, S. R., Champoux, M. C. and Nelson, R. J. (1994) RFLP mapping of genes conferring complete and partial resistance to blast in a durably resistant rice cultivar, *Genetics* **136**, 1421-1434.

Wang, Z., Mackill, D. J. and Bonman, J. M. (1989) Inheritance of partial resistance to blast in *indica* rice cultivars, *Crop Science* **29**, 848-853.

RICE SHEATH BLIGHT - PATHOGEN BIOLOGY AND DIVERSITY

S. BANNIZA and M. HOLDERNESS
CABI BioScience,
Bakeham Lane, Egham, Surrey TW20 9TY, UK.

1. Introduction

Rhizoctonia solani Kühn was originally reported causing sheath blight of rice in Japan at the beginning of this century (Miyake, 1910). However, rice sheath blight has developed into a major disease only during the past two decades. Nowadays, the disease is second only to, and often rivals, rice blast (*Pyricularia oryzae*) in terms of economic importance in many major rice-growing countries. The emergence of *R. solani* as a major rice pathogen has been attributed to the intensification of the rice-cropping system characterised by the development of new, short-statured, high-tillering, high-yielding varieties and an increase in nitrogen fertilisation inducing a favourable micro-climate for the pathogen. The rice sheath blight pathogen has been associated with isolates of one particular intraspecific group (ISG) of *Rhizoctonia solani*, anastomosis group (AG) 1 IA (Yamaguchi *et al.*, 1984; Vijayan and Nair, 1985; Borthakur and Addy, 1988; Jones and Belmar, 1989), but diversity within this ISG appears to be much greater than was previously assumed (Liu and Sinclair, 1993). Detailed knowledge of pathogen characteristics at this taxonomic level is still lacking and has hampered understanding of the patho-systems.

2. History and economic importance of sheath blight caused by *Rhizoctonia solani*

The first description of sheath blight appeared in Japan (Miyake, 1910) and the causal organism was subsequently identified as *Hypochnus sasakii* which had previously been described by Shirai (1906). In the Philippines, a similar disease was discovered by Reinking (1918) and Palo (1926) which they attributed to a fungus of the *Rhizoctonia* group. In Sri Lanka, Park and Bertus (1932) also associated sheath blight with a *Rhizoctonia* species, referring to it as *R. solani*. Two years later the disease was reported in China (Wei, 1934), but only after descriptions appeared in Brazil, Surinam, Venezuela, Madagascar and the USA was the initial assumption revised that sheath blight occurred only in Asia (Ou, 1985).

Sheath blight is considered to be of major importance in Sri Lanka, China, Taiwan and Japan, where it is second only to blast in terms of economic damage (Gangopadhyay and Chakrabarti, 1982). In China, the disease has affected more than 3.2 million hectares causing yield losses of over 200 million kg/year. In 1985-1990,

S. Sreenivasaprasad and R. Johnson (eds.), Major Fungal diseases of Rice Recent Advances, 201-211.
Prof. K. Manibhushan Rao Festschrift

sheath blight occurred in 47% of all paddy fields. In Malaysia, 15-20% of the total area planted to rice has been reported to be infected with the disease and in the 1993 season losses of 17-25% occurred. More than half of the total cultivated rice acreage has been infected with sheath blight in Korea and yield losses of up to 90% have been recorded. A dramatic increase in sheath blight incidence has also been reported from Vietnam where the area infected has increased tenfold from 21,000 ha in 1985 to about 200,000 ha in 1990 and 1991. Total yield losses have occurred in several hundred hectares planted to rice (Anonymous, 1993).

3. Disease development and host-pathogen interactions

3.1. DISEASE SYMPTOMS AND DEVELOPMENT

Symptoms on rice typically start to develop at or above water level, or above soil level in upland conditions. Pale, greenish-grey, ellipsoidal lesions appear which are initially about 2-3 cm long and have a dark brown margin, but then enlarge and coalesce in combination with a change in colour to fawn or off-white with a brown or purplish-brown margin. Lesions may be found anywhere on the leaf sheath, initially appearing on the outer sheath but then extending to the inner sheath, and under favourable conditions in humid warm climates they can even be detected on the leaf blade. Severe attack may lead to plant death (Gangopadhyay and Chakrabarti, 1982).

Favourable conditions for sheath blight development are low sunlight, high humidity (near 95%) and high temperatures between 28 and 32°C. Under these conditions, the fungus spreads rapidly by means of runner hyphae to upper plant parts as well as to adjacent plants. Usually disease development is most rapid in the early heading and grain filling stages.

Rice seedlings may be infected with *R. solani* in the nursery, if they are planted in infested soil. However, the importance of transmission by transplanting seedlings into the paddy field has not yet been established, in particular as the process includes a selection of the most viable seedlings (Gangopadhyay and Chakrabarti, 1982). It is suggested that sclerotia are the major source of infection (Damicone *et al.*, 1993). These are formed on or near the lesions, are easily detached, and survive for long periods in the soil (Ou, 1985; Damicone *et al.*, 1993). Upon coming into contact with a rice plant, they germinate and the emerging hyphae produce infection cushions on the exposed leaf sheath (Lee and Rush, 1983). From these infection cushions, haustoria grow and penetrate the host tissue and the typical lesions develop (Dodman and Flentje, 1970). Sclerotia are generally formed after the death of the invaded cells and are found on the tissue where fungal growth has been most prolific. However, the precise factors which initiate the formation of sclerotia are unknown (Parmeter and Whitney, 1970). Estimates and numbers of recovered sclerotia from soil vary widely in the literature. For example, Lee (1980) reported 216 to 701 sclerotia per litre of soil in the top 0-0.6 cm depth in fields with a history of heavy sheath blight infection, with viability ranging from 42.3 to 51.4%. At depths of 0.6 to 7.6 cm this was reduced to 27 to 87 sclerotia, with a viability range of 14.2 to 30.3%. It was also observed that sclerotia remained viable for up to two years. Initially, sclerotia are non-buoyant and sink to the ground but after about 30 days the exterior cells die and become void of cellular content with the consequence that sclerotia float on the water (Lee and Rush, 1983). If sclerotia happened to be buried in submerged paddy soil they were found to be non-viable after 14 days, a decline attributed to either anaerobic conditions or microbial factors (Mew and Rosales, 1986).

Lower counts of sclerotia have been observed in the tropics. Shahjahan *et al.* (1990) only detected 3.6 to 6.3 sclerotia in 0.5 litre of soil in the control plots of their experiments, and Cu *et al.* (1996) reported 1.23 sclerotia per litre of saturated paddy soil, suggesting that sclerotia may not be the dominant source of initial inoculum in tropical lowland rice. Other sources of initial inoculum are infected weeds, straw and stubble (Mew and Rosales, 1986), but no quantitative data on inoculum of this nature has been published to date.

3.2. HOST-PATHOGEN INTERACTIONS

Rice plants appear to be most susceptible to infection with *R. solani* at the tillering stage when the leaf sheath becomes discoloured (Gangopadhyay and Chakrabarti, 1982). Two main factors are thought to be responsible for this disease predominating during the second half of the crop. Young rice plants are more resistant to infection with this pathogen, but lose their resistance with maturation. Along with maturation, the encirclement of the leaf sheath around the stem loosens when the leaf matures, which physically facilitates penetration of the mycelium to the inner side of the sheath (Ou, 1985). Apart from changes in plant susceptibility with maturity, environmental factors influencing the micro-climate and in particular nitrogen supply have a major influence on disease development. A closed canopy in the later stage of rice plant development increases air humidity inducing a micro-climate conducive to the fungus. Savary *et al.* (1995) demonstrated that increased nitrogen supply (120 kg/ha) exacerbated the situation by increasing plant growth, thus increasing leaf-to-leaf and leaf-to-sheath contact and prolonging leaf surface wetness due to the higher relative humidity produced in the canopy where growth was denser. Heavily infected plants produce poorly filled grains and increased lodging and/or reduced tiller production due to death of the culm reduces yield (Lee and Rush, 1983).

Although rice plants are most susceptible at tillering, Cu *et al.* (1996) showed that infection only resulted in yield loss when the pathogen established during panicle initiation and flowering, causing a reduction in mean grain weight. In contrast, early infection proved not to affect yield, possibly because the plant compensated for diseased tissue by producing new tillers. Similarly, no effect on yield was observed in late infection at the grain filling stage when time was not sufficient for the fungus to progress significantly.

Apart from plant age, varietal characteristics appear to have a major influence on disease development. Various degrees of resistance of rice varieties have been associated with several different factors. Massaquoi and Rush (1987) investigated the amount of epicuticular wax on the leaf blade and the sheath surface by SEM. They observed a significant positive correlation between wax concentration and resistance level, associated with a lower number of infection cushions on varieties expressing a higher degree of resistance. It appears that a thicker wax layer could prevent contact between the pathogen and compounds produced by the rice plants which induce the formation of infection cushions, as was suggested previously (Flentje *et al.*, 1963; Marshall and Rush, 1980a . A lower number of infection cushions on varieties with a higher degree of resistance to *R. solani* was also observed by Groth and Nowick (1992) who used this character as one criterion for the selection of newly-bred rice varieties. A complete lack of these structures on resistant varieties was observed by Marshall and Rush (1980b), who only found lobate appressoria from where penetration was initiated. However, cells surrounding the invading hyphae collapsed thus slowing down its growth in the outer sheath of these varieties.

Rice plants respond to pathogen invasion with the production of new proteins which are commonly referred to as pathogenesis-related (PR) proteins (Lin *et al.*, 1995). Using two phytoalexin elicitors, Bera and Purkayastha (1995) artificially induced the production of PR proteins in rice plants and detected at least 5 similar types of proteins of this group in a susceptible rice variety. Treatment with one of the elicitors (Penicillin) induced a significantly higher number of proteins and reduced sheath blight disease after subsequent inoculation with the pathogen. Infection of moderately resistant rice varieties with *R. solani* resulted in higher levels of chitinases and ß-1,3-glucanases, two representatives of PR proteins, than were found in susceptible varieties (Anuratha *et al.*, 1996). Introduction of a constitutively-expressed chitinase gene into rice plants, increased resistance of the plants to infection by *R. solani* (Lin *et al.*, 1995) and appears to be a promising step forward to the development of rice varieties with a higher degree of resistance.

4. Functional and genetic diversity of the sheath blight pathogen

4.1. ANASTOMOSIS BEHAVIOUR

Diversity within the species *Rhizoctonia solani* is represented by anastomosis groups defined from hyphal anastomosis behaviour. To date, 12 AGs have been characterised (Ogoshi *et al.*, 1990). Within the AGs, other intraspecific groups have been defined on the basis of morphology, pathology, pectinase isoenzymes and DNA sequence homology (Vilgalys and Cubeta, 1994). Strains of AG-1 IA have been subdivided into three ISGs based on sclerotial form, cultural characteristics and DNA sequence homology (Sneh *et al.*, 1991).

Extensive studies on the anastomosis behaviour of isolates of *R. solani* from rice have been conducted in several major rice growing countries, e.g. Indonesia (Yamaguchi *et al.*, 1984), India (Borthakur and Addy, 1988; Vijayan and Nair, 1985) and USA (Jones and Belmar, 1989). The majority of isolates in these studies were found to belong to AG-1, and although rice was the major host plant, AG-1 isolates were also isolated from a number of other plant species, such as corn, sorghum, peanut, legumes and several vegetable species. Strains of this AG are known to exist on an extremely wide range of host plants (Baker, 1970) and have a wide distribution (Anderson, 1982) with the exception of arid areas where they have not yet been found (Parmeter *et al.*, 1969).

Diversity of plant-borne isolates of *R. solani* AG-1 in one paddy field was studied by Ogoshi and Ui (1983). By distinguishing between perfect fusion and imperfect fusion in their anastomosis testing, they were able to identify 41 clones in this paddy field. They could also demonstrate that isolates from the same culm showed a high degree of perfect fusion indicating that they were clones of the same strain, while the proportion of imperfect fusions with isolates of neighbouring culms and hills increased with distance.

4.2. MORPHOLOGICAL VARIATION AND PATHOGENICITY

The separation of three ISGs within AG-1 is mainly based on morphological characters, and we have in particular used the size and shape of sclerotia. Sclerotia of AG-1 IA are described as large (1-3 mm diam.) and relatively spherical, in contrast to those of strains of AG-1 IB which are said to be small and irregular-shaped. Small (0.2 - 0.8 mm) sclerotia are also found in AG-1 IC, but they are more spherical and such strains only appear to predominate in cooler climatic conditions (Sneh *et al.*, 1991). Sherwood (1969) gave the following morphological characters for AG-1 IA, which he called AG-1 Type 2 and considered to be the typical rice pathogen:

> Mycelium usually very light brown or whitish; coarsely radiate; occasionally moderately aerial; white patches scattered over the surface; sclerotia few to abundant; quite uniform in size, 0.7-3.0 mm diam, usually about 1.5 mm on agar surface; spherical with a slightly flattened bottom; scattered singly or sometimes joined laterally; surface smooth, firm, finely fuzzy and sometimes finely pitted where exudates occur; surface very dark brown (nearly black) at maturity.

A similar description was given by Vijayan and Nair (1985). Although attempts have been made to further separate morphological groups within AG-1 IA, several authors have concluded that isolates were indistinguishable in gross morphology (e.g. Kim *et al.*, 1988; Singh *et al.*, 1990; Foroutan and Rahimian, 1991). Similarly, pathogenicity testing did not allow the establishment of distinct groups (Jones and Belmar, 1989; Banniza *et al.*, 1996).

4.3. DIVERSITY IN INTRA- AND EXTRA-CELLULAR ENZYMES, AND PROTEINS

The characterisation of fungi by analysis of intracellular and extracellular enzymes is based on the phenomenon that most organisms contain several different genetically-determined forms of many of their active enzymes. These different forms are called isoenzymes or isozymes. They differ in amino acid sequence, but show the same enzymatic activity (Egger, 1992). Isozyme analysis has been increasingly used in the study of fungal pathogen diversity. In *R. solani* AG-1 IA, several studies have been conducted to illuminate diversity within this ISG. Variation in isozyme phenotypes was observed in some cases, but overall results based on isozyme analysis appear to be inconclusive.

The study of non-specific esterase patterns of isolates of AG-1, 2, 3 and 5 showed that each anastomosis group produced a distinct banding pattern (Matsuyama *et al.*, 1978). Within AG-1, isolates of AG-1 IA originating from rice had an almost identical banding pattern which was distinct from the banding pattern produced by isolates of AG-1 IB. Similarly, analysis of multiple isozyme band patterns of 10 enzymes for isolates of AG-1 showed that characteristic phenotypes were found for some of the 6 subgroups within AG-1 which were identified through the analysis of the ribosomal DNA (rDNA) internal transcribed spacer (ITS) region (Liu and Sinclair, 1993). However, some overlapping was also observed and isolates of two subgroups showed a diversity of phenotypes, leading to the conclusion that isozyme phenotypes alone could not be used to define subgroups.

Variation around one characteristic banding pattern was observed in zymograms of pectic enzymes of rice isolates (Banniza *et al.*, 1996). This major banding pattern appeared to be identical with the one observed by Cruickshank (1990) for an isolate of AG-1 IA, which was distinct from patterns of AG-1 IB and AG-1 IC. However, unlike in the case of AG-8 (Cruickshank, 1990), isolates of AG-1 IA could not be considered to show homogeneous banding patterns (Banniza *et al.*, 1996).

A higher degree of diversity was observed when mycelial proteins of five differentially-virulent isolates of *R. solani* from rice, obtained from different geographical regions, were subjected to electrophoresis (Zuber and Manibhushanrao, 1982). Although one band was expressed in all patterns, suggesting that it was characteristic for this group of isolates, the remaining bands varied in number, position and intensity. A similar degree of variation in banding patterns was observed in 3 out of 5 intra- and extra-cellular enzymes, while two enzymes were not expressed in all isolates.

4.4. DIVERSITY OF CELLULAR FATTY ACIDS

To date, the analysis of fatty acid profiles has been mainly used to differentiate between closely related strains of bacteria. Fatty acids are the most common ‘tail' of lipids in biological membranes. They are usually long chained carboxylic acids with an even number of carbon atoms (Rawn, 1983). This technique has been assessed for the characterisation of *R. solani* only recently by Johnk and Jones (1992, 1993, 1994), who studied the fatty acid composition of isolates of *R. solani* of AG-1 and AG-2.

Twelve fatty acids were detected and quantified from analysis of 10 isolates of AG-1 IA. Ten of these were used to create a library which enabled the rapid identification of other isolates (Johnk and Jones, 1992). Further investigations into AG-1, however, showed that qualitative composition of fatty acids was similar in all three subgroups of AG-1. Quantitative differences were only detected between AG-1 IC and the other two subgroups (Johnk and Jones, 1994), allowing only isolates of this subgroup to be identified with high precision.

4.5. GENETIC DIVERSITY

Molecular biology techniques have become a standard approach for the study of fungal relationships at inter- and intra-specific level (Bruns *et al.*, 1991). In the species *R. solani*, these methods have proved the taxonomic significance of the AG concept, showing that genetic groups were highly correlated with AG groups (Vilgalys and Cubeta, 1994). Increasingly, DNA based methods have also been used to investigate diversity within AG-1, to elucidate population structures with respect to particular pathosystems.

An intensive study into differentiation of ISGs within AG-1 using the rDNA - ITS region was conducted by Liu and Sinclair (1993) who were able to distinguish 6 ISGs (ISG 1A-1F). Twenty-four of the 61 isolates they studied had been isolated from rice. With the exception of one isolate which belonged to AG-1 IC, all of these had been identified as AG-1 IA isolates according to conventional concepts. Based on their analysis, however, these 23 AG-1 IA isolates belonged to 3 ISGs. Two ISGs IA and IF comprised the majority of strains. These groups did not exhibit a strong correlation with geographical origin. Isozyme analysis of the same set of isolates revealed that groups obtained by this approach did not correlate entirely with rDNA based groups, although several marker phenotypes could be identified in certain ISGs.

A high degree of variability in rDNA restriction patterns in AG-1 and its subgroups was also reported by Vilgalys and Gonzalez (1990). In particular, in AG-1 IA they observed a high degree of restriction fragment length polymorphism, which was attributed to divergence within this subgroup. In contrast, Jabaji-Hare *et al.* (1990), after digesting DNA of AG-1 isolates with *Eco*R1 and hybridising with different DNA clones, were not able to distinguish ISGs in AG-1 and suggested that this AG is composed of one homogeneous group with a single invariant RFLP pattern. Similar results were obtained in a study on the genetic relationships of 4 isolates of AG-1 IA and IB of *R. solani* by DNA/DNA hybridisation, which revealed a high degree of similarity within, but low levels of similarity between these ISGs (Kuninaga and Yokosawa, 1982). However, when analysing a larger set of isolates by DNA/DNA hybridisation, Vilgalys (1988) observed a considerable degree of variation in AG-1, which corresponded, at least in parts, with ISGs. Vilgalys explained the inconsistency between his study and the results presented in Kuninaga and Yokosawa (1982) by the different number of isolates, their geographic diversity and the differences in DNA hybridisation. The possibility of a process of microevolution within some AGs has also been suggested

A considerable degree of diversity was also observed when analysing isolates from rice-based cropping systems of various countries in SE Asia and West Africa (Banniza *et al.*, 1996; Banniza *et al.*, 1997). The majority of these isolates, although not all originating from rice plants, proved to belong to AG-1 IA, confirming that the typical pathogen of rice agro-ecosystems is mainly found in this ISG. Analysis of mitochondrial (mt) DNA RFLPs revealed that these AG-1 IA isolates were characterised by banding patterns which appeared to be variants of one major pattern (Banniza *et al.*, 1996). These patterns were distinct from those observed in non AG-1 isolates. Similarly, when analysing DNA of several of these isolates by variable number of tandem repeats (VNTR) PCR, the majority exhibited similar although distinct patterns, while a small group, including the non-AG-1 isolates had very different fragments (Banniza *et al.*, 1997). Both mt DNA RFLPs and VNTR-PCR profiles could not be correlated with geographical origin, and pathogenicity testing on rice showed that there was no significant difference in virulence among the majority of isolates.

5. Summary

A considerable effort has been made to increase understanding of the epidemiology of sheath blight and the host-pathogen interactions. It has been shown that modern production systems with new varieties and in particular an increase in nitrogen fertilisation have had a major impact on the rise of sheath blight as a major pathogen in rice production. However, despite this wealth of new insight into the disease, relatively little progress has been made in developing effective protection strategies. A major impediment to doing so has been a lack of knowledge of the diversity of the sheath blight pathogen and its dynamics in different agro-ecosystems.

A number of studies have been conducted with the objective of studying the diversity of isolates of *R. solani* using conventional criteria, and increasingly with modern molecular biology techniques. In the past, most of these studies have concentrated on clarifying the relationship between AGs, while during recent years more efforts have been made to shed light on the diversity of isolates present within various AGs. While some AGs, particularly AG-8 are well investigated, hardly anything is known about the population structure of other AGs.

The increase in economic importance of sheath blight has dramatically increased the interest in AG-1, and in particular AG-1 IA, and a number of systematic studies have been conducted as presented above. While it was previously assumed that this ISG within AG-1 was characterised by a high degree of homogeneity, recent studies have shown that isolates within this subgroup are much more diverse than expected. Further grouping has been attempted, but has not been conclusive so far. However, knowledge of the diversity of the rice sheath blight pathogens is essential for the understanding of the host-pathogen system, and for studying population structures and dynamics of this pathogen.

6. Acknowledgement

This work was funded by the UK Department for International Development as part of the research project R6643 'Investigation of the *Rhizoctonia* sheath disease complex in rice', commissioned through the Natural Resources Institute.

7. References

Anderson, N. A. (1982) The genetics and pathology of *Rhizoctonia solani*, *Annual Review of Phytopathology* **20**, 329-347.

Anonymous (1993) Country reports on crop loss due to sheath blight, in: *Rice sheath blight management workshop,. China National Rice Research Institute, Hangzhou, China, 11-15 October 1993*, IRRI, Philippines.

Anuratha, C.S., Zen-KuoChang, Cole, K.C., Mew, T., Muthukishnan, S. and Zen, K.C. (1996) Induction of chitinases and ß-1,3-glucanases in *Rhizoctonia solani*-infected rice plants: isolation of infection-related chitinase cDNA clone, *Physiologia-Plantarum* **97**, 39-46.

Baker, K.F. (1970) Types of *Rhizoctonia* diseases and their occurrence, in J.R. Parmeter Jr. (ed.), *Rhizoctonia solani: Biology and pathology*, University of California Press, Berkeley, Los Angeles.

Banniza, S., Rutherford, M.A., Bridge, P.D., Holderness, M and Mordue, J.E.M. (1996) Biological characterisation of *Rhizoctonia solani* in rice-based cropping systems. *Proceedings Brighton Crop Protection Conference*. Vol.**1**:399-404,The British Crop Protection Council, Farnham

Banniza, S., Bridge, P.D., Holderness, M., Rutherford, M.A. and Simons, S.A. (1997) Assessments of characteristics for determining pathogen populations of *Rhizoctonia solani* in rice-based cropping systems, in: *Proceedings of the 4th EFPP-Symposium on 'Diagnosis and Identification of Plant Pathogens'*.

Bera, S. and Purkayastha, R.P. (1995) Differential response of pathogenesis-related proteins to phytoalexin elicitors and its impact on sheath blight disease in rice, *Indian Journal of Experimental Biology* **32**, 902-905.

Borthakur, B.K. and Addy, S.K. (1988) Anastomosis grouping in isolates of *Rhizoctonia solani* causing sheath blight disease, *Indian Phytopathology* **41**, 351-354.

Bruns, T.D., White, T.J. and Taylor, J.W. (1991) Fungal molecular systematics, Annual Review of Ecological Systematics **22**, 525-564.

Cruickshank, R.H. (1990) Pectic zymograms as criteria in taxonomy of *Rhizoctonia*, *Mycological Research* **94**, 938-946.

Cu, R.M., Mew, T.W., Cassman, K.G. and Teng, P.S. (1996) Effect of sheath blight on yield in tropical, intensive rice production system, *Plant Disease* **80**, 1103-1108.

Damicone, J.P., Patel, M.V. and Moore, W.F. (1993) Density of sclerotia of *Rhizoctonia solani* and incidence of sheath blight in rice fields in Mississippi, *Plant Disease* **77**, 257-260.

Dodman, R.L. and Flentje, N.T. (1970) The mechanism and physiology of plant penetration by *Rhizoctonia solani*, in J.R.Parmeter Jr. (ed.), *Rhizoctonia solani, biology and pathology*, University of California Press, Berkley, pp. 149-160.

Egger, K.N. (1992) Analysis of fungal population structure using molecular techniques, in G.C Caroll and D.T. Wicklow (eds.), *The fungal community. Its organization and role in the ecosystem*, Marcel Dekker, New York.

Flentje, N.T., Dodman, R.L. and Kerr, A. (1963) The mechanism of host penetration by *Thanatephorus cucumeris*, *Australian Journal of Biological Science* **16**, 784-799.

Foroutan, A. and Rahimian, H. (1991) Distribution and characteristics of isolates of *R. solani* and *R. oryzae-sativae* in rice fields of Mazandaran, *Iranian Journal of Plant Pathology* **27**, 1-4.

Gangopadhyay, S. and Chakrabarti, N.K. (1982) Sheath blight on rice, *Review of Plant Pathology* **61**, 451-460.

Groth, D.E. and Nowick,E.M. (1992) Selection for resistance to rice sheath blight through number of infection cushions and lesion type, *Plant Disease* **76**, 721-723.

Jabaji-Hare, S.H., Meller, Y., Gill, S. and Charest, P.M. (1990) Investigation of genetic relatedness among anastomosis groups of *Rhizoctonia solani* using cloned DNA probes, *Canadian Journal of Plant Pathology* **12**, 393-404.

Johnk, J.S. and Jones, R.K. (1992) Determination of whole-cell fatty acids in isolates of *Rhizoctonia solani* AG-1 IA, *Phytopathology* 82, 68-72.

Johnk, J.S. and Jones, R.K. (1993) Differentiation of populations of AG-2-2 of *Rhizoctonia solani* by analysis of cellular fatty acids, *Phytopathology* **83**, 278-283.

Johnk, J.S. and Jones, R.K., (1994): Comparison of whole-cell fatty acid compositions in intraspecific groups of *Rhizoctonia solani* AG-1, *Phytopathology* **84**, 271-275.

Jones, R.K. and Belmar, S.B., (1989): Characterisation and pathogenicity of *Rhizoctonia* spp. isolated from rice, soybean, and other crops grown in rotation with rice in Texas, *Plant Disease* **73**, 1004-1010.

Kim, W.G., Kim, C.K. and Yu, S.H. (1988) Anastomosis grouping and cultural characteristics of isolates of *Rhizoctonia solani* Kühn from sclerotia overwintered in paddy fields, *Korean Journal of Plant Pathology* **4**, 136-142.

Kuninaga, S. and Yokosawa, R. (1982) DNA base sequence homology in *Rhizoctonia solani* Kühn I. Genetic relatedness within anastomosis group 1, *Annals of the Phytopathological Society of Japan* **48**, 659-667.

Lee, F.N., (1980): Number, viability and buoyancy of *Rhizoctonia solani* sclerotia in Arkansas rice fields, *Plant Disease* **64**, 298-300.

Lee, F. N. and Rush, M.C. (1983) Rice sheath blight: A major rice disease, *Plant Disease* **67**, 829-832. Lin, W., Anuratha, C.S., Datta, K., Potrykus, I., Muthukrishnan, S. and Datta, S.K. (1995) Genetic Engineering of rice for resistance to sheath blight, *Biotechnology* **13**, 686-691.

Liu, Z.L. and Sinclair, J.B. (1993) Differentiation of intraspecific groups within anastomosis group 1 of *Rhizoctonia solani* using ribosomal DNA internal transcribed spacer and isozyme comparison, *Canadian Journal of Plant Pathology* **15**, 272-280.

Marshall, D.S. and Rush, M.C., (1980a): Infection cushion formation on rice sheaths by *Rhizoctonia solani*. *Phytopathology*, 70:947-950.

Marshall, D.S. and Rush, M.C. (1980b) Relation between infection by *Rhizoctonia solani* and *R. oryzae* and disease severity in rice, *Phytopathology* 70, 941-946.

Massaquoi, R.C. and Rush, M.C., (1987): Relationship of quantity of epicuticular wax to resistance of rice to sheath blight. [Abstract] *Phytopathology*, **77**:1723

Matsuyama, N., Moromizato, Z., Ogoshi, A. and Wakimoto, S. (1978) Grouping *Rhizoctonia solani* with non-specific esterase zymogram, *Annals of the Phytopathological Society of Japan* **44**, 652-658.

Mew, T.W. and Rosales, A.M.(1986) Bacterization of rice plants for the control of sheath blight caused by *Rhizoctonia solani*, *Phytopathology* **76**, 1260-1264.

Miyake, I. (1910) Studien über die Pilze der Reispflanze in Japan, *Journal of the College of Agriculture, Imperial University of Tokyo* **2**, 237-276.

Ogoshi, A. and Ui, T. (1983) Diversity of clones within an anastomosis group of *Rhizoctonia solani* Kühn in a field, *Annals of the Phytopathological Society of Japan* **49**, 239-245.

Ogoshi, A., Cook, R.J. and Bassett, E.N. (1990) *Rhizoctonia* species and anastomosis groups causing root rot of wheat and barley in the Pacific Northwest, *Phytopathology* **80**, 784-788.

Ou, S.H. (1985) *Rice diseases*, Commonwealth Mycological Institute, Great Britain.

Palo, M.A. (1926) *Rhizoctonia* disease of rice: I. Study of the disease and of the influence of certain conditions upon the viability of the sclerotial bodies of the causal fungus, *Philippine Agriculturist* **15**, 361-375.

Park, A.R.C.S. and Bertus, L.S. (1932) Sclerotial diseases of rice in Ceylon. 1. *Rhizoctonia solani* Kühn, *Ceylon Journal of Science* **A11**, 319-331.

Parmeter, J.R.,Jr., Sherwood, R.T. and Platt, W.D. (1969) Anastomosis grouping of *Thanatephorus cucumeris*, *Phytopathology* **59**, 1270-1278.

Parmeter, J.R.,Jr., and Whitney, H.S. (1970) Taxonomy and Nomenclature of the imperfect state, in J.R. Parmeter, Jr (ed.), *Rhizoctonia solani: Biology and pathology*, University of California Press, Berkeley.

Rawn, J.D. (1983) *Biochemistry*, Harper & Row Publishers Inc.

Reinking, O.A. (1918) Philippine economic plant diseases, *The Philippine Journal of Science* **A13**, 165-274.

Savary, S., Castilla, N.P., Elazegui, F.A., McLaren, C.G., Ynalvez, M.A. and Teng, P.S. (1995) Direct and indirect effects of nitrogen supply and disease source structure on rice sheath blight spread, *Phytopathology* **85**, 959-965.

Shahjahan, A.K.M., Fabeller, N. and Mew, T.W., (1990): Effect of crop management practices on the sclerotia dynamics of *Rhizoctonia solani* in upland rice, *Bangladesh Journal of Plant Pathology* **6**, 19-23.

Sherwood, R. T. (1969) Morphology and pathology of four anastomosis groups of *Thanatephorus cucumeris*, *Phytopathology* **59**, 1924-1929.

Shirai, M. (1906) On *Hypochnus sasakii* n. sp., [Abstract] *Botanical Magazine, Tokyo* **20**, 319-323.

Singh, S.K., Satyanarayana, K. and Reddy, A.P.K. (1990) Studies on morphology, growth habit, hyphal anastomosis and virulence pattern of five isolates of sheath blight pathogen of rice, *Indian Phytopathology* **43**, 368-371.

Sneh, B., Burpee, L. and Ogoshi, A., (1991) Identification of *Rhizoctonia* species, APS Press, Minnesota.Vijayan, M. and Nair, C. M. (1985) Anastomosis grouping of isolates of *Rhizoctonia solani* Kühn (*Thanatephorus cucumeris* (Frank) Donk) causing sheath blight of rice *Current Science* **54**, 89-291.

Vilgalys, R. (1988) Genetic relatedness among anastomosis groups in *Rhizoctonia* as measured by DNA/DNA hybridization, *Phytopathology* **78**, 698-702.

Vilgalys, R. and Gonzalez, D. (1990) Ribosomal DNA restriction length polymorphisms in *Rhizoctonia solani, Phytopathology* **80**, 151-158.

Vilgalys, R. and Cubeta, M.A. (1994) Molecular systematics and population biology of *Rhizoctonia, Annual Review of Phytopathology* **32**, 135-155.

Wei, C.T. (1934) *Rhizoctonia* sheath blight of rice, *Bulletin of the College of Agriculture and Forestry, University of Nanking* **15**, 1-21.

Yamaguchi, T., Kardin, M.K., Wagiman and Tantera, D.M. (1984) Hyphal anastomosis groups of *Rhizoctonia solani* Kühn in Indonesia, *Penelitian Pertanian* **4**, 26-28.

Zuber, M. and Manibhushanrao, K. (1982) Studies on comparative gel electrophoretic patterns of proteins and enzymes from isolates of *Rhizoctonia solani* causing sheath blight disease in rice. *Canadian Journal of Microbiology* **28**, 762-771.

RELP ANALYSIS OF THE PCR-AMPLIFIED 28S RIBOSOMAL DNA FOR REVISION OF GENETIC RELATIONSHIPS IN *RHIZOTONIA* SPP.

MASARU MATSUMOTO and NOBUKAI MATSUYAMA
Laboratory of Plant Pathology,
Faculty of Agriculture, Kyushu University,
Fukuoka 812-81, Japan.

1. Introduction

Phytopathogenic fungi classified as *Rhizoctonia* spp. represent an admixture of taxonomically separate groups that vary in many significant characteristics, including their sexual stages (teleomorph), asexual stages (anamorph), and other characters (Tu and Kimbrough, 1978; Sneh *et al.*, 1991; Vilgalys and Cubeta, 1994). Understanding the genetic relationships is likely to aid in redefining the systematics of fungi described under the genus *Rhizoctonia.*

Casual agents of rice sheath blight disease and similar diseases, which have been identified as multinucleate *Rhizoctonia* spp. Such as *R.solani* anastomosis group (AG)1 IA and AG 2 2 IIIB (Ogoshi, 1976, 1987), *R.oryzae* (Oniki *et al.*, 1985) and binucleate *Rhizoctonia* spp. such as *R. fumigata* (Ogoshi *et al.*, 1979, 1983; Burpee *et al.*, 1980) and *R. oryzae - sativae* (Ogoshi *et al.*, 1979; Gunnell and Webster, 1987), have been previously classified according to their ecology, morphology and hyphal anastomosis (Watanabe and Matsuda, 1966; Ogoshi, 1976; Ogoshi *et al.*, 1979), and molecular taxonomic criteria (Vilgalys and Gonzalez, 1990; Damaj *et al.*, 1993; Duncan *et al.*, 1993).

Analysis of nuclear - encoded ribosomal DNA (rDNA) by restriction fragment length polymorphism (RELP) revealed that different AGs in *R. solani* and binucleate *Rhizoctonia* spp. Are genetically divergent (Cubeta *et al.*, 1991; Liu and Sinclair, 1992, 1993; Liu *et al.*, 1993; Matsumoto *et al.*, 1996a,b). However, genetic relationships between binucleate and multinucleate *Rhizoctonia* species are not known. Such variation would provide better understanding of the genetic divergence of *Rhizoctonia* spp., causal agents of rice diseases, because similar symptoms on rice sheath are difficult to differentiate by visual observation. In the present study, an attempt was made to investigate the genetic relationships in these *Rhizoctonia* spp., based on the results of RFLPs analysis of the polymerase chain reaction (PCR) amplified nuclear-encoded 28S rDNA.

S. Sreenivasaprasad and R. Johnson (eds.), Major Fungal diseases of Rice Recent Advances, 213-222.
Professor K. Manibhushan Rao Festschrift

2. Materials and Methods

Thirty isolates of *Rhizoctonia* species used in this study are listed in Table1. To prepare DNA samples for restriction analysis, three thin mycelial discs (4 mm in diameter) of each isolate cultured on PDA, were used to inoculate 50 ml of a liquid nutrient-broth medium (Difco) amended with 50mg/L of streptomycin sulfate in 100 ml flasks and incubated at 25° C for 7 days. The mycelial mat was harvested, washed repeatedly with water, lyoplised and then stored at 20° C until use. Extraction of genomic DNA was done according to the procedure of Vilgalys and Gonzalez (1990) with slight modifications. The lyophilised and ground mycelia (30 mg) were homogenised in 450μl of extraction buffer (50 mM Tris-HCl, pH 7.2, 50 mM EDTA, 1% sodium N-lauryl sarconsinate, 1% 2-Mercaptoethanol) and incubated at 65° C for 30min. The homogenate was thoroughly shaken using 450 μl of chloroform/phenol/isoamyl alcholol (24:25:1, v/v/v) mixture. After centrifugation for 15 min at 15,000 rpm, the water phase was collected and ammonium acetate was added to a final concentration of 2.5 M. DNA was precipated with 225 μl of isopanol and then collected by the centrifugation at 13,000 rpm for 15min. The pellet was washed with 70% aqueous ethanol solution, dired *in vacuo,* dissolved in 120 μl of TE buffer (10mM Tris-HCI, pH 8.0, 1 mM EDTA) and incubated at 37.5 C for 30 min for treatment with ribonuclease A (20mg/ml, DNase-free; Sigma, USA).

Two primers were synthesised for the amplification of a portion of 28S rDNA repeat homologous to positions 17-1, 448 within the *Saccharomyces cerevisiae* 25S ribosomal RNA according to the procedure of Vilgalys and Hester (1990). Ten μl of 100-fold dilution of DNA solution prepared by the above procedure was added to reaction mixture as recommended by the manufacture using *Tth* DNA polymerase (TOYOBO Biochemicals Co. Japan) in 100 μl volume. Thirty thermal cycles of 1 min at 94° C, 2 min at 50° C, 1 min ramp from 50-72° C and 3 min at 72° C were performed. After amplification, each sample was subjected to electrophoresis using a 1% agarose gel in TBE buffer (100 mM Tris-HCL, 20 mM EDTA, 100 mM boric acid, pH 8.3) After preparation of PCR-amplified rDNA, each PCR product was digested with 8 restriction enzymes (*Msp*I, *Hae*III, *HhaI, Bam*HI, *EcoRI, HinIII, TaqI* and *SauI*3AI) and the sample was subjected electrophoresis in a 3.5% agarose gel (NuSieve, FMC Bioproducts) with TBE buffer.

Relative similarities among isolates were estimated by the cluster analysis. A resemblance matrix based on Euclidean distance coefficients was computed from pairwise comparison of each isolate with every other isolate based on the number and size of digested fragments with each restriction enzyme. Principal component analysis (SYSTAT: Version 5.2.1, SYSTAT Inc.) was used to determine the degree of variability among *Rhizoctonia* spp. A dendrogram, for objective assessment of genetic relationships in *Rhizoctonia* spp., was constructed by an unweighted paired group method with arithmetic averaging (UPGMA) using the statistics software package SYSTAT 5.2.1.

TABLE 1. Isolates of *Rhizoctonia* species analysed in this experiment

Species	Isolate	AG	Host	Origin
Rhizoctonia solani	Cs-Ka	1 IA	Rice	ATCC[a)]76121
	C-325	1 IA	Rice	IFO[b)] 30935
	C-326	1 IA	Rice	IFO30936
	Cs-Gi	1 IA	Rice	AHU[c)]
	CS-IW	1 IA	Rice	AHU
	NI-9	1 IA	Rice	AHU[d)]
	TY-1	1 IA	Rice	AMU[c)]
	TY-2	1 IA	Rice	AKU
	C-96	2-2IIIB	Mat rush	ATCC 76155
	C-116	2-2IIIB	Mat rush	MAFF[f)]
	C-321	2-2IIIB	Mat rush	IFO 30943
	C-328	2-2IIIB	Rice	IFO 30944
	C-335	2-2IIIB	Rice	MAFF
	C-354	2-2IIIB	Rice	AKU
	KU-PI	2-2IIIB	Rice	AKU
	KU-P2	2-2IIIB	Rice	AKU
R. oryzae	Ro-0105	WAG-O	Rice	AMU
	M23	WAG-O	Rice	AMU
	R-1	WAG-O	Rice	AKU
	KAES-14	WAG-O	Rice	AKU
	C-505	WAG-O	Rice	ATCC 76155
	T-1101	WAG-O	Rice	AKU
	M.2122	WAG-O	Rice	AKU
	86TA13	WAG-O	Rice	AMU
R. oryzae-sativae	93Gi	Bb	Rice	AMU
	94S	Bb	Rice	AKU
	KS-T2-6	Bb	Rice	AKU
	C-445	Bb	Rice	ATCC76135

	STU-4	Bb	Rice	AKU
	SOS-1	Bb	Rice	AKU
R. fumigata	TO-7	Ba	Rice	AMU
	TM-2B	Ba	Rice	AMU
	KST1-3	Ba	Rice	AKU
	K-TK.T41	Ba	Rice	AKU

a) ATCC: American Type Culture Collection, USA
b) IFO: Insitute for Fermentation, Osaka, Japan
c) AHU: Faculty of agriculture, Hokkaido University, Hokkaido, Japan
d) AMU: Faculty of Agriculture, Meijyo University, Fukuoka, Japan
e) AKU: Faculty of Agriculture, Kyushu University, Fukuoka, Japan
f) MAFF: Ministry of Agriculture, Forestry and Fisheries, Ibaraki, Japan.

3. Results and Discussion

PCR amplification the 28S rDNA yielded a single 1.4 kb or 1.8 kb fragment, as detected by agarose gel electrophoresis (Figure 1). A single 1.4 kb fragment was obtained from two test isolates (*R. oryzae and R. oryzae-sativae*) and a 1.8-kb fragment was obtained from remaining isolates (*R. solani* AG 11A and AG 2-2 IIIB, and *R. fumigata).* Test isolates of *R.solani* AG 2-2 IIIB obtained from diseased mat rush and rice plants did not reveal the size variation of PCR-amplified products 28S rDNA (data not shown).

RFLP profiles obtained from digestion of PCR-amplified 28S rDNA represented specific fragment patterns (Figure .2 and 3, Table 2). Four enzymes, *HhaI, MspI, HaeIII* and *Sau*3AI, were used in differentiating different *Rhizoctonia* species. After digestion of PCR-amplified 28S rDNA with *Msp*I, all isolates produced a common 290-bp fragment. However, specific restriction fragments were obtained from isolates among four *Rhizoctonia* species (Figure 2, Table 2). Variation of restriction profiles in *R. solani* isolates was shown. Fragments of 310bp and 450 bp in *R. solani* AG 1IA and 280 bp and 500 bp in AG 2 2 III were obtained, respectively (Figure 2, Table 2). Digestion of the amplified 28SrDNA of *R. fumigata* and *R. oryzae-sativae* produced common 170 bp and 500 bp fragments and specific 330bp and 400 bp fragments, respectively (Figure 2, Table 2). Digestion with the enzymes *Hha*I, *Sau*3AI and *Hae*III revealed, a common 160 bp fragment with *Hha*I (Figure 3, Table 2), common 110bp and 800 bp fragments with *Sau*3AI (Table 2) and common 150 bp and 220 bp fragments with *Hae*III (Table2), respectively. The *Hha*I, *Sau*3AI and *Hae*III RELP profiles of AG 1IA and AG 2 2 IIIB in *R. solani* were identical. However, RFLP profiles with these enzymes presented significant variation for four *Rhizoctonia* spp. including the two *R. solani* groups stated above (Figure 2 and 3, Table 2)., All isolates of *Rhizoctonia* spp. produced a common 230 bp fragment after digestion of PCR amplified 28S rDNA with *Hha*I except for isolates of *R. oryzae* that produced unique sets of 260, 390 and 480 bp fragments

(Figure 3, Table 2) Digestion with *Sau*3AI and *Hae*III yielded unique sets of fragments among isolates of four *Rhizoctonia* spp. (Table 2). RELP profiles with *Sau*3AI were different among the isolates of four *Rhizoctonia* spp., whereas in the case of *R. fumigata* and *R. oryzae-sativae*, a common 450 bp fragment and unique sets of fragments were obtained by digestion with *Hae*III (Table 2).

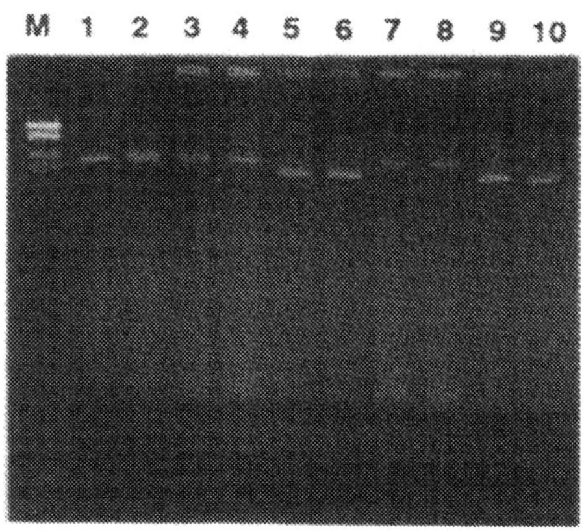

Figure1. Electrophoretic pattern of PCR - amplified 28S rDNA of *R. solani* AG 1 IA and AG 22 IIB, *R.solani, R oryzae, R.fumigata and R.oryzae-sativae* in a1.0% agarose gel. Lane 1, Cs-Ka; Lane 2, C-325; lane 3, C-96; Lane 4, C-354; Lane 3 C-96; Lane 4,C-354; Lane 5, C-505; Lane 6, Ro-0105; Lane 7,TO-7; Lane 8, KS-T1-3; Lane 9, 93Gi; Lane 10, K-T5 and M, size marker of lamda-DNA digested with *Eco* RI and *Hind*III (from top to bottom, 24756, co-migrating bands of 5148, 5148, 4937 and 4268, co-migrating bands of 2027 and 1904, 1584, 947, 831 and 564.

Principal component analysis showed distinct genetic relationships based on RELP profiles of PCR amplified 28S rDNA with *Hha*I, MspI, *Sau*3AI and *Hae*III digestion among isolates of four *Rhizoctonia* spp. including *R. solani* AG and AG 2-2 IIIB (Figure 4). The first two principal components (component 1 and 2) of this analysis accounted for 48.5% of the variation in the data. Principal component 1 accounted for 36.7% of the variability and principal component 2 accounted for 21.8% of the variability (Data not shown). Based on the patterns of digestion products of PCR amplified 28S rDNA with *Hha*I, *Msp*I, *Sau*3AI and *Hae*III, the relationships among four *Rhizoctonia* spp., including the two *R. solani* AGs were investigated by cluster analysis (Figure 5). The dendrogram generated revealed that the isolates of *R. oryzae* had divergent phylogenetic relationships. Two groups of *R.Solani*, AG 1IA and AG 2-2 IIIB were closely related in RFLP analysis, whereas *R. fumigata* and *R. oryzae-sativae* were genetically distinct.

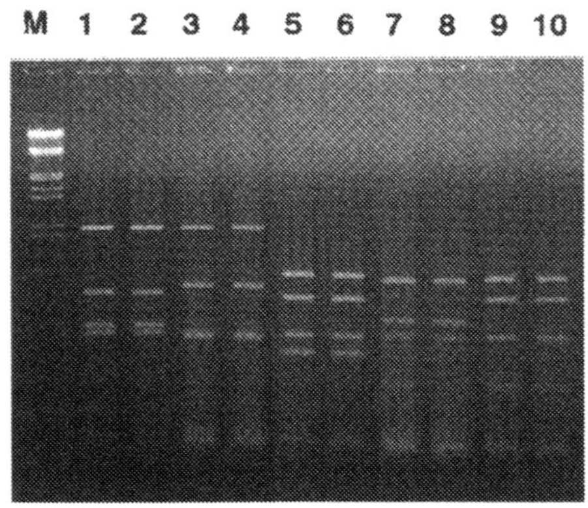

Figure 2. Electrophoretic pattern od PCR -amplified 28S rDNA of *R. solani* AG 1 IA and AG 22 IIB, *R.solani, R oryzae, R.fumigata and R.oryzae-sativae* in a 3.5% agarose gel. Lane 1, Cs-Ka; Lane 2, C-325; lane 3, C-96; Lane 4, C-354; Lane Lane C-96; Lane 4, C-354 5, C-505; Lane 6, Ro-0105; Lane 7,TO-7; Lane 8, KS-T1-3; Lane 9, 93Gi; Lane 10, K-T5 and M, size marker of lamda-DNA digested with *Eco* RI and *Hind*III (from top to bottom, 24756, co-migrating bands of 5148, 5148, 4937 and 4268, co-migrating bands of 2027 and 1904, 1584, 947, 831 and 564.

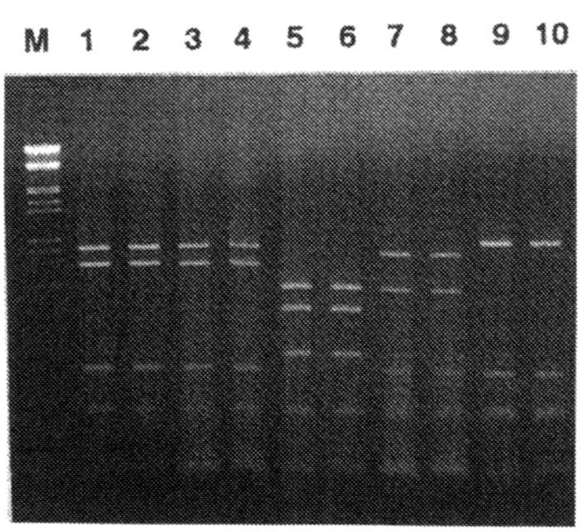

Figure 3. Electrophoretic pattern od PCR -amplified 28S rDNA of *R. solani* AG 1 IA and AG 22 IIB, *R.solani, R oryzae, R.fumigata and R. oryzae-sativae* in a 3.5% agarose gel. Lane 1, Cs-Ka; Lane 2, C-325; lane 3, C-96; Lane 4, C-354; Lane Lane C-96; Lane 4, C-354 5, C-505; Lane 6, Ro-0105; Lane 7,TO-7; Lane 8, KS-T1-3; Lane 9, 93Gi; Lane 10, K-T5 and M, size marker of lamda-DNA digested with *Eco*RI and *Hind*III (from top to bottom, 24756, co-migrating bands of 5148, 5148, 4937 and 4268, co-migrating bands of 2027 and 1904, 1584, 947, 831 and 564).

TABLE 2. DNA fragments (bp) generated by restriction of 28S rRNA gene from four *Rhizoctonia* spp., with four different enzymes

	Restriction enzyme			
Species[a)]	*MspI*	*Hha*I	*Sau*3AI	*Hae*II
R. solani AG 1-IA	290, 310, 450, 900	160, 230, 680, 850	110, 800, 900	150, 220, 250, 330, 450, 500
R. solani AG 2-2 IIIB	280, 290, 500, 900	160, 230, 680, 880	110, 800, 900	150, 200, 220, 330, 450, 500
R. oryzae	260, 290, 400, 550	160, 260, 390, 480	110, 200, 250, 800	100, 150, 220, 900
R. fumigata	170, 290, 330, 500	160, 230, 480, 880	110, 800, 850	150, 180, 220, 250, 450
R. oryzae-sativae	170, 290, 500	160, 230, 850	110, 800, 550	150, 220, 450, 550

a) Represent 12 AGs of *R. solani*, AGs of *R. oryzae* and 16 Japanese AGs of binucleate *Rhizoctonia spp.*

Analysis of the rDNA gene block for systematic studies is based on its usefullness as an evolutionary marker, considering that these genes contain, highly conserved sequences and variability over their entire length. It is possible to amplify and study DNA sequence variation occurring between PCR primer sites at almost any desired taxonomic level. Genetic analysis based on RFLP profiles by the digestion of PCR amplified 28S rDNA with *Hha*I, *Msp*I, *Sau*3AI *and Hae*III, revealed genetic relationships among four *Rhizoctonia* species including *R. solani* AG 1IA and AG 2-2 IIIB. The relationship among these *Rhizoctonia* species are important, because of the similar symptoms induced by them on rice plants. The isolates of *R.solani* AG1IA and AG 2-2 IIIB showed identical RFLP profiles after the digestion of the PCR amplified 28S rDNA with *Hha*I, *Msp*I, *Sau*3AI and *Hae*III, but showed different RFLP profiles with *Msp*I. Isolates belonging to four *Rhizoctonia* spp. *R. solani, R. oryzae, R. fumigata and R. oryza-sativae* showed different RFLP patterns and they were genetically divergent.

Recent studies based on RFLP analysis using PCR amplified rDNA have revealed genetic difference among different AGs in *R. solani* (Liu *et al.*, 1992, 1993; Liu and Sinclair, 1993; Matsumoto *et al.*, 1996a,b). RFLP profiles of PCR amplified rDNA indicated the existence of at lease three groups of isolates in AG 1 and five groups in AG 2. The results from this study supported the view that isolates of *R. solani* AG1 IA and AG2 2IIIB, causal agents of rice sheath diseases, are divergent based on RFLP profiles with *Msp*I. Isolates of *R. fumigata* and *R. oryzae-sativae*, predominant rice sheath pathogens, are morphologically similar and frequently anastomose (Ogoshi *et al.*, 1979, 1983) and but appear divergent based on RFLP analysis (Cubeta *et al.*, 1990, 1991).

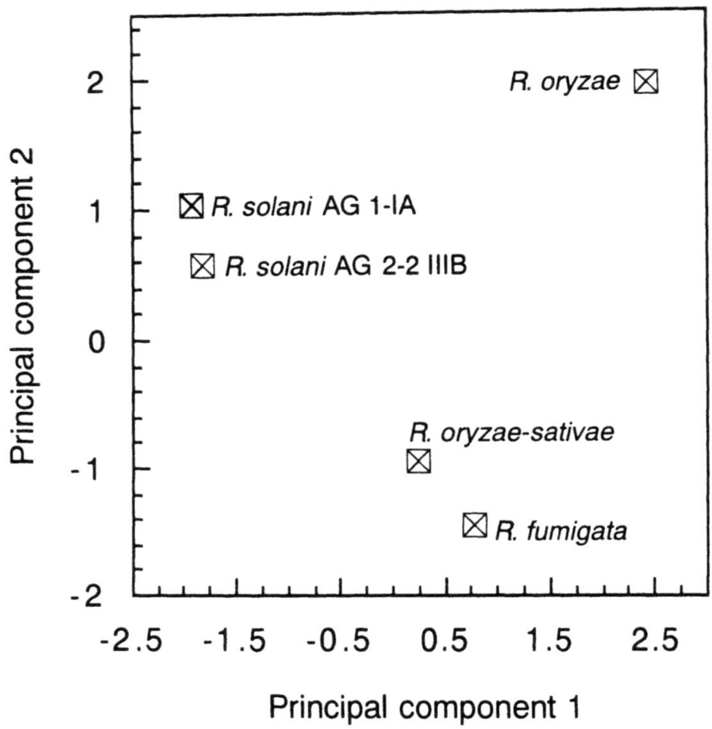

Figure 4. Variability in RFLP profiles of amplified 28S rDNA for *R.solani, R.oryzae, R. fumigata and R. oryzae-sativae*. Plot of the first two principal components (Principal Component 1 and 2) was derived from the digestion fragments with *Msp*I, *Hha*I, *Sau*3AI *and Hae*III

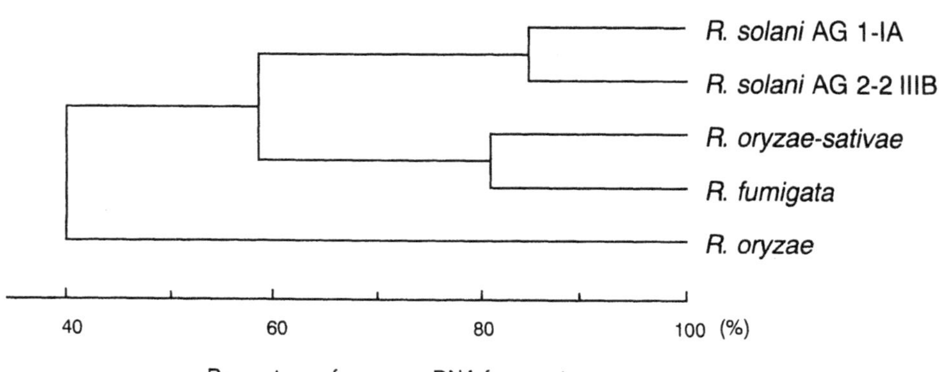

Figure 5. Dendrogram showing the genetic relationships among rDNA of *R. solani* AG 1 IA and AG 22 IIB, *R. solani, R. oryzae, R. fumigata and R. oryzae-sativae* based on the RFLP profiles.

Genetic divergence among isolates of *R. solani*, *R. oryzae*, *R. fumigata and R. oryzae-sativae*, was discerned by principal component and cluster analysis based on the RFLP profiles. Among multinclete *Rhizoctonia* spp., isolates of *R.solani* AG 1 IA and AG 2 2 IIB represented close genetic relationships, whereas isolates of *R. solani* and *R. oryzae* were genetically divergent. Isolates of binucleate *Rhizoctonia* spp. *R. fumigata* and *R. oryzae-sativae* were clearly divergent from multinucleate *Rhizoctonia* species such as *R. solani* and *R. oryzae* based on RFLP profiles.

4. Summary

We conclude that the genetic information enabled characterisation of relationships among isolates of *Rhizoctonia* spp., which are predominantly rice pathogens. Further investigation by comparison of sequence data from PCR amplified rDNA fragments will throw light on evolutionary relationships among different *Rhizoctonia* spp. It would also be useful to identify genetic markers that distinguish phylogenetically divergent *Rhizoctonia* spp. more clearly and rapidly.

5. References

Burpee, L.L., Sanders, P.L., Cole, H.Jr. and Sherwood, R.T. (180) Anastomosis groups among isolates of *Ceratobasidium cornigerum*, *Mycologia* **72,** 689-701.

Cubeta, M.A., Echandi, E., Abernethy, T. and Vilgalis, R. (1990) Characterisation of anastomosis groups of binucleate *Rhizoctonia* fungi using restriction analysis of ribosomal RNA genes, *Phytopathology* **80,** 1008 (abstract).

Cubeta M.A., E., Abernethy, T. and Vilgalis, R. (1991) Characterisation of anastomosis groups of binucleate *Rhizoctonia* species using restriction analysis of an amplified ribosomal RNA gene, *Phytopathology* **81,** 1390-1400.

Damaj, M., Jabaji-Hare, S.H. and Charest, P.M. (1993) Isozyme variation and genetic relatedness in binucleate *Rhizoctonia* spp., *Phytopathology* **83**, 864-871.

Duncan, S., Barton, E.J. and Brien, A.P. (1993). Analysis of variation in isolates of *Rhizoctonia solani* by random amplified polymorphic DNA assay, *Mycological Research* **97**, 1075-1082.

Gunnel, P.S. and Webster, R.K. (1987) *Ceratobasidium oryzae-sativae* sp.nov.,the teleomorph of *Rhizoctonia oryzae-sativae* and *Ceratobasidium setariae* comb. nov., the probabble teleomorph of *Rhizoctonia Fumigata* com.nov., *Mycologia* **79**, 731-736.

Liu, Z.L. and Sinclair, B.J. (1992) Genetic diversity of *Rhizoctonia solani* anastomosis group 2, *Phytopathology* **82,** 778-787.

Liu, Z.L. and Sinclair, B.J. 1993. Molecular differentiation of intrapsecific groups within anastomosis group I of *Rhizoctonia solani*, *Canadian Journal of Plant Pathology* **15,** 272-280.

Liu, Z.L. Domier,L.L. and Sinclair, B.J. 1993. ISG-specific ribosomal DNA polymorphism of th *Rhizoctonia solani* anastomosis group 2, *Mycologia* **85**, 795-800.

Mastumoto,M., Furuya, N. and Matsuyama. N. (1996a) PCR.RFLP analysis of amplified 23S ribosomal DNA for identification of *Rhizoctonia* spp., the causal agents of sheath diseases of rice plants, *Journal of Faculty of Agriculture Kyushu University* **41**, 39-44.

Matsumoot, M., Furuya, N., Takanami, Y. and Matsuyam, N.(1996) RFLP analysis of the PCR-amplified 28S rDNA in *Rhizoctonia solani*. *Mycoscience* **37**, 351-356.

Ogoshi, A. (1976) Studies on the grouping of *Rhizoctonia* solani Kuhn with hyphal anastomosis and on the perfect stages of groups, *Bulletin of National Institute of Agriculture Science Service* **C30**, 1-63.

Ogoshi, A. (1987). Ecology and pathogenicity of anastomosis and intraspecific groups of *Rhizoctonia* solani Kuhn, *Annual Review of Phytopathology* **25,** 125-143.

Ogoshi, A., Oniki, M.,Araki,T. and Ui, T. 91983). Studies on the anastomosis groups of binucleate *Rhizoctonia* and their perfect states, *Journal of the Faculty of Agriculture Hokkaido University* **61**, 244-260.

Ogoshi, A.Oniki, M., Sakai, R. and Ui, T. (1979). Anastomosis grouping among isolates of binucleate *Rhizoctonia, Transactions of the Mycological Society of Japan* **20**, 33-39.

Oniki, M., Ogoshi, A., Araki, T., Sakai, R. and Tanaka, S. (1985)., The perfect state of *Rhizoctoniap oryzae* and *R.. zeae* and the anastomosis groups of *Waitea circinata, Transactions of the Mycological Society of Japan* **20**, 33-39.

Sneh, B., Burpee, L.L. and Ogoshi, A., (1991) *Identification of Rhizoctonia species,* APS PRESS, St.Paul, MN.

Tu, C.C. and Kimbrough, J. W. (1978) Systematic and phylogeny of fungi in the *Rhizoctonia* complex, *Botanical Gazette* **139**, 454-466

Vilgalys, R. J.and Gonzakez, D. (1990) Ribosomal DNA restriction fragment length pollymorphisms in *Rhizoctonia solani, Phytopathology* **80**, 151-158.

Vilgalys, R. and Cubeta, M.A. (994). Molecular systematics and population biology of *Rhizoctonia*, *Annual Review of Phytopathology* **32**, 135-155.

Vilgalys, R. and Hester, M, (1990). Rapid genetic identification and mapping of enzymatically amplified ribosomal DNA from several *Cryptococcus* species, *Journal of Bacterialogy* **172**, 4238-4246.

Watanabe, B. and Matsuda, A. (1966). Studies on the grouping of *Rhizoctonia solani* Kuhn pathogenic to upland crops, *Bull. Appointed. Exp.* **7**, 1-131.

PROTOPLAST FUSION AND DNA PLASMID CHARACTERISATION IN *RHIZOCTONIA SOLANI*

T. HASHIBA, A. SASAKI and K. KATSURA
Faculty of Agriculture, Tohoku University
1-1 Tsutsumidori-Amamiyamachi Sendai 981, Japan.

1. Introduction

Protoplasts have been used routinely for physiological and biological investigations. More recently, protoplasts have been frequently used for genetic studies. Two methods have been developed for introducing favorable genes into microbial protoplasts (Figure 1). One method, protoplast fusion (Figure 1) serves as a convenient method for introducing genes from one fungal isolate to another. The protoplast fusion process has been shown to generate a large number of random recombinants. Furthermore, because the protoplast fusion is observed not only among different strains but also different species, this method of gene transfer may be advantageous to improve fungal strains. The other method, genetic transformation of protoplasts by the uptake of recombinant DNA (Figure 1) is widely used.

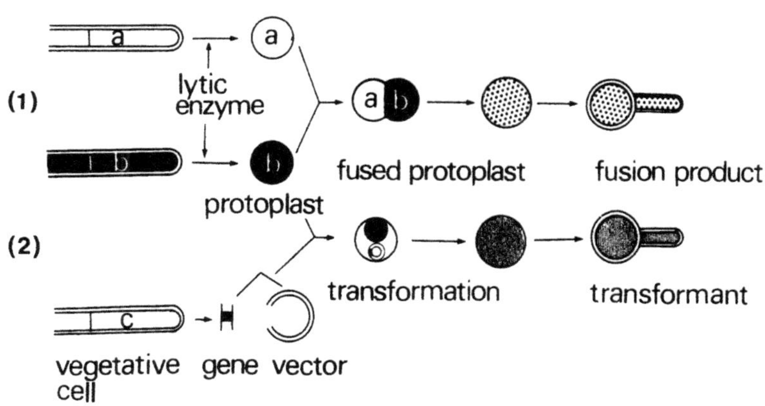

Figure 1. Two methods for introducing favorable genes into microbial protoplasts. (1) Genetic fusion of protoplasts. (2) Genetic transformation of protoplasts by DNA uptake.

S. Sreenivasaprasad and R. Johnson (eds.), Major Fungal diseases of Rice Recent Advances, 223-233.
Prof. K. Manibhushan Rao Festschrift

The development of an efficient transformation system is vital for genetic studies and strain improvement using the recombinant DNA technology. Because vectors are not readily available for the efficient transformation of *Rhizoctonia solani*, we have to open up the development of expression vectors. Many linear plasmids have been found in the species-complex *R. solani*, AGs 1-6 (Miyasaka *et al.*, 1990; Wako *et al.*, 1991; Jabaji-Hare *et al.*, 1994). Most of the plasmids are associated with mitochondria (Samac and Leong, 1989). The biological function of the plasmids is unknown but they occur in unusually high copy numbers in the host cytoplasm. These linear plasmids are likely to be useful for constructing transformation vectors with autonomous replication.

In this review, we describe improved methods for protoplast fusion, and the isolation and characterisation of the linear DNA plasmids from *R. solani*.

2. Protoplast fusion

2.1. FORMATION OF PROTOPLASTS

Various method for the production and isolation of protoplasts from different fungal groups have been described, and detailed discussion on protoplasts may be found in some reviews (Peberdy, 1979; Davis, 1985; Hashiba, 1992).

In this part, the modified method of Hashiba and Yamada (1982) for protoplast formation in *Rhizoctonia solani* is described (Figure 2) (Wako *et al.*, 1991). One gram (fresh weight) of the mycelium grown for 30-36 h in 20 ml of potato-sucrose liquid medium containing polypeptone was harvested on a 150 μm steel sieve, washed with distilled water. The mycelium was suspended in 10 ml pretreatment solution containing 5 mM EDTA (pH 8.0) and 0.2% 2-mercaptoethanol at room temperature. After 1 h the mycelia were washed with distilled water, and resuspended in 10 ml 0.6 M-mannitol containing a mixture of enzymes. The enzyme mixture comprised of cellulase 'Onozuka' RS (Yakult Biochemical Co.) 20 mg/ml, driserase (Kyowa Hakko Co.) 10 mg/ml, Novozyme 234 (Novo Enzyme Products) 0.4 mg/ml and chitinase T-1 (Asahi Industries Co.) 0.1 mg/ml. The age of the mycelium strongly affected the yield of protoplasts. The culture flask was placed on a reciprocal shaker set at 75 strokes per min at 34°C for 4 h. The culture was then filtered through a 150 μm steel sieve to remove mycelial fragments. The filtrate was centrifuged at least twice at 700 *g* for 5 min to remove the enzymes. The pellet was resuspended in 2 ml of 0.6 M mannitol.

High yield of protoplasts from filamentous fungi can be obtained by a combined enzyme system. The effect of the combinations of enzymes in enhancing protoplast release is presumably due to the synergistic action of the major components.

2.2. PURIFICATION OF PROTOPLASTS

The procedure for purification of protoplasts from *R. solani* is shown in Figure 2. The crude protoplast preparation obtained from the mycelial fragments of *R. solani* contained undigested tissues, broken protoplasts, and various cellular organelles in addition to protoplasts (Figure 2). Two ml of the protoplast suspension were layered onto 4 ml of 0.6 M sucrose, 1 mM $CaCl_2$, 5 mM HEPES-KOH (pH 7.0) and centrifuged in a swinging bucket rotor at 200 *g* for 5 min at room temperature. The intact protoplasts, which were located at the interphase of the two-phase system, were removed with a Pasteur pipette and resuspended in mannitol. Broken cells were found in the lower phase of the two-phase solution. The protoplasts were diluted with 0.6 M mannitol solution and centrifuged at 700 *g* for 5 min. The pellet containing the protoplasts was resuspended in the mannitol solution. An apparently homogeneous suspension of protoplasts was obtained by purifying the crude protoplast preparation with the liquid two-phase system of 0.6 M sucrose and 0.6 M mannitol (Figure 3).

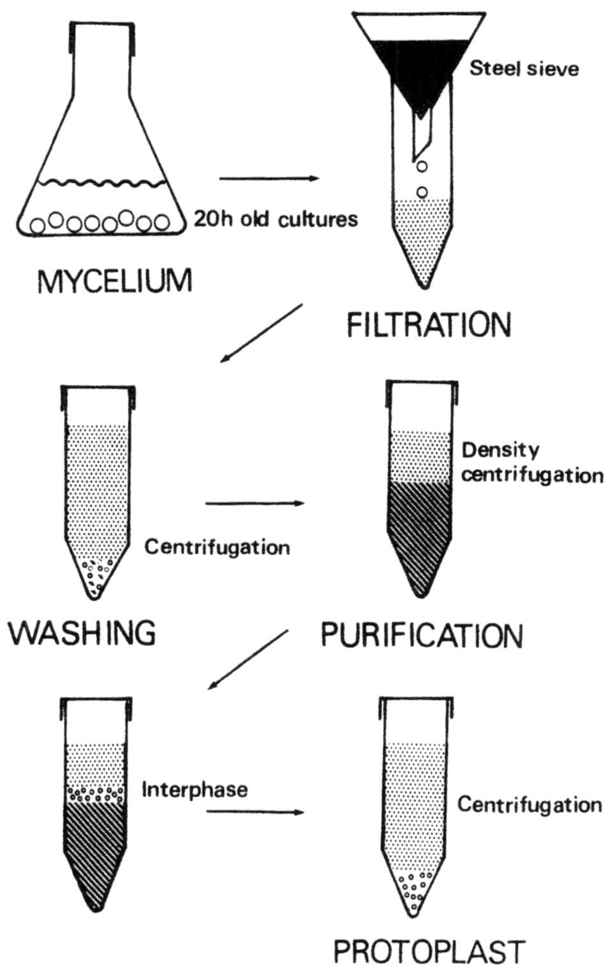

Figure 2. The procedure for isolation and purification of protoplasts from *R. solani*.

We found that intact protoplasts of *R. solani* have a strong tendency to partition at the interphase of the mannitol-sucrose two-phase system, leaving broken protoplasts in the lower phase; in the system with 0.5 M sucrose, both intact and broken protoplasts sedimented into the lower phase.

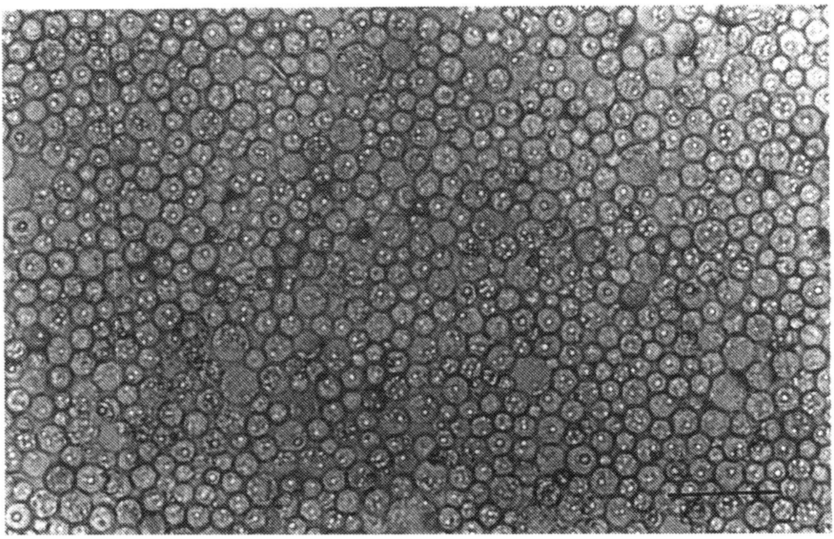

Figure 3. Light micrograph of intact protoplasts of *R. solani* after purification with the two-phase system. Bar represents 20 μm.

2.3. MUTAGENESIS AND ISOLATION OF AUXOTROPHIC MUTANTS

To develop a transformation system the first step is to obtain a selectable marker, such as auxotrophic and drug resistant mutants which can be used to screen fused protoplasts. Several difficulties were encountered in the attempt to isolate auxotrophic mutants of *R. solani*. For example, *R. solani* isolates do not produce spores on artificial media and vegetative cells are multinucleate.

We describe a procedure for the isolation of mutants in *R. solani* by using protoplasts (Hashiba and Yamada, 1984). This method appears to be more suitable for the production of various stable mutants.

Protoplasts of parental strain (5×10^6 protoplasts per ml) in 0.05 M phosphate buffer (pH 7.2) supplemented with 0.6 M mannitol were treated with N-methyl-N'-nitro-N-nitorosoguanidine (NTG) for 45-60 min at 25°C to give a 3-20% survival rate. The concentration of NTG used was 20 μg/ml. The mutagenised protoplasts were centrifuged and the pellet washed with mannitol buffer solution before being suspended in fresh mannitol buffer. The protoplasts were plated on complete medium (CM) supplemented with 0.6 M mannitol.

Colonies that appeared after incubation for 5 days at 28°C were transferred to fresh CM plates, and the resulting colonies were checked for growth on MM. Isolates showing growth on CM but not on MM were scored as auxotrophs. Specific nutritional requirements of auxotrophs were determined by the methods of Holliday (1956).

The reversion frequency of these mutants was about 10^{-3}. Thus, the proposed method may be broadly applicable for genetic analysis and manipulation of other multinucleate, filamentous, or non-spore producing fungi.

2.4. PROTOPLAST FUSION

The procedure for protoplast fusion of *R. solani* is based on modifications to the methods of Anne and Peberdy (1976) and Ferenczy *et al.* (1976). Protoplasts were prepared from the met$^-$ and arg$^-$ mutants. After being washed in 0.6 M mannitol, protoplasts obtained from the two different auxotrophic mutants to be fused were mixed (2×10^6 protoplasts per ml of each auxotroph) and centrifuged (3500 *g*, 5 min). The pelleted protoplasts were resuspended in 0.4 ml of a solution containing polyethylene glycol 4000 (PEG) and $CaCl_2$ in 10 mM Tris-HCl buffer and adjusted to pH 7.5 with 0.01 M NaOH. After incubation for 15 min, the suspension was centrifuged for 5 min at 3500 *g*. Protoplasts from fusion mixtures were plated directly on a selective regeneration medium lacking growth factors (methionine and arginine) required by each parental auxotrophic strain. Colonies that appeared after 5 days of incubation at 28°C were counted. The fusion frequency was calculated from the ratio of colonies growing on MM to colonies growing on CM.

The important factors affecting production of prototrophic colonies appear to be concentrations of PEG and $CaCl_2$ in the fusion mixtures. Ann and Peberdy (1976) and Ferenczy *et al.* (1975) carried out extensive studies of the conditions that influence the efficiency of PEG. They reported that PEG concentration, molecular weight of PEG, and treatment time were critical for increasing protoplast fusion frequency. The improved survival of *R. solani* at PEG levels higher than 30% may reflect a marked sensitivity of protoplasts of *R. solani* to osmotic damage.

2.5. CHARACTERISATION OF FUSED PROTOPLASTS

Prototrophic colonies that appeared on the MM were transferred to fresh MM and their properties were examined. As illustrated in Figure 4, prototrophic colonies differed morphologically from the colonies of the auxotrophic mutant strains and those of the parental field isolate. The parental field isolate produced sclerotia, but auxotrophic mutants produced no sclerotia at all. The colonies regenerated from the fused protoplasts, however, produced sclerotia (Figure 4). Most of them segregated sectors of parental auxotrophs. After the first recultivation of cells originating from fused protoplasts on MM, stable prototrophic progenies were obtained. Intact protoplasts were separated from mycelial fragments of stable prototrophic progenies. Then the fraction with uninucleate and enucleate protoplasts was plated on CM supplemented with 0.6 M mannitol. Colonies were transferred to fresh CM plates, and the resulting colonies were checked for growth on MM. All of the isolates grow on MM. The results indicated that the stable isolates that originated from fused protoplasts were prototrophic.

PEG treatment is an efficient method for the fusion of fungal protoplasts. In recent years, potential of the electrical fusion method in fungal protoplast fusion is being explored widely.

3. DNA plasmid

3.1. DETECTION AND PREPARATION OF DNA PLASMID.

DNA plasmid was detected using a slight modification of the procedure of Hirt (1967). Ten grams of lyophilized mycelium was ground to powder and suspended in 160 ml 0.1 M-citrate/phosphate buffer (pH 6.0) containing 0.1 M-EDTA and then lysed by adding 20 ml 10% (w/v) SDS solution. After the addition of 5 M NaCl to a final concentration of 1 M, this mixture was kept on ice overnight and then centrifuged at 15,000 *g* for 30 min. The mixtures were precipitated with 0.6 volumes of 2-propanol at 0°C , and the precipitates were collected by centrifugation. The total DNA precipitated was resuspended in 7 ml TE (20 mM-Tris, 0.25 mM EDTA, pH 8.0) buffer containing 0.4 % Sarkosyl, and purified by CsCl density-gradient centrifugation [7 g CsCl and 0.4 ml ethidium bromide (10 mg/ml)]. The samples were centrifuged at 95000 *g* for 48 h. The bands removed from the gradients were extracted four times with butanol, precipitated with ethanol, and dissolved in TE buffer. A portion of the DNA sample thus obtained was subjected to agarose gel electrophoresis in 0.7% agarose. The DNA plasmid band was excised and electroeluted. All isolates were independently analysed three times by gel electrophoresis for the presence of DNA plasmid.

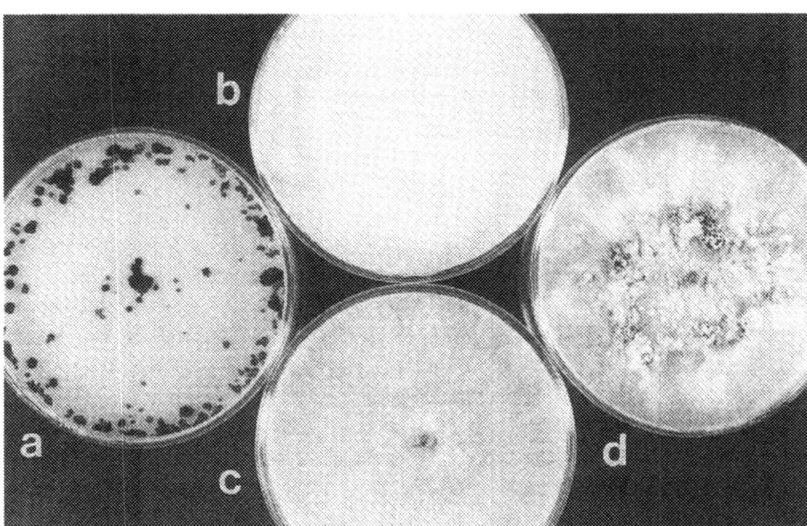

Figure 4. Cultural characteristics of *R. solani*. (a) Parent field isolate No. 131 grown on minimal medium (MM), (b) arg^- mutant grown on minimal medium supplemented with 20 μg arginine per ml, (c) met^- a mutant grown on MM supplemented with 20 μg methionine per ml. Both the mutants were normally unable to form sclerotia. (d) Prototrophic colonies resulting from intraspecific protoplast fusion grew on MM. Arrows indicate sclerotia.

TABLE 5. Southern hybridization analysis of plasmid DNA from *R. solani* AG 1 to 6 isolates probed with plasmid DNA from a representative isolate of each AG

				Source of probe:						
AG	ISG	Isolate	Size of plasmid (kb)	C-527 (AG1)	C-94 (AF2-2 IIIB)	H-16 (AG2-2 IV)	P-7-1 (AG3)	RI-64 (AG4	ST-8 (AG5)	HI-5-16 (AG6)
1		C-527	4.5	P[a]	-	-	-	-	-	-
2-2	III-B	C-6	5.2	-[b]	-	-	-	-	-	-
			3.2	-	+	-	-	-	-	-
		C-94	4.0	-	-	-	-	-	-	-
			2.0	-	P	-	-	-	-	-
			2.2	-	-	-	-	-	-	-
		C-125	4.3	-	-	-	-	-	-	-
			2.9	-	+	-	-	-	-	-
			2.2	-	-	-	-	-	-	-
		C-328	4.2	-	-	-	-	-	-	-
			2.9	-	+	-	-	-	-	-
		C-329	4.7	-	-	-	-	-	-	-
			2.9	-	+	-	-	-	-	-
2-2	IV	A-14-1	4.8	-	-	+	-	-	-	-
			2.7	-	-	-	-	-	-	-
		BV-6	4.0	-	+	+	-	-	-	-
			2.4	-	+	-	-	-	-	-
		C-616	3.0	-	-	+	-	-	-	-
		C-623	5.5	-	-	+	-	-	-	-
			2.9	-	-	+	-	-	-	-
		H-16	5.4	-	-	+	-	-	-	-
			4.8	-	-	P	-	-	-	-
			2.7	-	-	-	-	-	-	-
		SH-17-24-1	4.8	-	-	+	-	-	-	-
			4.2	-	-	+	-	-	-	-
			2.7	-	-	+	-	-	-	-
3		P-5	6.2	-	-	-	+	-	-	-
		P-7-1	5.2	-	-	-	P	-	-	-
		ST-3	6.2	-	-	-	+	-	-	-
4		GM-7	2.7	-	-	-	-	+		
		GM-11	2.7	-	-	-	-	+	-	-
		RI-64	2.7	-	-	-	-	P	-	-
5		ST-2-1	3.0	-	-	-	-	-	+	-
		ST-8	3.0	-	-	-	-	-	P	-
		W-3	3.0	-	-	-	-	-	+	-
		501	3.0	-	-	-	-	-	+	-
6		HI-5-16	5.2	-	-	-	-	-	-	P

a) P: the entire length of a plasmid obtained from the representative isolation of one AG was used as the hybridization probe.

b) +: considerable sequence homology was observed; -: no sequence homology was observed.

3.2. PROPERTIES OF DNA PLASMIDS

A survey of 114 isolates of *R. solani* from a wide range of plant hosts revealed that 48 isolates contained one to three plasmids (Hashiba *et al.*, 1984; Shimma *et al.*, 1988; Miyasaka *et al.*, 1990). We examined the sequence homology among DNA plasmids found in representative isolates by Southern blot analysis, using nick-translated DNA plasmids as probes (Miyasaka *et al.*, 1990; Chen *et al.*, 1992). Considerable sequence homology has been detected among DNA plasmids obtained from isolates within the same AG. This suggested that the distribution of DNA plasmids correlates with the AGs of *R. solani*.

Northern hybridisation analysis showed that RNAs hybridizing to the DNA plasmids are present in various *R. solani* isolates. Considerable sequence homology was obtained among plasmid-encoded transcripts obtained from isolates of the same AG, but no sequence homology was found among transcripts obtained from isolates of different AGs.

3.3. STRUCTURE

Three linear DNA plasmids were found in isolate RI-64 of anastomosis group 4 (AG 4) of *R. solani*. These plasmids, designated pRS64-1, -2, and -3, were of the same size (2.7 kb). Restriction mapping and Southern hybridisation analysis of pRS64-1, -2, and -3 revealed the presence of homologous regions. The DNA plasmids were resistant to both 3'-exonuclease and 5'-exonuclease even after treatment with proteinase K or alkali. The length of both terminal fragments that were generated by restriction endonuclease digestion was doubled under the denaturation condition, indicating that the linear DNA plasmids have hairpin loops at both termini (Miyashita *et al.*, 1990). A computer-based study of the pRS64-2 DNA folding at both termini predicted hairpin loop structures. The hairpin loops consisted of the left- and right-hand terminal 113 and 105 nucleotides, respectively, and had no sequence homology (Figure 5). They did not reveal flip-flop inversion. The hairpin formed cruciform base-paired structures (Katsura *et al.*, 1998).

3.4. REPLICATION

Replication of pRS64 DNAs may be initiated by site-specific nicking proximal to the loop structure and within the terminal homologous region. This results in the formation of a free 3'-hydroxyl end that can serve as a primer for DNA replication. If DNA replication proceeds from both ends simultaneously, mature molecules can be formed directly. If DNA synthesis is initiated from a single end, replication will proceed around the opposite hairpin, leading to the formation of a dimer. Southern blotting analysis of total DNA showed the presence of two types of dimeric forms of pRS64 DNA. One is a head-to-head dimer and the other is a tail-to-tail dimer (Miyashita *et al.*, 1990).

3.5. TRANSCRIPTION

Northern hybridisation analysis showed that poly $(A)^+$ RNA hybridising to pRS64 plasmids is present in the isolates of AG 4. Comparison of the nucleotide sequence of the cDNA derived from poly $(A)^+$ RNA and those of pRS64-1, -2, and -3 showed 100%, 73%, and 84% homology, respectively (Hongo *et al.*, 1994). This suggests that the poly

$(A)^{+}$ RNA hybridising to pRS64 represents the transcription products of pRS64-1, -2, and -3, and that the genetic information in pRS64 DNAs is expressed.

3.6. PROTEIN CODING

To determine potential coding regions, the nucleotide sequence of the cDNA clone pMHO4 was analyzed in all possible reading frames. In this analysis, the genetic code for mitochondrial systems of filamentous fungi was used. When plasmid DNA sequences were analyzed with GTG as the initiation codon, three open reading frames of 204 bp, one in each plasmid (ORF1-1, ORF2-1, and ORF3-1) were detected, which could encode polypeptide of 68 amino acid residues (Hongo *et al.*, 1994).

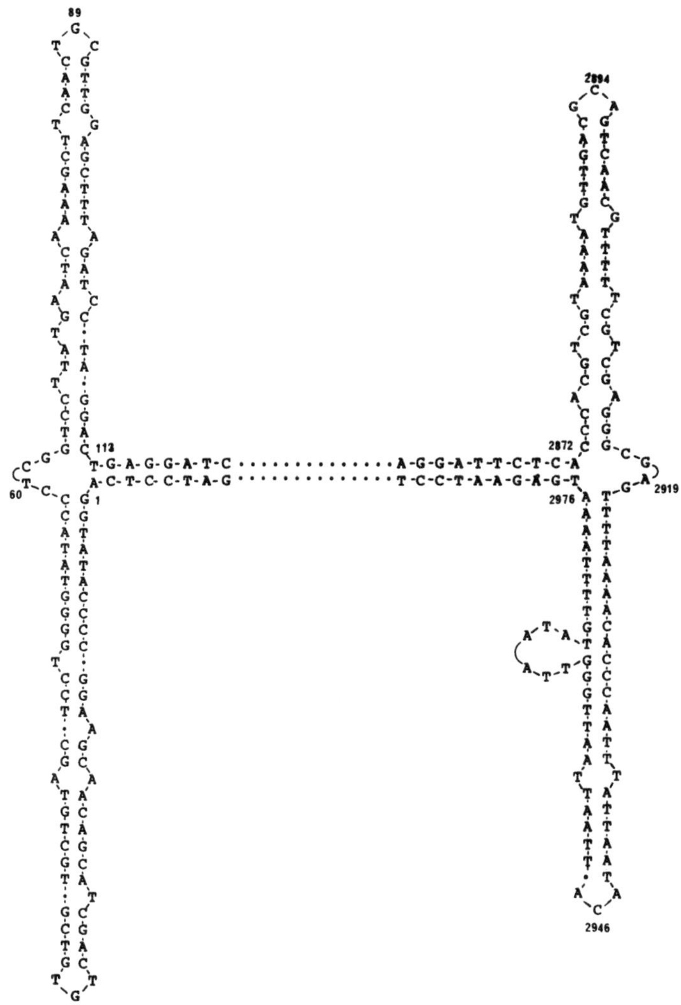

Figure 5. Computer-predicted secondary structure of both termini of pRS64-2. Nucleotide positions are indicated by numbers used in EMBL and GenBank accession. The analysis was made with the software DNASIS.

Based on the sequence analysis, the theoretical molecular weight of the putative protein is 7.2 kDa. The promoter consensus sequence of *E. coli*, TTGACA, is located 55 nucleotides upstream of the transcription initiation site (Figure 6). The consensus 'core enhancer' sequence, TGGAAAG, identified from higher eukaryotes, is located 33 nucleotides upstream of the transcription initiation site (Figure 6). Using the DNASIS software, the amino acid sequence of the putative polypeptide was compared with those of known proteins. The translation product of ORF1-1 did not show significant homology with known proteins in the data banks. Further studies are necessary to determine the function of ORF products encoded by linear DNA plasmids (pRS64) of *R. solani*.

```
961   ATTCAGCTTTGCTGAGTGTGCGGTGGCTCCTTGAGGACACCCCAATAGTGGCAGTCCCTG   1003
1104  AGACTTTATCTCAGGGGCCCCTATTGTGGGTGCTATTGACACCATTCAAGGGGGTAATGG   1063
1064  AAAGATCCGAGATAAAAGTGACTGCTTTATTCATCTGAAAATAGTAAGAGATCCTCTCTC   1123
1124  GGATCCGCAACTTCTATTAACGAAGGATCATAGCCATTACCCCAACCGGTGACATTGTTG   1183
                                                  M  T  L  L
1184  CCAAAGGCTGGCGCCTTTACAAACCGTCCAGACCGGAGGCTTCCGGGTGTCAATAGTGCT   1243
      P  K  A  G  A  F  T  N  R  P  D  R  R  L  P  G  V  N  S  A
1244  TCAGGCCGTAGCTGCATGAAGCTGCCGTGCAAATCGGAAGGCAATTGCCTCTCTATGCCG   1303
      S  G  R  S  W  M  K  I  P  C  K  S  E  G  N  C  L  S  M  P
1304  TGAGGCATAGTGCTAAGGAGTCACTCCCTTAGCGGCCTAGACCGTCTACCTTTGGAACCC   1363
      W  G  I  V  L  R  S  H  S  L  S  G  L  D  R  L  P  L  E  P
1364  CCAAAGGTAGCCTAGTAGGCCCTAGCATTTGAAAACTAGGGCCACATATAGGCCTGAGAT   1423
      P  K  V  A  *
```

Figure 6. Nucleotide sequence of the fundamental region of pMHO4 containing an open reading frame (ORF1-1). The presumed amino acid sequence encoded by ORF1-1 is shown from left to right below the nucleotide sequence. The consensus 'core enhancer' sequence, TGGAAAG, is located 33 bp upstream of the start (). The consensus promoter sequence, TTGACA, is shown by a bold line. Numbers at the beginning indicate the portion of the sequence from the terminal *Xho* I site on the left side.

4. Acknowledgement

The authors sincerely wish to thank Dr. T. Ishikawa, Teikyo University, Japan for helpful suggestions and critical reading of the manuscript, and Dr. Y. Ehara and Dr. T. Kobayashi for kind help.

5. References

Anne, J. and Peberdy, J. F. (1976) Induced fusion of fungal protoplasts following treatment with polyethylene glycol, *Journal of General Microbiology* **92**, 413-417.

Chen, C. L., Miyasaka, A., Miyashita, S., Ehara, Y. and Hashiba, T. (1992) Genetic relatedness of plasmid-like DNA in isolates from anastomosis group 4 of *Rhizoctonia solani*, *Annals of the Phytopathological Society of Japan* **58**, 286-291.

Davis, B. (1985) Factors influencing protoplast isolation, in J. F. Peberdy and L. Ferenczy (eds.), *Fungal protoplasts. Application in biochemistry and genetics*, New York, Marcel Dekker, Inc., pp.45-72.

Ferenczy, L., Kevei, F. and Szegedi, M. (1975) High frequency fusion of fungal protoplast, *Experimentia* **31**, 1028-1030.

Ferenczy, L., Kevei, F., Szegedi, M., Franko, A. and Rojik, I. (1976) Factors affecting high-frequency fungal protoplast fusion, *Experimentia* **32**, 1156-1158.

Hashiba, T. (1992) Isolation of fungal protoplast, in D. K. Arora, R. P. Elander, and K. G. Mukerji (eds.), *Handbook of applied mycology vol. 4. Fungal biotechnology*, New York, Marcel Dekker, Inc., pp.129-148.

Hashiba, T. and Yamada, M. (1982) Formation and purification of protoplasts from *Rhizoctonia solani*, *Phytopathology* **72**, 849-853.

Hashiba, T. and Yamada, M. (1984) Intraspecific protoplast fusion between auxotrophic mutants of *Rhizoctonia solani*, *Phytopathology* **74**, 398-401.

Hashiba, T., Homma, Y., Hyakumachi, M. and Matsuda, I. (1984) Isolation of a DNA plasmid in the fungus *Rhizoctonia solani*, *Jounal of General Microbiology* **130**, 2067-2070.

Hirt, B. (1967) Selective extraction of polyoma DNA from infected mouse cell cultures, *Journal of Molecular Biology* **26**, 365-369.

Holliday, R. (1956) A new method for the identification of biochemical mutants of micro-organisms, *Nature* **178**, 987.

Hongo, M., Miyasaka, A., Suzuki, F. and Hashiba, T. (1994) Expression of the linear DNA plasmid pRS64 in the plant pathogenic fungus *Rhizoctonia solani*, *Molecular General Genetics* **245**, 265-271.

Jabaji-Hare, S. H., Burger, A., Forget, L. and Lang, B. F. (1994) Extrachromosomal plasmids in the plant pathogenic fungus *Rhizoctonia solani*, *Current Genetics* **25**, 423-431.

Katsura, K., Suzuki, F., Miyashita, S., Nishi, T., Hirochika, H. and Hashiba, T. (1998) The complete nucleotide sequence and characterization of the linear DNA plasmids pRS64 from the plant pathogenic fungus *Rhizoctonia solani*, *Current Genetics.* (in press)

Miyasaka, A., Chen, C. L. and Hashiba, T. (1990) Detection and properties of plasmid-like DNA in isolates from nine anastomosis and intraspecific groups of *Rhizoctonia solani*, *Journal of General Microbiology* **136**, 1791-1798.

Miyashita, S., Hirochika, H., Ikeda, J. and Hashiba, T. (1990) Linear plasmid DNAs of the pathogenic fungus *Rhizoctonia solani* with unique terminal structures, *Molecular General Genetics* **220**, 165-171.

Peberdy, J. F. (1979) Fungal protoplasts: Isolation, Reversion, and Fusion, *Annual Review of Microbiology* **33**, 21-39.

Samac, D. A. and Leong, S. A. (1989) Mitochondrial plasmids of filamentous fungi: characteristics and use in transformation vectors, *Molecular Plant-Microbe Interactions* **2**, 155-159.

Shimma, Y., Uno, I., Hashiba, T. and Ishikawa, T. (1988) Characterization of a *Rhizoctonia solani* strain carrying plasmids, *Journal of General Applied Microbiology* **34**, 111-117.

Wako, T., Ishikawa, T. and Hashiba, T. (1991) Unique DNA plasmid pRS64 associated with chromosomal DNAs of the plant pathogenic fungus *Rhizoctonia solani*, *Journal of General Microbiology* **137**, 2817-2821.

PHYLLOPLANE YEASTS AS POTENTIAL BIOCONTROL AGENTS FOR RICE SHEATH BLIGHT DISEASE

A.K. M. SHAHJAHAN[1], M. C. RUSH[1] and D. E. GROTH[2]
[1]*Department of Plant Pathology and Crop Physiology, Louisiana AgriculturalExperiment Station, Louisiana State University Agricultural Center, Baton Rouge*
[2]*Rice Research Station, Crowley, Louisiana, USA.*

1. Introduction

Rice is an important cereal crop in the world. More than 50% of the world's population utilises rice for their main daily intake of carbohydrate. It is an important cash crop in several states in the United States. There are many constraints to rice production in the world with diseases being among the most significant due to the severity of losses and the lack of cost efficient control measures. Yield losses due to diseases range from 10% or more in the United States (Groth *et al*., 1991; Lee and Rush, 1983) to 20-25% in other countries (Ou, 1985; Rao, 1995; Shahjahan *et al*., 1986). Among the major fungal diseases of rice, sheath blight caused by *Rhizoctonia. solani*, is the most destructive necrotrophic disease (Lee and Rush, 1983; Ou, 1985; Rao, 1995; Rush and Lee 1992). The disease is endemic in nature and the pathogen overwinters as sclerotia in soil and in debris from the previous crop. The sclerotia and debris float to the surface of flood water and initiate infections on the lower leaf sheaths starting with the maximum tillering stage of growth. Under the conditions of high temperature and humidity prevailing during the panicle initiation to heading stages, sheath blight develops very rapidly. When the plants are heading, lesions may reach the flag leaf in as few as 10-15 days (Groth *et al*., 1991; Rush and Lee, 1992; Shahjahan and Mew, 1989). Most modern rice cultivars are susceptible to this disease. Commercial cultivars are susceptible or have low levels of partial resistance (Rush *et al*., 1995). Farmers presently use fungicides to control this disease in the United States (Groth *et al*., 1993) and in other countries (Ou, 1985), but this technology increases the cost of production as well as leaving potential pollutants in the environment. Fungicide use also affects the nontarget phylloplane microorganisms that naturally restrain pathogen development (Andrews and Kinnerly, 1978; Fokkema and de Nooij, 1981; Rush *et al*., 1991; Van Eeckhout *et al*., 1991). Because of its sustainable nature, scientists all over the world are now concentrating on developing biological methods of disease control.

Biological control using antagonistic fungi or bacteria has been used successfully with soilborne diseases of several crops (Campbell, 1989; Cook and Baker, 1983; Handelsman and Stabb, 1996; Lumsden *et al*., 1995). These antagonists were mostly

S. Sreenivasaprasad and R. Johnson (eds.), Major Fungal diseases of Rice Recent Advances, 235-252.
Prof. K. Manibhushan Rao Festschrift

isolated from the soil or rhizosphere. There has been limited success with foliar diseases. Recently, the use of antagonists inhabiting the phylloplanes of crop plants for biological control of foliar diseases is receiving more attention. These antagonists have the advantage of being naturally established on the plant organs where the pathogen also tries to establish for infection. This ensures *in vivo* interaction between the pathogen and the antagonist and results in the reduction of pathogen activity. Natural biological control by phylloplane antagonists may have a major effect on slowing disease development, but this effect is overcome when environmental conditions favour epidemics.

2. Biological control of foliar diseases with phylloplane antagonists

The phylloplane of plant aerial surfaces is an important ecological niche for pathogenic and saprophytic microorganisms. Included among the saprophytic microorganisms are antagonists that naturally restrict pathogens, and these may be used for artificial inoculation, or artificially enhanced for plant disease control (Andrews, 1992; Blakeman, 1985; Blakeman and Fokkema, 1982; Fokkema *et al.*, 1979; Lindow, 1985; Sackenheim *et al.*, 1994). The composition, population levels, and ecological succession of phylloplane microflora influence the growth of pathogens and thereby impact natural biological control of foliar diseases (Blakeman, 1985; Morris and Coleman, 1995; Sackenheim *et al.*, 1994; Williamson and Fokkema, 1985). Until recently not much attention was given to these phylloplane microorganisms as possible biocontrol agents.

Biological control mechanisms operating in nature are evidenced by the observation of *in vivo* interactions between the pathogen and the antagonists inhabiting plant phylloplanes (Akai and Kuramoto, 1968; Fokkema and Van der Meulen, 1976; Fokkema *et al.*, 1975, 1979). The increase in severity of infection by *Cochliobolus sativus* on rye (*Secale cereale* L.) and *R. solani* on rice were observed after spraying plants with Benlate 50 WP™ (benomyl) and Tilt™ (propiconazole), respectively, (Fokkema 1978, Rush et al. 1991, Van Eeckhout et al. 1991). The degree of disease severity on rye appeared to be related to the population level of saprophytes on the leaves that included the naturally occurring antagonists of the pathogen *C. sativus* (Fokkema, 1978; Fokkema *et al.*, 1975). Important among these phylloplane antagonists were yeasts. It has been reported that yeasts such as *Sporobolomyces roseus* Kluyver & Van Niel, *Cryptococcus* spp., and *Aureobasidium pullulans* (de Bary) Arnaud that inhabit the phylloplanes of wheat (*Triticum aestivum* L.), maize (*Zea mays* L.) and rye are potential biocontrol agents (Bashi and Fokkema, 1977; Fokkema, 1978; Fokkema and Van Der Meulen, 1976; Williamson and Fokkema, 1985). This suggested that these types of antagonists also occur on the phylloplanes of other crops. In rice this area has remained almost unexplored.

As early as in 1968, Akai and Kuramoto in Japan found that populations of *Candida* spp. were most abundant on rice leaves. Mixed inoculation of *Candida* spp.and *Cochliobolus miyabeanus* (Ito & Kur) Drech. reduced brown spot incidence on leaves by 50%. The *Candida* spp., however, was not able to inhibit the germination of conidia or prevent mycelial growth of the pathogen *in vitro*. Our studies at Louisiana State University show that yeasts are a major component of the phyllopane microbial community and that these yeasts can be used or manipulated to achieve biological control of rice sheath blight.

It has been speculated that the types of microflora present and the predominance of a particular type at any given time or stage of a plant's growth are dependent on the microenvironment and the nutrients available on the phylloplane (Andrews, 1992; Bashi and Fokkema, 1977; Dik *et al.*, 1984; Sackenheim *et al.*, 1994). Another important factor that contributes to predominance of a particular phylloplane microorganism is the interspecific 'competition' that occurs in nature (Shearer, 1995). The competition is usually for nutrients and space. Phylloplane microorganisms not only compete with each other for nutrients, but they also antagonize pathogens infecting the foliage by modes-of-antagonism which include antibiosis, competition, amensalism, and parasitism (Fokkema, 1984; Morris and Coleman, 1995; Shahjahan and Rush, 1996). Interspecific competition, which we have largely ignored in the past, results in natural biological control of diseases occurring on the leaf or aerial surfaces (Fokkema, 1978; Rush *et al.*, 1991). Fungal mycoflora that are primary colonizers of leaf surfaces include *Cladosporium*, *Fusarium*, *Curvularia*, *Alternaria*, and yeasts. Among yeasts, pink yeasts (*Sporobolomyces* spp.) predominate (Blakeman, 1985; Blakeman and Fokkema, 1982; Magan, 1986). These primary saprophytes multiply rapidly under favourable conditions and exhaust nutrients and space necessary for the pathogen to establish. Some even secrete toxic substances and arrest the growth or kill the pathogen (Blakeman, 1985; Blakeman and Fokkema, 1982; Lindow, 1985). These *in vivo* phenomena clearly show that there is potential for biological control of foliar diseases with phylloplane antagonists.

3. Biological control of foliar diseases with antagonistic yeasts

Among the phylloplane inhabiting microorganisms, yeasts have drawn the attention of many researchers because of their nature and growth habits. Yeasts grow well at temperatures below 32 C and at relative humidities ranging from 65 - 95%. Yeasts are resistant to desiccation, and easy to culture, multiply, and store. The field application of yeast is also relatively easy (Bashi and Fokkema, 1977; Dik *et al.*, 1992; Last, 1955; Sugar *et al.*, 1994). They are able to compete with other saprophytes and pathogens for both exogenous and endogenous nutrients and, therefore, are desirable antagonistic agents for use in biocontrol (Fokkema, 1984; Punja, 1997).

Biological control of foliar plant diseases with yeasts has not been explored extensively. Table 1 lists the names of the yeast antagonists that have been found effective in biological control of foliar diseases of field crops such as barley (*Hordeum vulgare* L.), maize, rye, wheat, cucumber (*Cucumis sativus* L.), bean (*Vicia fava* L.) and tomato (*Lycopersicon esculentum* Miller). The most common yeast antagonist appears to be *S. roseus*. Recently, yeasts have been found to be effective in biocontrol of postharvest diseases of fruits such as apple (*Mallus domestica* Borkh), pear (*Pyrus communis* L.), grape (*Vitis* spp.), peach (*Prunus persica* (L.) Batsch), avocado (*Persea americana* Miller), kiwi (*Actinidia chinensis* Planch), sweet cherry (*Prunus avium* (L.) L.) and citrus (Table 1). The ability of these yeasts to control preharvest or postharvest diseases can be enhanced through application of nutrients on the foliage or fruits (Fokkema *et al.*, 1979, Janisiewicz *et al.*, 1992, 1994).

TABLE 1. Studies on biological control of preharvest foliar diseases of crops and postharvest diseases of fruits with yeasts

Crop	Disease/pathogen	Yeast antagonist	Reference
Barley	Powdery mildew (*Erysiphe graminis*)	*Tilletiopsis pallescens*, *T. albicans*	Klecan *et al.*, 1989; Knudsen and Skau, 1993
Wheat	Brown spot (*Cochliobolus sativus*)	*Sporobolomyces roseus*	Bashi and Fokkema, 1977
	Leaf spot (*Septoria nodorum*)	*S. roseus*	Fokkema *et al*, 1979; Fokkema and Van der Muelen, 1976
Maize	Leaf spot (*Colletotrichum graminicola*)	*S. roseus*	Williamson and Fokkema, 1985
Rice	Brown spot (*C. miyabeanus*)	*Candida* spp.	Akai and Kuramoto, 1968
	Sheath blight (*R. solani*)	*S. roseus* and other spp.	Rush *et al*, 1998
Rye	Leaf spot (*C. sativus*)	*S. roseus*	Fokkema, 1973; Fokkema *et al.*, 1975
Rose	Powdery mildew (*Sphaerotheca pannosa*)	*T. pallescens*	Ng *et al.*, 1977
Tomato	Fruit rot (*Botrytis cinerea*)	Yeasts	Elad *et al.*, 1994
Cucurbits	Powdery mildew	*A. quisqualis*	McGrath and Shishkoff, 1996
	Powdery mildew (*E. graminis*)	*Sporothrix floculosa*	Hajlaoui and Belanger, 1993
Bean	Leaf spot (*B. cinerea*)	Yeasts	Elad *et al.*, 1994
Cucumber	Powdery mildew (*Sphaerotheca fuligena*)	*T. pallescens*, *T. minor*	Urquhart *et al.*, 1994; Urquhart and Punja, 1996
Fruits:			
Apple	Fruit rot (*B. cinerea*)	Yeasts	Filonow *et al.*, 1996
Apple, pear	Storage decay	*S. roseus*	Janisiewicz *et al.*, 1994; Chand-Goyal and Spotts, 1996
	Blue mold (*Penicillium expansum*)	*Kloeckera apiculata*, *Candida guillermondi*	Janisiewicz *et al.*, 1992; McLaughlin *et al.*, 1992
	Grey mold (*B cinerea*)	*Cryptococcus laurentii*	Roberts, 1990
Sweet cherry	Brown rot	Yeasts	Chand-Goyal and Spotts, 1996
Avocado	Fruit rot (*C. gloeosporioides*)	*Aureobasidium* spp.	Stirling *et al.*, 1995
Peach, grapes	postharvest rot	*Kloeckera apiculata*, *Candida guillermondi*	McLaughlin *et al.*, 1992
Citrus	Penicillium rot	Yeasts	Wilson and Chalutz, 1989
Kiwi	Storage rot (*B. cinerea*)	Yeasts	Cheah *et al.*, 1996

The efficacy of biocontrol can be further enhanced if we know the mode-of-antagonism of yeast antagonists against the target pathogen (Morris and Coleman, 1995). The modes-of-antagonism expressed among yeast antagonists so far identified include competition for nutrients (Finlow *et al.*, 1996; Fokkema, 1984; Fokkema *et al.*,

1979; Janisiewicz *et al.*, 1994), exclusion of the pathogen from the infection site (Roberts, 1990); reduction in growth of mycelium, reduced germination of propagules, interference with the formation of appressoria, and reduction in the size of lesions (Fokkema and Van der Meulen, 1976; Stirling *et al.*, 1995; Williamson and Fokkema, 1985); mycoparasitism (Wisniewski *et al.*, 1991); antibiosis or production of antimicrobial compounds (McCormack *et al.*, 1994); production of toxins or 'killer toxins') to lyse or kill pathogen propagules (Starmer *et al.*, 1987; Radler *et al.*, 1985), and induced resistance mediated through elicitors in the yeast cell walls (Reglinski *et al.*, 1994, 1995). For example, if the mechanism is competition for nutrients, then identification of a nutrient favouring yeast growth and amendment of the phylloplane environment with that nutrient should enhance the yeast's competitive ability and give the biocontrol agent an advantage over the pathogen, resulting in increased disease control compared to nonamended conditions.

4. Biological control of sheath blight disease of rice

Although the potential of yeasts as biocontrol agents has been studied for foliar diseases of some crops (Table 1), there have been no published studies on rice phylloplane yeasts and their potential in controlling rice foliar diseases. Sheath blight is considered second in importance among the rice foliar diseases worldwide. The disease is difficult to control without the use of fungicides. Studies on the biological control of sheath blight were initiated in the eighties at the International Rice Research Institute, Phillipines (Mew and Rosales, 1986). Subsequently, pathologists in other countries also initiated programmes on biological control of this disease (Lin *et al.*, 1992; Rabindran and Viddyasekaran, 1996; Sarker *et al.*, 1992). The antagonists were all isolated from soil or the rice rhizosphere and tested against *R. solani*. In all cases, the selected antagonists were bacteria and were identified as *Pseudomonas fluorescens* (Mew and Rosales, 1986; Rabindran and Viddyasekaran, 1996; Sarker *et al.*, 1992). However, the technology has not yet reached farmers, although it has been reported in India that a peat based formulation of this bacterium has been developed for use by growers (Rabindran and Viddyasekaran, 1996). Our studies at Louisiana State University focused on antagonists inhabiting the rice phylloplane (Shahjahan and Rush, 1996; Shahjahan *et al.*, 1996, 1997; Rush *et al.*, 1998). This ecological niche was found to be colonised by both saprophytic and pathogenic microorganisms belonging to the actinomycetes, bacteria, and fungi including yeasts (Shahjahan and Rush, 1996; Shahjahan *et al.*, 1997; Rush *et al.*, 1998). Yeasts appeared to be an integral natural component of this phylloplane community and have drawn our attention because of their nature and biocontrol potential.

5. Potential of yeasts as biocontrol agents for rice sheath blight

The rationale behind using antagonists inhabiting the phylloplane of the target crop for controlling foliar diseases is that the antagonists are already adapted to the environment and establish quickly. As a result the antagonists may derive a competitive advantage over the pathogen in *in vivo* interactions under normal environmental conditions. Also if the antagonist is indigenous to the crop's phylloplane and ubiquitous, then the antagonist's population may be manipulated through nutrient amendment or environmental modifications rather than by application of the antagonist to foliage. We, therefore, undertook a program to identify and classify rice phylloplane microorganisms and to explore the potential of those microbes for biological control of rice sheath blight.

The population of yeasts on the rice phylloplane varied with plant age, the parts of the plant sampled, and the cultivar grown. The yeast population was found to be higher at the maximum tillering stage than at maturity. At maximun tillering, the yeast population was higher on the leaf sheath than on the leaf blade. The reverse was true at maturity (Table 2). This could be due to differences in nutrient status or to leaf wetness period and moisture availability at these growth stages. However, plants were not as prone to infection by *R. solani* at early growth stages as they were during the later stages suggesting a possible relationship between populations of antagonists on the phylloplane and development of sheath blight.

We have isolated from the phylloplanes of different rice cultivars 280 yeast isolates by the contact plating method and by the washing and dilution method on potato dextrose agar (PDA), acidified PDA, and malt extract yeast extract agar media (Figure 1). We found that at least 18 different types of yeasts were present on rice leaf blades and sheaths (Table 3). Most frequently occurring among them are pink yeasts belonging to the genus *Sporobolomyces.* The *in vitro* interactions of these yeasts and *R.solani* were

TABLE 2. Population[1] (mean cfu/cm^2 of leaf blade or sheath surface) of yeasts on the phylloplanes at the maximum tillering and maturity stages of five rice cultivars grown in the field at the Rice Research Station, Crowley, LA, USA in 1996

Cultivar	Maximum tillering stage		Maturity stage	
	Leaf blade	Leaf sheath	Leaf blade	Leaf sheath
Bengal	35.0	99.8	5.8	1.0
Cypress	29.7	66.6	0.6	0.0
Lemont	9.8	41.4	0.6	0.2
Mars	18.3	17.2	2.1	0.0
Teqing	37.0	86.0	0.6	0.8
Mean	26.5	62.2	2.3	0.4

[1]population of yeasts was determined on acidified potato dextrose agar medium by contact plating method. Values are means of three random tiller samples.

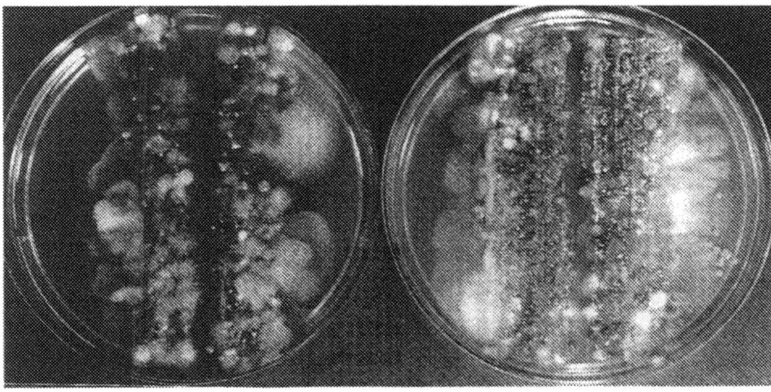

Figure 1. Contact plates of leaf blades of the rice cultivar Bengal (left) and the cultivar Teqing (right) selected at maturity and plated on potato dextrose agar medium showing the fungi (filamentous) and yeasts (round, convex, butyrus colonies) 5 days after incubation at 28 C. Note the varietal differences in yeast population.

competition, reverse mycoparasitism, or in rare cases antibiosis (Table 3). Isolates selected for field evaluation exhibited the potential to control *R. solani* in the series of tests listed in Table 4. In dual culture tests the yeast isolate and *R. solani* were challenged on PDA plates for antagonistic interactions such as antibiosis (producing a clear zone of inhibition between the pathogen and the antagonist), competition (the pathogen and the antagonist grow towards each other, meet, and then stop growing), and mycoparasitism (either the pathogen or the antagonist overgrows the other affecting further growth and survival). The Magenta vessel test was conducted on 2-3 week old sterile rice seedlings grown in Magenta vessels. The vessels contained 40 ml of sterile half strength Murashige and Skoog medium (1962). Seedlings were inoculated with a mycelial plug of *R. solani* one day before spraying with the antagonist (Figure 2). Sheath blight development on seedlings was measured on a rating scale of 0-5 where 0 meant no disease and 5 meant all seedlings were blighted or dead. In the cut leaf test sclerotia of *R. solani* were dipped into yeast cell suspension (10^{4-6} cfu/ml) for 10 seconds and placed on one edge of cut leaf pieces laid flat in Petri dishes lined with moist filter paper. Sclerotial germination was recorded and sheath blight development was measured after 10 days of incubation at 28 C on a rating scale of 0-5 where 0 meant no disease and 5 meant the cut leaf was completely covered with blighting symptoms. The test in the greenhouse was conducted on plants grown in pots. The field test was conducted at the Rice Research Station, Crowley, LA. In both tests all plants except the untreated were inoculated with *R. solani* inocula, prepared on Rice hull : rice grain (2:1) medium, at the first joint elongation stage and sprayed with the 24-hour-old yeast suspension (10^{4-6} cfu/ml) 7 days after inoculation and again at 14 days. Disease development was measured at maturity on a rating scale of 0-9, where 0 meant no disease and 9 meant leaf blade and sheath area of all tillers blighted (Groth *et al.*, 1990).

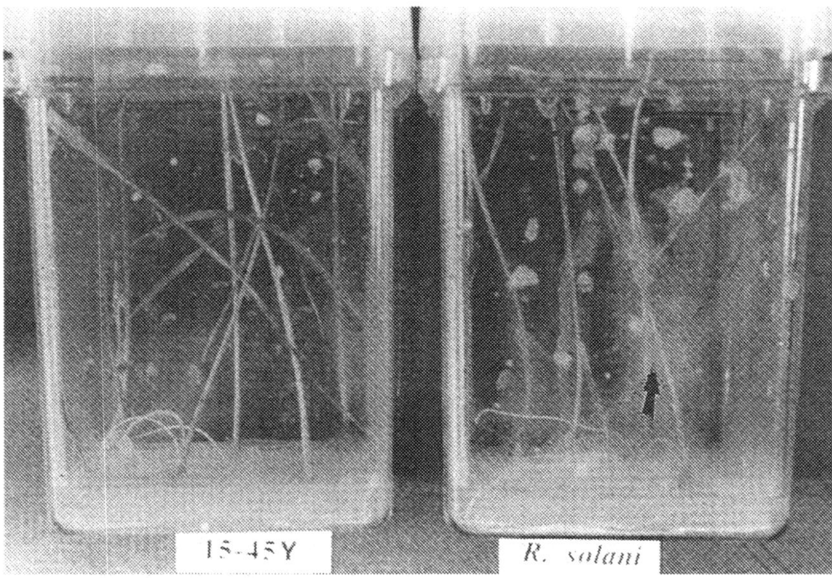

Figure 2. Magenta vessels showing 18 day old sterile rice seedlings inoculated with *R. solani* (right) and *R. solani* followed by spraying with a yeast antagonist (left). Note the growth of *R. solani* mycelium (arrows) on the seedlings in the control vessel (right) compared to almost no growth of *R. solani* on the yeast treated seedlings.

TABLE 3. Cultural characteristics of yeasts isolated from the rice phylloplane and their *in vitro* interactions with *R. solani* causing sheath blight disease of rice

Isolate No.[1]	Cultural Characteristics				Interactions[6]
	Color[2]	Cell[3]	Ballist[4]	Cyclo[5]	
LR951172	pinkish	arthrospores	-	+	comp-over
LR951438	off-white	arthrospores	-	+	over
LR951552	brown	cylindrical	-	-	comp-over
LR951564	peach	cylindrical	-	-	comp
LR951565	pink	oval	+	+	comp
LR951566	beige	cylindrical	-	-	comp-over
LR951755	cream	round	-	-	over
LR951906	almond	round	-	+	over
LR952001	almond	oval	-	+	over
LR952006	cream	round	-	-	over
LR952009	tangerine	oval	-	+	comp-over
LR952051	brown	cylindrical	-	+	antibios-comp
LR952058	pink-orange	oval	-	+	comp-over

LR952069	peach	cylindrical	-	-	comp-over
LR952115	orange	oval	-	-	comp-over
LR952116	off-white	oval	-	+	over
LR952124	peach	cylindrical	-	-	over
LR952145[a]	pink	oval	+	-	comp
LR952145[p]	pink	oval	-	-	comp
LR952175	orange	oval	+	+	comp
LR952185	white	slender	-	+	comp-over
LR952193	white	round	-	-	comp
LR952200	pink	oval	-	+	comp
LR952216	cream	cylindrical	-	+	comp-over
LR952220	off-white	arthrospores	-	+	over
LR952229	cream	oval	-	-	over
LR961663	straw	oval	+	-	over
LR961703	grey	oval	+	-	over

[1]The stock number of the isolates used in the test. [2]Colour of the colony on potato dextrose agar medium, [3]Shape of individual cells, [4]Ballistospore production (+ =observed, - = not observed), [5]Growth in presence of cyclohexamide (+= grows, - = does not grow), [6]Interaction between yeast and *R. solani* after dual culture on PDA (antibios = antibiosis, comp = competition, over =*R. solani* overgrows yeast- possible reverse mycoparasitism).

TABLE 4. *In vitro* and *in vivo* tests for selection of yeast isolates for field evaluation against the sheath blight disease of rice, Louisiana State University, Baton Rouge, LA., USA

Type of test[1]	Number of isolates tested	Number of Isolates selected
Dual culture test[1]	280	60
Magenta vessel test[2]	32	5
Cut leaf test[3]	27	9
Greenhouse test[4]	27	13
Field bioassay[5]	40	7

[1] In dual culture tests the yeast isolate and *R. solani* were challenged on PDA plates for antagonistic interactions.
[2]Magenta vessel test was conducted on 2-3 week old sterile rice seedlings grown in Magenta vessels containing 40 ml of sterile half strength Murashige and Skoog medium (1962). Seedlings were inoculated with a mycelial plug of *R. solani* one day before spraying with the antagonist.
[3]In the cut leaf assay the sclerotia of *R. solani* were dipped into yeast cell suspension (10^{4-6} cfu/ml) for 10 seconds and placed at one edge of cut leaves laid flat in a Petri dish lined with moist filter paper. Germination of sclerotia and development of sheath blight lesions was noted after 10 days incubation at 28 C.
[4 & 5]The test in the greenhouse was conducted on plants grown in pots. The field test was conducted at the Rice Research Station, Crowley, LA. In both tests all plants except the untreated were inoculated with *R. solani* inocula, prepared on Rice hull : rice grain (2:1) medium, at the first joint elongation stage and sprayed with the yeast suspension 7 days after inoculation and again at 14 days.

Sheath blight ratings for plants or plots treated with yeasts in four screening tests are presented in Table 5. The performance of some of the yeasts was not consistent in all the tests. However, Pearson's correlation analysis of the disease severity values of plants sprayed with these yeasts in different tests showed moderate to high correlations. Highest correlation was obtained between the greenhouse and field results (Table 6).

TABLE 5. Efficacy of selected yeast isolates for biocontrol of rice sheath blight in three screening tests conducted in the laboratory and greenhouse of the Department of Plant Pathology and Crop Physiology, Louisiana State University, Baton Rouge, and in a field bioassay conducted at the LSU Rice Research Station, Crowley LA., USA[1]

Isolates tested[2]	Magenta vessel test[3] ShB (0-5)	Cut leaf test[4] ShB (0-5)	Greenhouse test[5] ShB (0-9)	Field bioassay[5] ShB (0-9)
LR951172	4.2 0ab	4.00 bc	5.00 a	7.50 a-d
LR951438	4.50 ab	4.40 b	5.00 a	7.67 abc
LR951552	2.50 de	3.75 cde	4.50 b-f	7.00 a-g
LR951564	4.75 a	3.80 cde	4.00 def	7.33 a-e
LR951565	3.75 a-d	3.75 cde	4.5 b-f	6.50 d-h
LR951566	4.50 ab	3.10 fg	5.25 abc	6.00 gh
LR951755	5.00 a	4.25 bc	4.00 def	7.00 a-g
LR951906	4.50 ab	3.00 fg	3.50 fg	6.83 b-g
LR952001	4.50 ab	3.40 ef	4.75 b-e	6.50 d-h
LR952006	2.00 d	3.40 ef	4.25 c-g	7.33 a-e
LR952009	2.50 de	3.10 fg	3.50 fg	7.00 a-g
LR952051	2.50 de	3.40 ef	4.25 c-g	6.50 e-h
LR952058	5.00 a	3.25 g	5.50 ab	6.83 b-g
LR952069	3.50 a-d	4.50 b	3.75 efg	7.50 a-d
LR952115	5.00 a	3.40 ef	3.25 g	8.00 a
LR952116	4.00 abc	3.00 fg	4.00 def	6.33 e-h
LR952124	4.00 abc	4.40 b	3.25 g	6.67 c-h
LR952145[a]	2.50de	3.00 fg	4.25 c-g	7.00 a-g
LR952145[p]	3.00 bcd	3.80 cde	4.50 b-f	6.67 c-h
LR952175	4.25 ab	3.40 ef	4.25 c-g	7.00 a-g
LR952185	5.00 a	2.60 g	4.50 b-f	7.33 a-e
LR952193	3.75 abc	4.40 b	4.5 b-f	6.83 b-g
LR952200	4.75 a	4.40 b	4.50 b-f	7.00 a-g
LR952216	5.00 a	2.70 g	4.50 b-f	6.83 b-g
LR952220	4.25 ab	3.40 ef	5.50 ab	7.00 a-g
LR952229	4.50 ab	1.75 g	4.50 b-f	7.33 a-e
LR961663	4.50 ab	4.00 bcd	4.00 d-f	-
LR961703	4.25 ab	3.50 def	6.00 a	-
LR172(*R.solani*)	5.00 a	5.00 a	6.00 a	8.00 a
Untreated	0.00 e	0.00 i	0.00 h	1.50 i

[1]Data from a larger test with entries not shown

[2]Refers to the stock number of the isolates tested.

[3]This test was conducted on 2-3 weeks old sterile rice seedlings grown in Magenta vessels containing 40 ml of sterile half strength Murashige and Skoog medium (1962). Seedlings were inoculated with a mycelial plug of *R. solani* one day before spraying with the antagonist. Sheath blight (ShB) was measured 10 days after inoculation on a rating scale of 0-5 where 0 meant no disease and 5 meant all seedlings blighted or killed.

[4]In this test sclerotia of *R. solani* were dipped in yeast cell suspension (10^{4-6} cfu/ml) for 10 seconds and placed at one edge of the cut leaf laid flat in a petri dish lined with moist filter paper. Lesion development was measured on a rating scale of 0-5 where 0 meant no disease and 5 meant the whole leaf area blighted.

[5]In the greenhouse and field tests all plants except the untreated were inoculated with *R. solani* inocula prepared in rice hull : rice grain (2:1) medium at the first joint elongation stage and sprayed with the yeast suspension 7 days after inoculation and again at 14 days. Disease development in the greenhouse test was measured 30 days after inoculation and in the field test at maturity on a rating scale of 0-9 where 0 meant no disease and 9 meant leaf blade and sheath area of all tillers blighted. All data are means of 3 replications. Means followed by a common letter did not differ significantly at the 5% level based on Duncan's multiple range test.

TABLE 6. Pearson correlation coefficients between the sheath blight disease ratings of plants treated with antagonistic yeasts by different testing methods

Method	Magenta vessel	Cut leaf	Greenhouse	Field
Magenta vessel	1.00	0.46*	0.60*	0.64*
Cut leaf		1.00	0.56*	0.71**
Greenhouse			1.00	0.72**
Field				1.00

**= significant at the 1% level and *= significant at the 5% level.

This suggests that selection based on any one method may be adequate to identify an effective antagonist for biocontrol against sheath blight, although greenhouse tests appeared to be the best for selecting antagonists. Five isolates that we selected performed well compared to the *R. solani* inoculated control in most of these tests and they reduced sheath blight development and increased yield in the field bioassay. These isolates were LR951565, LR951566, LR952001, LR952051, and LR952116 (Table 5). Four of these isolates were evaluated in the field in 1996 and 1997 in 4X16 m plots (Table 7). Compared to the *R. solani* inoculated control, all four isolates reduced sheath blight severity and increased yield in the field evaluation tests. The results were generally significant in 1996 because the climate was favourable for sheath blight development. In 1997, due to dry weather during the time that disease developed, disease pressure was low and consequently the differences were not statistically significant. It is, however, evident from all of the tests that these four isolates have the potential for biological control of sheath blight, especially when the disease pressure is high. One of these isolates, LR951565, is a *Sporobolomyces* sp. This isolate gave consistent *in vitro* and *in vivo* results and is a candidate for developing formulations for commercial release.

TABLE 7. Field evaluation of selected yeast antagonists isolated from the rice phylloplane for biological control of rice sheath blight disease caused by *Rhizoctonia solani*, Rice Research Station, Crowley, LA, USA

Isolate [1]	1996		1997	
(Treatments)	DS (0-9)[2]	Yield (lbs/A)[3]	DS (0-9)[2]	Yield (lbs/A)[3]
LR951565	6.50 b	3878.2 b	3.08 ab	8354.1 b
LR951566	6.00 b	3842.5 b	2.88 abc	8296.9 b
LR952051	6.30 b	3660.5 b	2.63 bc	8214.6 b
LR952116	6.30 b	3345.6 b	2.38 c	8255.1 b
LR172(R.solani)	8.00 a	2429.3 c	3.25 a	7965.3 b
Untreated	1.50 c	5709.8 a	1.00 d	8933.1 a

[1]Stock number of each isolate. They were selected based on their performance in other tests. Isolate LR951565 is a *Sporobolomyces* sp.

[2]Disease severity (DS) of the treated plots was measured at maturity on a scale of 0-9 where 0 meant no disease and 9 meant leaf blade and sheath area of all tillers blighted. All plots except the untreated were inoculated at the first joint elongation stage with *R. solani* inocula grown in rice hull : rice grain (2:1) medium.

[3]Yield is in lbs/A at 12% moisture. 1996 values are means of 3 replications and 1997 values are means of 4 replications. Means followed by a common letter do not differ significantly at the 5% level based on Duncan's multiple range test

These four isolates were further evaluated in preliminary greenhouse and field tests by applying them alone or in combination with carbon sources; such as lactose, sorbitol or sucrose at 2% concentration. Lactose and sorbitol were found to be poorly utilized by *R. solani in vitro*. Lactose was, on the other hand, better utilised by isolate LR951552 and sorbitol by LR961565. Sucrose was utilised equally well by the pathogen and the antagonists. Applications were made twice at 10-day intervals starting 10 days after inoculation with *R. solani* inocula grown in rice hull : rice grain (2:1) medium The isolate LR951552 when applied with lactose or by itself significantly decreased sheath blight severity in the greenhouse test. This treatment also decreased sheath blight severity and increased yield in the field test compared to the disease control plots. The yield performance of this treatment was better than the lactose or LR951552 treatment (Table 8). It has been reported that yeasts antagonize pathogens through nutrient competition. Yeasts exhaust the nutrients in short supply and restrict the growth of the pathogen (Andrews, 1992; Blakeman, 1985; Fokkema, 1984). On the other hand enriching the phylloplane with a nutrient that the pathogen can not utilize helps the antagonist to multiply rapidly and occupy the space needed for the pathogen to grow

TABLE 8. Effect of application of two yeast antagonists alone or in combination with a carbon source on sheath blight development on plants grown in the greenhouse or in the field at Rice Research Station, Crowley, LA, USA

Treatment[1]	Sheath blight severity (GH) (0-9 scale)	Sheath blight severity (Field) (0-9 scale)[2]	Grain yield (Field) (lbs/A)[3]
Lactose 2%	3.8 a-d	2.38 a-d	8235.8 abc
Sorbitol 2%	3.7 a-d	2.62 abc	8229.8 abc
Sucrose 2%	4.0 abc	1.88 d	8365.8 ab
LR951552	3.2 bcd	2.43 bcd	8638.5 ab
LR951565	3.2 bcd	2.25 bcd	8720.8 ab
LR951552+Lactose	3.2 bcd	2.28 bcd	8710.0 ab
LR951565+Sorbitol	4.0 abc	2.00 cd	8337.3 ab
LR951665+Sucrose	4.0 abc	2.75 ab	7698.5 c
LR172 (R. solani)	4.7 a	3.00 a	8339 ab
Untreated (healthy)	0.0 e	0.00 e	8879.5 a

[1]All plots except the untreated were inoculated with *R. solani* inocula grown in rice hull : rice grain (2:1) medium. Spraying with the carbon source (2%), yeast suspension (10^{4-6} cfu/ml) or mixture of yeast and carbon source as shown was done twice at 10-day intervals starting 10 days after inoculation.
[2]Sheath blight severity was measured on a rating scale of 0-9 where 0 means no disease and 9 means all tillers blighted completely.
[3]Grain yield is in lbs/A at 12% moisture. Greenhouse data are means of 3 replications and field data are means of 4 replications. Means followed by a common letter do not differ significantly at the 5% level based on Duncans multiple range test.

(Janisiewicz *et al.*, 1992). The result of the greenhouse and field tests, although not significant due to low disease pressure, gave an indication that by amending the phyllosphere with a carbon source it may be possible to enhance the biocontrol efficacy of the yeast antagonist.

Although fungicides are usually effective in controlling sheath blight, sometimes disease develops even after fungicidal sprays. This could be due to the effect of the fungicide on the phylloplane antagonists that provide natural biological control (Rush *et al.*, 1991). We screened phylloplane yeast isolates against the fungicides commonly used to control sheath blight (Groth *et al.*, 1993). Isolate LR951565 (*Sporobolomyces* sp.) was tolerant to Rovral™ (iprodione) at 6000 ppm formulated. We sprayed this isolate alone and in combination with iprodione to determine the effect on biocontrol of sheath blight disease in greenhouse and field tests. The plants in the greenhouse were grown in pots. In both greenhouse and field tests, plants were inoculated at the first joint elongation stage with *R. solani* grown on rice hull : rice grain (2:1) medium. Plants were sprayed twice at 10 day intervals starting 10 days after inoculation with yeast suspension (10^{4-6} cfu/ml), Rovral 50 WP (at 16 oz/A/15 gal water) or with a mixture of yeast and Rovral. Sheath blight development was measured 30 days after inoculation from the greenhouse test and at maturity from the field test on a 0-9 rating scale, where 0 meant no disease and 9 meant all tillers were completely blighted. All three treatments

TABLE 9. Application of yeast (*Sporobolomyces* sp.) alone or in combination with the fungicide Rovral on greenhouse and field grown plants on biocontrol efficacy of the yeast against rice sheath blight and effect on microbial and yeast populations on the phylloplane

Treatment[1]	DS (0-9)[2] Greenhouse	DS (0-9)[2] Field	Yield (lbs/A)[3] Field	Microbial population[4] leaf bl Field	Microbial population[4] leaf sh	Yeast population[4] leaf bl Field	Yeast population[4] leaf sh
Untreated	0.0 e	0.00 c	8830 a	51.91	89.51	1.07	0.23
R. solani	4.0 a	4.67 a	7715 c	77.69	114.81	0.12	0.06
Yeast (LR951565)	3.2 ab	3.50 b	7860 c	49.19	36.00	0.59	0.10
Rovral	1.5 cd	3.17 b	8176 b	71.04	72.02	0.37	0.05
Yeast+Rovral	2.0 cd	3.33 b	8006 c	68.76	51.40	2.92	0.90

[1]All plots except the untreated were inoculated with *R. solani* inocula as previously described at the first joint elongation stage. Spraying with the yeast suspension (10^{4-6} cfu/ml), Rovral, or yeast and Rovral mix was conducted twice at 10-day intervals starting 10 days after inoculation.
[2]Sheath blight severity was measured 30 days after inoculation in the greenhouse and in the field at maturity on a 0-9 rating scale where 0 means no disease and 9 means leaf blades and sheaths of all tillers blighted.
[3]Grain yield in lbs/A at 12% moisture level. Greenhouse data are means of 3 replications and field data are means of 4 replications. Means followed by a common letter do not differ significantly at the 5% level based on Duncan's multiple range test.
[4]Microbial (actinomycetes, bacteria, and fungi including yeasts) and yeast populations were determined at maturity on potato dextrose agar, acidified PDA and King's medium B and by the contact plating method of isolation. Data are means of three contact platings of randomly selected leaf blade and sheath per medium.

reduced sheath blight development and increased yield in both greenhouse and field tests. The effects of Rovral and the mixture of Rovral and yeast on sheath blight reduction and of Rovral on yield were significantly different from the inoculated control (Table 9). The effect of Rovral and yeast mix on yield was not synergistic or statistically significant due to low disease pressure in 1997. Microbial populations on leaf sheaths of untreated and *R. solani* inoculated plants were higher than on leaf sheaths of yeast, Rovral or yeast and Rovral mix treated plants. The yeast populations on leaf blade and sheath were higher in Rovral and yeast treated plants than on leaf blades and sheaths in other treatments reflecting tolerance to Rovral and survival in presence of this fungicide (Table 9). Phylloplane microorganisms are affected by fungicidal sprays on the aerial surfaces (Andrews and Kinnerly, 1978; Fokkema *et al.*, 1979; Fokkema and De noij, 1981) The effect may be differential on species of yeasts inhabiting the phylloplane (Blakeman, 1985). Recent studies show that yeasts applied with low doses of fungicide are effective in reducing postharvest diseases of apple, pear and sweet cherry (Chand-Goyal and Spotts 1996a, b). These results suggest that there is potential for yeasts to be used as biocontrol agents either alone or in combination with fungicides, like Rovral, for controlling sheath blight on rice.

6. Conclusion

Sheath blight caused by *Rhizoctonia solani* Kuhn is a major constraint to rice production in many countries of the world including the United States. Because of the lack of a high level of sheath blight resistance in commercial cultivars, rice growers have no alternatives but to use fungicides for its control. To reduce grower's dependence on fungicides, a method that is cost effective and that can easily be incorporated into the integrated management system needs to be developed. Biological methods of control using antagonists inhabiting the rice phylloplane appear to be an alternative solution for sheath blight management. Among the antagonistic microorganisms inhabiting the rice phylloplane environment, yeasts show potential as biological agents. They are naturally adapted to the phylloplane where the pathogen also must establish for infection. This is important because the efficacy of biocontrol of foliar diseases will depend on survival of the antagonists and interactions with the pathogen on the phylloplane.

Antagonists inhabiting the rice phylloplane have not been screened or developed as biological control agents for rice diseases. Our studies showed that yeasts are capable of reducing sheath blight development and increasing yield in the field. They are tolerant to commonly used fungicides, which means that they can be used alone or in combination with fungicides, if necessary, to achieve effective control of the disease.

The microbial inhabitants of the rice phylloplane may vary in different environments. Therefore, we need to isolate and select phylloplane antagonists from the environment where they will be used to ensure easy adaptation and survival of the antagonist for *in vivo* interactions against the target pathogen, as well as, other indigenous microorganisms for effective management of the disease. This concept has helped us to identify yeasts inhabiting the rice phylloplane as antagonists of *R. solani* and as potential biocontrol agents for rice sheath blight in Louisiana.

7. Acknowledgement

We thank Dr. Salliana R. Stetina, our colleague, for her critical review and valuable comments on the manuscript.

8. References

Akai, S. and Kuramoto, T. (1968) Microorganisms existing on leaves of rice plant and the occurrence of brown leaf spot, *Annals of the Phytopathological Society of Japan* **34**, 313-316.

Andrews, J. H. (1985) Strategies for selecting antagonistic microorganisms from the phylloplane, in C. E. Windels and S. E. Lindow (eds.), *Biological control on the phylloplane*, The American Phytopathological Society, St. Paul, MN.

Andrews, J. H. (1992) Biological control in the phyllosphere, *Annual Review of Phytopathology* **30**, 603-635.

Andrews, J. H. and Kinnerly, C. M. (1978) The effect of pesticide programs on nontarget microbial populations of apple leaves, *Canadian Journal of Microbiology* **24**, 1058-1072.

Bashi, E. and Fokkema, N. J. (1977) Environmental factors limiting growth of *Sporobolomyces roseus*, an antagonist of *Cochliobolus sativus* on wheat leaves, *Transactions of the British Mycological Society* **68**, 17-25.

Blakeman, J. P. (1985) Ecological succession of leaf surface microorganisms in relation to biological control, in C. E. Windels and S. E. Lindow (eds.), *Biological control on the phylloplane*, APS, St. Paul, MN, pp6-30.

Blakeman, J. P. and Fokkema, N. J. (1982) Potential for biological control of plant pathogens on the phylloplane, *Annual Review of Phytopathology* **20**, 167-192.

Campbell, R. (1989) *Biological control of microbial plant pathogens*, Cambridge University Press, Cambridge, UK.

Chand-Goyal, T. and Spotts, R. A. (1996a) Postharvest biological control of blue mold of apple and brown rot of sweet cherry by natural saprophytic yeasts alone or in combination with low doses of fungicides, *Biological Control* **6**, 253-259.

Chand-Goyal, T. and Spotts, R. A. (1996b) Control of postharvest pear diseases using natural saprophytic yeast colonists and their combination with a low dosage of thiobendazole. *Postharvest Biology & Technology* 7, 51-64.

Cheah, L. H. , Wilson, C. L. and Marshal, A. P. (1996) Rapid screening for antagonist against *Botrytis* storage rots using leaf and fruit tissue, *Postharvest Biology and Technology* **8**, 223-228.

Cook, R. J. and Baker, K. F. (1983) *The nature and practice of biological control of plant pathogens*, APS press, St. Paul, MN.

Dik, A. J., Fokkema, N. J. and Van Pelt, J. A. (1992) Influence of climatic and nutritional factors on yeast population dynamics in the phyllosphere of wheat. *Microbial Ecology 23*,41-52

Elad, Y., Kohl, J. and Fokkema, N. J. (1994) Control of infection and sporulation of *Botrytis cinerea* on bean and tomato by saprophytic yeasts, *Phytopathology* **84**, 1193-1200.

Filonow, A. B.,Vishniac, H. S., Anderson, J. A. and Janisiewicz, W. J. (1996) Biological control of *Botrytis cinerea* in apples by yeasts from various habitats and their putative mechanisms of antagonism, *Biological Control* **7**, 212-220.

Fokkema, N. J. (1973) The role of saprophytic fungi in antagonism against *Drechslera sorokiniana* (*Helminthosporium sativum*) on agar plates and on rye leaves with pollen, *Physiological Plant Pathology* **3**, 195-205.

Fokkema, N. J. (1978) Fungal antagonism in the phyllosphere, *Annals of Applied Biology 89*, 115-142.

Fokkema, N. J. (1984) Competition for endogenous and exogenous nutrients between *Sporobolomyces roseus and Cochliobolus sativus, Canadian Journal of Botany* **62**, 2463-2468.

Fokkema, N. J. and de Nooij, M. P. (1981) The effect of fungicide on the microbial balance in the phyllosphere, *EPPO Bulletin* **11**, 303-310.

Fokkema, N. J., den Houte, J. G., Koselman, Y. J. C. and Nelis, A. L. (1979) Manipulation of yeasts on field grown wheat leaves and their effect on *Cochliobolus sativus*, and its susceptibility to benomyl, *Netherlands Journal of Plant Pathology* **85**, 176-186.

Fokkema, N. J., Van der Laar, J. A., Nelis-Blomberg, A. L. and Schippers, B. (1975) The buffering capacity of the natural mycoflora of rye leaves to infection by *Cochliobolus sativus*, and its susceptibility to benomyl, *Netherlands Journal of Plant Pathology* **81**, 176-186.

Fokkema, N. J. and Van der Meulem, F. (1976) Antagonism of yeast like phylloplane fungi against *Septoria nodorum* on wheat leaves, *Netherlands Journal of Plant Pathology* **82**,13-16.

Gokulapalan, C. and Nair, M. C. (1992) Phylloplane fungi of rice as mycoparasites of *Rhizoctonia solani, Indian Phytopathology* **45**, 367-368.

Groth, D. E., Rush, M. C. and Lindberg, G. D. (1990) Foliar fungicides for control of rice diseases in the United States. in B. T. Grayson, M. B. Green, and L. G. Copping (eds.), *Pest Management in Rice*. Elsevier Applied Sci., NY, pp. 31-52.

Groth, D. E., Rush, M. C. and Hollier, C. A. (1991) *Rice Diseases and Disorders in Louisiana, Lousiana Agricultural Experimental Station Bulletin* **No. 828**, 37.

Groth, D. E., Rush, M. C., Geisler, G. G. and Hollier, C. A. (1993) *Foliar fungicides for use in the management of rice diseases, Lousiana Agricultural Experimental Station Bulletin* **No. 840**, 37.

Handelsman, Jo and Stabb, E. V. (1996) Biocontrol of soilborne plant pathogens, *The Plant Cell* **8**, 1855-1869.

Hajlaoui, M. R. and Belanger, R. R. (1993) Antagonism of the yeast like phylloplane fungus *Sporothrix floculosa* against *Erysiphe graminis* var *tritici, Biocontrol Science and Technology* **3**, 427-434.

Hijwegen, T. (1986) Biological control of cucumber mildew by *Tilletiopsis minor, Netherlands Journal of Plant Pathology* **92**, 93-95.

Janisiewicz, W. J. and Bors, B. (1995) Development of a microbial community of bacterial and yeast antagonists to control wound invading post harvest pathogens of fruits, *Applied and Environmental Microbiology* **61**, 3261-3267.

Janisiewicz, W. J., Usall, J. and Bors, B. (1992) Nutritional enhancement of biocontrol of blue mold on apples, *Phytopathology* **82**, 1364-1370.

Janisiewicz, W. J., Peterson, D. L. and Bors, B. (1994) Control of storage decay of apples with *Sporobolomyces roseus, Plant Disease* **78**, 466-470.

Klecan, A. L., Hippe, S., and Somerville, S. C. (1989) Reduced growth of *Erysiphe graminis* f.sp. *hordei* induced by *Tilletiopsis pallescens, Phytopathology* **80**, 325-331

Knudsen, I. M. B. and Skou, J. P. (1993) The effectivity of *Tilletiopsis albescens* in the biological control of powdery mildew, *Annals of Applied Biology* **123**, 173-185.

Last, F. T. (1955) Seasonal incidence of *Sporobolomyces* on cereal leaves, *Transactions of the British Mycological Society*. **38**, 221-239.

Lee, F. N. and Rush, M. C. (1983) Rice sheath blight: a major disease, *Plant Disease* **67**, 829-832.

Lin, B. R., Wu, S. Z., Xu, Y. M., Mew, T. W. and Rosales, A. M. (1992) Screening of antagonistic bacteria for biocontrol of rice sheath blight, *Chinese Journal of Rice Science* **6**, 77-82.

Lindow, S. E. (1985) Foliar antagonists: status and prospects, in *Biological Control in Agricultural IPM systems*, Academic Press, NY, pp. 395-413.

Lumsden, R. D., Lewis, J. A. and Fravel, D. R. (1995) Formulation and delivery of biocontrol agents for use against soilborne plant pathogens' in F. R. Hall and J. B. Barry (eds.) *Biorational Pest Control Agents: Formulation and Delivery*, ACS Symposium Series No. 595, ACS, Washington D.C., pp.166-182.

Magan, N. and Lacy, J. (1986) The phylloplane microflora of ripening wheat and effect of late fungicide applications, *Annals of Applied Biology* **109**, 117-128.

McCormack, P. J., Wildman, H. G. and Jeffries, P. J. (1994) Production of antibacterial compounds by phylloplane inhabiting yeasts and yeast like fungi, *Applied and Environmental Microbiology* **60**, 927-931.

McGrath, M. T. and Shishkoff, N. (1996) Evaluation of AQ10 (*Ampelomyces quisqualis*) for cucurbit powdery mildew under field conditions, *Phytopathology* **86**, 53 (Abstr.)

McLaughlin, R. J., Wilson, C. W., Droby, S., Ben-Arie, R. and Chalutz, E. (1992) Biological control of postharvest diseases of grapes, peach, and apple with yeasts *Kloeckera apiculata* and *Candida guilliermondii, Plant Disease* **76**, 470-473.

Morris, J. J. and Coleman, P. M. (1995) Ecological succession and biological control in the phyllosphere, *Canadian Journal of Botany* **73**(Suppl.1), 76-82.

Mew, T. W. and Rosales, A. M. (1986) Bacterization of rice plants for control of sheath blight caused by *Rhizoctonia solani, Phytopathology* **76**, 1260-1264.

Murashige, T. and Skoog, F. (1962) A revised medium for rapid growth and bioassay with tobacco tissue culture, *Physiologia Plantarum* **15**, 473-497.

Ng, K. K., MacDonald, L. and Punja, Z. K. (1977) Biological control of rose powdery mildew with the antagonist yeast *Tilletiopsis pallescens, HortScience* **32**, 262-266

Ou, S. H. (1985) *Rice Diseases,* The Commonwealth Mycological Institute, Kew, England.

Punja, Z. K. (1997) Comparative efficacy of bacteria, fungi, and yeasts as biocontrol agents for diseases of vegetable crops, *Canadian Journal of Plant Pathology* **19**, 315-323.

Rabindran, R. and Viddyasekaran, P. (1996) Development of a formulation of *Pseudomonas fluorescens* PfALR2 for management of rice sheath blight, *Crop Protection* **15**, 715-721.

Radler, F., Pfeiffer, P. and Dennert, M. (1985) Killer toxins in new isolates of the yeasts *Hanseniaspora uvarum* and *Pichia kluyveri, FEMS Microbiology Letters* **29**, 269-272.

Rao, K. M. (1995) *Sheath Blight Disease of Rice*, Daya Publishing House, Delhi, India.

Reglinski, T., Lyon, G. D. and Newton, A. C. (1994) Induction of resistance mechanisms in barley by yeast derived elicitors, *Annals of Applied Biology* **124**, 509-517.

Reglinski, T., Lyon, G. D. and Newton, A. C. (1995) The control of *Botrytis cinerea* and *Rhizoctonia solani* on lettuce using elicitors extracted from yeast cell walls, *Journal of Plant Diseases and Protection* **102**, 257-266.

Roberts, R. G. (1990) Postharvest biological control of gray mold of apple by *Cryptococcus laurentii, Phytopathology* **80**, 526-530.

Rush, M. C. and Lee, F. N. (1992) Sheath blight, in R. K. Webster and P. S. Gunnel (eds.) *Compendium of Rice Diseases,* APS, St. Paul, MN, pp.22-23.

Rush, M. C. Prabhu, A. S. and Van Eeckhout, E. (1991) Natural biological control of sheath blight in rice. *Proceedimgs of the XII International Plant Protection Congress,* Rio de Janeiro, Brazil.

Rush, M. C., Pan, X. B., Sha, X. Y., Linscombe, S. D., Xie, Q. J. and Erwin, S. R. (1995) Development of sheath blight resistance in rice, in *87th Annual Research Report*, Rice Research Station, Crowley, LA, pp. 380-390.

Rush, M. C., Shahjahan, A. K. M., Groth, D. E. and Jones, J. P. (1998) Biological control of sheath blight disease of rice (abstract) *27th Rice Technical Working Meeting*, Reno, NV. March 1-3, 1998.

Sackenheim, R., Weltzien, H. C. and Kast W. K. (1994) Effects of microflora composition in the phyllosphere on biological regulation of grape vine fungal diseases, *Vitis* **33**, 235-240.

Sarker, D. K., Sharma, N. R. and Shahjahan, A. K. M. (1992) Antagonistic soil bacteria for biocontrol of rice sheath blight disease caused by *Rhizoctonia solani* Kuhn, *International Rice Research Newsletter* **17**, 22-23.

Shahjahan, A. K. M. and Mew, T. W. (1989) Analysis of sheath blight development in Bangladesh, *Bangladesh Journal of Plant Pathology* **5**, 47-52.

Shahjahan, A. K. M., Sharma, N. R., Ahmed, H. U. and Miah, S. A. (1986) Yield loss in modern rice varieties due to sheath blight in Bangladesh, *Bangladesh Journal of Agricultural Research* **11**, 82-90.

Shahjahan, A. K. M. and Rush, M. C. (1996) Identification of phylloplane microorganisms in Louisiana rice and potential for biological control of rice diseases, *Proceedings of the 26th Rice Technical Working Group meeting*, San Antonio, TX, p. 31.

Shahjahan, A. K. M., Rush, M. C., Groth, D. E. and Jones, J. P. (1997) Potential for biological control of rice sheath blight disease (ShB) with phylloplane fungi, *Phytopathology* **87** (abstract) Supplement, S88.

Shahjahan, A. K. M., Rush, M. C., and Groth, D. E. (1996) Rice phylloplane microorganisms and their potential for biocontrol of sheath blight disease of rice, in *88th AnnualResearch Report,* Rice Research Station, Crowley, LA, pp. 468-476.

Shearer, C. A. (1995) Fungal Competition, *Canadian Journal of Botany* **73,** Supplement 1, S1259-S1264.

Starmer, W. T., Ganter, P. F., Aberdeen, V., Lachance, M. A. and Paff, H. J. (1987) The ecological role of killer yeasts in natural communities of yeasts, *Canadian Journal of Microbiology* **33**, 783-796.

Stirling, A. M., Coates. L. M., Pegg, K. G. and Hayward, A. C. (1995) Isolation and selection of bacteria and yeasts antagonistic to preharvest infection of avocado by *Colletotrichum gloeosporioides, Australian Journal of Agricultural Research* **46**, 985-995.

Sugar, D., Roberts, R. G., Hilton, R. J., Righetti, T. L. and Sanchez, E. E. (1994) Integration of cultural methods with yeast treatment for control of postharvest fruit decay in pear, *Plant Disease* **78**, 791-795.

Urquhart, E. J. and Punja, Z. K. (1996) Epiphytic growth and survival of *Tilletiopsis pallescens*, a potential biological control agent of *Sphaerotheca fuligena*, on cucumber leaves, *Canadian Journal of Botany* **75**, 892-901.

Urquhart, E. J., Menzies, J. G. and Punja, Z. K. (1994) Growth and biological control activity of *Tilletiopsis* species against powdery mildew (*Sphaerotheca fuligena*) on greenhouse cucumbers, *Phytopathology* **84**, 341-351.

Van Eeckhout, E., Rush, M. C. and Blackwell, M. (1991) Effects of rate and timing of fungicide application on incidence and severity of sheath blight and grain yield of rice, *Plant Disease* **75**, 1254-1261.

Williamson, M. A. and Fokkema, N. J. (1985) Phyllosphere yeasts antagonize penetration from appressorium and subsequent infection of maize leaves by *Colletotrichum graminicola, Netherlands Journal of Plant Path*ology **91**, 265-276.

Wilson, C. L. and Chalutz, E. (1989) Postharvest biological control of *Penicillium* rots of citrus with antagonistic yeast and bacteria, *Sci. Hortic.* **40**, 105-112.

Wisneiwski, M., Biles, C., Droby, S., McLaughlin, R., Wilson, C. and Chalutz, E. (1991) Mode of action of the postharvest biocontrol yeast, *Pichia guilliemondi*. 1. Characterization of attachment to *Botrytis cineara, Physiological Plant Pathology* **39**, 245-258.

BIOCONTROL STRATEGIES FOR THE MANAGEMENT OF RICE SHEATH BLIGHT DISEASE

U.I. BABY
Division of Plant Pathology, UPASI Tea Research Institute,
Nirar Dam B.P.O, Valparai, Tamil Nadu, India, 642 127.

1. Introduction

Sheath blight (ShB) disease of rice incited by *Rhizoctonia solani* Kuhn was first reported from Japan in 1910 by Miyaki and subsequently from Philippines (Reinking, 1918), Srilanka (Park and Bertus, 1932) and China (Wei, 1934). The occurrence of the disease in India was first reported from Gurudaspur, Punjab (Paracer and Chahal, 1963). Since then it has become one of the most common and destructive diseases in many rice growing tracts of India (Sridhar *et al.*, 1989). The disease has gained importance in the tropics as it causes yield loss up to 50% (Tsai, 1976). Comprehensive treatises on the disease are available (Manibhushanrao *et al.*, 1979; Gangopadhyay and Chakrabarti, 1982; Ou, 1985; Dath, 1990; Dasgupta, 1992; Manibhushanrao, 1995).

The pathogen *R. solani* is a versatile soilborne saprophyte with high competitive saprophytic ability and wide host range. It survives in soil as sclerotia or thick walled mycelia. Sclerotia represent the primary source of inoculum (Leu and Yang, 1985) and they remain viable in soil for several months over a wide range of temperature and moisture (Park and Bertus, 1932; Manian and Manibhushanrao, 1990). The disease also spreads through airborne basidiospores (Kozaka, 1970; Lee and Rush, 1983) and reports are available on their seedborne nature (Kannaiyan and Prasad, 1978). The disease appears at seedling, tillering and booting stage of the plant and the damage to crop is most severe when the disease appears at later stages. Due to the peculiar versatility of the pathogen, its seedborne nature, competitive saprophytic ability and prolonged survival of the propagules in soil, control of the disease has become very difficult. Chemical fungicides are costly with non-target effects, environmentally hazardous and are not very successful in controlling the disease. In this scenario, biocontrol strategies are gaining importance in controlling the disease. A glimpse on various methods and their efficacy are discussed.

2. Organic soil amendments

The impact of organic amendments on soilborne plant pathogen has been reviewed (Huber and Watson, 1970; Lumsden *et al.*, 1983; Linderman, 1989). A large number of diseases on a variety of crops have been successfully controlled through organic

S. Sreenivasaprasad and R. Johnson (eds.), Major Fungal Diseases of Rice Recent Advances, 253-269.
Prof. K. Manibhushan Rao Festschrift

amendments such as green manures (Muehlchen *et al.*, 1990), plant residues (Baker and cook, 1974; Sumner *et al.*, 1978), oilseed cakes (Lumsden *et al.*, 1983; Dutta and Deb, 1986) and composted organic wastes (Schuler *et al.*, 1989; Gorodecki and Hadar, 1990). Organic amendments effectively controlled rice ShB in greenhouse and field conditions (Rajan and Menon, 1975; Rajan, 1980; Kannaiyan and Prasad, 1981). The effect of various organic amendments such as oilseed cakes, green leaf manures and agro-industrial wastes on the physico-chemical and biological characteristics of the soil with respect to the survivability of the pathogen and disease suppression was critically studied (Manibhushanrao *et al.*, 1989a; Baby, 1991; Baby and Manibhushanrao, 1993a).

2.1. PHYSICO-CHEMICAL PROPERTIES

Incorporation of organic amendments improved the soil porosity, water holding capacity and organic matter status. Macronutrients increased initially and reduced in later stages, while micronutrients showed a linear increase throughout. The leaching of N in amended soil was slow and a steady supply was noticed due to their slow release (Baby *et al.*, 1996).

2.2. BIOLOGICAL PROPERTIES

The soil augmentation of organic substances resulted in a rapid stimulation of soil microflora in the non-rhizosphere as well as rhizosphere zone. The type of amendment as well as the period of incubation had a significant influence on microbial population in both the zones. In the non-rhizosphere soil, fungal and bacterial population was higher in early stages of decomposition of the amendments (upto 30 days) while actinomycetes population increased when the other groups started declining. On the other hand, amendments significantly increased rhizosphere microflora with the increasing age of the plant. The rhizosphere fungi and bacteria responded better to oilseed cakes and agro-industrial wastes, while the actinomycetes to that of green manures (Baby and Manibhushanrao, 1996a). Further, the microbial population was high in the rhizosphere of susceptible cv. TKM9 than the resistant cv. IET 5656 (Figure 1). The population level of the residential *Trichoderma* spp. was also triggered by organic amendments (Baby and Manibhushanrao, 1993a). In general, the population was higher between 20 and 50 days after soil augmentation (Figure 2). In addition, production of antifungal substances toxic to the pathogen and enhanced production of CO_2 were also observed in amended soil (Baby and Manibhushanrao, 1997). The survivability of the *R. solani* propagules reduced as the decomposition of the amendment progressed. After 60 days of incubation, the per cent survival was significantly reduced with all the amendments tested (Baby and Manibhushanrao, 1993a). Moreover, periodical monitoring indicated that the viability as well as infectivity of the sclerotia was drastically reduced in amended soil (Table 1). The ShB incidence was reduced significantly in susceptible as well as resistant cvs. under field condition, with the addition of organic substrate. The protection achieved was maximum with green leaf manures followed by oilseed cakes and agro-industrial wastes (Baby and Manibhushanrao, 1993 a). In addition, there was a significant increase in the growth and yield of the plant (Table 2). Based on the results of the study it is clear that disease control achieved with organic soil amendments is through complex processes and interactions. The organic amendments

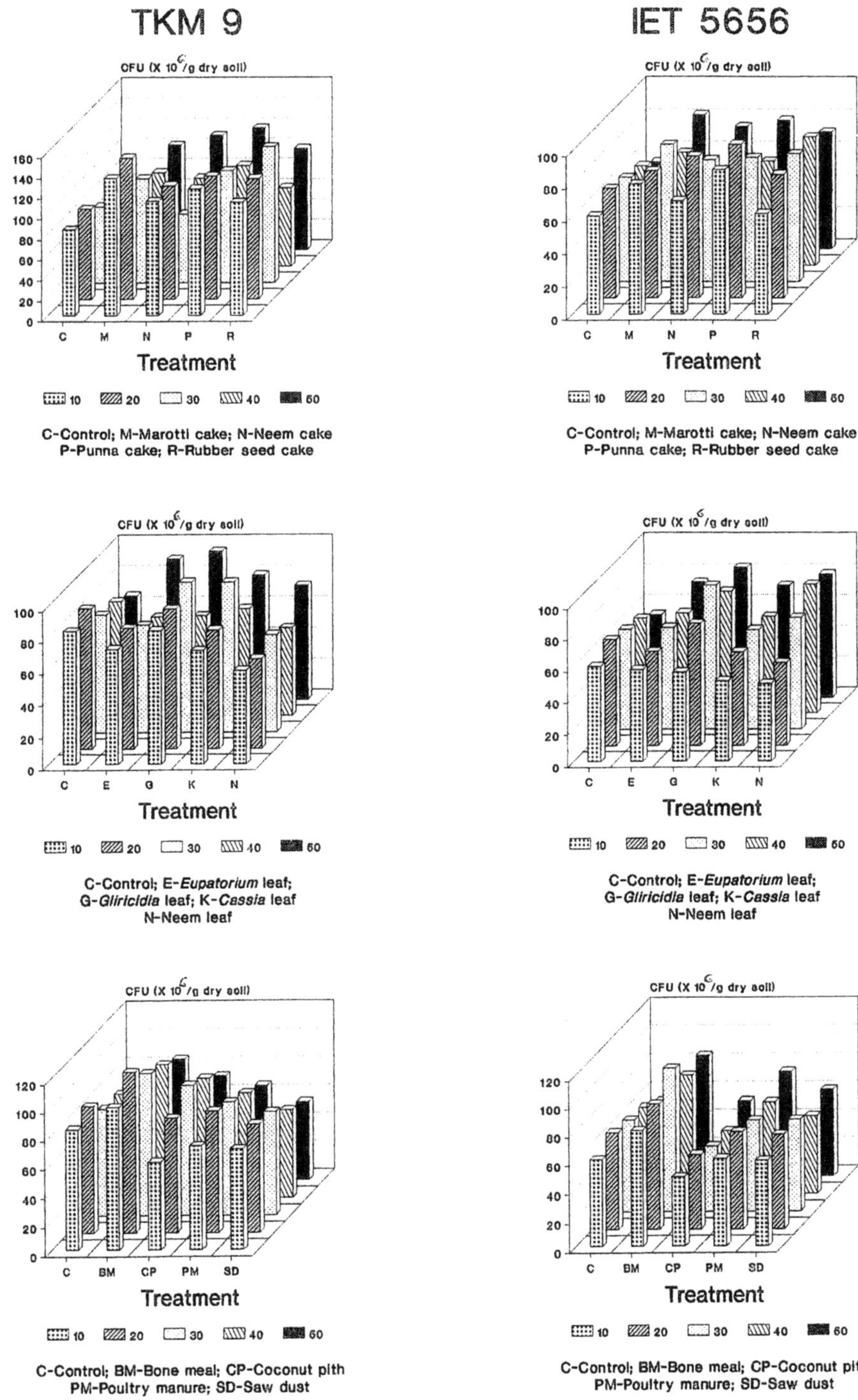

Figure 1. Influence of organic amendments on rhizosphere microflora of susceptible (TKM 9) and resistant (IET 5656) rice cvs.

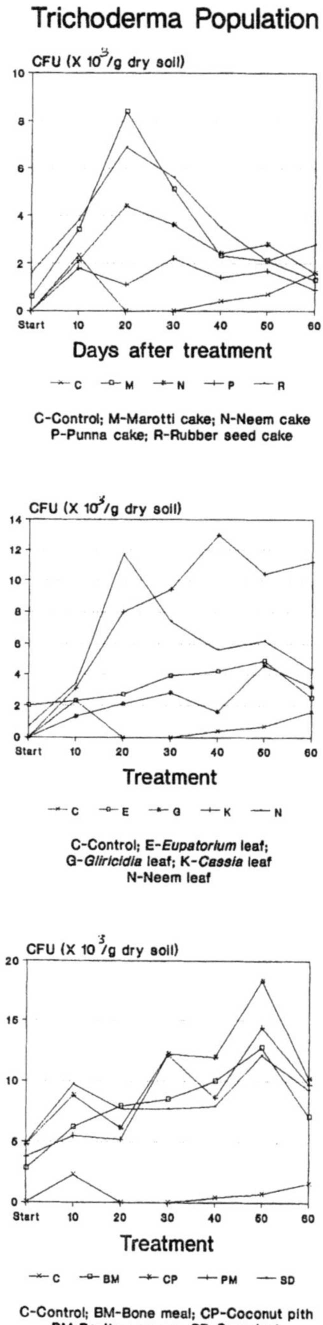

Figure 2. Effect of organic amendments on *Trichoderma* population in non-rhizosphere soil.

can have a range of effects on the pathogen and these effects are strongly related to the kind of substrate added and its stage of decomposition. Enhancement of microbial population results in competition and high level of fungistasis and propagule lysis (Lumsden *et al.*, 1983; Singh and Singh, 1984). Organic amendments favour the build up of antagonistic microbes that can decrease the inoculum potential of the pathogen (Linderman, 1989). In this context, the increased population of *Trichoderma* spp. in amended soil (Baby and Manibhushanrao, 1993a) is significant. Further, production of antifungal substances in amended soil (Chun and Lockwood, 1985) and also the sensitivity of *R. solani* to CO_2 (Papavizas and Davey, 1962) have been reported. as possible mechanisms of action of soil amendments in the rice ShB system (Figure 3).

3. Biocontrol through antagonists

Biocontrol through the use of residential and introduced antagonists has gained importance in recent years and proved successful with many crops. Among the many potentially antagonistic soil inhabitants, members of the genus *Trichoderma* , *Gliocladium*, *Bacillus* and *Pseudomonas* have gained considerable importance (Cook and Baker, 1983; Papavizas, 1985; Schippers *et al.*, 1987). Biological control of plant pathogens by seed bacterisation has been found to be successful in many crops (Weller and Cook, 1983; Colyer and Mount, 1984). Seed bacterisation and spray application of bacterial antagonists suppressed rice ShB and increased grain yield and also the fluorescent bacteria were detected in the roots, coleoptile and first and second leaves of the seedlings raised from bacterised seeds (Mew and Rosales, 1992). Vasanthadevi *et al.* (1989) observed drastic reduction in the ShB incidence on various rice cvs under greenhouse and field conditions. Mycoparasitism of species of *Trichoderma* (Rosales and Mew, 1982; Roy and Sayre, 1984) and *Gliocladium* (Gokulapalan and Nair, 1986) on rice ShB pathogen *R. solani* is known. Manibhushanrao *et al.* (1989b) isolated *T. longibrachiatum* and *G. virens* from rice field soils and established their mycoparasitic potential against ShB pathogen. The role of wall lytic enzymes in mycoparasitism (Howell, 1987; Manocha, 1987), production of antibiotics such as trichodermin, suzukacillin and alamethicin by *T. harzianum* (Dennis and Webster, 1971a,b) and gliovirin by *G. virens* (Howell and Stipanovic, 1983) have been reported. Kumaresan and Manibhushanrao (1991) studied the mycoparasitic potential of two *T. harzianum* isolates against *R. solani* and established the strainal variation in their mycoparasitic and biocontrol potential. Elavarasan (1989) established the biocontrol potential as well as growth promotive effect of *G. virens* under greenhouse condition and Baby and Manibhushanrao (1993b) confirmed their potency under field condition.

Genetic manipulation of biocontrol agents to improve their efficacy is emphasised (Baker, 1989). Fungicide mutant strains of *T. longibrachiatum* and *T. harzianum* were raised by UV irradiation (Viji *et al.*, 1993). The mutants were tolerant to bavistin and expressed higher mycoparasitic activity and competitive saprophytic ability than their parent strains *in vitro*. Under greenhouse condition, the mutants suppressed ShB disease more effectively than the parent strains (Baby, 1997). Mrinalini and Lalithakumari

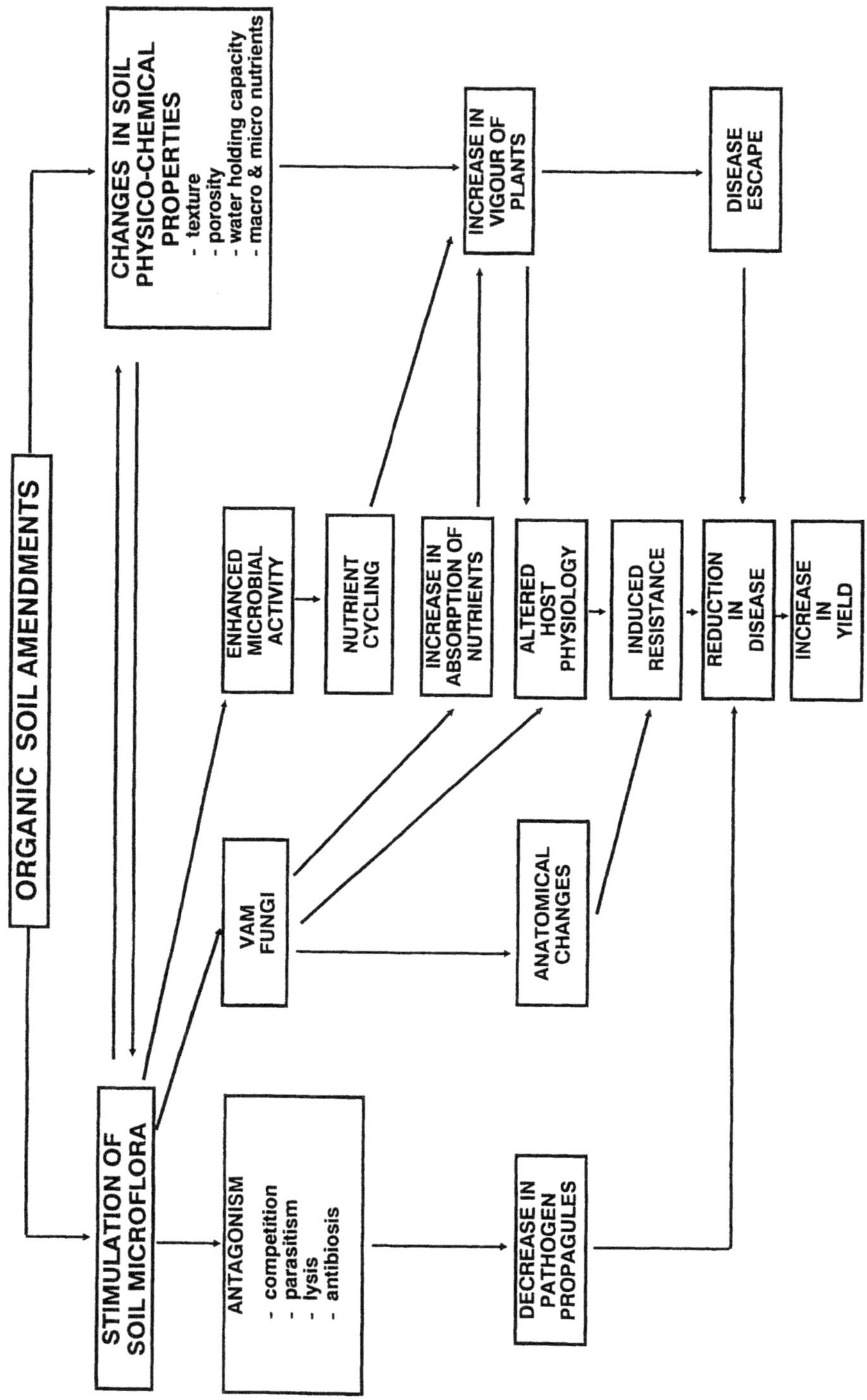

Fig. 3 **Possible mechanisms of action of organic amendments against rice sheath blight disease**

TABLE 1. Effect of organic amendments on the viability and infectivity of *R. solani* sclerotia

	Days of incubation															
	15		30		60		90		120		150		180		210	
Amendment	VBI	DSI	VBI	DSI	VBI	DSI	VBI	DSI	VBI	DSI	VBI	DSI	VBI	DSI	VBI	DSI
Control (unamended soil)	82.5	4.4	78.2	3.8	70.6	4.4	72.7	4.3	68.2	4.5	65.8	4.3	68.4	4.6	62.4	3.5
Marotti cake	65.7	0.8	70.1	2.2	60.3	2.1	50.8	3.3	40.1	3.0	30.3	1.0	24.6	0.8	6.3	0.0
Neem cake	80.3	1.8	71.7	4.4	48.1	2.8	46.0	3.1	45.8	3.3	32.2	2.8	28.7	2.0	26.6	2.0
Punna cake	81.8	1.0	73.7	0.2	57.3	1.3	31.3	0.9	31.9	2.4	34.3	2.8	23.1	1.9	16.7	1.8
Rubber seed cake	83.5	3.8	69.1	4.1	50.1	2.0	38.9	1.1	37.6	1.1	25.3	1.5	30.2	1.8	22.3	1.7
Eupatorium leaf	48.4	0.3	58.2	0.7	46.2	3.1	44.2	1.9	40.3	2.3	37.1	2.3	40.5	1.2	34.0	1.3
Gliricidia leaf	84.5	1.2	57.7	2.0	40.6	1.3	36.8	2.8	38.1	3.2	35.9	3.0	28.2	1.7	20.2	0.7
Cassia leaf	80.5	2.2	64.9	1.2	51.9	2.9	26.7	2.6	30.4	2.9	23.1	2.9	30.6	1.6	25.3	1.8
Neem leaf	78.7	0.9	74.4	3.2	48.6	3.1	40.7	3.2	43.5	3.3	25.1	2.8	26.3	2.0	23.3	1.9
Bonemeal	81.0	1.5	77.5	2.4	52.8	2.1	50.5	2.8	46.7	3.0	30.1	1.5	20.1	0.5	9.6	0.0
Coconut pith	66.7	2.3	63.1	0.8	43.3	1.4	20.3	0.9	22.6	1.3	28.3	1.8	25.3	2.0	16.0	1.0
Poultry manure	85.4	3.0	62.6	4.1	42.6	2.7	40.2	1.5	42.6	2.5	45.8	3.0	43.3	3.0	40.5	2.6
Saw dust	83.3	4.2	76.6	4.3	55.5	2.8	51.7	4.9	55.6	4.5	50.3	4.5	48.5	3.4	45.3	2.1

VBI- Viability Index of of sclerotia
DSI- Disease Severity Index

TABLE 2. Effect of organic amendments on sheath blight incidence, growth and yield

Amendment	Disease incidence		Growth response	
	TKM 9	IET 5656 (Booting stage)	Height	Yield (T/ha)
Control (unamended soil)	18.1 -	10.5 -	54.8	5.51
Marotti cake	10.3 (43.1)	5.0 (52.4)	63.3	5.75
Neem cake	10.4 (42.6)	4.9 (53.8)	59.5	5.40
Punna cake	9.4 (47.8)	5.2 (50.5)	62.9	4.81
Rubber seed cake	7.1 (60.7)	3.4 (67.6)	60.5	5.92
Eupatorium leaf	9.6 (46.9)	4.3 (59.5)	57.7	4.80
Gliricidia leaf	6.9 (61.7)	3.0 (71.4)	56.5	5.06
Cassia leaf	9.1 (49.8)	3.5 (66.7)	58.7	3.91
Neem leaf	7.5 (58.4)	3.7 (64.8)	50.3	4.33
Bonemeal	11.0 (39.1)	5.6 (46.7)	60.6	6.05
Coconut pith	9.2 (49.2)	4.7 (55.2)	55.3	5.42
Poultry manure	9.3 (48.8)	4.3 (59.1)	56.7	5.22
Saw dust	10.1 (44.3)	4.8 (54.3)	54.3	4.31

Values in parentheses represent per cent protection

(1996) developed improved strains of *Trichoderma* by fusing protoplasts of *T. harzianum* and *T. longibrachiatum* which are antagonistic toseveral rice pathogens including *R. solani* and found that the hybrid strains (fusants) controlled ShB disease and increased the growth of rice plants more effectively than the parents under greenhouse conditions.

4. Vesicular arbuscular mycorrhizae (VAM)

VAM fungi, a component of rhizosphere microflora, form symbiotic association with higher plants. (Baltruschat and Dehne,1988; An *et al.*, 1993). Enhanced nutrient uptake, especially immobile elements like P, Cu, Zn etc. (Tinker, 1975; Clarkson, 1985), plant growth promotion (Cooper, 1983; Manjunath and Bagyaraj, 1986), stress tolerance (Allen and Boosalis, 1983; Marx and Schenck, 1983) and resistance to soilborne plant pathogens (Schenck, 1981; Caron, 1989) are some of the important attributes of VAM fungi. Association of VAM fungi with upland (Gangopadhyay and Das, 1982; Ammani *et al.*, 1985) and low land (Sivaprasad *et al.*, 1990) rice is known. The VAM fungal species associated with upland rice has been reported (Brown et al., 1988). Baby and Manibhushanrao (1992) surveyed the indigenous VAM fungi in the rice fields of Kerala state, India. The areas surveyed included fields below the mean sea level. It has been found that species like *Glomus occultum*, *G. fasiculatum*, *G. aggregatum*, *G. ambisporum* were dominant; *Acaulospora*, *G. mosseae*, *G. microaggregatum*, *Gigaspora gigantea* and *G. albida* were common and many other species of *Glomus*, *Sclerocystis* sp. and *Scutellospora* sp. were rare forms in rice field soils. The spore density appeared to be more closely related to pH and soil type than nutrient status. Biological control potential of VAM fungi (VAMF) has been reported with many crop plants (Graham and Menge, 1982; Kaye *et al.* 1984; Rosendahl, 1985).

Baby and Manibhushanrao (1996b) studied the response of rice plants to VAMF infection with reference to growth and ShB control, under greenhouse condition. The plants were inoculated with VAMF before, after and simultaneously with *R. solani* and the results (Table 3) showed that the soil augmentation with VAMF inoculum after the establishment of the pathogen influenced neither growth nor ShB incidence. Simultaneous application had a significant interaction, while prior inoculation with VAMF was significantly superior in promoting growth and suppressing the disease. The influence of organic amendments namely, oil cakes, green leaf manures and agro-industrial wastes on the VAMF infection and spore density as well as ShB incidence were critically studied (Baby and Manibhushanrao, 1996c). VAMF spore density and per cent infection of rice root were influenced by the type of amendment, their stage of decomposition and age of the plant (Figure 4 and Table 4). It was also observed that some of the amendments (marotti and punna cakes, eupatorium, cassia and neem leaves, bonemeal, coconut pith and saw dust) favoured arbuscular development. Assessment of mycorrhizal colonisation in the roots of healthy and *R. solani* infected rice plants revealed that the spore density and per cent infection were significantly higher in healthy plants than in diseased plants (Table 5). Induced resistance, reduced disease incidence and increased growth response by the VAMF explore the possibilities of exploiting this natural symbiont in the ShB disease management strategies.

TABLE 3. Effect of VA mycorrhizal fungi on the growth and sheath blight incidence under greenhouse condition

Treatment	Height (cm)	Disease incidence (%)
Control (R. solani alone)	15.2 c	41.0 c
R. solani + VAMF (Post application)	15.0 c	40.0 c (2.0)
R. solani + VAMF (Simultaneous application)	16.0 b	30.0 b (25.0)
VAMF + R. solani (Prior application)	18.6 a	27.0 a (33.0)

In column, means followed by the same letters are not significantly different (P= 0.001) according to Duncan's multiple range test.
Values in parentheses indicate per cent protection.

5. Cross protection and Induced resistance

Cross-protection is the prevention of disease by the use of a biotic agent which could be an avirulent or mildly virulent pathogen or a non-pathogen. Induced resistance, on the other hand, is a type of phenomenon where the host defence mechanism is triggered to meet the threat posed by the pathogen. Cross-protection may, therefore, be by the induced resistance or by other mechanism(s) of antagonism. An avirulent strain of *R. solani* (R7) from potato was tested *in vitro* against rice ShB (Manibhushanrao *et al.*, 1988). The incompatible interaction on excised rice leaves reduced the ShB spread considerably. The magnitude of resistance was clearly visible in rice leaves prior inoculated with R7, yielding 81% against differentially virulent isolates of *R. solani*.

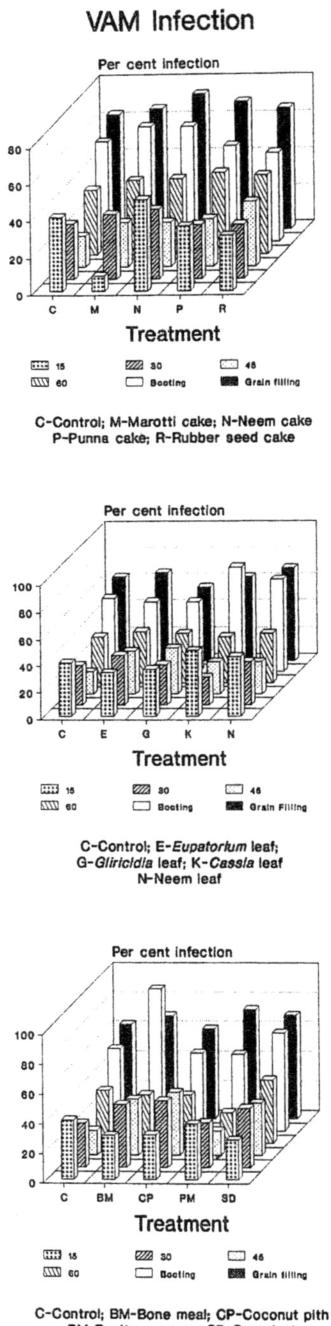

Figure 4. Influence of organic amendments on arbuscular mycorrhizal fungal infection of field grown rice plants.

TABLE 4. Influence of organic amendments on VAM fungal spore density under field conditions

	Plant age (days)					
Amendment	15	30	45	60	Booting	Grain filling
Control (unamended soil)	249c	407b	380k	253j	323i	370j
Marotti cake	170d	383d	452j	480d	396j	363j
Neem cake	118g	247f	713c	546b	560b	385i
Punna cake	133f	235g	734a	526c	549c	368j
Rubber seed cake	162e	172i	308l	370g	425h	394h
Eupatorium leaf	92i	211h	723b	566a	680a	805a
Gliricidia leaf	89i	266e	550f	327h	446g	626c
Cassia leaf	108h	249f	530g	466e	498e	590d
Neem leaf	95i	131j	578e	425f	515d	698b
Bonemeal	250c	234g	584e	551b	379j	570e
Coconut pith	307a	396c	597d	310i	387j	448g
Poultry manure	292b	507a	517h	331h	482f	632c
Saw dust	290b	228g	472i	329h	353k	496f

The results are expressed as number of spores per 100g dry soil.

Means followed by the same letter in a given column do not differ significantly.

TABLE 5. Mycorrhizal development and response of rice plant to sheath blight incidence at different stages of growth

	Booting			Grain filling		
	Healthy	Infected	t(P<0.01)	Healthy	Infected	t(P<0.01)
No. of spores/100g dry soil	203	186	3.02	400	340	10.47
Per cent infection	60	45	16.63	74	15	6.40
Intensity of infection	2-3	1-2	-	2-3	1-2	-
No. of vesicles per root segment	35	18	7.71	56	32	12.37

The ShB pathogen *R. solani* produces phytotoxins, Phenyl acetic acid (PAA) and its hydroxy derivatives (*m*-HPAA, *o*-HPAA and *p*-HPAA). These metabolites are toxic to plant tissue at higher concentrations (>100 ppm) and produce lesions comparable to those caused by the pathogen. But at low concentrations (PAA- 1.0%, HPAA- 0.1%), these toxins induced 25- 30% protection from the disease on excised rice leaves (Waheeta *et al.*, 1987). These metabolites were neither phytotoxic nor fungitoxic. The acquired resistance may presumably be due to the physical and/ or chemical barriers elicited in the host. Plant immunisation using inducer chemicals is well established. Manibhushanrao *et al.* (1990) tested an array of chemical elicitors namely, lithium sulphate, sodium azide, sodium selenite, thioglycollic acid, cycloheximide and penicillin on rice ShB. The seed and root dip treatment of the chemicals yielded protection up to 60 %. Further, the leaf diffusate bioassay against *R. solani* revealed a positive correlation between the disease reduction and the enhanced fungitoxicity level implicating the involvement of phytoalexin type substances.

6. Integrated control

Integrated control is the combined use of chemicals and biocontrol agents to control disease. It is proved that a combination of two or more provide high level of control than that could be achieved with each disease management system. Rajan and Alexander (1988) studied the effect of combination of soil amendments, antagonistic fungus and fungicide on rice ShB. The treatments consisted of soil amendments (rice husk, punna cake, glyricidia leaf, lime and gypsum along with an untreated control) as main components; and *T.viride* and carbendazim along with a control as sub-components.

TABLE 6. Effect of biocontrol agents and amendments individually and in combination on the growth response and sheath blight intensity in field grown rice plants

Treatment	Height (cm)	Grain yield (T/ha)	*Disease intensity(%)	Per cent protection
Control (Untreated)	44.8c	3.00f	15.11a	-
G. virens (GV)	47.4bc	3.94de	9.27b	38.6
T. longibrachiatum (TL)	48.8abc	4.44cde	9.47b	37.3
Gliricidia (G)	46.5bc	3.62ef	8.71b	42.4
Neem (N)	49.8abc	4.60bcd	8.99b	40.5
GV + G	55.5ab	.92a	57.64b	49.4
TL + G	56.7a	5.73a	6.66b	55.9
GV + N	57.8a	5.40ab	8.15b	46.1
TL + N	55.1ab	5.33abc	7.29b	51.8

Values followed by the same letters are not significantly different (P= 0.01) according to DMRT.
*At booting stage.

The results showed that the disease incidence was least in all the combination treatments with a subsequent reduction in *R. solani* population and an increase in general microflora (fungi and bacteria). The efficacy of the combination of organic amendments and fungal antagonists in controlling rice ShB was studied under field condition (Manibhushanrao and Baby, 1991; Baby and Manibhushanrao, 1993b). Two fungal antagonists (*G. virens* and *T. longibrachiatum*) and two organic amendments (*Gliricidia* and neem cake) were tested. Integration of these systems significantly enhanced the efficacy over their individual effect (Table 6).

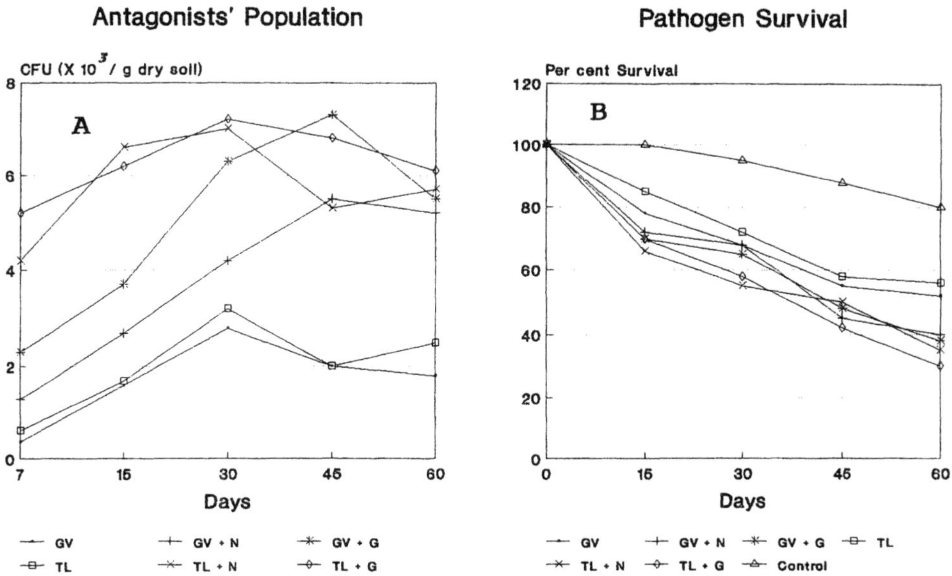

Figure 5. Population dynamics of antagonistic fungi and *Rhizoctonia solani* in soil with various treatments. A. Survival of antagonistic fungi; B. Survival of *R. solani*.

The growth as well as grain yield also were increased by varying degrees in different treatments. When the antagonists were supplemented with organic substrates, an increase in their colony forming units and subsequent reduction in the pathogen population were noticed (Fig 5). Viji *et al.* (1991) observed a difference in the response of ShB pathogen and their fungal antagonists namely, *G. virens*, *T. harzianum* and *T. longibrachiatum* their response to different groups of systemic fungicides. The pathogen *R. solani* was highly sensitive to both benzimidazole and organophosphorus fungicides. On the otherhand, the fungal antagonists were highly sensitive to benzimidazole fungicides, while tolerant to organophosphorus fungicides. The result indicated the possibility of integrating fungal antagonists with chemicals which are less toxic to the antagonistic/ beneficial organisms. Further, Papavizas *et al.* (1982) and Abd- El Moity *et al.* (1982) have shown that new biotypes of *T. harzianum* can be selected for tolerance or resistance to fungicides. Fungicide resistant biotypes were found to be superior to their parents as biocontrol agents (Cook and Baker, 1983; Baby, 1997).

7. Conclusion

In the present environmental protection scenario, it is absolutely necessary to exploit the biological/ natural resources in disease management strategies to minimise the load of chemical pesticides for a clean and safe environment. Extensive studies on the rice - ShB system with various non- chemical methods proved the prospects in this line.

Organic wastes as soil amendments effectively controlled the ShB disease through a complex biotic and abiotic interactions, including the stimulation of saprophytic residential microflora. Experiments with biocontrol agents indicate the possibilities of exploiting the bacterial and fungal antagonists and also improving their bioefficacy through genetic manipulation. Vesicular arbuscular mycorrhizae as biocontrol agents, cross protection through avirulent isolates and induced resistance have also proved to be promising. The superiority of integrating various disease control methods is also evident from the study.

8. References

Abd-El Moity, T.H., Papavizas, G.C and Shatla, M.N. (1982) Induction of new isolates of *Trichoderma harzianum* tolerant to fungicides and their experimental use for control of white rot of onion, *Phytopathology*, **72**, 396- 400.

Allen, M.F. and Boosalis, M.G. (1983) Effects of two species of VA mycorrhizal fungi on drought tolerance of winter wheat, *New Phytologigt* **93**, 67- 76.

Ammani, K., Venkateswarlu, K. and Rao,A.S. (1985) Development of vesicular- arbuscular mycorrhizal fungi of upland rice, *Current Science* **59**, 1120- 1122.

An ZO, J.W., Hendrix, D.E., Hershman., R.S., and Henson, G.T. (1993) The influence of crop rotation and soil fumigation on a mycorrhizal fungal community associated with soybean, *Mycorrhiza* **3**, 171- 182.

Baby, U.I. (1991) *Studies on the control of rice sheath blight disease with organic soil amendments*. Ph. D. Thesis, University of Madras, Madras.

Baby, U.I. (1997) Biocontrol potential of fungicide resistant mutants of *Trichoderma* spp., *Indian Journal of Microbiology* (In Press)

Baby, U.I. and Manibhushanrao, K. (1992) Vesicular arbuscular mycorrhizal fungi of rice field soils, *Acta Botanica Indica* **20**, 10-15.

Baby, U.I. and Manibhushanrao, K. (1993a) Control of rice sheath blight disease by organic manures, in R. Ravikumar (ed.), *Proceedings of the fifth Kerala Science Congress*, Organized by State Commission on Science, Technology and Environment, Government of Kerala, Thiruvananthapuram, pp. 158- 161.

Baby, U.I. and Manibhushanrao, K. (1993b) Control of rice sheath blight through the integration of fungal antagonists and organic amendments, *Tropical Agriculture*. (Trinidad) **70**, 240- 244.

Baby, U.I. and Manibhushanrao, K. (1996a) Effect of organic amendments on soil microflora and plant growth, *Soil Biology and Fertility* (in press).

Baby, U.I. and Manibhushanrao, K. (1996b) Fungal antagonists and VA mycorrhizal fungi for biocontrol of *Rhizoctonia solani*, the rice sheath blight pathogen. in K. Manibhushanrao and A. Mahadevan (eds.), *Recent developments in biocontrol of plant pathogens*, Current trends in life sciences **Vol. XXI**, Today and Tomorrow's Printers and Publishers, New Delhi, pp. 1-9.

Baby, U.I. and Manibhushanrao, K. (1996c) Influence of organic amendments on arbuscular mycorrhizal fungi in relation to rice sheath blight disease, *Mycorrhiza* **6**, 201- 206.

Baby, U.I. and Manibhushanrao, K. (1997) Suppression of rice sheath blight disease in organic manure amended soil, *Agriculture Ecosystem and Environment* (in press).

Baby, U.I., Manibhushanrao, K. and Elango, L. (1996) Effect of organic amendments on physico- chemical properties of rice field soil, *Communications in Soil Science and Plant Analysis* (in press).

Baker, R. (1989) Some perspectives on the application of molecular approaches to biocontrol problems, in J. M. Whipps and R.D. Lumsden (eds.), *Biotechnology of fungi for improving plant growth*, Cambridge Univ. Press, Cambridge, pp. 220-223.

Baker, K.F. and Cook, R.J. (1974) *Biological control of plant pathogens*, Freeman, Sanfransico, U.S.A.

Baltruschat, H. and Dehne, H.W. (1988). The occurrence of vesicular arbuscular mycorrhizae in agro-ecosystems. Influence of nitrogen fertilization and green manure in continuous monoculture and in crop rotation on the inoculum potential of winter wheat, *Plant and Soil* **107**, 279- 284.

Brown, M.B., Quimio, T.H. and Castro, A.M. (1988). Vesicular arbuscular mycorrhizas associated with upland rice (*Oryza sativa* L.), *Philppine Agriculture* **71**, 317- 332.

Caron, M. (1989) Potential use of mycorrhizae in control of soil-borne diseases, *Canadian Journal of Plant Pathology* **14**, 177-179.

Chun, D. and Lockwood, J.L. (1985) Reduction of *Pythium ultimum*, *Thielaviopsis basicola* and *Macrophomina phaseolina* populations in soil associated with ammonia generated from urea, *Plant Disease* **69**, 154- 158.

Clarkson, D.T. (1985) Factors affecting mineral nutrient acquisition by plants, *Annual Review of Plant Physiology* **36**, 77- 115.

Colyer, P.D. and Mount, M.S. (1984) Bacterization of potatoes with *Pseudomonas putida* and its influence on post harvest soft rot disease, *Plant Disease* **68**, 703- 706.

Cook, R.J. and Baker, K.F. (1983) *The nature and practice of biological control of plant pathogens*, American Phytopathological Society, St. Paul, Minnesota, U.S.A.

Cooper, K.M. (1983) Mycorrhizal fungi can improve growth of horticultural crops, *Orchardist* (N. Z.) **56**, 410-413.

Dasgupta, M.K. (1992) Rice sheath blight: The challenge continuous, in U.S. Singh, A.N. Mukopadhayay, J. Kumar and H.S. Chanbe (eds.), *Plant diseases of International Importance* **Vol. I**, Printice Hall, New Jersey, U.S.A., pp. 130- 157.

Dath, A.P. (1990) *Sheath blight disease of rice and its management*, Associate Publishing Co., New Delhi.

Dennis, C. and Webster, J. (1971a) Antagonistic properties of species-groups of *Trichoderma*. 1. Production of non-volatile antibiotics, *Transactions of British Mycological Society* **57**, 25- 39.

Dennis, C. and Webster, J. (1971 b) Antagonistic properties of species-groups of *Trichoderma*. 2. Production of volatile antibiotics, *Transactions of British Mycological Society* **57**, 41- 48.

Dhillion, S.S. (1992) Host endophyte specificity of vesicular arbuscular mycorrhizal colonisation of *Oryza sativa* L. at the pretransplant stage in low or high phosphorus soil, *Soil Biology Biochemistry* **24**, 405-411.

Dutta, B.K. and Deb, P.R. (1986) Effect of organic and inorganic amendments on the soil rhizosphere microflora in relation to the biology and control of *Sclerotium rolfsii* causing foot rot of soybean, *Journal of Plant Disease and Protection* **93**, 163-171

Elavarasan. A. (1989) *Studies on the biocontrol potential of Gliocladium virens against rice sheath blight disease*, M. Phil. Dissertation, University of Madras, Madras.

Gangopadhyay, S. and Chakrabarti, N.K. (1982) Sheath blight of rice, *Review of Plant Pathology* **61**, 451-460.

Gangopadhyay, S. and Das, K.M. (1982) Occurrence of vesicular arbuscular mycorrhizae in rice in India, *Indian Phytopathology* **35**, 83-85.

Gokulapalan, C. and Nair, M.C. (1986) Mycoparasites of *Rhizoctonia solani* and control of sheath blight of rice, (abstract) *Fourteenth Annual meeting of the Mycological Society of India and Seminar on Applied Mycology*, Thanjavur, India, p.18.

Gorodecki, B. and Hadar, Y. (1990) Suppression of *Rhizoctonia solani* and *Sclerotium rolfsii* diseases in container media containing composted separated cattle manure and composted grape marc, *Crop Protection* **9**, 271- 274.

Graham, J.H. and Menge, J.A. (1982) Influence of vesicular arbuscular mycorrhizae and soil phosphorus on take all disease of wheat, *Phytopathology* **72**, 95- 98.

Howell, C.R. (1987) Relevance of mycoparasitism in the biological control of *Rhizoctonia solani* by *Gliocladium virens*, *Phytopathology* **77**, 992- 994.

Howell, C.R. and Stipanovic, R.D. (1983) Gliovirin, a new antibiotic from *Gliocladium virens*, and its role in the biological control of *Pythium ultimum*, *Canadian Journal of Microbiology* **29**, 321- 324.

Huber, D.M. and Watson, A.D. (1970) Effect of organic amendment on soil-borne plant pathogens, *Phytopathology* **60**, 22- 26.

Kannaiyan, S. and Prasad, N.N. (1978) Seed-borne nature of sheath blight pathogen *Rhizoctonia solani* in rice, *International Rice Research Newsletter* **3**, 10.

Kannaiyan, S. and Prasad, N.N. (1981) Effect of organic amendments on seedling infection of rice caused by *Rhizoctonia solani*, *Plant and Soil* **62**, 131- 133.

Kaye, J.W., Pfleger, F.L. and Stewart, E.L. (1984) Interaction of *Glomus fasciculatum* and *Pythium ultimum* on greenhouse grown *Poinsettia*, *Canadian Journal of Botany* **62**, 575- 579.

Kozaka, T. (1970) *Pellicularia* sheath blight of rice plant and its control, *Japan Agriculture Research* **5**, 12-16.

Kumaresan, S. and Manibhushanrao,K. (1991) Studies on the biological control of rice sheath blight disease, *Indian Journal Plant Pathology* **9**, 64- 70.

Lee, F.N. and Rush, M.C. (1983) Rice sheath blight: a major rice disease, *Plant Disease* **67**, 829- 832.

Leu, L.S. and Yang, H.C. (1985) Distribution and survival of sclerotia of rice sheath blight fungus, *Thanatephorus cucumeris*, in Taiwan, *Annals of Phytopathological Society Japan* **51**, 1-7.
Linderman, R.G. (1989) Organic amendments and soil-borne diseases, *Canadian Journal of Plant Pathology* **11**, 180- 183.
Lumsden, D.R., Lewis, J.A. and Papavizas, G.C. (1983) Effect of organic amendments on soilborne plant diseases and pathogen antagonists, in W. Lockeretz (ed.), *The environmentally sound agriculture*, Praeger press, New York, U.S.A., pp. 51- 70.
Manian, S. and Manibhushanrao, K. (1990) Influence of some factors on the survival of *Rhizoctonia solani* in soil, *Tropical Agriculture* **67**, 207- 208.
Manibhushanrao, K. 1995. *Sheath blight disease of rice*, Daya Publishing House, New Delhi, India.
Manibhushanrao, K. and Baby, U.I. (1991) Management of rice sheath blight using fungal antagonists and organic amendments, *International Rice Research Newsletter* **16**,19-20.
Manibhushanrao, K., Baby, U.I. and Joe, Y (1989) Effect of organic amendments on the saprophytic survival of rice sheath blight pathogen and the soil microflora, *Oryza* **26**, 71- 89.
Manibhushanrao, K., Joe, Y. and Madathiammal, P (1990) Elicitation of resistance in rice to sheath blight disease, *International Journal of Tropical Plant Disease* **8**, 193- 197.
Manibhushanrao, K., Manian, S. and Zuber, M. (1979) Sheath blight disease of rice in East Asia, *Journal of Plant Disease and Protection* **86**, 499-509.
Manibhushanrao, K., Sreenivasaprasad, S., Baby, U.I. and Joe, Y. (1989b) Susceptibility of rice sheath blight pathogen to mycoparasites, *Current Science* **58**, 515- 518.
Manibhushanrao, K., Sreenivasaprasad, S., Chitralekha, R.S. and Kalaiselvi, K. (1988) Cross protection in rice to sheath blight, *Journal of Indian Botanical Society* **67**, 97- 100.
Manjunath, A. and Bagyaraj, D.J. (1986) Response of blackgram, chickpea and mungbean to vesicular arbuscular mycorrhizal inoculation in unsterile soil, *Tropical Agriculture* **63**, 33-35.
Manocha, M.S. (1987) Cellular and molecular aspects of fungal host-mycoparasite interaction, *Journal of Plant Disease and Protection* **94**, 431- 444.
Marx, D.H. and Schenck, N.C. (1983) Potential of mycorrhizal symbiosis in agricultural and fungal productivity, in T. Kommendahl and P.H. Williams (eds.),*Challenging problems in plant health*, 75th Annual publication of the Ammerican Phytopathological Society, St. Paul, Minnesota, U.S.A.
Mew, T.W. and Rosales, A.M. (1992) Control of *Rhizoctonia* sheath blight and other diseases of rice by seed bacterization, in E.J. Tjamos *et al*., (eds.), *Biological control of plant diseases*, Plenum press, pp. 113-123.
Miyake, I. (1910) Studien uber die Plze der Reispflanze in Japan, *Journal of College of Agriculture Imperial University Tokyo* **2**, 237-276.
Mrinalini, C. and Lalithakumari, D. (1996) Protoplast fusion: A biotechnological tool for strain improvement of *Trichderma* spp. in K. Manibhushanrao and A. Mahadevan (eds.), *Recent developments in biocontrol of plant pathogens*, Current trends in life sciences **Vol. XXI**, Today and Tomorrow's Printers and Publishers, New Delhi, pp.133-146.
Muehlchen, A.M., Rand, R.E. and Parke, J.L. (1990) Evaluation of crucifer green manures for controlling *Aphanomyces* root rot of peas, *Plant Disease* **74**, 651- 654.
Ou, S.H. (1985) *Rice diseases*, CABI, Slough, U.K.
Papavizas, G.C. (1985) *Trichoderma* and *Gliocladium*: biology, ecology and potential for biocontrol, *Annual Review of Phytopathology* **23**, 23- 54.
Papavizas, G.C. and Davey, C.B. (1962) Activity of *Rhizoctonia* in soil as affected carbon di oxide, *Phytopathology* **52**, 759-766.
Papavizas, G.C., Lewis, J.A. and Abd-ElMoity, T. (1982) Evaluation of new biotypes of *Trichoderma harzianum* for tolerance of benomyl and enhanced biocontrol capabilities, *Phytopathology* **72**, 126- 132.
Paracer, C.S. and Chahal, D.H. (1963) Sheath blight caused by *Rhizoctonia solani* Kuhn- A new record in India, *Current Science* **32**, 328- 329.
Park, M. and Bertus, L.S. (1932) Sclerotial diseases of rice in Ceylon, 1. *Rhizoctonia solani* Kuhn, *Ceylon Journal of Science* **11**, 319- 331.
Rajan, K.M. (1980) Soil amendments in plant disease control, *International Rice Research Newsletter* **5**, 515.
Rajan, K.M. and Alexander, S. (1988) Management of sheath blight disease of rice with *Trichoderma viride* and some soil amendments in relation to the population of pathogen in soil, *Journal of Biological Control* **2**, 36- 41.
Rajan, K.M. and Menon, R.M. (1975) Effect of organic amendments on plant growth and intensity of sheath blight of rice, *Agricultural Research Journal of Kerala* **13**, 179- 181.

Reinking, O.A. (1918) Philippine economic plant diseases, *Philippines Journal of Science and Agriculture* **13**, 165-274.

Rosales, A.M. and Mew, T.W. (1982) Antagonistic effects of soil microorganisms on rice sheath blight pathogen, *International Rice Research Newsletter* **7**, 12- 13.

Rosendahl, S. (1985) Interactions between the vesicular arbuscular mycorrhizal fungus *Glomus fasciculatum* and *Aphanomyces euteiches* root rot of peas, *Journal of Phytopathology* **114**, 31-40.

Roy, A.K. and Sayre, R.M. (1984) Electron microscopical studies of *Trichoderma harzianum* and *T. viride* and mycoparasitic activity of the former on *Rhizoctonia solani* f. sp. *sasakii*, *Indian Phytopathology* **37**, 710- 712.

Schenck, N.C. (1981) Can mycorrhizae control root disease?, *Plant Disease* **65**, 230- 231.

Schippers, B., Bakker, A.W. and Bakker, P.H.A.M. (1987) Interaction of deleterious and beneficial rhizosphere microorganisms and their effect on the cropping of potatoes, *Annual Review of Phyto Pathology* **25**, 339- 358.

Schuler, C., Biala, J., Bruns, C., Gottschall, R., Awers, S. and Vogtmann, H. (1989) Suppression of root rot on peas, beans and beet roots caused by *Pythium ultimum* and *Rhizoctonia solani* through the amendment of growing media with composted organic household waste, *Journal of Phytopathology* **127**, 227-238.

Singh, N. and Singh, R.S. (1984) Significance of organic amendment of soil in biological control of soilborne plant pathogens, in K.G. Mukerji, V.P. Agnihotri and R.P. Singh (eds.), *Progress in microbial ecology*, Print House, Lucknow, India, pp. 303- 323.

Sivaprasad, P., Sulochana, K.K., Salam, M.A. (1990) Vesicular arbuscular mycorrhizae (VAM) colonization in low land rice roots and its effect on growth and yield, *International Rice Research Newsletter* **15**, 14- 15.

Sridhar,R., Manibhushanrao, K. and Sinha, A.K. (1989) Innovative approaches to management of some major diseases of rice, *Journal of Scientific and Industrial Research* **48**, 181- 185.

Sumner, D.R., Johnson, A.W., Glaze, N.C. and Dowler, C.C. (1978) Root diseases of snap bean and southern pea in intensive cropping systems, *Phytopathology* **68**, 955- 961.

Tinker, P.B. (1975) Soil chemistry of phosphorus and mycorrhizal effects on plant growth, in F.E. Sanders, B. Mosse and P.B. Tinker (eds.), *Endomycorrhizas*, Academic Press, London, pp. 353-371.

Tsai, W.H. (1976) The influence of rice sheath blight on yield at different inoculation stages, *Annual Report of the Taiwan Agriculture Research Institute*, **1975**, pp. 115- 116.

Vasanthadevi, T., Malarvizhi, R., Sakthivel, N.and Gnanamanickam, S.S. (1989) Biological control of sheath blight of rice in India with antagonistic bacteria, *Plant and Soil* **119**, 325-330.

Viji, G., Baby, U.I. and Manibhushanrao, K. (1991) Efficacy of systemic fungicides to sheath blight pathogen *Rhizoctonia solani* and its fungal antagonists, *Symposium on New Frontiers in Chemical Control*, University of Madras, Madras, India.

Viji, G., Baby, U.I. and Manibhushanrao, K. (1993) Induction of fungicidal resistance in *Trichoderma* spp. through UV irradiation, *Indian Journal of Microbiology* **33**, 125-129.

Waheeta, A., Sreenivasaprasad, S. and Manibhushanrao, K. (1987) Induced resistance in rice to sheath blight disease, *Current Science* **56**, 486- 489.

Wei, C.T. (1934) *Rhizoctonia* sheath blight of rice, *Bulletin of College of Agriculture and Forestry* (University of Nanking), **No.15**, p.21.

Weller, D.M. and Cook, R.J. (1983) Suppression of take-all of wheat by seed treatment with fluorescent Pseudomonads, *Phytopathology* **73**, 463- 469.

SHEATH ROT DISEASE OF RICE: CURRENT STATUS AND CONTROL STRATEGIES

N. SAKTHIVEL
Center for Biotechnology
Pondicherry University
Kalapet, Pondicherry 605014
India.

1. Introduction

Rice (*Oryza sativa* L.) is a versatile food crop of the World, providing staple food for almost 2 billion people. More than 90% of the World's rice is grown and consumed in Asia. Rice crop is attacked by several bacterial, fungal, viral pathogens and pests. Among them blast, bacterial blight, tungro virus are considered as major diseases. In recent years, sheath rot disease caused by *Sarocladium oryzae* (Sawada) Gams and Hawksworth has become a severe production constraint. This paper describes the present level of knowledge on the biology, pathology, epidemiology of *S. oryzae* and strategies for the management of sheath rot disease.

2. Sheath rot: a potential constraint of rice grain yield and quality

Sheath rot has been a common problem on densely planted, nitrogen-responsive, high yielding semi-dwarf as well as tall rice cultivars in all rice growing areas in India (Nair, 1976; Raina and Singh, 1980; Vasantha Devi, 1991 Phookan and Hazarika, 1992). In Tamil Nadu and Pondicherry states of Southern India, all high yielding rice cultivars were susceptible to sheath rot (Lakshmanan, 1984; Sakthivel, unpublished). In Punjab, sheath rot has threatened the most popular rice cultivar PR106 (Kang and Singh, 1987). High yield losses due to sheath rot have been reported in all Asian countries.

The disease was first identified by Sawada in 1922 and later it became a common problem in Japan (Kawamura, 1940), South and Southeast Asia (Agnihothrudu, 1973; Amin *et al.*, 1974; Chin 1974; Nair, 1976), Kenya, Nigeria (CABI, 1980) and Brazil (Mathur, 1981). The disease is considered as a major problem in upland rice in West Africa on rice cultivars introduced from Asia (Ou, 1985). Sheath rot was also reported in the U.S.A. (Shahjahan *et al.*, 1977). Depending on the conditions, yield losses are vary from 20 to 85%. In Taiwan, the extent of damage ranged from 3 to 20% and sometimes the losses were as high as 85% (Chen, 1957). Sheath rot incidence, in some

S. Sreenivasaprasad and R. Johnson (eds.) Major Fungal Diseases of Rice Recent Advances, 271-283.
Prof. K. Manibhushan Rao Festschrift

instances, has led to a total destruction of the crop in Thailand (Surin and Disthaporn, 1977). Mild to severe crop losses were reported from Brazil (Mathur, 1981). Losses of 52.8% were reported in the Philippines (Estrada *et al.*, 1984) and 10 to 85% yield losses have been reported in India (Amin *et al.*, 1974; Prabhakaran *et al.*, 1974).

Sheath rot infection results in chaffy, discoloured grains and affects the viability and nutritional value of seeds (Vidhyasekaran *et al.*, 1984; Misra *et al.*, 1990). The protein (Vidhyasekaran *et al.*, 1984) and starch contents (Mohan and Subramanian, 1979) of sheath rot infected seeds are decreased. Prevalence of *S. oryzae* in different parts of rice grains was reported (Mukerjee and Yadav, 1989; Shamsi *et al.*, 1995) and the average incidence of sheath rot in seed coat, endosperm and embryo was reported as 59.3%, 21.8% and 4.2% respectively (Sachan and Agarwal, 1995).

3. The causal organism

The fungus was first described as *Acrocylindrium oryzae* (Sawada, 1922) and its name was changed later as *Sarocladium oryzae* (Sawada) Gams and Hawksworth (Gams and Hawksworth, 1975). The genus *Sarocladium* was divided into two species, *S. attenuatum* (Agnihothrudu, 1973) and *S. oryzae* primarily distinguished by the conidial shapes and size (Gams and Hawksworth, 1975). *S. attenuatum* contains verticillate conidiophores and conidia are longer, and narrower than the round ended conidia of *S. oryzae*. The elongated nature of the conidia is not an useful differentiating character (Bridge *et al.*, 1989). There is no consistency in esterase isozyme pattern of *S. oryzae* (Bridge *et al.*, 1989) and therefore, based on biochemical and morphological characters recognition of *S. oryzae* and *S. attenuatum* as distinct species appears to be incorrect. Therefore, *S attenuatum* can be considered as a synonym of *S. oryzae* (Bridge *et al.*, 1989). There is confusion on the taxonomy of the fungus and surprisingly, symptoms induced by both species are extremely similar. It is always not clear which reports refer to which species of the *Sarocladium* (Ou, 1985). With the advent of DNA technology, the DNA markers may distinguish the population of *Sarocladium*. Mycelial colonies of *S. oryzae* on the host are pale pinkish or whitish and sporulation is abundant in the center. *S. oryzae* produces verticilliate condiophore with 1 or 2 phialides which contain fusiform, often curved, hyaline, single cell conidia of 3.5 to 9 X 1 to 2.5 μm (Ou, 1985).

4. Symptomatology

Sheath rot disease is typically characterised by the production of greyish brown lesions on upper leaf sheaths enclosing panicles. Lesion colour varies from grey brown to purple brown depending upon the cultivars (Shahjahan *et al.*, 1977). The initial lesions are 0.5 to 1 cm long and 0.2 to 0.5 cm wide, chocolate brown in colour, whereas the healthy sheath around the lesions remain green (Amin *et al.*, 1974). The lesions become oblong to linear in shape and have irregular margins. At a later stage, they enlarge, coalesce and cover the entire sheath. White, powdery masses of conidia are contained in the center of each lesion (Agnihothrudu, 1973). Panicles from affected plants do not emerge or partially emerge and are compressed inside the sheath (Agnihothrudu, 1973; Amin *et al.*, 1974). The fungus causes browning of grains and also induces sterility (Chien and Huang, 1979; Ou, 1985) and empty heads (Chao-Yan Hsieh, 1977).

5. Epidemiology and disease cycle

Epidemiology and disease cycle of *S. oryzae* are presented in Figure 1. The fungus severely infects the high yielding dwarf cultivars under water stressed conditions in both rainfed and upland ecosystem (Gnanamanickam and Mew, 1991; Ou, 1985). The pathogen attacks both semi dwarf as well as traditional tall cultivars (Sakthivel and Gnanamanickam, 1986a,b; Sakthivel and Gnanamanickam, 1987).

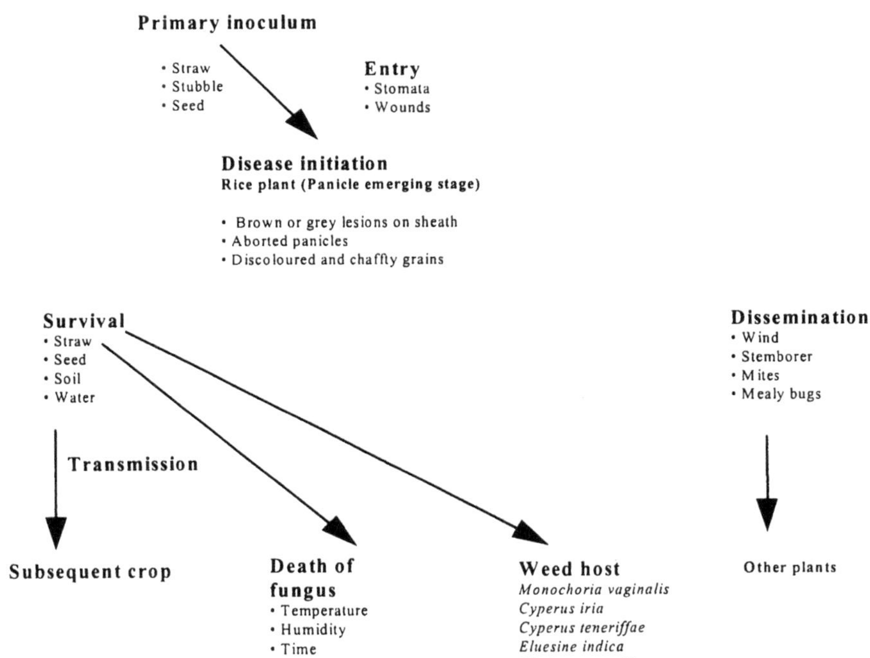

Figure 1. Disease cycle and epidemiology of *Sarocladium oryzae*

The fungus survives as mycelium in infected seeds, plant residues (Chakravarthy and Biswas, 1978; Misra *et al.*, 1990; Singh and Raju, 1981), soil on the left over stubble and straw and on weed hosts. Conidia are easily carried over long distances by wind (Amin *et al.*, 1974). *S. oryzae* survives atleast 4 months in seed and 7 months in sheath at room temperature and as long as 10 months in leaf sheaths in the field. The fungus infects at least five weed hosts, *Echinochloa colona* (Rahman *et al.*, 1982), *Eluesine indica, Monochoria vaginalis, Cyperus teneriffae* and *C. iria* (Balakrishnan and Nair, 1981) and also attacks bamboo (Boa and Brady, 1987; Bridge *et al.*, 1989).

However, only isolates from weeds have been shown to infect rice and the bamboo isolates are not pathogenic to rice (Boa and Brady, 1987). *S. oryzae* enters through stomata or injuries and ramifies intercellularly in the vascular bundles and mesophyll tissues (Shahjahan *et al.*, 1973). The disease has been associated with injuries caused by mites, stem borers, mealy bugs or viral infection (Chin, 1974; Ou, 1985) and its biological association with tarsonemid mites which transmit the spores has been confirmed (Surin and Disthaporn, 1977; Lo and Ho, 1979; Chow *et al.*, 1980; Liang, 1980; Lewin and Vidhyasekaran, 1987). A rice mealy bug, *Brevennia rehi* has been reported as a vector for sheath rot (Lakshmanan, 1992). *S. oryzae* infection occurs after the plant has been weakened by other diseases and infestations of stem borers. The bore holes of *Chilo* sp. at panicles are ideal infection sites for the fungus. *S. oryzae* is associated with other fungi such as *Fusarium roseum* (Sawada, 1922; Raju and Singh, 1981), *F. moniliforme* (Kang and Rattan, 1983) and *Verticillium* (Gnanamanickam and Mew, 1992). In laboratory assays, *S. oryzae* suppressed the growth of fungal pathogens of rice such as *Sclerotium oryzae* (stem rot pathogen), *Gaeumannomyces graminis* var. *graminis* (crown sheath rot pathogen), *Pyricularia oryzae* (blast pathogen) and *Rhizoctonia solani* (sheath blight pathogen) (Gnanamanickam and Mew, 1991). In greenhouse tests, *S. oryzae* suppressed the symptom development and disease severity of stem rot and crown sheath rot pathogens (Gnanamanickam and Mew, 1991). Such interactions with other stem-attacking pathogens of rice is believed to have helped *S. oryzae* in its emergence as a dominant rice disease. Although the fungus infects rice plants at all stages, the disease is visible primarily on flag leaf sheaths (Shahjahan *et al.*, 1977) at panicle emergence. The reason for this has not been established yet. Hot and humid weather favours the disease development. Temperature of 20 to 30 °C and relative humidity in the range of 65 to 85% favour the development of sheath rot (Upadhyay and Diwakar, 1984). Although the fungus is seedborne and is seed transmitted (Shahjahan *et al.*, 1977; Chuke, 1983; Milagrosa, 1987; Agarwal *et al.*, 1989; Mew and Misra, 1994), the initiation of infection from the infected seed has not been clearly demonstrated (Rahman *et al.*, 1982). This is because of the lack of sensitive methods.

6. Physiology

S. oryzae is a very slow-growing fungus (Shahjahan *et al.*, 1977). The optimum growth occurs at 20 to 48 °C and a pH of 6.4 (Tasugi and Ikeda, 1956). Based on the nutritional physiology, maize agar or rice straw agar, sodium nitrate, potassium nitrate and sucrose are found suitable for the growth of the fungus (Shahjahan *et al.*, 1977; Chien and Huang, 1979). The fungus produces cellulolytic, pectinolytic, proteolytic, oxidative enzymes (Singh, 1982; Malathy, 1984), the steroid antibiotic helvolic acid (Dachne *et al.*, 1968) and non-steroid antibiotic cerulenin (Turner and Alridge, 1983;

Berdy, 1986; Sakthivel and Gnanamanickam, 1986a). Cutinase enzyme is not produced by *S. oryzae* (Sakthivel and Gnanamanickam, unpublished). Isoenzymes of esterase are produced inconsistently in *S. oryzae* (Bridge *et al.*, 1989) Biochemical constituents of differentially virulent *S. oryzae* isolates showed a positive correlation with their degree of pathogenicity (Joe and Manibhushanrao, 1995). Very little published information was available on phytotoxic fungal metabolites of *S. oryzae*. For the first time, cerulenin was identified as a toxin involved in the induction of sheath rot symptoms and virulence of the fungus (Sathivel and Gnanamanickam, 1986c; Sakthivel, 1987; Vasantha Devi and Gnanamanickam, 1993). Variation in cerulenin and helvolic acid production has been identified in *S. oryzae* isolates collected from a single field (Sakthivel and Mrinal, unpublished). Induction of sheath rot symptoms by cerulenin was confirmed through the development of cerulenin negative mutants isolates that did not produce the rotting symptoms (Vasantha Devi and Gnanamanickam, 1993).

In a recent investigation, (Sakthivel and Muthukrishnan, unpublished), the simultaneous production of cerulenin and helvolic acid in culture fluids, infected tissues and grains was studied. Selective bioassays for cerulenin and helvolic acid were developed based on their toxicity towards *Candida albicans*-ATCC2140 and *Clavibacter michiganensis* (Figure 2). Both cerulenin and helvolic acid mimic sheath rot symptoms similar to those of fungus. These metabolites were characterised through their biological properties and confirmed by their profiles of thin-layer-chromatography (TLC) and High-performance-liquid-chromatography (HPLC) (Figure 3). According to a recent study on the induction profiles of chitinase enzyme, these metabolites selectively induced only the 28 kDa protein of chitinase (Figure 4) and not the 35 kDa protein that is normally induced by the fungus in rice plants (Sakthivel and Muthukrishnan, unpublished).

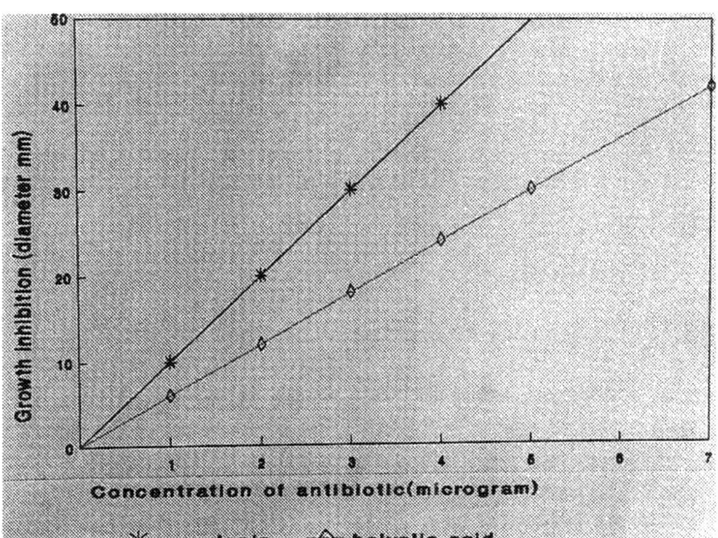

Figure 2. Calibration curve is based on the selective toxicity of cerulenin and helvolic acid towards *Candida albicans* ATCC2140 and *Clavibacter michiganensis* respectively.

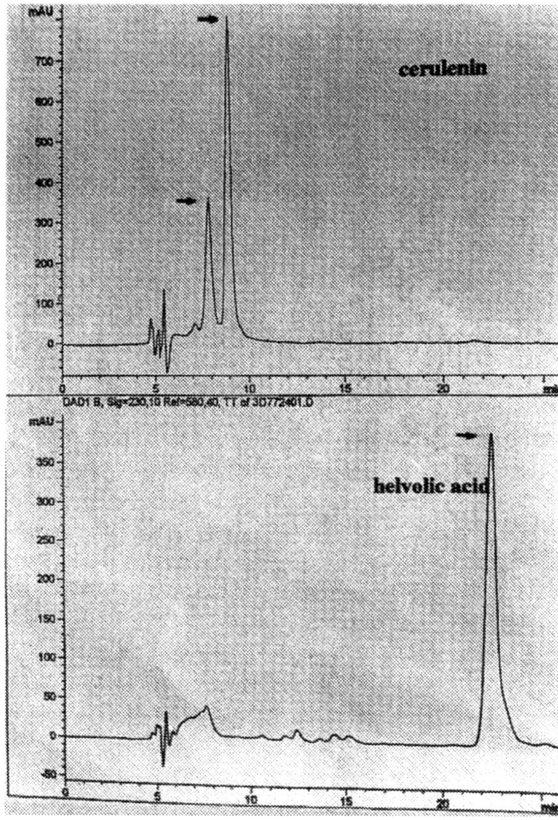

Figure 3. High-performance-liquid-chromatography (HPLC) profilesof cerulenin and helvolic acid. Cerulenin and helvolic acid were isolated from the infected grains, purified through silica gel column and used for HPLC analysis. Cerulenin gave two major peaks at 7.8and 8.8 min. at 210 nm. These two peaks were coeluted with the cerulenin standard which also gave two peaks due to its unstable nature. Helvolic acid gave one major peak at 22.4 min. at 230 nm.This peak was coeluted with helvolic acid standard.

7. Disease control strategies

7.1. RESISTANT CULTIVARS AND CULTURAL PRACTICES

All high-yielding international rice cultivars are highly susceptible to sheath rot disease. Therefore, considerable effort needs to be made to screen for resistant cultivars. Few cultivars such as Te-tep, ARC 7117, Ramtulasi, Intan, and Zenith are considered as resistant (Amin, 1976). Field sanitation, crop residue management, control of weeds and pathogen-free seed may enhance the effectiveness of sheath rot control. Since several weeds are known to spread the *S. oryzae* fungus (Figure 1), the alternate weed hosts such as *E. colona, E. indica, M. vaginalis, C. teneriffae* and *C. iria* should be destroyed around the rice fields (Rahman *et al*, 1982). According to several reports the sheath rot disease was more prevalent in densely planted fields (Agnihothrudu, 1973; Amin *et al.*, 1974; Amin, 1976; Agarwal *et al.*, 1989) than with wider spacing. Significant reduction in disease spread and increase in grain and straw yields were observed with the spacing of 15 X 15 cm (Kanagasabai, 1980).

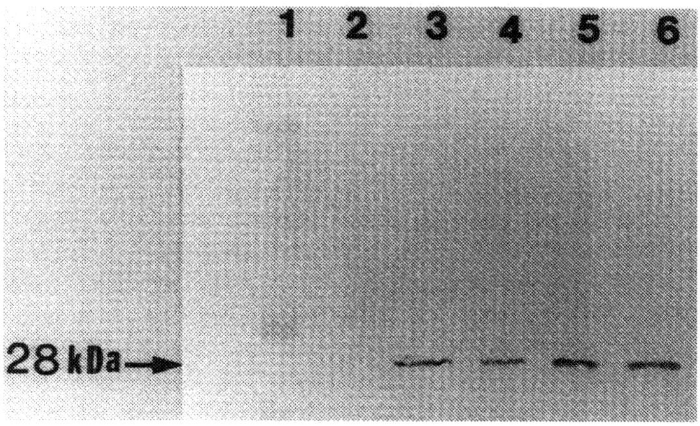

Figure 4. Induction of chitinase by cerulenin and helvolic acid. Western blot showing the 28 kDa band of chitinase due to ceruleninand helvolic acid. The 35 kDa chitinase band is absent. Leaf sheaths were inoculated with sterile paper discs loaded with metabolites (5 μg). Sheaths were sampled at 3, 6, 9 days aftertreatment and used for Western blot analysis. 1 = Standard marker;2 = control (ethanol 5 μl); 3 = cerulenin treatment for 3 days; 4 = helvolic acid treatment for 3 days; 5 = cerulenin treatment6 days; 6 = helvolic acid treatment for 6 days.

7.2. FUNGICIDES, BALANCED NUTRIENTS, GROWTH SUBSTANCES AND ORGANIC AMENDMENTS

Chemical fungicides such as dithane M-45, benlate, carbendazim, tridemorph, bavistin, brestanol, dithane, hinosan , diflotaton and gypsum have been used to control sheath rot (Rajan and Nair, 1978; Mathur, 1981; Raju and Singh, 1981; Narashimhan *et al.*, 1994). Seed treatments with such chemicals do not kill the fungi present inside the glumes and therefore, fungicides are not very effective in sheath rot control. Impact of these fungicides on environment and induction of resistance towards chemicals should be assessed. Plant extracts are well documented for antibacterial and antifungal activities. Toxicity of plant extracts towards several rice fungi including *S. oryzae* has been reported (Miah *et al.*, 1990). Neem, punna and rubber cakes have been used to control sheath rot.

Resistance in crop plants can be induced by providing a balanced application of fertilisers. Since high disease incidence was reported in all high nitrogen-responsive cultivars due to their conduciveness to the development of sheath rot (Misra and Mathur, 1983; Akanda *et al.*, 1984), application of heavy doses of nitrogenous fertilisers should be avoided. Sheath rot incidence is significantly lowered with application of potassium (Kanagasabai, 1980; Muralidhar, 1982) and the use of ammonium chloride (Alagarsamy and Bhaskaran, 1986). Application of nitrogen fertilisers increased the susceptibility of the host by decreasing the phenolic content. Potassium application augmented the synthesis of phenols and led to a decrease in sheath rot incidence. Foliar spray of different micronutrients have reduced the sheath rot incidence and increased the grain yield significantly (Kanagasabai, 1980; Alagarsamy and Bhaskaran, 1986). Sheath rot incidence was reduced also due to the applications of CaSo4, ZnSo4, manganese, ferrous sulphate and borax (Kanagasabai, 1980).

7.3. BIOLOGICAL CONTROL USING BACTERIAL ANTAGONISTS

Heterotrophic rhizobacteria such as *Pseudomonas fluorescens* were reported as antagonists of *S. oryzae* (Figure 5) (Sakthivel and Gnanamanickam, 1986a,b; Sakthivel *et al.*, 1986). In field tests selected strains of *P. fluorescens* were used as seed inoculants and they effected a 20 to 42% reduction in sheath rot incidence along with increases in plant height, tiller number, and grain yields of 3 to 160% (Sakthivel and Gnanamanickam, 1987). The strain *P. fluorescens*-Pfcp used in these studies produced a phenazine antibiotic. Several secondary metabolites of *Pseudomonas* spp. showed antifungal activities aganist *R. solani* (Rosales *et al.*, 1995). Similarly antibiotic production by other effective strains of *P. fluorescens* has been shown to mediate biological control of rice blast disease caused by *Pyricularia grisea* (Gnanamanickam and Mew, 1992; Valasubramanian, 1993).

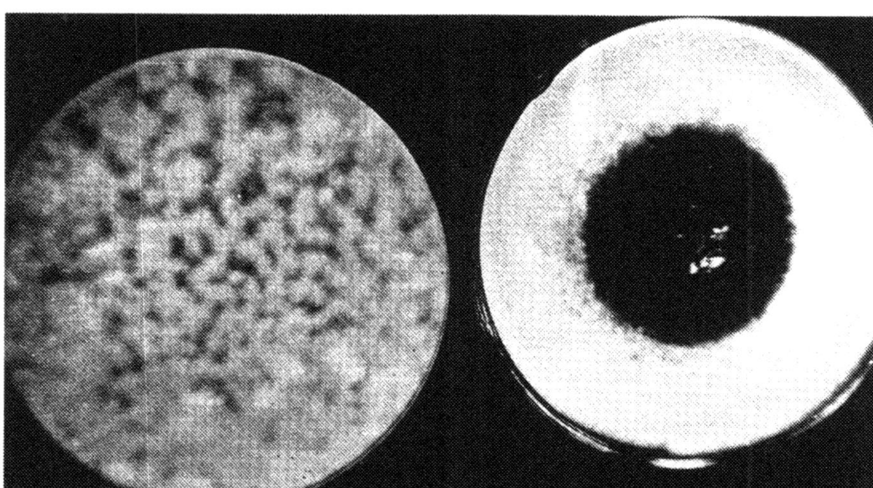

Figure 5. Antibiosis assay showing the growth inhibition of *Sarocladium oryzae* due to *Pseudomonas fluorescens*-Pfcp. The control potato dextrose agar (PDA) plate on the right showing the plateful growth of *S. oryzae*. The plate on the left showing the growth inhibition of *S. oryzae*. The plate was inoculated in the centre with agar plug containing *P. fluorescens*-Pfcp after the surface of plate was spread with conidial suspension of *S. oryzae*.

These antagonistic bacteria have been shown to have broad spectrum inhibitory activity against several other rice fungi such as *R.. solani, F. roseum, Nigrospora* spp. and *Curvularia* spp. and therefore, seed bacterisation with these antagonists may promote seedling vigour, plant growth and reduce pathogen infection (Sakthivel and Gnanamanickam, 1986b Sakthivel *et al.*, 1986; Sakthivel, 1987; Rosales *et al.*, 1995).

7.4 . TRANSGENIC RICE

Transgenic rice plants that express a rice chitinase gene constitutively have been developed. Transgenic rice plants with high levels of chitinase enzyme exhibit increased resistance to *R. solani* (Lin *et al.*, 1995). Currently these transgenic plants are being evaluated for resistance to sheath rot and stem rot diseases [Sakthivel and Muthukrishnan, unpublished] through a rice biotechnology project offered by the Rockefeller Foundation, New York, U.S.A.

8. Summary

Sheath rot disease was once considered as a minor disease, but in recent years it has become a major production constraint in all the rice-growing countries particularly in Asia. Lack of effective and affordable chemical control for the resource-poor farmers of the Asian tropics and the lack of resistance in rice that could be exploited for breeding for sheath rot resistance are matters of concern.

Recently, transgenic rice plants which expressed chitinase enzyme constitutively were developed. Development of new transgenic rice with thaumatin-like protein is in progress (Muthukrishnan and Velazhahan, pers. comm.). Preliminary results showed reduction in sheath rot disease incidence at greenhouse level Field-level disease responses of transgenic plants aganist sheath blight, sheath rot and stem rot diseases are being evaluated.. Biological control of sheath rot as an alternative to chemical control appears to be an effective strategy for disease management, for the resource-poor farmers of Asia. This aspect needs to be pursued and efficient antagonistic bacterial strains identified for making biological formulations that could be applied in the field.

Little is known about the pathogenic and genetic diversity of *S. oryzae*. Variabilty in the production of phytotoxic metabolites such as cerulenin and helvolic acid in *S. oryzae* has been identified (Vasantha Devi, 1991; Vasantha Devi and Gnanamanickam, 1993; Sakthivel and Muthukrishnan ,unpublished). However, the role of these metabolites in sheath rot induction or virulence has to be verified perhaps by making genetic analysis of *S. oryzae*. Mutants of *S. oryzae* through transposon mutagenesis might provide valuable insight into the question concerning how important are these metabolites as determinants of sheath rot disease. New research initiatives are required to gain a thorough understanding of pathogen diversity. A comprehensive global collection of *S. oryzae* isolates will be vital for a successful outcome. Development of molecular markers will be of immense value for checking the presence of seedborne *S. oryzae* in rice germplasm which moves across international borders. Recently, collaborative research has been initiated between the International

Mycological Institute (IMI) in U.K. and the International Rice Research Institute (IRRI) in the Philippines to develop polymerase chain reaction (PCR) based techniques to detect *S. oryzae* in rice seed (David Pearce, pers. comm.). PCR based diagnostic studies to detect and differentiate *S. oryzae* have been initiated recently in the Center for Biotechnology, Pondicherry University, India. Such efforts are likely to improve our understanding of the influence of seed inocula of *S. oryzae* on disease transmission, epidemic development and rice yields.

9. Acknowledgements

I thank the University Grants Commission, Government of India, New Delhi for the award of a research fellowship. Part of work presented in this paper was done in the Center for Advanced Studies in Botany, University of Madras, India when I was a graduate student during 1984 to 1987 at Prof. S. S. Gnanamanickam's Laboratory. I thank Prof. S. S. Gnanamanickam who encouraged me to work on rice diseases. Studies on characterization of phytotoxic fungal metabolites and their effect on the induction of chitinase enzyme and the evaluation of transgenic rice expressing chitinase for sheath rot resistance have been conducted at Prof. S. Muthukrishnan's Laboratory, Department of Biochemistry, Kansas State University, U.S.A. through a collaborative research program. I thank the Rockefeller Foundation, New York, U.S.A. for the award of a rice biotechnology career fellowship program. Microbial cultures, *Candida albicans* ATCC2140 and *Clavibacter michiganensis* were supplied by Ms. Susan Porter, Peterson's Clinical Laboratory, Manhattan, U.S.A and Mr. Bruce Raymundo, Plant Pathology Department, Kansas State University, U.S.A. respectively. PCR primer sequences for *S. oryzae* were kindly supplied by Dr. David Pearce International Mycological Institute (IMI), U.K.

10. References

Agnihothrudu, V. (1973) *Acrocylindrium oryzae* Sawada-Sheath rot on paddy, *Kavaka* **1**, 68-71.

Agrawal, P. C., Mortensen, C. N. and Mathur, S. B. (1989) *Seedborne diseases and Seed Health Testing of Rice*, CAB International Mycological Institute, Kew, Surrey, U.K.

Akanda, S. I., Shajahan, A. K. M. and Miah, S. A. (1984) Nitrogen fertilization and sheath rot (ShR) development in rice, *International Rice Research Newsletter* **9**, 12-13.

Alagarsamy, G. and Bhaskaran, R. (1986). Effect of foliar spray of macro and micronutrients on sheath rot (ShR) incidence and rice grain yield, *International Rice Research Newsletter* **11**, 18.

Amin, K. S. (1976) Sources of resistance to *Acrocylindrium oryzae* - sheath rot of rice, *Plant Disease Reporter* **60**,72-73.

Amin, K. S., Sharma, B. D. and Das, C. R. (1974) Occurrence of sheath rot of rice in India caused by *Acrocylindrium, Plant Disease Reporter* **58**, 358-360.

Balakrishnan, B. and Nair, M. C. (1981) Weed hosts of *Acrocylindrium oryzae* Saw. sheath rot pathogen of rice, *International Rice Research Newsletter* **6**, 13.

Berdy, J. (1986) Further antibiotics with practical application, in H. pape and Rehm, H. J. (eds.), *Biotechnology*, Microbial Products II, **vol. 4**, 456-507.

Bridge, P. D., Hawksworth, D. L., Kavishe, D. F. and Farnell, P. A. (1989) A revision of the species concept in *Sarocladium*, the causal agent of sheath rot in rice and bamboo blight, based on biochemical and morphometric analyses, *Plant Pathology* **38**, 239-245.

Boa, E. R. and Brady, B. I. (1987) *Sarocladium oryzae* associated with blight of bambusa species in Bangladesh, *Transaction of the British Mycological Society* **89**, 161-166.

CAB International Mycological Institute. (1980) *Description of Pathogenic Fungi and Bacteria No.673.*
Chakravarthy, D. K. and Biswas, S. (1978) Estimation of yield loss in rice affected by sheath rot, *Transaction of the British Mycological Society* **62**, 226-227.
Chao-Yan Hsieh (1977). "Empty head" a serious problem in Taiwan, *International Rice Research Newsletter* **2**, 5.
Chen, M. J. (1957) Studies on sheath rot of rice plant, *Journal of Agriculture* (Taiwan) **6**, 84-102.
Chien, C. C. and Huang, C. H. (1979) The relation between sheath rot and the sterility of rice plant, *Journal of Agricultural Research in China* **29**, 7-16.
Chin, K. M. (1974) Sheath rot disease of rice, *MARDI Research Bulletin* **2**, 9-12.
Chow, Y. S., Tzean, S. S., Chang, C. S. and Wang, C. W. (1980) A morphological approach of the tarsonemid mite, *Steneotarsonemus spinkii* (Tarsonemidiae) as a rice (*Oryza sativa*) spontonea, *Plant Pest Acta Arachn*. Tokyo **29**, 25-42.
Chuke, K. C. (1983) *Pathological and physiological studies on sheath rot caused by Sarocladium oryzae (Sawada) W. Gams and Hawksw.* M. S. Thesis. University of Philippines, Philippines.
Dachne, W. V., Lorch, H. and Godtfredsen, W. O. (1968) Microbiological transformation of fusidane- type antibiotics- a correlation between fusidic acid and helvolic acid, *Tetraheadran Letters* **47**, 4843-4846.
Estrada, B. S. Torres, C. Q. and Bonmann, J. M. (1984) Effect of sheath rot on some yield components, *International Rice Research Newsletter* **9**, 14.
Gams, W., and Hawksworth, D. L. (1975) The idendity of *Acrocylindrium oryzae* Sawada and a similar fungus causing sheath rot of rice, *Kawaka* **3**, 57-61.
Gnanamanickam, S. S. and Mew, T. W. (1991) Interactions between *Sarocladium oryzae* and stem attaking fungal pathogens of rice, *Plant and Soil* **138**, 213-219.
Gnanamanickam, S. S., and Mew, T. W. (1992) Biological control of blast disease of rice (*Oryza sativa*, L.) with antagonistic bacteria and its mediation by a *Pseudomonas* antibiotic, *Annals of the Phytopathological Society of Japan* **58**, 380-385.
Joe, Y. and Manibhushanrao, K. (1995). Biochemical constitutents of differentially virulent isolates of *Sarocladium* spp. causing sheath rot disease of rice. *Journal of Plant Disease and Protection* **102**, 291-298.
Kanagasabai, R. (1980). *Studies on an integrated approach for the control of sheath rot disease of rice caused by Acrocylindrium oryzae Sawada.* M.Sc. (Ag) thesis, Annamalai University, India.
Kang, M. S. and Rattan, G. S. (1983) Sheath rot in the Punjab, India, *International Rice Research Newsletter* **8**, 7-8.
Kang, M. S. and Singh, I. (1987) Some pathological and physiological diseases of rice in Punjab, *International Rice Research Newsletter* **12**, 24.
Kawamura, E. (1940) Notes on the sheath rot plant with special reference to its causal organism, *Acrocylindrium oryzae* Saw., *Annals of the Phytopathological Society of Japan* **10**, 55-59.
Lakshmanan, P. (1992) *Brevennia rehi* Lindinger, vector for the sheath rot (ShR) pathogen, *International Rice Research Notes* **17**, 23.
Lakshmanan, P. (1984) Effective control of sheath rot (ShR) disease,. *International Rice Research Newsletter* **9**, 14.
Lewin, H. D. and Vidhyasekaran, P. (1987) Controlling sheath rot (ShR) in rice, *International Rice Research Newsletter* **12**, 17.
Liang, W. J. (1980) Rearing method for tarsonemid mites (*Stenetarsonemus spinki*) free of rice sheath rot (*Acrocylindrium oryzae*), *Plant Protection Bulletin*, Taiwan **22**, 23-30.
Lin, W., Anuratha, C. S., Datta, K., Potrykus, I., Muthukrishnan, S. and Datta, S. K. (1995) Genetic engineering of rice for resistance to sheath blight, *Bio/Technology* 13, 686-691.
Lo, K. and Ho, C. C. (1979) Studies on the cause of empty head of rice in Taiwan *Journal of Agricultural Research in China* **28**, 193-197.
Malathy, A. (1984) *Physiological studies on Acrocylindrium oryzae, the causal organism of sheath rot of rice*, M.Phil. Dissertation, University of Madras, India.
Mathur, S. C. (1981) Observations on diseases of dryland rice in Brazil, *International Rice Research Newsletter* **6**, 11-12.
Mew, T. W. and Misra, J. K. (1994) *A Manual of Rice Seed Health Testing,* International Rice Research Institute, Los Banos, Laguna, Philippines.
Miah, M. A. T., Ahmed, H. U., Sharma, N. R., Ali, A. and Miah, S. A. (1990). Antifungal activity of some plant extracts, *Bangladesh Journal of Botany* **19**, 5-10.
Milagrosa, S. P. (1987) *Transmission of Sarocladium oryzae (Sawada) W. Gams and Hawksw. through seed,* Ph. D. Dissertation, University of the Philippines, Los banos, Laguna, Philippines.

Misra, J. K., Gergon, E., and Mew, T. W. 1990. Organism causing rice seed discoloration and their possible effect on germinability, *Rice Seed Health Newsletter* **1**, 6.
Misra, A. K. and Mathur, S. C. (1983). Sheath rot incidence in cultivars grown at different nitrogen levels, *International Rice Research Newsletter* **8**, 5.
Mohan, R. and Subramanian, C. L. (1979) Yield losses due to sheath rot disease of rice caused by *Acrocylindrium oryzae* Sawada, *Madras Agricultural Journal* **66**, 195-197.
Mukerjee, P. and Yadav, A. S. (1989). Histopathology and mode of infection of sheath rot (*Sarocladium attenuatum* Gams and Hawksw.) affected rice (*Oryza sativa* L.), *Indian Journal of Plant Pathology* **7**, 85-90.
Muralidhar, B. (1982). *Studies on certain physiological and biochemical changes in rice plants as influenced by potassium fertilization and Acrocylindrium oryzae inoculation,* M.Sc. thesis, Annamalai University, India.
Narasimhan, V., Ramadoss, N., Sridhar, V. V. and Kareem, A. A. 1994. Using gypsum to manage sheath rot in rice, *International Rice Research Newsletter* **19**, 27-28.
Nair, R. (1976) Incidence of sheath rot in rice-a potential problem for Sambalpur, Orissa, *International Rice Research Newsletter* **1**, 19.
Ou, S. H. (1985) *Rice Diseases*, CAB International Mycological Institute, Kew, Surrey, U. K.
Phookan, A. K., and Hazarika, D. K. (1992) Distribution of sheath rot (ShR) in six agroclimatic zones of Assam, India, *International Rice Research Newsletter* **17**, 16.
Prabakaran, J. Ranganathan, V. and Prasad, N. N. (1974) Occurrence of sheath rot disease of rice caused by *Acrocylindrium oryzae* Sawada, *Annamalai University Agricultural Researc Journal* **5**, 182-183.
Rahman, M. M., Shajahan, A. K. M. and Miah, S. A. (1982) *Echinochloa colona* an alternate host of *Acrocylindrium oryzae* causing sheath rot of rice, *International Rice Research Newsletter* **7**, 16-17.
Rajan, K. M. and Nair, P. V. (1978) Fungicidal control of sheath rot and glume discolouration in rice, *Indian Journal of Mycology and Plant Pathology* **8**, 85-87.
Raina, G. L., and Singh, G. (1980) Sheath rot outbreak in Punjab, *International Rice Research Newsletter* **5**, 16.
Raju, C. A. and Singh, R. A. 1981. Development of inoculation technique and varietal reaction to sheath rot of rice, *Philippine Phytopathology* 17, 67-73.
Rosales, A. M., Thomashow, L., Cook, R. J. and Mew, T. W. (1995). Isolation and identification of antifungal metabolites produced by rice associated antagonistic *Pseudomonas* spp., *Phytopathology* **85**, 1028-1032.
Sachan, I. P. and Agarwal, V. K. (1995) Seed discolouration of rice: location of inoculum and influence of nutritional value, *Indian Phytopathology* **48,** 14-20.
Sakthivel, N. (1987) *Biological control of Sarocladium oryzae (Sawada) Gams & Hawksworth, sheath rot pathogen of rice by bacterization with Pseudomonas fluorescens Migula*, Ph. D. Dissertation, University of Madras, India
Sakthivel, N., Sivamani, E., Unnamalai, N. and Gnanamanickam, S. S. (1986) Plant growth-promoting rhizobacteria in enhancing plant growth and suppressing plant pathogens, *Current Science* **55**, 22-25.
Sakthivel, N. and Gnanamanickam, S. S. (1986a) Toxicity of *Pseudomonas fluorescens* towards rice sheath rot pathogen, *Acrocylindrium oryzae* Saw., *Current Science* **55**, 106-107.
Sakthivel, N., and Gnanamanickam, S. S. (1986b) Bacterization of rice with *Pseudomonas fluorescens* reduces sheath rot infection, *International Rice Research Newsletter* **11**, 17-18.
Sakthivel, N. and Gnanamanickam, S. S. (1986c) Isolation of and assay for cerulenin produced by sheath rot pathogen, *Sarocladium oryzae* (Saw) Gams., *Current Science* **55**, 988-989.
Sakthivel, N. and Gnanamanickam, S. S. (1987) Evaluation of *Pseudomonas fluorescens* for suppression of sheath rot disease and for enhancement of grainyield in rice (*Oryza sativa* L.), *Applied Environmental Microbiology* **53**, 2056-2059.
Sawada, K. (1922) Descriptive catalogue of Formosan fungi. *II Report, Government Research Institute, Department of Agriculture* (Formosa) **2**, 135.
Shajahan, A. K. M., Harahap, Z. and Rush, M. C. (1977) Sheath rot of rice caused by *Acrocylindrium oryzae* in Lousiana, *Plant Disease Reporter* **61**, 307-310.
Shamsi, S., Khan, A. Z. M. N. A., Shajahan, A. K. M. and Miah, S. A. (1995). Prevalence of *Sarocladium oryzae* (Sawada) W. Gams and D. Hawksw. in different parts of rice grains, *Bangladesh Journal of Botany* **24**, 217-219.
Singh, N. (1982) Carbon and nitrogen nutrition of *Acrocylindrium oryzae* and production of pectic and cellulolytic enzymes, *Transaction of the British Mycological Society* **78**, 355-357.
Singh, R. A. and Raju, C. A. (1981) Studies on sheath rot on rice, *International Rice Research Newsletter* **6**, 11-12.

Surin, A. S. and Disthaporn, S. (1977) Rice abortion of sheath rot-a serious rice disease in Thailand, *International Rice Research Newsletter* **2**, 16-17.

Tasugi, H. and Ikeda, Y. (1956) Studies on the sheath rot of rice plants caused by *Acrocylindrium oryzae* Sawada, *Bulletin of Wantn. Institute of Agriculture* Tokyo **Ser. 6**, 151-166.

Turner ,W. B. and Aldridge, D. C. (1983) *Fungal metabolites II,* Academic Press, London.

Upadhyay, R. K. and Diwakar, M. C. (1984) Sheath rot (ShR) in Chhatisgarh, Madhya Pradesh, India, *International Rice Research Newsletter* **9**, 6.

Valasubramanian, R. (1993) *Biological control of rice blast,* Ph.D. Dissertation, University of Madras, India.

Vasantha Devi, T. (1991) Sheath rot disease of rice, Ph. D. Dissertation, University of Madras, India.

Vasantha Devi, T. and Gnanamanickam, S. S. (1993) Toxin produced by *Sarocladium oryzae* involved in inducing sheath rot (ShR) symptoms in rice, *International Rice Research Newsletter* **18**, 30-31.

Vidhyasekaran, P., Ranganathan, B., and Radhakrishnan, J. (1984) Quality of rice grains from sheath rot affected plants, *International Rice Research Newsletter* **9**, 19.

SPECIES CONCEPT IN *SAROCLADIUM*, THE CAUSAL AGENT OF SHEATH ROT IN RICE AND BAMBOO BLIGHT

D. A. PEARCE, P. D. BRIDGE and D. L. HAWKSWORTH
CABI BioScience,
Bakeham Lane, Egham, Surrey TW20 9TY, UK.

1. Introduction

Sarocladium is a genus of hyphomycetes resembling *Verticillium* and *Gliocladium*. It is the causal organism of sheath rot of rice and blight in bamboo. *Sarocladium oryzae* was once considered to be a minor disease of rice, but is becoming increasingly destructive and widely distributed. It infects rice plants at various stages of growth, entering through stomata or wounds, but is most destructive after booting, and frequently associated with the presence of insect damage or *Pseudomonas* infection. The management of sheath rot of rice by cultural, biological, and chemical methods has been reviewed (Ramabadran *et al.*, 1990), although no effective control methods are currently available, seed treatment with fungicides such as Dithane M-45 and Benlate effectively eliminates seedborne inocula.

2. Taxonomy

The genus *Sarocladium* W. Gams & D. Hawksw. was introduced in 1975 to accommodate the two species *S. oryzae* (Sawada) W. Gams & D. Hawksw. (syn. *Acrocylindrium oryzae* Sawada) and *S. attenuatum* W. Gams & D. Hawksw., both of which were reported as causal agents of sheath rot of rice (*Oryza sativa*). The two species were differentiated primarily based on the more regularly verticillate conidiophores and long narrow acerose conidia produced by *S. attenuatum*. The species concepts at that time were based on rather few isolates as the disease was then starting to become more widespread. The species concept in the genus was subsequently revised by Bridge *et al.* (1989) when more strains had become available. On the basis of 50 strains, it was concluded that conidial size and shape variations between the originally recognised two species represented a continuum, and that *S. attenuatum* should be treated as a synonym of *S. oryzae*. This view was consolidated by a variety of enzyme and substrate utilization tests. Bridge *et al.* (1989) included in their study a number of isolates described as *S. oryzae* that had been isolated from bamboo blight in Bangladesh (Boa and Brady, 1987). They treated these as conspecific, but noted that the bamboo isolates differed from those from rice in some physiological and biochemical properties.

S. Sreenivasaprasad and R. Johnson (eds.) Major Fungal Diseases of Rice Recent Advances, 285-292.
Prof. K. Manibhushan Rao Festschrift

3. Pathology

Sarocladium oryzae was first described (as *Acrocylindrium oryzae*) from rice in Taiwan (Sawada, 1922). The organism causes sheath-rot of rice which affects the uppermost leaf sheaths enclosing the young panicles (Ou, 1972; CMI, 1985). The disease starts as oblong or irregular lesions with brownish margins and greyish centres 0.5-1.5 cm long, which later enlarge and coalesce. Whitish mycelium may be seen in the centre of the lesions and occasionally inside the rotted sheaths. The disease can result in shrivelled grains, brown discolouration, a reduced number of spikelets per panicle, and in extreme cases sterility. Sheath rot was originally considered as a relatively minor disease of rice, originally reported from China. It was first recognised in India by Agnihothrudu (1973) and Amin *et al.* (1974) and it is now widespread, occurring throughout rice growing countries in Asia, Africa and the Americas. Losses in yield due to sheath rot can vary widely and while losses of between 3 and 26% have been reported, losses up to 50% and 85% can result. In general, greater losses are seen most commonly in lowland environments, and particularly during the rainy season in both rainfed and upland systems. It is thought that *S. oryzae* is distributed as conidia which enter rice plants through stomata or wounds, and infection may also be associated with insect damage. This is supported by reports of severe infections in densely planted fields where stem borers have infested, and in the recovery of pure cultures of *S. oryzae* from the bodies and eggs of tarsonemid mites. The fungus enters the plant through stomata or wounds. Secondary infections may be windbourne, the fungus entering the host through injured tissue (Amin *et al.*, 1974). Seed borne transmission was first established by Chuke (1983), and this mode of transmission is the subject of considerable ongoing research. *Sarocladium oryzae* being frequently detected in 1-9 % of seeds and with moderate epidemic potential.

In addition to causing sheath rot on rice, *S. oryzae* has also been isolated from natural infections on various grasses and sedges found growing in association with rice. Pathogenicity studies have confirmed that it is possible to infect rice with the isolates from grasses and sedges and it is possible that these grass populations provide a natural reservoir for the pathogen in the environment (Balakrishnan and Nair, 1981). *S. oryzae* has also been shown to infect *Oryza rufipogon, Cyperus difformis, Echinochloa colonum, E. crus-galli, Eleusine indica, Hymenachne assamica, Leresia hexandra, Monochoria vaginalis and Panicum walense* (Deka and Pookan, 1992; Balakrishnan and Nair, 1981).

S. oryzae has also been reported as the causal agent of a bamboo blight in Bangladesh, where it has been isolated from several bamboo species including *Bambusa balcoa, B. polymorpha, B. tulda, B. vulgaris* and *Meloccana baccifera* (Boa and Brady, 1987). Pathogenicity testing has suggested the development of blight by isolates from bamboo (Rahman and Krishna, 1984), but cross pathogenicity testing with isolates from rice on bamboo has been inconclusive (Boa and Brady, 1987). The disease is believed to be restricted to Bangladesh, but has now been observed in India (E. Boa pers. comm.).

4. Morphology

Sarocladium oryzae grows slowly on potato dextrose agar producing colonies of 1.9 cm diameter in 5 days at 30 ° C. Aerial mycelium is sparse and pale orange, and the colony reverse is generally a darker orange. On oat agar colonies are whitish to pale pink or saffron at 21 ° C. Colony consistency varies with the isolate, but is usually cottony and sometimes powdery (Boa and Brady, 1987). The branched septate hyphae can be up to 2.5 µm thick, although gnarled wider hyphae also occur. The conidiophores are abundant and irregularly branched. The branches arise in whorls at one level and are strongly appressed, with secondary branches often exceeding the primary ones in length. More often the branches arise either singly in dense rows or laterally. The conidiogenous cells may develop singly on vegetative hyphae or on slender conidiophores in a fascicular manner. Smooth, aseptate, hyaline conidia are formed in slimy masses at the tips of the phialides. The conidia are subcylindrical (3.5-7 x 1-1.5 µm), occasionally curved, and have round to acerose ends (Brady, 1980; Bridge *et al.*, 1989).

5. Physiology and biochemistry

The optimum growth temperature for *Sarocladium oryzae* is 30-31° C, with little growth at 37° C and none at 13° C. However, temperature responses may vary between isolates and depend upon the culture medium used. Some authors have found the maximum temperature for growth to be 37° C, while in another study it was reported as in the range 37-44° C (Bridge *et al.*, 1989). Similarly the reported pH optima differ, for example pH optima of 5.5 for growth and sporulation (Zhuge *et al.* 1985) and a maximum dry weight yield at pH 8 (Giri and Swain, 1994). Isolates are also reported to grow and sporulate well at relative humidity levels of 80-100% (Giri and Swain, 1994).

Extracellular cellulases (as carboxymethyl cellulase), invertases, phenolases, nucleases and proteases have been reported, and the relative production of some of these enzymes has been linked to differential levels of pathogenicity on rice. The production of extracellular nucleases has also been linked with specific strains, with this ability being recorded for many isolates from rice, while isolates from bamboo failed to produce the enzyme (Bridge *et al.*, 1989). A wide range of carbon and nitrogen sources are utilised, and again responses may be strain specific, although sucrose and glucose have been reported as the preferred carbon sources. When isolates are grown on different nitrogen sources, the pigmentation of the colony may be altered. Bridge *et al.* (1989) described two agar media which contained either sodium nitrite or ammonium oxalate as the sole sources of nitrogen. Isolates of *Sarocladium oryzae* from bamboo produced very strong yellow and brown pigments when grown on these media, while isolates from rice remained unpigmented. This pigmentation was tentatively suggested as being due to a flavonoid-containing compound, but no further analysis has been undertaken to our knowledge.

Very little information is available on the secondary metabolites produced *by S. oryzae* and only the hexaketide molecule cerulenin and the antibiotic helvolic acid have been widely reported in the literature. However, Bridge *et al.* (1989) detected more than 15 distinct spots by thin layer chromatography, so it is likely that other secondary compounds are also produced.

6. Infraspecific variation

Considerable variation is observed in the degree of pigmentation and the conidial size of isolates of *S. oryzae.* The range of conidial size was discontinuous and the long narrow conidium isolates were described as *S. attenuatum.* However, Bridge *et al.* (1989) examined some 50 strains comprising both living cultures and dried reference material and observed that the conidial size differences represented two extremes in a continuous range. Their decision to place *S. attenuatum* in synonymy with *S. oryzae* was partly based on this observation, but was also supported by the general levels of similarity seen for other physiological and biochemical characters. These authors considered the bamboo isolates to be a distinct group within the species, but did not make an explicit taxonomic separation.

Variation has also been reported in virulence of the isolates on rice, and these differences have been associated with differing levels of extracellular enzymes. The first report of infraspecific variation that gave some indications as to the genetic variability of isolates was by Bridge *et al.* (1989), who analysed the electrophoretic properties of intracellular esterases. They found that isolates produced multiple bands and detected 7 different isoenzyme patterns in 13 isolates. There were no consistent correlations between these band patterns and the known histories and activities of the isolates.

Recent studies on genetic diversity within *Sarocladium oryzae* have involved polymerase chain reaction (PCR) based amplification and fingerprinting techniques. These techniques have the ability to differentiate very closely related, or perhaps individual isolates (Pearce and Holderness, 1996; Bridge *et al.*, 1998). At IMI, a number of *Sarocladium oryzae* cultures have been obtained from different geographical locations around Luzon in the Philippines and also in Nueva Ecija, San Mateo and Iguig in October, 1994. DNA was extracted from each isolate taken and a selection of primers based on simple repetitive sequences were used to generate DNA profiles following PCR amplification (Figures 1 and 2). Subsequent work undertaken at IMI has shown that strains producing identical symptoms and isolated from different plants can be differentiated at both the regional, field and panicle scale.

In addition to this work on rice isolates, four isolates from bamboo blight have also been characterised. These isolates all showed identical band patterns with the simple sequence repeat primers used. The patterns largely comprised of bands that had been seen among rice isolates, but one band that was present in the bamboo isolates was not seen in any of the rice isolates.

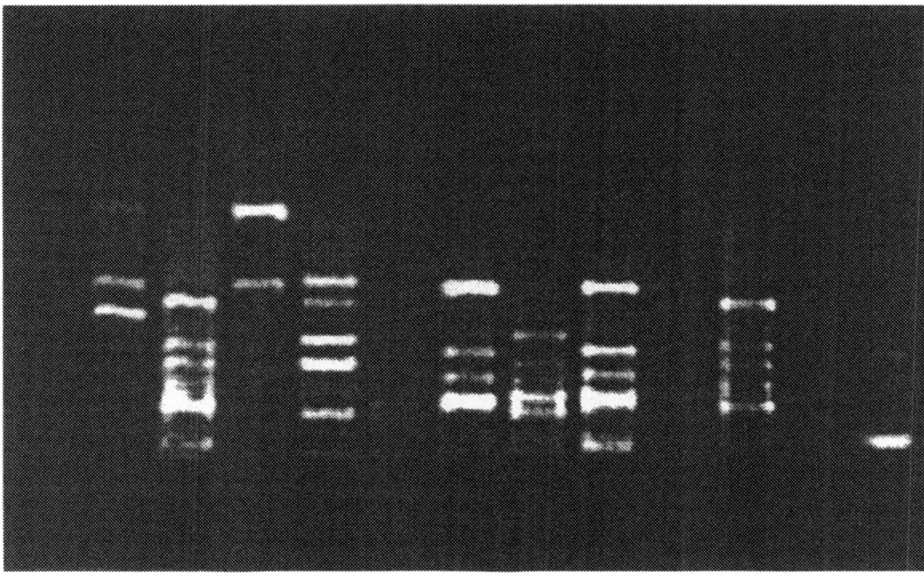

Figure 1. DNA profiles obtained from *Sarocladium oryzae* from three different geographical locations within the Philippines using two different RAPD primers.

Lanes 1 – 12, respectively are

1	5a	Nueva Ecija plot 1 (primer 70)
2	5a	Nueva Ecija plot 1 (primer 71)
3	24	San Mateo plot 1 (primer 70)
4	24	San Mateo plot 1 (primer 71)
5	23	Isolate from the same panicle as 24 (primer 70)
6	23	Isolate from the same panicle as 24 (primer 71)
7	4s	Iguig plot 1 (primer 70)
8	4s	Iguig plot 1 (primer 71)
9	2-2	San Mateo plot 2 (primer 70)
10	2-2	San Mateo plot 2 (primer 71)
11	W	Water
12	KB	Marker

The taxonomic status of the bamboo isolates is therefore not clear; they show less infra-population variation than comparable collections of rice isolates in their SSREP-PCR banding patterns, which indicates that all of the bamboo isolates are possibly very closely related. This is supported by the earlier work of Bridge *et al.* (1989), who found that all of the bamboo isolates could be differentiated from those from rice by their failure to produce extracellular nucleases and their distinctive pigmentation when grown on nitrite and ammonium oxalate containing media. It is possible that the bamboo isolates may prove to merit species or subspecies rank. However, the formulation of this hypothesis through the current findings would be premature in the absence of comparable information from an expanded range of rice, grass, and sedge isolates from a wide geographic area and hosts.

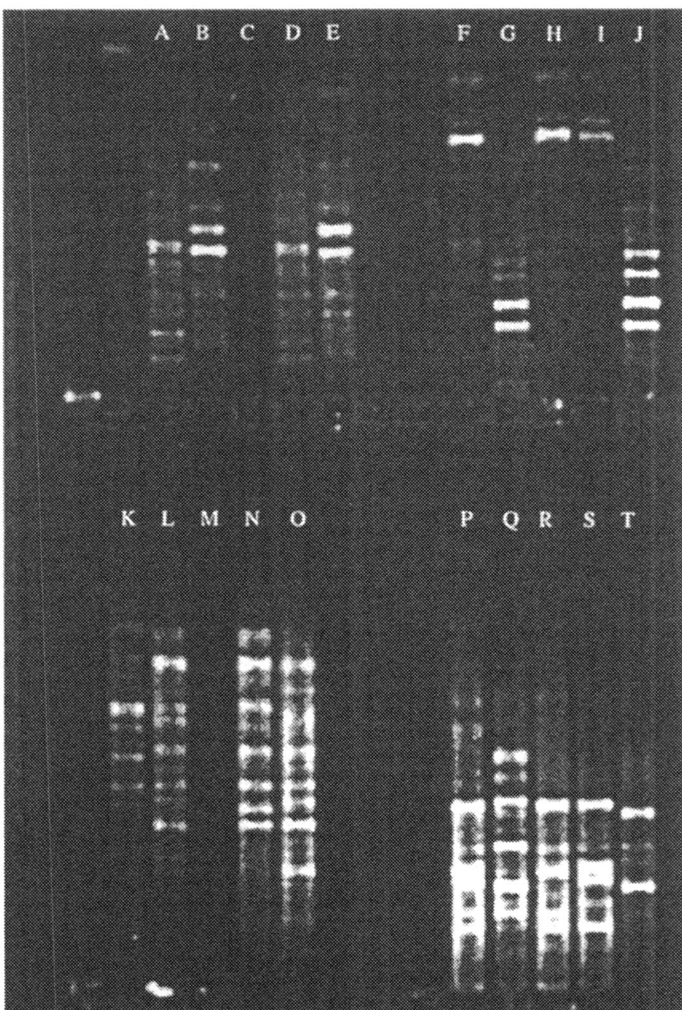

Figure 2. DNA profiles from *Sarocladium oryzae* isolates from two different geographical locations in the Philippines using two different VNTR primers and two different RAPD primers.

Lanes A – T, respectively are

Primer 70 (RAPD)

A	2-2	San Mateo plot 2
B	5b	Nueva Ecija plot 2
C	5a	Nueva Ecija plot 1
D	24	San Mateo plot 1
E	23	Isolate from the same panicle as 24

Primer 71 (RAPD)

F	2-2	San Mateo plot 2
G	5b	Nueva Ecija plot 2
H	5a	Nueva Ecija plot 1
I	24	San Mateo plot 1
J	23	Isolate from the same panicle as 24

Primer RY (VNTR)

K	2-2	San Mateo plot 2
L	5b	Nueva Ecija plot 2
M	5a	Nueva Ecija plot 1
N	24	San Mateo plot 1
O	23	Isolate from the same panicle as 24

Primer GF (VNTR)

P	2-2	San Mateo plot 2
Q	5b	Nueva Ecija plot 2
R	5a	Nueva Ecija plot 1
S	24	San Mateo plot 1
T	23	Isolate from the same panicle as 24

7. Summary

The genus *Sarocladium* consists of a single species *Sarocladium oryzae*. Isolates of this species cause sheath rot in rice, a disease which is becoming increasingly destructive and widespread. In addition, *S. oryzae* has been reported as the causal agent of a bamboo blight, and isolates of the species have also been collected from field weeds and sedges. Biochemical and molecular markers appear to have the potential for defining pathogen groups.

8. References

Agnihotrudu, V. (1973) *Acrocylindrium oryzae* Sawada - sheath rot of paddy, *Kavaka* **1**, 69-71.

Amin, K.S., Sharma, B.D. and Das, C.R. (1974) Occurrence in India of sheath rot of rice caused by Acrocylindrium, *Plant Disease Reporter* **58**, 358-360.

Balakrishnan, B. and Nair, M.C. (1981) Weed hosts of *Acrocylindrium oryzae* Saw., a sheath rot pathogen of rice, *International Rice Research Newsletter* **6**, 13.

Brady, B.L.K. (1980) *CMI Descriptions of Pathogenic Fungi and Bacteria*, **No. 673**, Commonwealth Agricultural Bureaux, Slough, UK.

Bridge, P.D., Hawksworth, D.L., Kavishe, D.F. and Farnell, P.A. (1989) A revision of the species concept in *Sarocladium*, the causal agent of sheath-rot in rice and bamboo blight, based on biochemical and morphometric analyses, *Plant Pathology* **38**, 239-245

Bridge, P.D., Pearce, D.A., Rivera, A. and Rutherford, M.A. (1998) VNTR derived oligonucleotides as PCR primers for population studies in filamentous fungi, Letters in Applied Microbiology (in press).

Boa, E.R. and Brady, B.L. (1987) *Sarocladium oryzae* associated with blight of *Bambusa* species in Bangladesh, *Transactions of the British Mycological Society* **89**, 161-166

Chuke, K.C. (1983) *Pathological and physiological studies of sheath rot of rice caused by Sarocladium oryzae*, M.S. Thesis, University of the Philippines at Los Banos, College, Laguna.

Commonwealth Mycological Institute. (1985). *Sheath rot of rice* (*Sarocladium oryzae*). Annotated-Bibliography, Commonwealth Agricultural Bureaux, Plant Series. **No.M.21**, p 12

Deka, A.K. and Phookan, A.K. (1992) Some common weed hosts of *Sarocladium oryzae* in Assam, India, *International Rice Research Newsletter* **17 (6),** 25.

Giri, S.K. and Swain, N.C. (1994) Effect of pH, temperature and relative humidity on growth and sporulation of *Sarocladium oryzae* causing sheath-rot disease of rice, *Orissa Journal of Agricultural Research* **7**, 31-34

Ou, S.H. (1972) *Rice Diseases*, Commonwealth Mycological Institute, CAB, 368 pp.

Pearce, D.A. and Holderness, M. (1996). *Methods of identification for fungi*. Planning Workshop on Seed Health for Disease Management, 22-25 January, 1996, International Rice Research Institute, Los Banos, Philippines, p 36.

Rahman, M.A. and Khisha, S.K. (1984). Bamboo blight with particular reference to *Acremonium strictum*, *Bano Biggyan Patrika* **10**, 81-93.

Ramabadran, R., Velazhahan, R. and Vidhyasekaran, P. (1990) Management of sheath rot of rice, in P. Vidhyasekaran (ed.), *Basic research for crop disease management*, Daya Publishing House; New Delhi; India, P.367-375.

Sawada, K. (1922) *Descriptive catalogue of Formosan fungi II*. Report of the Government Research Institute, Department of Agriculture, Formosa 2, 1-135.

Zhuge, G.H., Lapis, D.B. and Chuke, K.C. (1985) Studies on rice sheath rot disease II. On the physiology of the pathogen *Sarocladium oryzae* (Sawada) Gams and Hawksworth, *Acta Phytopathologica Sinica* **15**, 81-86.

EPIDEMIOLOGY AND DISEASE MANAGEMENT OF BROWN SPOT OF RICE IN INDIA

N.K.CHAKRABARTI
B/12/320 Kalyani 741 234, Nadia Dt., West Bengal, India.
Former Joint Director, Central Rice Research Institute Cuttack, India.

1. Introduction

Rice (*Oryza sativa* L.) suffers from a number of fungal diseases and many of them occur at times in fairly severe form and cause substantial loss to the crop yield. Among the major fungal diseases of rice, brown spot occupies not only an important position, but also historical interest. The disease is prevalent throughout the rice growing tracts of the world. It is found to be pendemic in heavy rainfall areas viz. the Himalayas, Assam, Malabar coast and West Bengal. Its occurrence has been reported in Asia, Africa, South America and U. S. A. The disease is generally severe in soils of low pH, low available potassium (K_2O) and also deficient in essential and trace elements.

2. Disease incidence

In India, brown spot was first recorded in 1922 and the causal fungus was described as *Helminthosporium oryzae*, a weak parasite under normal conditions. Later, Padmanabhan *et al.* (1948) while confirming the earlier report, observed that, under epiphytotic conditions, the damage could be enormous. Two epiphytotics due to brown spot disease of rice, one in 1918-19 in the Krishna-Godavari delta in the southern part of India and another in 1942 in Bengal (before partition) have been recorded. Loss in grain yield in Bengal during the 1942 epiphytotic ranged from 50 to 90% in several cases. This was considered to be one of the major contributing factors to the Bengal famine in 1943, as reported by the Famine Enquiry Commission headed by Sir John Woodhead in 1945 (Padmanabhan *et al.*, 1948; Padwick, 1950; Ghosh *et al.*, 1960; Padmanabhan, 1963, 1973). Some of the rice workers are unwilling to correlate the severe occurrence of this disease in 1942 in Bengal, to the causes of the famine, in the following year. It appears, however, that the findings of the Famine Enquiry Commission and post dated analysis by Padmanabhan and other workers could not be ruled out totally in this context. Padmanabhan (1942) analysed the meteorological conditions prevalent in some parts of Bengal during that period and found that, the weather conditions highly favoured the severe incidence of the brown spot disease. He also found considerable loss in the grain yield when compared with non-epidemic years (Padmanabhan *et al.* 1948). Also in the personal experience of the author, the local farmer confirmed the serious occurrence of both the foliage and grain spotting of rice

S. Sreenivasaprasad and R. Johnson (eds.), Major Fungal diseases of Rice Recent Advances, 293-306.
Prof. K. Manibhushan Rao Festschrift

which were typical of the brown spot disease. Apparently, the rice obtained from the seriously infected grains, when cooked was not suitable for human consumption due to bitter taste and unfit even as cattle feed.

Bedi and Gill (1960) reported loss of grain weight to the tune of 4.28 - 29% and Chattopadhyay *et al.* (1975) estimated the loss ranging between 26.2 to 51.8% due to heavy grain infection. According to Vidhyasekharan and Ramadoss (1973) the disease reduced the number of tillers and there was inhibition of root and shoot elongation. Under severe infection reduction in yield was recorded by Aluko (1975) in Nigeria, where under moderate infection loss was estimated to be 12%. Early senescence due to brown spot infection was reported from Surinam (Klomp, 1977). Various workers from Japan, however, observed that, the disease was rarely found on rice plants growing in normal soil, but if the soil condition was poor a soil with "akiochi" (a physiological disorder) the disease was severe (Baba, 1958; Sato, 1964-65;Takahashi, 1967). Average grain yield loss of 8.2-23% due to attack of *Cochliobolus miyabeanus* (perfect state) was observed in the mangrove swamps of north-west of Sierra Leone (west Africa). The disease incidence was found to be linked with unfavourable soil and favourable meteorological conditions with long dew period and moderate temperature (Fomba *et al.*, 1990)

3. Pathogen biology

3.1. SYMPTOMATOLOGY

Small brown lesions are ellipsoidal or oval, most conspicuous (around 2.8 mm x 0.5 mm) on the aerial parts namely, coleoptiles, leaf blades, leaf sheaths and glumes. Lesions with a light-brown or greyish centre and a dark-reddish brown margin were observed on foliage at maturity. In cases of severe infection, spotting of grains and discolouration of seeds are observed and the affected grains some times get shrivelled.

3.2. THE CAUSAL ORGANISM

The pathogen was first described as *Helminthosporium oryzae* Breda de Haan (Mitra, 1931). The perfect stage of the fungus was obtained in culture by Ito and Kuribayashi in 1927 and described as *Ophiobolus miyabeanus*. Drechsler (1934) and Dastur (1942) described the fungus as *Cochliobolus miyabeanus* under a new genus *Cochliobolus.* In 1955 it was suggested that the name *Helminthosporium oryzae* be changed to *Bipolaris oryzae*, considering the bipolar nature of germination of the conidia. However, it was observed by various workers that the conidia of the fungus germinated from both the ends and from one or more of intercalary cells also. Later, Subramanian and Jain (1966) accommodated the fungus (anamorph of *C. miyabeanus*) as *Drecshlera oryzae* Jain and Sub. The teleomorph of the fungus is at present known as *Cochliobolus miyabeanus* (Ito et. Kurib) Dreschsler ex Dastur.

3.3. TAXONOMY

The fungal hyphae are dark-brown to olivaceous in colour and 8.15 µm or more in diameter. The sporophores arise as lateral branches which measure 150-600 µm x 4-8. µm The gemiculations are not always conspicuous. Conidia are straight or curved to one side with 5-14 septa generally. They are broadest at the middle or slightly below, and tapering into blunt ends. Conidial walls are thick, but thinner in the terminal cells and conidia are generally hyaline, sometimes filled with granular contents. Chattapadhyay and Chakrabarti (1952) reported variation in conidial size, and divided the conidia into different types namely, type I with 24-156 µm x 8-22 µm, type II with 45-89 µm x 14-22µm. Chattopadhyay and Dasgupta (1959) obtained two different types of isolates, one of which was a morphological variant of *Drechslera rostrata* and the other was a typical one (*D. oryzae*), both being pathogenic on rice leaves. Earlier, Atkins (1972) also isolated *D. rostrata* from rice plants in the U. S. A. Successful inoculation and isolation of the fungus on rice was reported by Nelson and Kline (1961). Wei (1957), as quoted by Ou (1985), reported that, conidia of *D. oryzae* (*C. miyabeanus*) measure differently in various countries. For example, a) 90 µm x 16 µm in Java, b) 15-132 µm x 10-26 µm in Japan, c) 45-106 µm x 14-17 µm in India, d) 24-122 µm x 7-23 µm in China and e) 35-170 µm x 11-17 µm in the U. S. A. Though Nisikada and Miyake (1922) reported the existence of two different types of strains as Japanese and American strains, the Indian type as described above by Chattopadhyay and Chakrabarti (1952) may be considered a distinctive one.

3.4. SPORULATION

Studies carried out on conidial production by *D. oryzae* have shown that, a temperature range of 21-26°C was optimum and a relative humidity (RH) of 92% and above was associated with higher production of conidia with maximum at 100% RH (Chattopadhyay and Dasgupta, 1965). Suitable medium for growth and sporulation of the fungus is Richard's medium containing starch and glucose or starch alone, whereas glucose was considered a poor carbon source. Production of conidia was higher with KNO_3, peptone, glutamine acid and asperagin (Padmanabhan and□ Chakrabarti, 1985). Soprulation was induced or increased by the exposure of the fungus to near ultra violet radiation (Yadav, 1963). High humidity with periodic showers, rather than prolonged rainy season favoured conidial release and dissemination. The maximum number of conidia were observed during cooler months of October to December (Chakrabarti, 1955). Such observation was also recorded by Sreeramulu and Seshavataram(1962), Chandwani *et al.* (1963) and Kulkarni *et al.* (1982). Existence of different strains of the fungus was reported by Nishikado (1926), which was later confirmed by various other workers including Nawaz and Kausar (1962). However, Padmanabhan (1953) opined against the existence of physiological specialisation in the fungus. Misra and Chatterji (1963) reported that sporulating ability and pathogenicity were found to differ between two isolates of the pathogen. Chakrabarti and Mohanty (1973) detected two races from 14 isolates of the fungus, using rice cv. AC 2045 as one of the differentials, but other workers could not identify any physiological races.

3.5. TOXIC METABOLITES

The fungus has been reported to produce various toxic metabolites in culture : i) orgastaril (Murty and Subramanian, 1959), ii) a toxin capable of reducing seed germination in cereals and inhibiting the growth of roots and shoots of rice plants (Rangaswamy and Pandurangan, 1962) and iii) production of proteolytic enzymes and possible correlation with the tryptophane production and later conversion to IAA (Mahadevan and Chandramohan, 1966). Further Oku (1960) reported the presence of a toxic substance ophiobolin from diseased leaves, which caused wilting of plants. Capability of copper oxychloride to completely inactivate the toxin was reported by Narain and Simhachalam (1976).

3.6. SPORE DISPERSAL

The larger brown spots that are mostly produced on the second and third leaves from top give rise to the production of more conidia at about 95% RH. Potassium deficiency in pot cultures was found to influence the formation of large lesions similar to those occurring in muck soil (Sato, 1964, 1965; Dasgupta and Chattopadhyay, 1977). The rate of dispersal of conidia and disease spread are influenced by the interaction between the period of incubation and a period of active sporulation of a lesion (Van der Plank, 1963). The airborne conidia of *D. oryzae* were analysed to determine their relationship with the meteorological factors in West Bengal (Chakrabarti, 1955) and Orissa (Chandwani *et al.*, 1963 and a double peak of spore liberation was observed during a 24 h period. A temperature range of 25-35°C is congenial for the growth of the fungus, while 25-30° C is optimal for disease development. The conidial production is very high under conditions of high humidity (RH 93% or more) and at 25-30° C temperature. The prevalence of airborne conidia is influenced mainly by humidity, dew deposition and rainfall, along with congenial temperature. Under the conditions prevailing in West Bengal, presence of conidia was very high from early September to November. Periodic showers (intermittent), warm temperature (about 25-30°C), high humidity and dew deposition were observed during this period which are congenial for higher production and dispersal of conidia of *D. oryzae*.

The conidiophores of various species of *Drechslera* are discharged violently. A violent spore discharge involves an exogenous surface electrical phenomenon, which initiates repulsion between the spores and sporophores. The charge becomes static under low RH weakening the connection between the spores and sporophores and eventually breaking it, propelling electrostatic spores into air. The source of charge is evaporation and condensation of water (Leach, 1976). However, any such forcible release of conidia could not be observed in *D. maydis* (Aylor and Day, 1976). Various workers have observed that, in most of the phytopathogenic fungi liberation of spores are passive in nature. In a study to determine the effect of water drops on the vertical dispersal of conidia of *D. oryzae*, using glass beads, it was observed that, the role of water drops was more important than the strike off effect of the glass beads (Gangopadhyay and Chattopadhyay, 1973).

3.7. DISEASE CYCLE

The disease is primarily seed borne in nature (Ganguly, 1946; Padmanabhan, 1949; Chattopadhyay, 1951; Mohanty and Chakrabarti, 1976) appears in two phases namely, a) the primary infection at seedling stage and b) the secondary infection at post-tillering and later stages of growth. Padmanabhan *et al.* (1953) found that the brown spot pathogen remained viable only in seeds. Chattopadhyay and Chakrabarti (1954), however, observed that, the fungus survived in the stubbles in the fields of the infected crop of the wet (Kharif) season, harvested in November/December till next June, but does not survive in the soil.

3.8. HOST RANGE

Leersia hexandra, a common weed in the rice fields of India was observed to act as an alternative collateral host of the pathogen in nature (Chattopadhyay and Chakrabarti, 1953). At the Central Rice Research Institute (CRRI), Cuttack, India, *Echinocloa colonum* was also found to occur as a collateral host in nature. Twenty two species of wild rice were inoculated with isolates of the pathogen *D. oryzae*, obtained from cultivated rice and wild type *Oryza longiglumis*. Four host species were observed to be resistant to the isolates of the pathogen from cultivated rice and *O. longiglumis* (Dath and Chakrabarti, 1973). Subramaniam and Ramakrishnan (1973) also reported the occurrence of the pathogen on few types of wild rice. Kernkamp *et al* (1976) reported severe blight on *Zizania aquatica*, a cultivated wild rice in USA due to attack of the fungus. Earlier Japanese and other workers were able to artificially infect, several species of cereals of wheat, oat, maize and grasses like *Eleusine coracana* and *Setaria italica* by the fungus (e.g. Nishikado and Miyakee, 1922).

4. Host physiology

The influence of soil conditions on brown spot infection has not been studied thoroughly. Intensity of brown spot of rice increased at low soil temperature and was also highly influenced by the nutritional conditions of soil (Thomas, 1941). The disease intensity was higher both at lower and higher levels of N application (Chattopadhyay and Dickson 1960a; Chattopadhyay and Chakrabarti, 1965). Purkayastha and Mukhopadhyay (1979) also recorded similar observations and found that, fungitoxicity detected in the diffusates of leaves decreased both under excess and deficient N manuring. Accumulation of greater amount of soluble N in the host tissue was associated with increased susceptibility (Gangopadhyay and Chattopadhyay, 1975). The seedling leaves from high NH_4 level developed larger epidermal motor cells and mesophyll cells with thinner walls (Chattopadhyay and Dickson, 1960b). Application of phosphatic fertiliser in the field was not found to have any effect either on leaf or grain infection by *D. oryrae* (Chakrabarti, 1956). However, Dasgupta and Chattopadhyay (1977) observed that, the harmful effects of unfavourable doses of N was reduced by application of very high doses of phosphatic fertiliser. Kaur *et al.* (1986) reported that application of P at lower (50 ppm) dosage was associated with the reduction in disease intensity, which, however, increased with further increase in P application. Matsuo

(1948), Ono (1953) and Akai (1962) reported that, with K deficiency, the growth of rice plants was severely affected and the number and size of diseased spots were comparatively large without a clear margin. Conidia of *D. oryzae* produced on rice plants grown under the absence of deficiency of K were having greater virulence. Reduced incidence of the disease was associated with increase in concentration of K in the nutrient solution or in soil. Padmanabhan *et al.* (1962a, b) and Chattapadhyay and Chakrabarti (1965) observed higher disease incidence with lower K content in the nutrients. Orsenigo (1957) and Paik (1975) however, reported that, there was no influence of K on the incidence of brown spot disease. Abdel Hak *et al.* (1973) could not notice reduction in disease incidence when K_2SO_4 was applied in the field. Addition of K in the leached soil along with nitrogen had lower infection than those receiving N alone (Chakrabarti and Padmanabhan, 1962). Russel (1978) suggested that application of K and P at higher levels could help to minimise the disease incidence. According to Kaurav and Samnantaray (1982) the disease was most severe in soil with high pH, low organic content, low N and K levels and low cation exchange capacity. The disease was found to be severe under low availability of plant nutrients and dry conditions. It has also been found that, deficiency of micronutrients namely, Mn, Zn, Fe was associated with higher incidence of brown spot. A negative correlation was observed between calcium and disease severity in clay and loamy soil and positive one in sandy loam soil.

The wet season crop (July-November i.e. *Kharif*) transplanted by mid-August under Indian conditions had lower infection to those transplanted at a later date. In the dry season (December/January - April i.e. *rabi*) crop transplanted during the early part had higher disease incidence. Plants transplanted at closer spacing had higher level of disease incidence when compared to those grown under wider spacing (Chattopadhyay and Chakrabarti, 1965; Chakrabarti and Mohanty, 1975). Plants grown under dry conditions without additional water supply had higher level of infection (Hemmi and Suzuki, 1931; Aluko, 1969, 1975; Herrera and Seidal, 1978). Higher disease intensity has also been reported in the crop grown under rainfed upland soil than in irrigated flooded fields. This might be due to low nutrient uptake in the former situation as high air humidity and low soil moisture inhibit the absorption of Si, K and also reduce SiO_2 and K_2O content of leaves (Baba and Takahashi, 1957) under transplanted condition. Infection on both leaves and grains was observed to be higher in comparison to that under broadcast sowing in experimental fields (Chattopadhyay and Chakrabarti, 1957a).

5. Disease resistance

Chattapadhyay and Chakrabarti (1957b), while studying morphological basis of disease resistance, observed that the thickness of cuticle and the epidermis and the number of silicated epidermal bulliform cells were higher in resistant cvs. However, no difference in SiO_2 content of leaves of either resistant or susceptible cvs. was observed. Rice plants were observed to be most susceptible to brown spot at a) seedling stage (2-3 weeks old) and b) between the booting and flowering stages (Chattopadhyay and Chakrabarti, 1955). Further, Padmanabhan and Ganguly (1954), Klomp (1977) and Dhanpal and Prasad (1977) observed that the susceptibility of rice plant against brown spot increases with the increase in the age of the plant. Second leaf from the top was found to be most susceptible to the disease, while the youngest leaf was most resistant (Dasgupta and Chattopadhyay, 1975).

5.1. BIOCHEMICAL STUDIES

Accumulation of starch around the diseased spots on the leaves was observed due to a decreased activity of α-amylase and increased activity of invertase near the spot (Akai *et al.*, 1958; Tanaka and Akai, 1960). Soluble and insoluble N, reducing sugar and other soluble carbohydrates decreased in plants infected artificially, grown both in normal and `Akiochi' soil (in Japan), whereas insoluble carbohydrate increased (Asada, 1957). Inoculation of susceptible plants resulted in increase of Glutamic acid content after 2 days (Shishiyana *et al.*, 1969), followed by a decrease after 4 days. After inoculation with *D. oryzae* aspartic acid content was found to decrease. While studying the mechanism(s) of disease development, it was observed that, accumulated phenolics were oxidised to quinones by the polyphenol oxidase secreted by the fungus *in situ*. These quinones are polymerised into brown pigment very rapidly through enzymatic action. The deposited brown polymer spreads into necrotic lesions to form typical brown spot symptom. While the concentration of bound phenolic substances were higher in susceptible host types, the concentration of free phenols, free phenolic acids and total phenolic substances were higher in brown spot resistant cvs. Chattopadhyay and Samaddar (1976, 1980), however, did not observe any significant difference in total free phenols among healthy or inoculated rice plants in either resistant or susceptible cvs. Glycosides and wall bound phenols, however, decreased in both the types of cvs. during the later stages of infection.

Spraying with some amino acid solutions considerably increased the symptoms caused by *D. oryzae*, while at higher concentration some of the amino acids decreased the symptoms moderately (Sinha and Bandopadhyay, 1974). Spraying of rice seedlings with glutamic acid and glutamine was associated with decrease in resistance of the top most leaves to brown spot pathogen. It has been suggested that, the role of amino acids may not be entirely in lower (second and third from the top) leaves (Dasgupta, 1980). Greater peroxidase activity was observed in resistant cvs. when compared to that of susceptible cv. Benibhog (Chattopadhyay and Bera, 1979). Presence of ascorbic acid in a greater quantity in resistant cvs. T 141 and CO 20 was reported by Pushpanaden Padmabai (1957). Seedling leaves of resistant cv. FR 43B had highest content of ascorbic acid, but no significant difference could be observed in the adult leaves of different cvs. in this respect. Extracts from resistant rice cvs. were found to check conidial germination and length of germ tube growth. During spore germination of *D. oryzae*, production of toxins (Ophiobolin A and Ophiobolin B) at a very low concentration was observed. The spore germination fluids also caused susceptibility inducing phytotoxicity (Xico *et al.*, 1991) Further, Vidyasekharan *et al.* (1990) reported that, brown spot resistant rice plants could be regenerated from toxin resistant calluses. The resistance was found to be heritable.

6. Disease Management

A number of strategies such as application of suitable cultural practices, use of fungicides, introduction of resistant or moderately resistant cvs., use of tissue culture techniques and biocontrol measures have been attempted for the management of brown spot of rice. Padmanabhan *et al.* (1962a, b) reported that, occurrence of brown spot disease was considerably higher in the plants which had not received K in their nutrient, when compared with the plants which received 10 ppm and 20 ppm K_2O. In leached soil the disease was observed to be higher than in the plants grown in unleached soil. Plants that received N alone after leaching, showed higher infection than those grown in unleached soil; plants that received N, P and K after leaching showed lower infection. Soil amelioration with minor elements namely, Mn, Mg, Fe and/or Ca after soil testing would be helpful in reducing the incidence of the disease. Late planting beyond mid - August during the main *kharif* season and before February in *rabi* and application of N in heavy doses, particularly when not in combination with P_2O_5 or K_2O are to be avoided. Removal and burning of infected stubbles of the previous (infected) rice crop left over in the field after harvest has also been suggested for adoption, specially in brown spot endemic areas. Application of N in slow release form has been found to minimise the disease incidence (Vidhyasekharan *et al.*, 1983). Infection was also found to decrease with the application of coaltar coated urea at transplanting along with the increase in yield in cvs. IR 20 and ADT 36 (Mohanty and Chakrabarti 1982; Vidyasekharan *et al.*, 1986).

6.1. CHEMICAL CONTROL

Pre-sowing treatment of seeds, incubation in hot water (53-54°C for 10-12 min) or fungicide seed treatment (copper/ organomercurials/ dithiocarbamates/ antibiotics) was found to be helpful in reducing the primary seed borne infection. Application of benzimidazoles/ carbandezim was also found to minimise the disease incidence. Conidia of *D. oryzae* failed to germinate in media containing 1000 or 10,000 ppm of compounds containing Cu, Zn and Hg (Haware, 1967). Application of fungicides as foliar spray has been observed to be effective in reducing the brown spot incidence. Cu based fungicides were found effective in checking secondary air borne infection. It was also reported that, thiocarbamates were promising in checking the disease progress. Application of organophosphatic fungicides, dithiocarbomates (Maneb complex), Aureofungin with copper sulphate (50 ppm + 25 ppm) were reported to lower the disease incidence (Chattopadhyay and Chakrabarti, 1961; Mukherji and Bagchi, 1964; Mishra and Singh, 1971; Chakrabarti and Mohanty, 1975). Application of bavistin (Carbendezim) at 1g/l of water as spray is also recommended. Application of certain herbicides were found to reduce the disease incidence along with improvement in grain quality in terms of protein and amylose content (Chakrabarti *et al.*, 1975; Chakrabarty and Sen, 1976). Application of edifenphos (hinosan) at 0.1% and biloxazol was found to be effective in reducing brown spot disease. Since various strains tolerant to edifenphos have been detected both from the laboratory and field tests, if required biloxazol can be used as a substitute. In a laboratory experiment, Lalithakumari and Annamalai (1988) identified a tolerant mutant to bitartonol (Baycor) and mancozeb (DITHANE M-45). However, this has not yet become a practical field problem.

6.2. BIOCONTROL AND INDUCED HOST RESISTANCE

Biocontrol of rice brown spot was attempted (Chakrabarti and Padmanabhan, 1976; Chakrabarti and Chaudhari, 1992) using *Bacillus mycoides* as the antagonist. Studies on the efficacy of biocontrol need to be intensified not only from the point of view of economy, but also to achieve environmentally friendly disease control.

Induction of resistance in rice against brown spot disease by spraying conidia of a wild race of the fungus has been reported. Phytoalexin inducers like ferric chloride, nickel nitrate, sodium molybdate, DL-methionine, sodium selenite and thioglycollic acid provided fair degree of protection against the disease when sprayed on rice seedlings (Sinha and Trivedi, 1978; Sinha and Hait, 1982). Development of disease resistant cvs. was attempted through screening of large collection of germplasm, both at national/ international levels. Rice cvs. CH 13, CH 45, T.141, CO 2O, BAM 10 were considered resistant in India by Padmanabhan *et al.* (1948-73). Rice cvs. such as TN 1 and Mucy Nwang in Thailand and CI 9515 in U.S.A. were found resistant to the disease (Ou, 1985), and were used as donor parents in breeding for disease resistance. Much information is however, not available on inheritance of resistance to brown spot disease. Nagai and Hara (1930) observed that, the resistance seems to be a dominant character. However, Adair (1941) was of the opinion that, resistance was recessive involving multiple genes.

7. Summary

The brown spot disease of rice which earlier used to prevail in widespread form causing considerable damage to the crop, is at present occurring sporadically and is generally observed in mild to moderate form at certain localities. As such, it would be wise to keep a constant vigil over its reappearance and spread, especially in developing countries. In India for example, the disease is still prevalent in a fairly moderate form in certain pockets,namely, Bihar, Uttar Pradesh, Tamil Nadu and Kerala. The high yielding, semi-dwarf, nitrogen responsive cvs. that are being introduced from abroad or through large scale breeding programmes, should be screened thoroughly for their susceptibility to brown spot. The possibility of new pathotypes of *D. oryzae* appearing naturally cannot also be overlooked.

8. References

Abdel-Hak, T.M., Sirry, A.R., Ashour, W.A. and Kamel, S.M. (1973) Effect of different environmental factors on blast and brown spot diseases of rice in A.R.E., *Agricultural Research Review* **51**, 45-62.

Adair, C.R. (1941) Inheritance in rice of reaction to *Helminthosporium oryzae* and *Cercospora oryzae*, *Technical Bulletin* U. S. D. A. **772**, p.18.

Akai, S. (1962) Potash application and occurrence of *Helminthosporium* leaf spot on rice, *Potash Review* **23**, 27.

Akai, S., Tanaka, H. and Noguchi, K. (1958) On the mechanism of starch accumulation in tissues surrounding spots in leaves of rice plants to the attack of *Cochliobolus miyabeamnus*. I. Observation on starch accumulation in tissues surrounding spots, *Annals of Phytopathological Society of Japan* **23**, 111-116.

Aluko, M.O. (1969) Relative prevalence of blast and brown leaf spot on upland rice in Nigeria, *Plant Disease Reporter* **53**, 875-877.

Aluko, M.O. (1975) Crop losses caused by brown leaf spot disease of rice in Nigeria, *Plant Disease Reporter* **59**, 609-613.

Asada, Y. (1957) Changes of nitrogen compound, carbohydrates, reducing ascorbic acid and respiration accompanied by the infection of *Cochliobolus miyabeanus* and existence of hyphae in diseased spots, *Annals of Phytopathological Society of Japan* **22**, 103-106.

Atkins, J.G. (1972) An unreported *Helminthosporium* disease of rice for the United States, *Plant Disease Reporter* **46**, 26.

Aylor, D.E. and Day, P.R. (1976) Conidial release in *Helminthosporium*, *Phytopathology* **66**, 537.

Baba, I. (1958) Nutritional studies on the occurrence of *Helminthosporium* leaf spot and `akiochi' of the rice plant, *Bulletin of National Institute of Agricultural Sciences* **D7**, 1-157.

Baba, I. and Takahashi, Y. (1957) Growth and susceptibility to *Helminthosporium* leaf spot disease of rice plants grown in different soils under different water temperature, *Journal of Agricultural Methodology* **13**, 41-44.

Bedi, K.S. and Gill, H.S. (1960) Losses caused by the brown leaf spot disease of rice in the Punjab, *Indian Phytopathology* **13**, 161-164.

Chakrabarti, N.K. (1955) *Studies on certain aspects of Helminthosporium disease of paddy with particular reference to conditions in West Bengal*, Doctoral thesis, University of Calcutta, India, p.116.

Chakrabarti, N.K. (1956) Effect of phosphatic fertilizer on the incidence of infection of *Helminthosporium oryzae* Breda de Haan on paddy, *Proceedings of 43rd Indian Science Congress* **Part III**, p.215 (abstract).

Chakrabarti, N.K. and Padmanabhan, S.Y. (1962) Studies on *Helminthosporium* disease of rice VI : Nutritional factors and disease expression, - Effect of leaching of soil on incidence, *Proceedings of 49th Indian Science Congress* **Part III,** p.479 (abstract).

Chakrabarti, N.K. and Mohanty, S.K. (1973) Relationship between the different dates of transplanting and incidence of brown spot disease in high-yielding rice varieties grown in *rabi* season, *Proceedings of 60th Indian Science Congress* **Part III** (abstract).

Chakrabarti, N.K. and Mohanty, S.K. (1975) Control of brown spot disease of rice through application of fungicides, *Science and Culture* **41**, 539-541.

Chakrabarti, N.K., Mohanty, S.K., Das, A.B. and Lodh, S.B. (1975) Chemical control of brown spot of rice and its effect on the nutritional constituents of rice grains, *Plant Disease Reporter* **59**, 1028-1032.

Chakrabarti, N.K. and Padmanabhan, S.Y. (1976) Recent trend in control of rice diseases, *Proceedings of National Academy of Sciences, India* **46(B)**, 137.

Chakrabarti, N.K. and Chaudhari, R (1992) Brown spot of rice in U.S.Singh *et al* (eds.), *Plant Diseases of International Importance*, Vol. I, Sec. ed., Prentice-Hall Inc., U. S. A., pp. 116-129.

Chakrabarty, D.K. and Sen, C. (1976) Nature of action of herbicides in controlling brown spot of paddy caused by *Helminthosporium oryzae*, *Journal of the Society of Experimental Agriculture* **1**, 48-50.

Chandwani, G.H., Balakrishnan, M.S. and Padmanabhan S.Y. (1963). *Helminthosporium* disease of rice. V. A study of the spore population of *Helminthosporium oryzae* over rice fields, *Journal of the Indian Botanical Society* **42**, 1-14.

Chattopadhyay, A.K. and Samaddar, K.R. (1976) Effect of *Helminthosporium oryzae* infection and ophiobolin on the cell membrane of host tissues, *Physiological Plant Pathology* **8**, 131-139.

Chattopadhyay, A.K. and Samaddar, K.R. (1980) Effects of *Helminthosporium. oryzae* infection and ophiobolin on the phenol metabolism of host tissues, *Journal of Phytopathology* **98**, 193-202.

Chattopadhyay, S.B. (1951) *Helminthosporium* disease of paddy - seed as source of infection, *Proceedings of the 38th Indian Science Congress* **Part III**, p.142 (abstract).

Chattopadhyay, S.B. and Bera, A.K. (1979) Peroxidase activity in rice leaves infected with *H. oryzae* in relation to lesion maturation, *Phytoparasitica* **7**, 11-15.

Chattopadhyay, S.B., Chakrabarti, N.K. (1952) Variation in conidial character of *Helminthosporium oryzae* Breda de Haan, *Proceedings of the 39th Indian Science Congress* **Part III**, pp.33-34 (abstract).

Chattopadhyay, S.B. and Chakrabarti, N.K. (1953) Occurrence in nature of an alternative host (*Leersia hexandra* SW) of *Helminthosporium oryzae* Breda de Haan, *Nature* **172**, 550.

Chattopadhyay, S.B. and Chakrabarti, N.K. (1954) Survival of Helminthosporium oryzae Breda de Haan in the field in nature, *Proceedings of the 41st Indian Science Congress* **Part III**, p.123 (abstract).

Chattopadhyay, S.B. and Chakrabarti, N.K. (1955) Variability in resistance and susceptibility of different varieties of paddy to infection of *Helminthosporium oryzae* Breda de Haan, *Proceedings of the 42nd Indian Science Congress* **Part III**, (abstract).

Chattopadhyay, S.B. and Chakrabarti, N.K. (1957a) Effect of methods of cultivation on the incidence of `Brown spot' disease in paddy, *Indian Agriculture* **1**, 45-52.

Chattopadhyay, S.B. and Chakrabarti, N.K. (1957b) Relationship between anatomical characters of leaf and resistance to infection of *Helminthosporium oryzae* in paddy, *Indian Phytopathology* **10**, 130-132.

Chattopadhyay, S.B. and Chakrabarti, N.K. (1961) Application of fungicides in the control of secondary air-borne infection of *Helminthosporium oryzae* Breda de Haan, *Indian Phytopathology* **14**, 88-92.

Chattopadhyay, S.B. and Chakrabarti, N.K. (1965) Effect of agronomic practices on susceptibility of rice plants to infection of *Helminthosporium oryzae*, *Rice News Letter(ICAR)* **13**, 39-44.

Chattopadhyay, S.B., Chakrabarti, N.K. and Ghosh, A.K. (1975) Estimation of loss in yield of rice due to infection of brown spot incited by Helminthosporium oryzae, *International Rice Community. Newsletter* **24**, 67-70.

Chattopadhyay, S.B. and Dasgupta, C. (1959) *Helminthosporium rostratum* Drechs. on rice in India, *Plant Disease Reporter* **43**, 124-134.

Chattopadhyay, S.B. and Dasgupta, C. (1965) Factors affecting conidial production of *Helminthosporium oryzae, Indian Phytopathology* **18**, 160-167.

Chattopadhyay, S.B. and Dickson, J.G. (1960a) Relation of nitrogen to disease development in rice seedlings infected with *Helminthosporium oryzae*, *Phytopathology* **7**, 434-438.

Chattopadhyay, S.B. and Dickson, J.G. (1960b) Induced mutation and cytological investigation in *Helminthosporium oryzae*, *Phytopathology* **50**, 439-442.

Dasgupta, M.K. (1980) Effect of some amino acids and amines on the development of brown spot in rice, *Oryza* **17**, 125-128.

Dasgupta, M.K. and Chattopadhyay, S.B. (1975) Variation in susceptibility of rice leaves to the brown spot, *Indian Phytopathology* **28** 17-21.

Dasgupta, M.K. and Chattopadhyay, S.B. (1977) Effect of environmental factors on the development of brown spot of rice caused by *H. oryzae, Journal of the Society of Experimental Agriculture* **2**, 24-27.

Dastur, J.F. (1942) Notes on some fungi isolated from `black point' affected wheat kernels in the central provinces, *Indian Journal of Agricultural Science* **12**, 731-742.

Dath, A.P. and Chakrabarti, N.K. (1973) Wild rice species, the alternative hosts of *Helminthosporium oryzae* Brede de Haan, *Science and Culture* **39**, 394-397.

Dhanpal, N. and Prasad, N.N. (1977) Effect of different levels of nitrogen on the development of rice brown spot disease *H. oryzae* Brede de Haan, *Madras Agricultural Journal* **64**, 347-348.

Fomba, S.N., Singh, N. (1990) Crop losses caused by rice brown spot disease in mangrove swamps of north-western Sierra Leone, *Tropical Pest Management* **36**, 387-393.

Gangopadhyay, S. and Chattopadhyay, S.B. (1973). Vertical dispersal of the conidia of *Helminthosporium oryzae in vitro*, *Indian Phytopathology* **26**, 249-256.

Gangopadhyay, S. and Chattopadhyay, S.B. (1975) Weather condition and brown spot disease development of rice in West Bengal, *Indian Journal of Mycology and Plant Pathology* **4**, 207-210.

Ganguly, D. (1946) *Helminthosporium* disease in Bengal, *Science and Culture* **112**, 220-223.

Ghosh, R.L.M., Ghatge, M.B. and Subramanyan, V. (1960) *Rice in India* (Revised ed), Indian Council of Agricultural Research , New Delhi, p.474.

Haware, M.P. (1967) Effect of certain copper, zinc and mercury compounds on *Sclerotium oryzae* Calt. and *Helminthosporium oryzae* Breda de Haan, *Research Journal of the Jawaharlal Nehru Krishi Viswavidyalaya*, Jabalpur **1**, 89-90.

Hemmi, T. and Suzuki, H. (1931) On the relation of soil moisture to development of *Helminthosporium* disease of rice seedlings, *Forschn Geb. Pollkrankh* **1**, 90-98.

Herrera, L. and Siedel, D. (1978) On the injurious effect of *Cochliobolus miyabeanus* (Ito and Kuribayashi) Drechsler ex Dastur, in "Rice growing in Cuba", *Archiv fur Phytopathologic Pflanzenscluntz* **14**, 285.

Ishaque, M. and Talukdar, M.J. (1962) Studies on the effect of soil moisture on the development of leaf spot disease of rice caused by *Helminthosporium oryzae*, *Pakistan Journal of Biological and Agricutural Sciences* **5**, 10-17.

Ito, S. and Kunibayashi, K. (1927) Production of ascigerous stage in culture of *Helminthosporium oryzae, Annals of the Phytopathological Society of Japan* **2**, 1-8.

Kaur, P., Kaur, S. and Padmanabhan, S.Y. (1986) Effect of calcium on the development of brown spot disease of rice, *Indian Phytopathology* **39**, 57-61.

Kaurav, L.P. and Samnantaray, R.N. (1982) Occurrence of brown spot in rice in relation to nutrient soil status, *International Rice Research Newsletter* **7**, 16.

Kernkamp, M.F., Kroll, R. and Woodruff, S.C. (1976) Diseases of cultivated wild rice in Minnesota, *Plant Disease Reporter* **60**, 771-775.

Klomp, A.O. (1977) Early senescence of rice and *Drechslera oryzae* in Wageningen Polder, *Surinam Agricultural Research Report*, Wageningen, No.**859**.
Kulkarni, S., Ramakrishnan, K. and Hegde, R.K. (1982) Epidemiology and control of brown leaf spot of rice in Karnataka. IX-Atmospheric variation in spore load of *Drechslera oryzae*, *Indian Phytopathology* **35**, 80-82.
Lalithakumari, D and Annamali P. (1988) Observation of the field performance of edifenphos against *Drechslera oryzae* (Brown spot of rice*), Chemical Control Newsletter* **10**, 26.
Leach, C.M. (1976) An electrostatic theory to explain violent spore liberation by *Drechslera teurica* and other fungi, *Mycologia* **68**, 63-86.
Mahadevan, A. and Chandramohan, D. (1966) Auxin production by rice leaf blight fungus, *Helminthosporium oryzae* and other species of *Helminthosporium*, *Bulletin of the Indian Phytopathological Society* **3**, 91-96.
Matsuo, T. (1948) On the effect of potassium deficiency in the soil upon the occurrence of `Helminthosporiose' of rice plant, *Annals of the Phytopathological Society of Japan* **13**, 10-13.
Misra, A.P. and Chatterya, A.K. (1963) Comparative study of two isolates of *Helminthosporium oryzae* Breda de Haan, *Indian Phytopathology* **16**, 275-281.
Misra, A.P. and Singh, T.B. (1971) Fungi toxicity of certain copper and organic fungicides to mycelial growth of *A. tenius* and *H. oryzae*, *Indian Journal of Mycology and Pathology* **1**, 60-63.
Mitra, M. (1931) A comparative study of species and strains of *Helminthosporium* on certain Indian cultivated crops, *Transactions of the British Mycological Society* **15**, 254-293.
Mukherji, S.K. and Bagchi, B.N. (1964) Control of secondary air borne infection of *Helminthosporium* disease of paddy, *Rice Newsletter* (ICAR) **12**, 103-105.
Mohanty, S.K. and Chakrabarti, N.K. (1976) Occurrence of seed borne fungi in high yielding rice varieties, *Journal of Experimental Agriculture* **1**, 49-50.
Mohanty, S.K. and Chakrabarti, N.K. (1982) Effect of slow release forms of nitrogen application on the incidence of brown spot disease of rice, *Oryza* **18**, 241.
Murty, T.K. and Subramanian, S.S. (1959) A note on the metabolic products of *Helminthosporium oryzae*, *Current Japanese Journal of Genetics* **28**, 282-283.
Nagai, I. and Hara, S. (1930) On the inheritance of variation in a strain of rice plant, *Japanese Journal of Genetics* **5**, 140-144.
Narain, A. and Simhachalam, G. (1976) *Helminthosporium oryzae* toxin and its inactivation by fungicides, *Indian Phytopathology* **29**, 232-237.
Nawaz, M. and Kausar, A.G. (1962) Cultural and pathogenic variation in *Helminthosporium oryzae*, *Biologia* **8**, 35-48.
Nelson, R.R. and Kline, D.M. (1961) The pathogenicity of certain species of *Helminthosporium* to species of Graminae, *Plant Disease Reporter* **45**, 644-648.
Nishikado, Y. (1926). Comparative studies of *Helminthosporium* disease of rice in the Pacific regions, *Proceedings of the 3rd Paso-Pacific Science Congress*, Tokyo, p.2113 (Abstract).
Nishikado, Y. and Miyake, C. (1922) Studies on the helminthosporiose of the rice plant (cited by Ou, 1985).
Oku, H. (1960) Fungicidal action of polyphenols and the role of polyphenoloxidase of the fungus, *Journal of Phytopathology* **38**, 342-354.
Ono, K. (1953) Morphological studies on blast disease and sesame leaf spot of rice plant, *Journal of Hokuriku Agriculture* **2**, 75.
Orsenigo, M. (1957) Estrazione purificazione della cochliobolina, una tomino prodotte de *Helminthosporium oryzae*, *Journal of Phytopathology* **29**, 189-196.
Ou, S.H. (1985) *Rice Diseases,* The Common Wealth Agricultural Bureaux, Farnham Royal, U. K., p.380.
Padmanabhan, S.Y., Chowdhury, K.R.R. and Ganguly, D. (1948) Nature and extent of damage caused by *Helminthosporium* disease of rice, *Indian Phytopathology* **1**, 34-47.
Padmanabhan, S.Y. (1949) Occurrence of fungi inside rice kernels, *Current Science* **18**, 442-443.
Padmanabhan, S.Y. (1953) Specialization in pathogenicity of *Helminthosporium oryzae* and *Pyricularia oryzae*, *Proceedings of the 49th Indian Science Congress*, **Part IV**, p18, (Abstract).
Padmanabhan, S.Y. (1963) *Heliumuthosporium* disease of rice. VII - A study of the meteorological factors associated with epiphytotics of 1942 in Bengal, *Oryza* **1**, 101-110.
Padmanabhan, S.Y. (1973) The Great Bengal famine, *Annual Review of Phytopathology* **11**, 11-26.
Padmanabhan, S.Y. and Ganguly, D. (1954) Relation between the age of rice plant and its susceptibility to *Helminthosporium* and blast diseases, *Proceedings of the Indian Academy of Sciences 29*, 44-50.
Padmanabhan, S.Y. and Chakrabarti, N.K. (1985) Diseases of rice in India (1) Fungal Diseases in *Rice Research in India*, ICAR, New Delhi, pp.459-506.

Padmanabhan, S.Y., Ganguly, D. and Balakrishnan, M.S. (1953) Source and development of seedling infection in *Helminthosporium* disease of rice, *Indian Phytopathology* **6**, 95-115.

Padmanabhan, S.Y., Chakrabarti, N.K. and Kameswar Rao, K.V.S.R. (1973) Forecasting and control of rice diseases, *Proceedings of the Indian Academy of Sciences*, Part B, **37**, 423-429.

Padmanabhan, S.Y., Abichandrani, T.G., Chakrabarti, N.K. and Patnaik, S. (1962a) Studies on *Helminthosporium* disease of rice VI. Nutritional factors and disease expression - 2. Effect of potassium, *Proceedings of the Indian Science Congress*, **Part III**, p.480 (abstract).

Padmanabhan, S.Y., Kameswar Rao, K.V.S.R., Naha, P.M. and Veeraraghavan, J. (1962b) Prospects of control of blast and helminthosporium disease with antibiotics, *2nd All-India Rice Research Workers' Conference* (ICAR), September, 1962.

Padwick, G.W. (1950) *Manual of rice diseases*, Commonwealth Mycological Institute, Kew, U. K., p.198.

Paik, S.B. (1975). The effects of silicate, nitrogen, phosphorus and potassium fertilizers on the chemical components of rice plants and incidence of blast disease of rice, *Korean Journal of Plant Protection* **14**, 97-109.

Purkayastha, R.P. and Mukhopadhyay, R. (1979) Level of amino acids and post-infectional formation of antifungal substances in relation to susceptibility of rice plants against *H. oryzae* at different nitrogen supply, *Journal of Plant Disease and Protection* **83**, 221-228.

Pushpananden Padmabai (1957) Ascorbic acid and helminthosporiose of *Oryza sativa* L., *Current Science* **26**, 26.

Rangaswamy, G. and Pandurangam, S.V. (1962) Studies on the helominthosporiose of some graminaceous hosts in South India. I. Fungal morphology, identity and pathogenicity, II. Physiological properties of fungi, *Journal of the Annamalai University* **24**, 138-55; 156-76.

Russel, G.E. (1978) *Plant breeding for pest and disease resistance studies in the agricultural and food science*, Butterworths, London, U. K., p.485.

Sato, K. (1964-65) Studies on the blight diseases of rice plant, *Bulletin of the Institute of Agricultural Research*, Tohoku University **15**, 199-237; 239-342; **16**, 1-54.

Shishiyama, J., Egawa, H., Mayama, E. and Akai, S. (1969) Role of amino acids in the development of *Helminthosporium* blight disease of rice plant and some enzyme activities relating to amino acid metabolism in the host-parasite interaction, *Memoirs of the College of Agriculture*, Kyoto University, Japan **95**, 7-34.

Sinha, A.K. and Bandopadhyay, A.K. (1974) Increase of resistance of rice seedlings to *H. oryzae* infection following treatment with some amino acids, *Journal of Phytopathology* **80**, 209-214.

Sinha, A.K. and Hait, G.N. (1982) Host sensitization as a factor in induction of resistance in rice against *Dreschlera* by seed treatment with phytoalexin inducers, *Transactions of the British Mycological Society* **79**, 213.

Sreeramulu, T. and Seshavataram, V. (1962) Spore content of air over paddy fields. I. Change in a field near Pentapadu from September 21 to December 31, 1957, *Indian Phytopathology* **15**, 61-74.

Subramanian, C.V. and Jain, B.L. (1966) A revision of some Graminicolus Helminthosporia, *Current Science* **35**, 352-355.

Subramanian, C.V. and Ramakrishnan, G. (1973) The occurrence of brown spot *Helminthosporium oryzae* Breda de Haan, on some wild species of *Oryza*, *Current Science* **42**, 327.

Takahashi, Y. (1967) Nutritional studies on the development of *Helminthosporium* leaf spot, in *Proceedings of the Symposium on Rice diseases and their control by growing resistant varieties and other measures*, Forest and Fisheries Council, Tokyo, pp.157-170.

Tanaka, H. and Akai, S. (1960) The formation of artificial spots and inhibition of starch decomposition in rice leaves by various chemical compounds, *Annals of the Phytopathological Society of Japan* **25**, 156-164.

Thomas, K.M. (1940) *Detailed administration report of the Government Mycologist, Madras, for the year 1939-40*, p.18.

Van der Plank (1963) *Plant Diseases. Epidemics and control*, Academic Press, N. Y., p.349.

Vidhyasekharan, P. and Ramadoss, N. (1973) Quantitative and qualitative losses in paddy due to Helminthosporiose epidemic, *Indian Phytopathology* **26**, 479-484.

Vidhyasekharan, P., Rajamanickam, B. and Lewin, H.D. (1986) Effect of coaltar coated urea on brown spot, *International Rice Research Newsletter* **11**, 20.

Vidhyasekharan, P., Ranganathan, K., Palaniappan, S.P. and Ramasamy, S. (1983) Effect of slow release nitrogen fertilizers on rice brown spot disease, *International Rice Research Newsletter* **8**, 11.

Vidhyasekharan, P., Ling, D.H., Borromeo, E.S., Zapate, F.J. and Mew, T.W. (1990) Selection of brown spot resistant rice plants *from Helminthosporium oryzae* toxin resistant calluses, *Annals of Applied Biology* **117**, 515-523.

Wei, C.T. (1957) *Manual of rice pathogens*, Science Press, Beijing, p.267.
Xico, J.Z., Trude, M., Doke, N. and Nishimura, S. (1991) Phytotoxin produced by germinating spores of *Bipolaris oryzae*, *Phytopathology* **81**, 58-64.
Yadav, B.S. (1963) Effect of ultra violet and visible radiation on the sporulation of species of *Helminthosporium*, *Disease Abstracts* **24**, 1786-1787.

Prof. K.Manibhushanrao to whom this volume is dedicated started his career as Research Assistant at the Central Rice Research Institute, Orissa, India during 1962-63, where I had the privilege of working in the same department. I wish him a happy retired life.

RICE LEAF SCALD: PATHOGEN BIOLOGY AND DIVERSITY

H. C. TURNER and R. BLACK
Natural Resources Institute
University of Greenwich, Central Avenue, Chatham Maritime, Chatham, Kent ME4 4TB, UK.

1. Introduction

Rice leaf scald is caused by the fungal pathogen *Monographella albescens* (Thüm) Parkinson, Sivanesan and Booth (anamorph *Gerlachia oryzae* (Hashioka and Yokogi) W. Gams). It is a disease that can be found throughout virtually all of the rice growing regions of the world (International Mycological Institute, 1996). Under favourable climatic and cultural conditions, *M. albescens* can pose a serious threat to rice production and is often ranked in the first three or four fungal constraints for a region, along with blast (*Magnaporthe grisea*), sheath blight (*Rhizoctonia* spp.) and brown spot (*Helminthosporium oryzae*) (Jones *et al.*, 1991; Alam *et al.*, 1985; Cuevas-Perez and Gaona, 1988; Kempf, 1983; Prabhu and Sitarama-Prabhu, 1989). Incidence of infection in an area can reach high levels under favourable conditions, resulting in significant loss of yield, such as the 23.4% yield reduction reported in India by Srinivasan (1981).

Much of the literature relating to this disease comprises first reports of the pathogen's presence in a given region, or revisions of the taxonomy of the anamorph and teleomorph. Variations in symptomology and host tissues affected have given rise to a variety of common names for the disease, for example leaf scald, brown leaf blight, leaf tip blight and leaf tip drying (Ou, 1985). A succinct overview of the taxonomy of the pathogen was produced by Ou (1985), while Parkinson *et al.* (1981) have provided good evidence for the current classification, with *Monographella albescens* superseding *Metasphaeria albescens* (Thum) for the teleomorph, and *Gerlachia oryzae* superseding *Rhynchosporium oryzae* (Hashioka and Yokogi) for the anamorph.

Despite the wide geographical distribution of *M. albescens*, this pathogen has been the subject of remarkably little research, being overshadowed by the more damaging rice blast pathogen *Magnaporthe grisea* (anamorph *Pyricularia oryzae*). However, as blast-resistant rice varieties have been developed and introduced into general use around the world, it is worth bearing in mind that leaf scald is a near-ubiquitous disease that can become a severe problem when favourable conditions are prevalent. In light of this, some attention should be devoted to the biology of the causal agent and particularly to

S. Sreenivasaprasad and R. Johnson (eds.), Major Fungal diseases of Rice Recent Advances, 307-319.
Prof. K. Manibhushan Rao Festschrift

the levels of genetic and phenotypic variability within and between populations. Current understanding of these aspects is therefore discussed below.

2. Morphology and Culture

M. albescens is a uninucleate, filamentous fungus (Hyponectriaceae, Ascomycota) which is most often observed causing characteristic light-and-dark banded brown lesions which extend down the leaf from the tip or inwards from leaf margins, drying the leaf tissue as the lesion develops (Figure 1, arrowhead). Scald infection can occur in conjunction with other diseases, such as blast (Figure 1, arrow). *M. albescens* has also been shown to infect the leaf sheath, seeds, spikelets and panicles (Ribeiro, 1983).

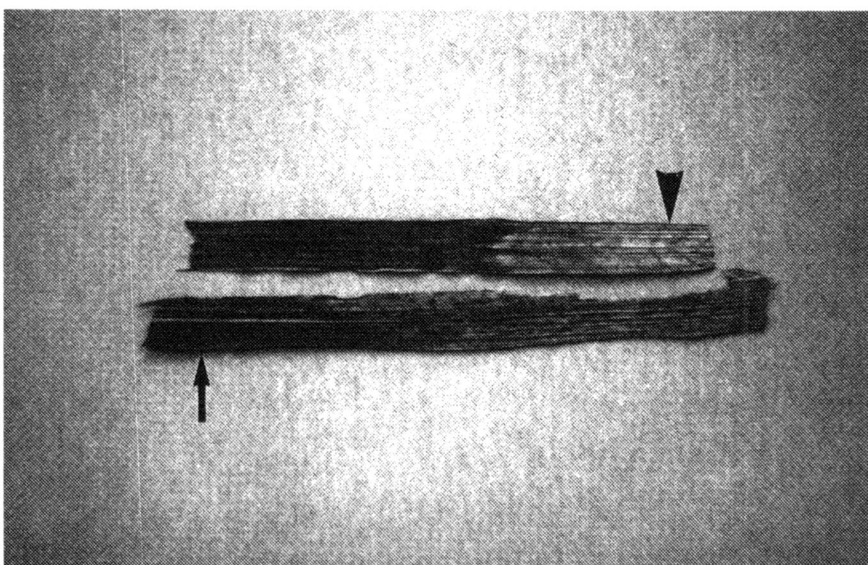

Figure 1. Rice leaves showing typical leaf scald symptoms. Note how the lesion, spreading down from the leaf tip, shows a distinctive banding pattern (arrowhead). These leaves also show signs of infection by the rice blast pathogen *Magnaporthe grisea* (arrow).

The pathogen forms falcate, predominantly bitunicate conidia on the leaf surface (Figure 2); the conidiogenous structures often emerge from stomata. The conidia are 9-14 μm long by 3-4.5 μm wide (Ou *et al.*, 1978). Perithecia, containing colourless paraphyses and sets of eight ascospores per ascus, develop within the mesophyll layer of infected leaves.

M. albescens can be cultured on potato dextrose agar (PDA) where it produces abundant pale pink or cream mycelium (Parkinson, 1980). Conidia form on ageing

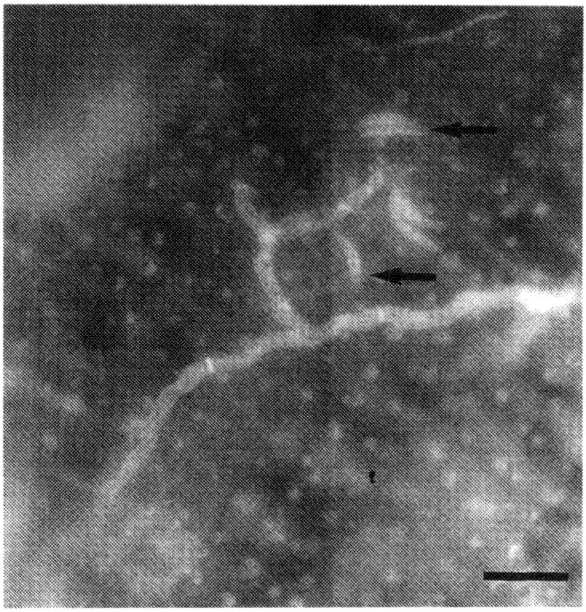

Figure 2. Formation of bitunicate conidia (arrows) at the leaf surface. This figure shows fungal material on the surface of a rice leave. The infected tissue was cleared in methanol, stained with fluorescent brightener and observed under UV illumination. Scale bar represents 20μm.

cultures in orange-pink gelatinous patches, particularly if the medium is supplemented with an aqueous extract of rice leaves. However, prolonged maintenance *in vitro* results in attenuation of the isolates' pathogenicity and sporulation capacity (Turner and Black, 1996). Production of perithecia outside host tissues has not generally been achieved. However, successful perithecial formation has been induced in cultured isolates, by crossing isolates on rice leaf sheaths placed on 1.7% agar containing Sach's medium, but whether the pathogen was homothallic or heterothallic could not be determined.

3. Sources of Infection

M. albescens is primarily a foliar pathogen, usually invading host tissues from the leaf tips and margins. There is some debate as to the major sources of inoculum in the field. A number of papers have been published describing the isolation of the pathogen from rice seeds and implicating seed transmission as an important mechanism. However, some reports suggest minimal incidence of *M. albescens* on seeds collected from heavily infected rice fields and that plants raised from infected seeds remained disease-free. Clearly this is an important point which requires further clarification if effective disease control, by seed-treatment or by improved agronomic practices, is to be developed.

The possibility that viable pathogen inoculum persists in crop detritus has been examined. It appears that viable inoculum persisted for no more than a few months on leaves buried in the soil, although this period would be long enough to provide inoculum where rice is cropped on a bimodal basis. However, scald is more commonly a problem in non-irrigated cropping systems where rice is grown on an annual basis. It has been reported that the pathogen did not survive the dry season (November - May) under field conditions, but survived in harvested seeds, stored at 26-28°C, for over a year. At NRI, we have found that viable isolates can be obtained from dried rice leaves at least a year after collection from the field (Turner and Black: unpublished data) although some loss of virulence may have occurred. Also, laboratory conditions for leaf preservation were far more favourable than those likely to prevail in the field.

Alternate hosts are another potential source of inoculum for this disease. Several host species have been identified amongst graminaceous weeds commonly found around the margins of rice fields, for example *Echinocloa crus-galli, Chloris* sp., *Eleusine* sp., *Melinis* sp., *Rhynchelitrum* sp., *Setaria* sp. and various *Oryza* spp. (Das, 1994). The transmissibility of *M. albescens* between these potential hosts and cultivated rice would bear further investigation, as would the phenotypic and genetic similarities between isolates found on rice and non-rice hosts.

4. The Infection Process

Conidia are transmitted by water-splash and germinate on the host surface. Singh and Gupta (1983) have shown that rice leaf exudates and extracts stimulate conidial germination and germ-tube growth. Interestingly, this stimulation was found to be more pronounced with extracts from susceptible as opposed to resistant rice cultivars.

The germ-tubes that emerge from the conidia often anastomose (Ou, 1985) and it has been observed that infection by anastomosis complexes was significantly more likely to be successful than infection originating from solitary conidia. Some host-derived carbohydrates are known to stimulate anastomosis in *M. albescens*.

Once the conidia have germinated, the hyphae start to spread over the surface of the rice leaf. We have found that, at the leading edge of the mycelium, the hyphae often run parallel to the longitudinal contours of the leaf surface (Figure 3a). Lateral branches develop from these exploratory hyphae. Where stomata are encountered, the hyphae form swollen, appressorial infection structures and penetrate the host through the stomatal apertures (Figure 3b). Behind the leading edge of the lesion, the mycelium continues to develop, forming a complex mass of inter-connected hyphae in which no preferred orientation can be discerned (Figure 3c).

It should be noted that, although stomatal penetration is the most frequent method of infection, *M. albescens* is also capable of direct penetration of host tissues. This is particularly the case for tissues lacking the heavy deposits of cuticular waxes common to leaf surfaces. Our own observations have confirmed that direct penetration

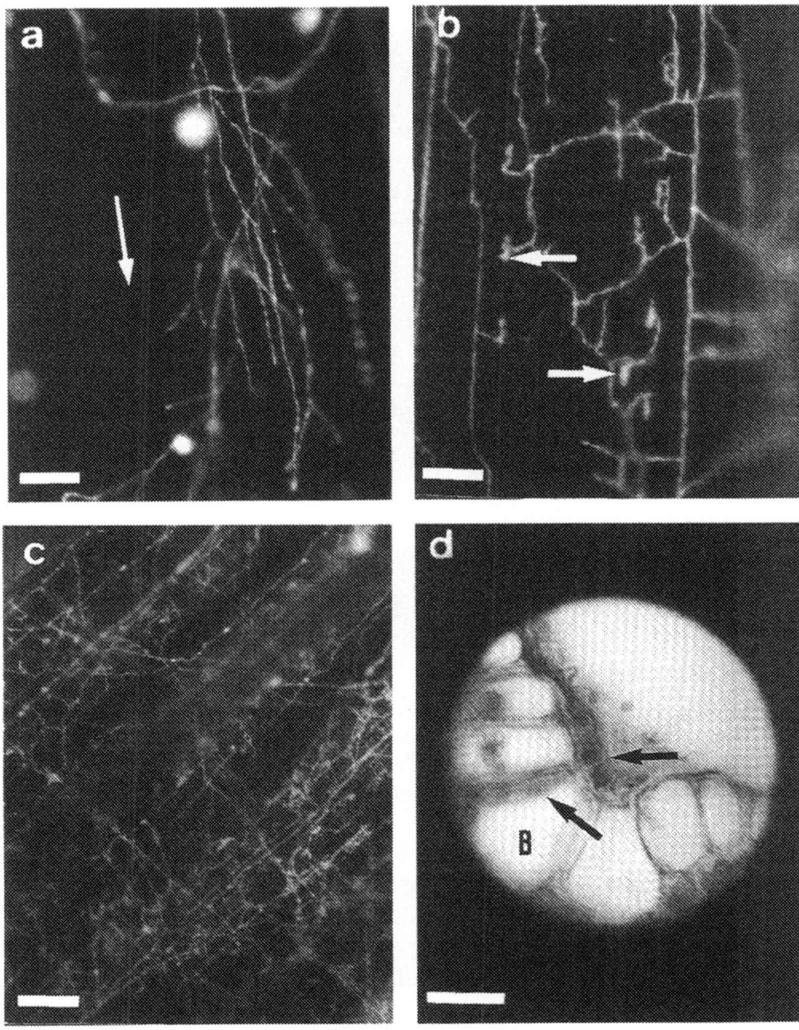

Figure 3. Stages of rice leaf infection by *M. albescens*. (a) Hyphae at the leading edge of a lesion show a preferred orientation that parallels the direction (arrowed) of the longitudinal ridges of the leaf surface. Scale bar represents 40μm. (b) Behind the leading edge of the mycelium, side-branches develop. Where stomata are encountered appressoria (arrow) are formed and the stomatal aperture penetrated. Scale bar represents 40μm. (c) Deeper within the lesion, proliferation of hyphal branching and anastomosis result in a complex hyphal mat over the leaf surface. Scale bar represents 20μm. (d) Direct penetration of the leaf surface, via the bulliform cells (B). Surface and internal hyphae indicated by arrows. Scale bar represents 10μm. (a) to (c) show surface fungal material on methanol-cleared rice leaves, stained with fluorescent brightener and observed under UV illumination. (d) shows a transverse section of a methanol-cleared, infected leaf stained with 0.05% toluidine blue.

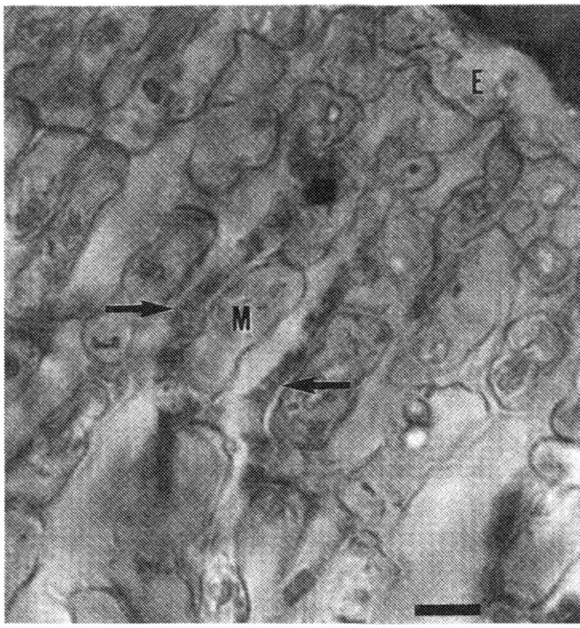

Figure 4. Transverse section of an infected rice leaf, showing the intercellular ramification of *M. albescens* hyphae (arrows) between the mesophyll cells (M). The epidermis (E) is visible at the top right. Infected leaf material was cleared in methanol, sectioned on a Bright Cryostat microtome and stained with 0.05% toluidine blue. Scale bar represents 5μm.

can occur, even into leaves with their heavily-waxed cuticles, notably via the bulliform cells (Figure 3d), although stomatal penetration was still the predominant method of entry.

Once the host leaf has been penetrated, the hyphae ramify between the mesophyll cells (Figure 4). This penetration and intercellular ramification begins in the green region of the leaf ahead of the expanding necrotic zone. Some penetration of the hosts mesophyll cells occurs as the intercellular network develops and the leaf tissues begin to die off. Phenolic compounds are deposited and it becomes harder to distinguish the fungal material present in this necrotic zone. Viable hyphae do persist, however, for it is in the necrotic zone that perithecia are formed (Figure 5).

M. albescens, grown in liquid medium (potato dextrose broth, supplemented with aqueous rice extract), produces one or more compounds that are toxic to rice plants, inducing chlorosis, wilting and necrosis (Singh and Gupta, 1983). This toxicity was found to be specific to rice plants, and may comprise an important mechanism contributing to the successful invasion of the host by the pathogen. Research within our own laboratory has demonstrated that conidia held in aqueous suspension for 48 hours can also release one or more host-specific toxic components into the water (Turner and Black, 1996) and that these components are dependent on light for their activity (Noor

Azman, 1994). Elucidation of the conditions for maximum toxin production and of the nature of the toxin(s), however, was hampered by loss of virulence in the pathogen, and by attenuation of toxin production, during maintenance of the pathogen in culture.

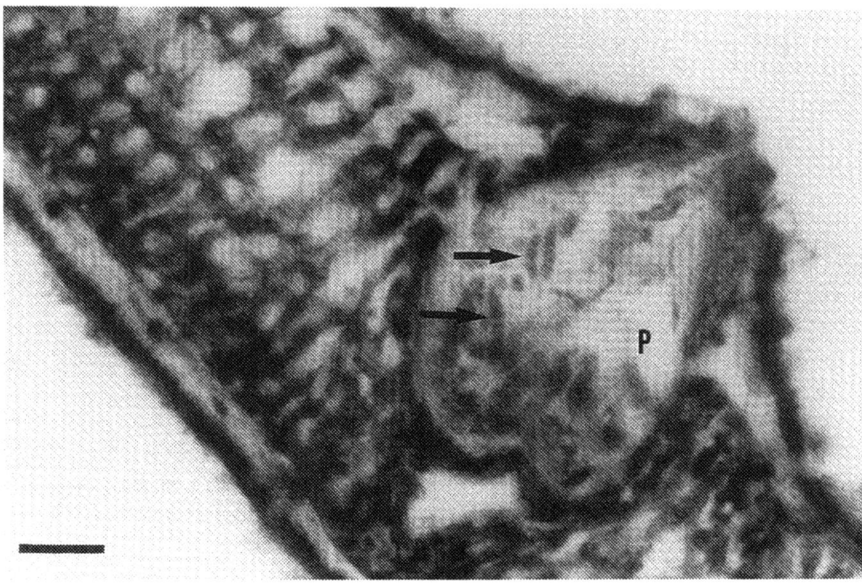

Figure 5. Longitudinal section of the necrotic zone of an *M. albescens* lesion, showing a perithecium (P) and asci (arrows). The infected leaf originated from Nigeria. The leaf material was cleared in methanol, sectioned on a Bright Cryostat microtome and stained with 0.05% toluidine blue. Scale bar represents 20μm.

5. Host Resistance

In a number of regions where leaf scald presents a significant constraint to rice production, screening of cultivars for resistance has been introduced. For the most part, this has been done through field-trials, usually using the International Rice Research Institute (IRRI) standard evaluation system to determine relative disease severity. The proportion of tested rice lines identified as showing high levels of resistance, according to these standards, has often been disappointingly low. For example, only 5 selections out of 437 showed high resistance, with another 33 selections showing moderate resistance in India. More recently, Saifulla (1993a) found only one scald-resistant entry out of 490 accessions tested at Karnal. Under lowland conditions in Indonesia, only 10 out of 100 cultivars showed good resistance. Saifulla (1993b) also reported an accession that showed relatively high levels of resistance to both scald and blast.

Ribeiro (1983) reported that the incidence of severe scald disease episodes has increased since the introduction of semi-dwarf rice varieties to Brazil. This is a worrying trend, given that improved introductions are commonly of semi-dwarf stock. However, these data also suggest that the traditional varieties may well prove to be a valuable source of scald-resistance traits that could be of use in future breeding programs, especially if these traits can be clearly identified.

Thus far, no specific mechanisms for host resistance have been identified, nor any genetic basis for resistance, making it difficult for breeding programs to fix resistance traits into agronomically useful rice lines. Generally speaking, even 'resistant' cultivars can be infected by *M. albescens*, but will suffer only limited lesion development and little if any loss of yield. Bonman *et al.* (1990) tested 288 rice accessions for resistance to *M. albescens* and found that, rather than distinct groupings of resistant, tolerant and susceptible accessions, there was a continuum from low to high levels of resistance, with no accessions showing complete resistance to the pathogen. This suggests that the interactions between *M. albescens* and rice plants is not dominated by major-gene resistance, unlike the resistance traits identified in barley cultivars against barley leaf scald, *Rhynchosporium secalis* (Graner and Tekauz, 1996; Schweizer *et al.*, 1995; Abbott *et al.*, 1995). Clearly, more work needs to be done if the relationship between *M. albescens* and its host is to be understood and this understanding used to improve the disease resistance of rice cultivars developed in the future.

It has been noted on several occasions that the application of nitrogenous fertiliser, or growth of rice on nitrogen-rich soils, results in a general increase in the susceptibility of rice plants to leaf scald (Das, 1994; Misra and Mathur, 1982; Mondal *et al.*, 1986; Singh and Gupta, 1983). Interestingly, Winslow (1992) reported that supplementing the levels of soluble silicon available to *indica* rice plants in Nigeria significantly reduced the severity of leaf scald disease. Improved resistance may have been conferred through increased deposition of silicon at the leaf surface. However, Winslow noted that upland *japonica* types, which had up to 100% higher levels of silicon in their leaves than the lowland *indica* types, proved to be equally susceptible to leaf scald, which argues against this hypothesis. Also, the propensity of *M. albescens* for entry into the leaf via stomata, rather than by direct penetration of the epidermis as effected by the blast pathogen, would seem to provide a ready method for bypassing such a structural defence.

6. Biological and Molecular Diversity of the Pathogen

The phenotypic and genotypic variability of *M. albescens* are areas that have been almost completely neglected to date. Below we have summarised the information currently available and outlined some of the work presently in progress to improve on this situation.

6.1. CULTURAL DIVERSITY

Singh and Gupta (1985) published an analysis of variation among six isolates of the pathogen obtained from various regions of India, noting that these isolates showed differences in total conidial biomass production, when grown in liquid culture. They also conducted some assessment of the isolates' pathogenicity, by spraying seedlings of a susceptible cultivar (Jaya) and a resistant cultivar (Dular) with conidial suspensions. Differences in pathogenicity were found to be more pronounced on the susceptible Jaya seedlings.

Parkinson (1980) described four main groups of fungal morphology for *M. albescens* isolates grown in culture, distinguished by the abundance of aerial hyphae and conidial masses produced and the intensity of the mycelial pigmentation. Fungal morphology *in vitro* should, however, be treated with caution. We have observed significant changes in the production of aerial hyphae and conidial masses by isolates held in culture over a period of several months. No relationship has been established between morphological type of an isolate in culture and its pathogenicity, but we have observed that isolates of *M. albescens* held in culture do suffer attenuation of pathogenicity over a period of several months which can be associated with a loss of sporulation capacity.

6.2. PATHOGENIC DIVERSITY

Laboratory experiments designed to assess the resistance of various cultivars to leaf scald disease have almost invariably utilised only one or two isolates of the pathogen. For example, Bonman *et al.* (1990) used an *in vitro* method, as well as whole-plant inoculation, to determine the relative susceptibility of 288 rice accessions to *M. albescens*. However, only two isolates, both from the Philippines, were used to challenge the host plants. Even then, some variation in the isolates' pathogenicity towards the various rice accessions was noted and led the authors to suggest that this was due to pathogenic specialisation. The authors also speculated that such specialisation may account for disparities within the literature reporting on the relative susceptibility of certain cultivars to leaf scald.

In collaboration with the West Africa Rice Development Association (WARDA), we have begun to assess the pathogenic variability of *M. albescens* isolates. *In vitro* analyses, following the detached leaf protocol described by Bonman *et al.* (1990), are being used to investigate the pathogenicity of selected West African isolates of *M. albescens* against a range of rice cultivars. Segments of fully expanded leaves from two month old rice plants are placed on 2% water agar plates, and inoculated with 3mm-diameter mycelial plugs taken from the margins of *M. albescens* cultures grown on PDA (Oxoid). The length of developing lesions is determined at 3 days after inoculation. Results of one such experiment, with 12 replicates per isolate/cultivar combination, are shown in Figure 6.

As can be seen in Figure 6, the pattern of virulence for each isolate towards the various rice cultivars is distinctive. For example, while both isolates 20.13 and 20.20 do well on cultivar ITA 306, and isolate 20.19 shows only intermediate lesion development, this situation changes when cultivar SIPI is used: both 20.13 and 20.19 do well and it is isolate 20.20 that shows limited lesion development. Characteristic patterns of pathogenicity, even within this small set of isolates, are becoming apparent. These findings go some way towards supporting the suggestion of pathogenic specialisation made by earlier workers (Singh and Gupta, 1985; Bonman *et al.*, 1990). The potential existence of *M. albescens* populations geographically distinguishable by pathogenic specialisation would have significant implications for local and international breeding programs aiming to incorporate scald-resistance traits into rice germplasm.

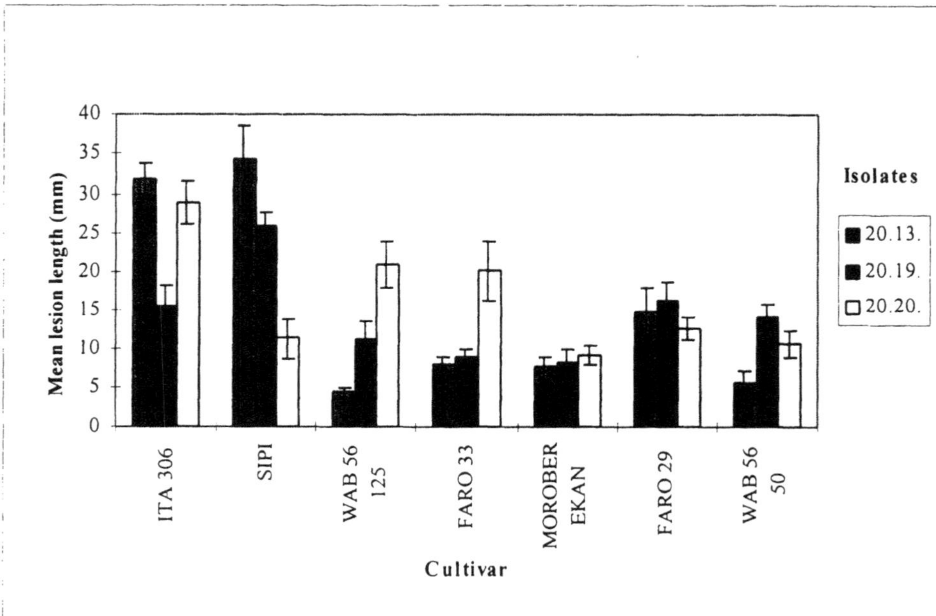

Figure 6. Mean lesion length (in mm +/- 95% confidence limits) at three days after inoculation of detached rice leaves with plugs of *M. albescens*, following the protocol of Bonman *et al.* (1990). Isolates 20.13, 20.19 and 20.20 were isolated from leaf material collected from a WARDA trial site in the Ivory Coast, and inoculated onto the rice varieties indicated right. Control replicates, where rice leaves were inoculated with PDA only, did not form lesions.

In vivo analyses are also in progress at NRI, in parallel with the detached-leaf protocol outlined above. Clip-inoculation of seedlings at the 5-leaf stage is carried out by cutting the tip off fully-expanded rice leaves with mycelium-coated scissors. The inoculated plants are then enclosed in a plastic bag, in order to maintain the high level of humidity required for consistent levels of infection. At 6 days after inoculation, the leaf tips are collected into methanol for clearing. After 24 hours, the leaf tips are stained with fluorescent brightener (Sigma Chemical Co.) and viewed under ultra-violet illumination with an Olympus BH-2 microscope. Where infection has been successful, the pathogen can be clearly seen growing over both abaxial and adaxial leaf surfaces. Although more labour-intensive, this method does more closely simulate field conditions, where whole-plant responses may be important to the plants' resistance to, or tolerance of, leaf scald infection.

6.3. GENETIC DIVERSITY

Given the limited research that has been directed towards phenotypic variability of *M. albescens*, it is not surprising that the genetic variability of this pathogen on different continents, or among different agro-ecological systems in a country, remains unknown. However, we have now begun studies, in conjunction with the phenotypic work outlined above, on genetic variability among and within *M. albescens* populations from S. E. Asia, West Africa and South America. Isolates of *M. albescens* obtained from diseased leaf material collected from these regions, with the assistance of IRRI, WARDA and the Centro Internationale de Agricultura Tropicale (CIAT) respectively,

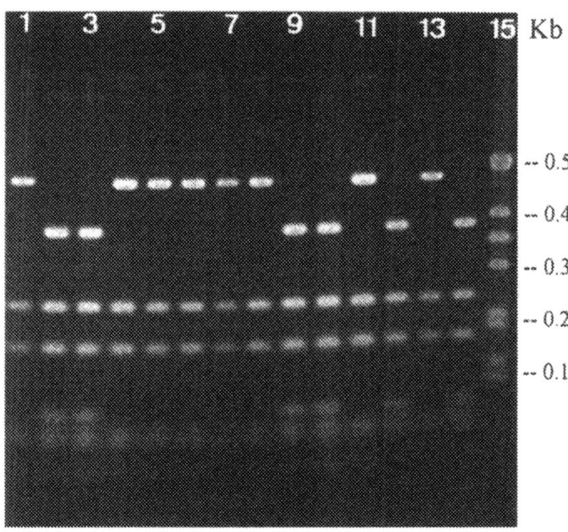

Figure 7. DNA amplified from the ITS region of the genome, using the polymerase chain reaction, was digested with *Hha* I. All isolates in this figure were isolated from material collected in the Ivory Coast. Isolates are as follows: Lane 1 = F37-3, lane 2 = DP2, lane 3 = MOR1, lane 4 = MOR2, lane 5 = SHIN2, lane 6 = F37-3b, lane 7 = W5650-2, lane 8 = DY3, lane 9 = IR5-3, lane 10 = IDSA6-1, lane 11 = IDSA6-2, lane 12 = MOR3, lane 13 = M22-1, lane 14 = NP2, lane 15 = molecular weight standard (Kb ladder, Gibco BRL).

were grown in potato dextrose broth and the DNA extracted according to a protocol adapted from that of Lodhi *et al.* (1994). A region of the ribosomal DNA (rDNA) was then amplified by polymerase chain reaction (PCR) using the primers ITS1 and R635. These primers amplify the variable internal transcribed regions 1 and 2, the 5.8S gene and part of the 28S gene. Aliquots of the amplified product were then digested with selected restriction endonucleases, including *Hinf* I, *Hae* III, *Rsa* I, *Hpa* II, *Taq* I and *Hha* I. The digestion fragments were separated on a 2% agarose gel, using 0.5x TBE buffer and the restriction fragment patterns were visualised on a UV trans-illuminator after staining the gel with ethidium bromide.

The region of rDNA analysed showed little variability. Isolates could be divided into two groups only, based on the restriction digest patterns from digestion with *Hha* I (Figure 7). Both groups were represented in isolate collections from each of the countries of origin examined. Sequencing of PCR amplified ITS region in selected isolates has shown that the two *Hha* I-derived AFLP groupings result from a single base variation within an *Hha* I restriction site in the second internal transcribed spacer region (ITS 2). Analysis of the rDNA-intergenic spacer (IGS) region of *M. albescens* is in progress at NRI and a greater degree of variability has already begun to emerge in this region. In time, we should be able to determine any correlation with geographic source or virulence of the isolates. Information on the genetic diversity of this pathogen will provide valuable clues to the history and biology of *M. albescens*.

7. Summary

M. albescens, the causal agent of the rice leaf scald disease, is an agronomically significant pathogen in many rice-growing regions of the world. Various authors have described the infection process, but many aspects of this process are still not clearly understood. For example, the contribution of host-specific toxins to the pathogen's success before and after penetration of the host is not known. Infected seeds, detritus from previous crops and alternate hosts have each been suggested as probable sources of primary inoculum under field conditions, but no consensus has been reached on their relative importance generally, or even under specific cropping systems. Field trials have indicated that some rice varieties have useful levels of resistance or tolerance to local populations of *M. albescens*, but the mechanisms of resistance have not been identified. The existence of pathogenic specialisation among *M. albescens* populations and even of physiological races has been suggested, but remains unconfirmed. Essentially, *M. albescens* is a little-studied pathogen, but work is in progress to better understand its epidemiology and genetic variability, with a view to developing more effective control of rice leaf scald.

8. Acknowledgements

This publication is an output from a research project funded by the Department for International Development of the United Kingdom who can accept no responsibility for any information provided or views expressed [DFID project code R6738; Renewable Natural Resources Research Strategy, Crop Protection Programme.]. The authors also wish to acknowledge Dr. A. Sy and Mr. M. K. Akator of the West African Rice Development Association for their assistance in collecting material and conducting infection assays; Dr. G. McKay of the Queen's University of Belfast for DNA sequence analyses.

9. References

Abbott, D. C., Lagudah, E. S. and Brown, A. H. D. (1995) Identification of RFLPs flanking a scald resistance gene on barley chromosome 6, *Journal of Heredity* **86**, 152-154.

Alam, M. S., John, V. T. and Zan, K. (1985) Insect pests and diseases of rice in Africa, in *Rice improvement in Eastern, Central and Southern Africa* (9-19 April 1984, Lusaka, Zambia), International Rice Research Institute, Manila, pp.67-82.

Bonman, J. M., Mackil, A. O. and Glaszmann, J. C. (1990) Resistance to *Gerlachia oryzae* in rice, *Plant Disease* **74**, 306-309.

Cuevas-Perez, F. and Gaona, J. S. (1988) Disease selection in rice in Colombia and Central America, *International Rice Research Newsletter* **13**, 14-15.

Das, S. R. (1994) Effect of nitrogenous fertilisers on leaf scald severity of rice, *Orissa Journal of Agricultural Research* **7**, 108-110.

Graner, A. and Tekauz, A. (1996) RFLP mapping in barley of a dominant gene conferring resistance to scald (*Rhynchosporium secalis*), *Theoretical and Applied Genetics* **93**, 421-425.

International Mycological Institute (1996) *Monographella albescens* (Thum.) V.O. Parkinson, Sivan. & C. Booth, *Distribution Maps of Plant Diseases*, Map No. 492, edition 3,CAB International, Wallingford, UK.

Jones, M. P., Jeutong, F. and Tchatchoua, J. (1991) Diseases of rice in Cameroon, *International Rice Research Newsletter* **16**, 19-20.

Kempf, D. (1983) Disease control, *Lavoura Arrozeira* **37**, 28-32.

Lodhi, M. A., Ye, G-N., Weeden, N. F. and Reisch, B. (1994) A simple and efficient method for DNA extraction from grapevine cultivars and *Vitis* species, *Plant Molecular Biology Reporter* **12**, 6-13.

Misra, A. K. and Mathur, S. C. (1982) Scald susceptibility of cultivars grown at different nitrogen levels, *International Rice Research Newsletter* **7**, 6.

Mondal, A. H., Ahmed, H. U. and Miah, S. A. (1986) Effect of nitrogen on the development of leaf scald disease of rice, *Bangladesh Journal of Botany* **15**, 213-215.

Ou S. H. (1985) *Rice Diseases*, CAB International, Wallingford, UK.

Ou, S. H., Nuque, F. L. and Vergel de Dios, T. I. (1978) Perfect stage of *Rhynchosporium oryzae* and the symptoms of rice leaf scald disease, *Plant Disease Reporter* **62**, 524-528.

Parkinson, V. O. (1980) Cultural characteristics of the rice leaf scald fungus, *Rhynchosporium oryzae*, *Transactions of the British Mycological Society* **74**, 509-514.

Parkinson, V. O., Sivanesan, A. and Booth, C. (1981) The perfect state of the rice leaf scald fungus and the taxonomy of both the perfect and imperfect states, *Transactions of the British Mycological society* **76**, 59-69.

Prabhu, A. S. and Sitarama-Prabhu, A. (1989) Control of the principal diseases of rice in dry lands, *Informe Agropecuario Belo Horizonte* **14**, 58-63.

Ribeiro, A. S. (1983) Occurrence of leaf scald on irrigated rice in Rio Grande do Sul, *Fitopatologia Brasileira* **8**, 195-199.

Saifulla, M. (1993a) Source of resistance in rice to blast, leaf scald and narrow brown leaf spot, *Agricultural Science Digest Karnal* **13**, 169-170.

Saifulla, M. (1993b). Screening of rice cultures against blast, brown spot and leaf scald diseases. *Agricultural Science Digest Karnal* **13**, 171-172.

Singh, S. A. and Gupta, P. K. S. (1983) Effect of nitrogen fertilisation on the severity of leaf scald of rice caused by *Rhynchosporium oryzae*, *Beitrage zur Tropischen Landwirtschaft und Veterinarmedizin* **21**, 89-93.

Singh, S. A. and Gupta, P. K. S. (1985) Variation among some isolates of *Rhynchosporium oryzae* - the rice leaf scald fungus, *Indian Phytopathology* **38**, 535-537.

Srinivasan, S. (1981) Yield losses due to leaf scald disease, *International Rice Research Newsletter* **6**, 15.

Schweizer, G. F., Baumer, M., Daniel, G., Rugel, H. and Roder, M. S. (1995) RFLP markers linked to scald (*Rhynchosporium secalis*) resistance gene Rh2 in barley, *Theoretical and Applied Genetics* **90**, 920-924.

Turner, H. C. and Black, R. (1996) *Rice scald (*Monographella albescens*): Host-pathogen relationships, pathogenicity and resistance*, Final Technical Report R5336/A0208, Chatham Maritime, UK, Natural Resources Institute.

Winslow, M. D. (1992) Silicon, disease resistance, and yield of rice genotypes under upland cultural conditions, *Crop Science* **32**, 1208-1213.

Zaimah Bt. Noor Azman, N. (1994) Investigation of pathogenicity of *Monographella albescens* towards rice and other Gramineae. MSc Thesis, Wye College, University of London.

FUNGAL DISEASES OF RICE IN GHANA: KNOWLEDGE AND RESEARCH NEEDS

S. K. NUTSUGAH[1] and J. K. TWUMASI[2]
[1] *Savanna Agricultural Research Institute (SARI), P. O. Box 52, Tamale, Ghana*
[2] *Crops Research Institute (CRI), P. O. Box 3785, Kumasi, Ghana.*

1. Introduction

Rice is an important food and cash crop in Ghana. Domestic consumption is rapidly increasing and likely to reach 762,000 mt of paddy equivalent in 2006. The current production barely exceeds 221, 300 mt (PPMED, 1995). Large scale mechanised rice production prevalent during the 1970s collapsed once Government subsidies were withdrawn. However, small scale rice farming is now on the raise. With the increase in consumption, the rice supply is more dependent on imports (170,000 mt/yr on average, i.e. $^2/_3$ of the requirement). The total production has indeed increased over the past 10 years, but is still far below domestic demand. In the inland valleys, the area cropped under water controlled system is currently estimated at less than 2,000 ha i.e. 2% of the cultivated area (MOFA & CFD Draft Report, 1997). Rice acreage increased from 76,100 to 99,900 ha between 1986 and 1995 with most of the cultivation under upland/lowland conditions (PPMED, 1995). While the scope for increasing the irrigated production is limited, there is clearly a large potential for developing rice in the inland valleys, which are presently under utilised, with only 99,900 ha of rice of a possible 400,000 ha.

Fungal diseases are one of the major production constraints in most rice-growing areas in Ghana. Some diseases are major and others are minor, some being location-specific. In Ghana, there is paucity of information on the major fungal diseases of rice. Rice research started in the early 1950s with the introduction of *Oryza sativa* species to replace low yielding *O. glaberrima* species in Ghana (Dekuku *et al.*, 1986). Rainfed upland/lowland rice research is carried out in Nyankpala with additional testing sites at Salaga, Yendi and Manga (all in the Northern Savanna zone where about 70% of Ghana's rice is cultivated). Irrigated rice research is conducted at Kpong Agricultural Research Station of the University of Ghana in the Accra plains (Dekuku *et al.*, 1986), while in-land valley rice research is carried out at the Crops Research Institute's sites at Aframso (forest-savanna transition zone) and Besease (deciduous rain forest ecology) in Ashanti Region. Rice producers in and around Nyankpala have suffered major losses and even total crop failure to leaf blast pathogen, *Magnaporthe grisea* (anamorph *Pyricularia grisea*) (NAES Annual Report, 1985). The same disease devastated a

S. Sreenivasaprasad and R. Johnson (eds.), Major Fungal diseases of Rice Recent Advances, 321-329.
Prof. K. Manibhushan Rao Festschrift

research field at Hohoe (Volta Region) in 1997 (Twumasi, unpublished data). In 1996, a preliminary survey of diseases/pests of upland/hydromorphic rice was conducted in some rice-growing areas in northern Ghana. Table 1 presents a summary of the relative importance of the most common diseases of rice. Research work on the control of these diseases continues under the sponsorship of the National Agricultural Research Project (NARP) Rice Programme at Savanna Agricultural Research Institute (SARI), Nyankpala and Crops Research Institute (CRI), Kumasi. Work is in progress on protecting rice against pests and diseases, but there are still challenges ahead in gaining increased understanding of rice diseases and in devising methods to control or minimise their effects on the crop.

TABLE 1. Relative occurrence and severity of some of the most prevalent fungal diseases of rice in Ghana[1]

Disease	Relative severity[2]
Brown spot	3+++
Blast	2+++
Narrow brown leaf spot	2+++
Grain discoloration	2+++
Leaf scald	2++
False smut	1+
Sheath blight	1+
Seedling blight	1+

[1] Adapted from Nutsugah, 1997 a,b
[2] Coding for disease severity and occurrence is as follows:
1 = of minor importance, 2 = of moderate importance,
3 = of major importance, + = occasionally present (sporadic),
++ = commonly present, +++ = always present.

In this chapter, fungal diseases of current and potential importance on rice in Ghana are discussed and the research needs identified. The objective is to initiate and enhance international collaborative research on themes which are of relevance to farmers.

2. Present knowledge

A systematic survey was conducted in farmers' fields and experimental plots to determine the prevalence of rice diseases in northern Ghana during the 1996-97 rainy seasons. The results obtained (Nutsugah, 1997a,b) revealed that fungi cause more diseases of rice in Ghana than any other group of pathogens. Similar surveys conducted in southern Ghana (Twumasi and Adu-Tutu, 1995; Twumasi, 1996) appear to support this view. However, many rice diseases occur sporadically causing severe local damage but having little impact on national production. In this paper, diseases caused by fungi are grouped into foliar, grain and seedling diseases.

2.1. FOLIAR DISEASES

The principal foliar diseases occurring on rice in Ghana are brown spot, blast, narrow brown leaf spot, leaf scald and sheath blight. Under the NARP rice protection programme, three research activities were initiated during 1996 rainy season. The objectives of these activities were to (1) characterise the genetic diversity of rice blast and scald pathogens, (2) conduct field level varietal screening evaluations of selected germplasm/lines against major diseases such as brown spot, rice blast, narrow brown leaf spot and leaf scald and (3) define the extent of disease incidence in different types of rice-growing ecologies to determine potential interaction of environmental and cultural conditions on disease.

Initially, an assessment of the disease incidence and severity was made in different rice-growing areas. Collections of the rice blast and scald pathogens were obtained from the different areas, preserved and mailed to the UK for comparative studies of the molecular genetic components with known isolates. These activities began during August-October 1996 and repeated in 1997 rainy season. The survey locations included Bontanga, Kadia, Kpalsugu, Kpinginga, Kukpehi, Loagri, Nabogu, Nasia, Nyankpala, Salaga, Savelugu, Taha, Tamale and Yepeligu (Northern Region), Bolgatanga, Fumbisi, Manga, Navrongo and Tono (Upper East Region) (Table 2). In southern Ghana, the locations included Nobewam, Besease, Kwadaso, Aframso, Adugyama (Ashanti Region), Kobina-Annor Krom, Wasa Simpa (Western Region), Dawhenya (Greater Accra Region), Hohoe and Afife (Volta Region).

Assembly of germplasm and lines for evaluation was done in April-May with subsequent seeding in June-July. Disease incidence observations were made during the crop progression with emphasis on hot spots for disease. For the characterisation of populations of Ghanaian type rice blast and scald, rice entries were evaluated in microplots. Entries were bordered with spreader rows to ensure disease development. Disease levels were recorded at least once a week and leaf blast and scald samples were collected, preserved and mailed to CABI Bioscience, Horticultural Research International and Natural Resources Institute, UK for characterisation.

The important diseases recorded during the surveys were brown spot, leaf blast, leaf scald, glume discoloration and sheath blight. The distribution and severity of fungal diseases of rice in northern Ghana are summarised in Table 2. Disease incidence and severity arising from natural infection in the fields varied markedly among the different rice-growing areas surveyed in the regions. Brown spot was the most widely prevalent disease in the areas surveyed. The incidence was, however, more apparent in the Northern Region (NR) than in the Upper East Region (UER) (Table 2). It wiped out a whole field belonging to some co-operative farmers at Aframso. Leaf scald was quite widespread at Nyankpala. The disease severity reached an alarming proportion at Nyankpala to the extent that a promising rice variety (Mashuri) proposed for on-farm evaluation had to be dropped from trials. Leaf blast occurrence during the 1996 rainy season was observed to be far lower than leaf scald and brown spot. However, observations revealed that there were leaf blast, nodal blast, neck blast and panicle blast, suggesting that the disease is a potential constraint to rice production in Ghana. Sheath blight was detected at three locations, Nyankpala and Yepeligu in the NR and at Dawhenya in the Greater Accra Region.

TABLE 2. Distribution and severity of fungal diseases of rice in northern Ghana

Location	Disease						
	Brown spot	Leaf blast	Narrow brown leaf spot	Leaf scald	Glume discoloration	False smut	Sheath blight
Northern Region							
Bontanga	L[1]	L	-[2]	L	-	-	-
Golinga	L	L	-	M	-	-	-
Kadia	L	L	-	-	-	-	-
Kpalsugu	L	L	-	L	-	-	-
Kpinginga	M	L	-	L	-	-	-
Kukpehi	L	L	-	M	-	-	-
Loagri	M/H	L	M/H	0	L/M	0	0
Nabogu	L	L	L	0	L	0	0
Nasia	M	M	M	0	L	0	0
Nyankpala	H	M	M	M/H	M	L	L
Salaga	L	0	M	0	L	L	0
Savelugu	0	0	0	L/M	L	0	0
Taha	L	L	0	0	L	0	0
Yepeligu	L	L	L	L	L	0	L
Upper East Region							
Bolgatanga	M/H	M/H	M/H	L	L	0	0
Fumbisi	-	-	-	-	-	-	-
Manga	M	L	M	L	L	0	0
Navrongo	M	L/M	M	M	M	0	0
Tono	M	M	0	M	L	0	0

[1] L = Low (damage less than 10% of affected leaf area or infected florets or florets with discoloured- glumes),
M = Moderate (damage between 10 and 50% of affected leaf area or infected florets or florets with discoloured glumes),
H = High (damage above 50% of affected leaf area or infected florets or florets with discoloured glumes),
0 = Disease not recorded,
[2] Data not collected.

The study of genetic and phenotypic variability of *Magnaporthe grisea* and *Monographella albescens* populations within Ghana is on-going. MGR 586 fingerprinting of the *M. grisea* isolates and identification of the lineages are in progress. Limited molecular variability was observed in the ITS region of the *M. albescens* genome and preliminary data indicated a higher level of variability in the IGS region (Banniza and Holderness, 1997; Sreenivasaprasad and Chipili, 1997; Turner, 1997, 1998; Chipili *et al.*, 1998; Sreenivasaprasd *et al.*, 1998).

Incidence and severity of leaf blast, brown spot and leaf scald on 15 rice lines screened at Nyankpala during 1996 rainy season are shown in Table 3. The disease pressure in the blast nursery was generally low. In 1997, the disease pressure in the blast nurseries can be described as moderate to high with the exception of leaf scald and false smut (Table 4, 5 and 6). Across the three screening sites, narrow brown leaf spot seemed to be the most frequently occurring disease followed by brown spot, glume discoloration and leaf scald in that order. Apart from entries Empsac 102 and CT 9506-48-1-1R which had a mean score of 4 across the locations, some of the entries screened performed better than the check, GR18, GR19 and GR21 against the disease recorded in the nurseries. Rice variety GR 18 exhibited stable resistance across the screening sites. Furthermore, entries SIK 55-263-2, IR 64, Centa A1 and BD 2 showed multiple resistance against the major fungal diseases. It is worthwhile to further evaluate the performance of these accessions in advanced agronomic trials.

TABLE 3. Major diseases of rice accessions screened in disease observation nursery at Nyankpala, 1996 rainy season

	Score (1 – 9)[1]		
Entry	Leaf blast	Brown spot	Leaf scald
CT 8837-1-17-1P			
Bouake 189	1	3	2
SIK 55-263-2	3	4	2
IR 64	1	1	2
Centa A1	2	4	2
BD 2	1	2	2
CT 9799-9-2-1-1	2	2	2
Empasc 102	3	3	2
CT 9506-48-1-1P	1	2	3
PNA 1126F_4-150	1	2	3
GR 19	2	2	2
GR 21	2	2	1
GRUG-7	2	2	3
SIK 55-263-5	1	2	3
GR18	1	1	2
Mean[2] (15 Entries)	1.6	2.2	2.1
LSD (0.05)	0.8	1.2	1.3
CV (%)	36	38	43

[1]Visual scores of disease incidence and severity were recorded using the IRRI standard evaluation system,
[2] Mean of four replicates.

2.2. GRAIN DISEASES

The grain diseases prevalent in Ghana are grain discoloration and false smut. Grain discoloration occurs in all the rice-growing areas in the country. Seed mycoflora studies conducted in Southern Ghana on 19 discolored paddy rice varieties during 1993 rainy season revealed the presence of a large number of fungi (Twumasi, 1993, 1995). These include *Aspergillus flavus, A. fumigatus, A. terreus, Alternaria padwickii, Cladosporium herbarum, Curvularia lunata, C. inaequalis, C. pallescens, Cochliobolus miyabeanus, Fusarium moniliforme, F. semitectum, F. solani, F. dimerum, Nigrospora oryzae, Penicillium* sp., *Phoma sorghina, Rhizopus* sp., *Trichoderma* sp. and *Trichoconis padwickii* (Table 7). Addison (1971) had earlier encountered several seedbone fungi on one rice variety. These fungi are possibly responsible for the panicle disease complex symptoms observed on the grains. They were found to affect the seed germination adversely (Twumasi, 1995). Poor stand due to low germination is a major problem in all the rice-growing areas. Grain mold probably poses a potential problem for seed production, but does not appear to be a limiting factor for farmers. False smut caused by *Ustilaginoidea virens* was reported at two screening sites (Nyankpala and Salaga) in northern Ghana during the 1997 rainy season. In a survey in the Volta Region of Ghana (Twumasi, 1996), heavy false smut incidence was observed (GRUG-7 variety) in a co-operative farmers' field. In general, it is a disease of minor importance.

TABLE 4. Major diseases of rice accessions screened in disease observation nursery at Manga, 1997 rainy season

	Score (1 – 9)[1]			
Entry	Brown spot	Narrow brown leaf spot	Glume Discoloration	Leaf scald
CT 8837-1-17-1P	5.0	3.0	3.0	0.0
Bouake 189	4.0	6.0	3.0	0.0
SIK 55-263-2	4.0	6.0	3.0	2.0
IR 64	5.0	5.0	2.0	0.0
Centa A1	4.0	4.0	2.0	0.0
BD 2	4.0	5.0	2.0	0.0
CT 9799-9-2-1-1	4.0	6.0	2.0	1.0
Empasc 102	7.0	8.0	6.0	1.0
CT 9506-48-1-1P	6.0	7.0	5.0	1.0
PNA 1126F_4-150	3.0	6.0	3.0	0.0
GR 19	5.0	6.0	3.0	0.0
GR 21	5.0	6.0	4.0	0.0
GRUG-7	5.0	6.0	5.0	0.0
SIK 55-263-5	4.0	6.0	2.0	1.0
GR18	3.0	4.0	2.0	0.0
Mean[2] (15 Entries)	4.5	5.6	3.1	0.4
CV (%)	34.3	29.6	50.4	-

[1] Visual scores of disease incidence and severity were recorded using the IRRI standard evaluation system,
[2] Mean of four replicates.

2.3. SEEDLING BLIGHT

Seedling blight or damping-off is a disease complex caused by several seedborne pathogens. *Alternaria padwickii, Cochliobolus miyabeanus, Pyricularia grisea* and *Fusarium moniliforme* are among the seedborne fungi causing reduced germination, seed rotting, lesions on roots and coleoptiles, stunting, yellowing and seedling blighting in water-seeded rice. These fungi also cause foliage diseases and/or grain discoloration as mentioned above.

TABLE 5. Major diseases of rice accessions screened in disease observation nursery at Nyankpala, 1997 rainy season

	Score (1 – 9)[1]			
Entry	Brown spot	Narrow brown leaf spot	Glume Discoloration	Leaf scald
CT 8837-1-17-1P	3.0	4.0	4.0	0.0
Bouake 189	4.0	3.0	4.0	0.0
SIK 55-263-2	3.0	2.0	3.0	1.0
IR 64	3.0	2.0	2.0	0.0
Centa A1	3.0	2.0	2.0	0.0
BD 2	2.0	3.0	4.0	0.0
CT 9799-9-2-1-1	4.0	3.0	4.0	1.0
Empasc 102	3.0	4.0	5.0	0.0
CT 9506-48-1-1P	3.0	4.0	6.0	0.0
PNA 1126F$_4$-150	3.0	4.0	6.0	0.0
GR 19	4.0	3.0	6.0	0.0
GR 21	4.0	4.0	7.0	1.0
GRUG-7	3.0	4.0	6.0	0.0
SIK 55-263-5	3.0	3.0	5.0	0.0
GR18	3.0	3.0	3.0	0.0
Mean[2] (15 Entries)	3.2	3.2	4.5	0.2
CV (%)	36.8	40.8	37.3	-

[1] Visual scores of disease incidence and severity were recorded using the IRRI standard evaluation system
[2] Mean of four replicates.

TABLE 6. Major diseases of rice accessions screened in disease observation nursery at Salaga, 1997 rainy season

	Score (1 – 9)[1]			
	Brown spot	Narrow brown leaf spot	Glume Discoloration	Leaf scald
CT 8837-1-17-1P	3.0	4.0	4.0	0.0
Bouake 189	3.0	3.0	2.0	0.0
SIK 55-263-2	4.0	6.0	4.0	1.0
IR 64	5.0	5.0	3.0	0.0
Centa A1	5.0	6.0	4.0	0.0
BD 2	4.0	5.0	4.0	0.0
CT 9799-9-2-1-1	4.0	5.0	3.0	2.0
Empasc 102	5.0	5.0	4.0	1.0
CT 9506-48-1-1P	4.0	5.0	5.0	1.0
PNA 1126F$_4$-150	5.0	5.0	5.0	2.0
GR 19	5.0	3.0	3.0	0.0
GR 21	4.0	4.0	4.0	0.0
GRUG-7	5.0	4.0	4.0	0.0
SIK 55-263-5	4.0	6.0	5.0	1.0
GR18	5.0	4.0	4.0	0.0
Mean[2] (15 Entries)	4.3	4.7	3.9	0.5
CV (%)	41.6	36.9	41.5	-

[1] Visual scores of disease incidence and severity were recorded using the IRRI standard evaluation system,
[2] Mean of four replicates.

TABLE 7. Isolation of seedborne fungi from rice accessions at Aframso, 1993 rainy season

Variety	Mean score (1 – 9)[1]	Microorganisms isolated
Tox 3108-56-4-2-2-2	1.0	*Nigrospora oryzae*
BR 153-23-10-1-3	2.0	*N. oryzae*
IR 19660-46-1-3-2-2	0	-
AR 43	0	-
Tox 3142-1-1-3-1	5.0	*Cochliobolus miyabeanus, Pyricularia grisea*
AR 50	7.0	*Co. miyabeanus, P. grisea, N. oryzae, Curvularia pallescens*
AR 70	1.0	*C. miyabeanus, P. grisea*
ITA 239	4.0	*N. oryzae, Fusarium solani, P. grisea, Rhizoctonia solani, N. oryzae, Co. miyabeanus*
AR 31	0	-
MR. MORE	8.0	*P. grisea, R. solani, N.oryzae, Co. miyabeanus, Cu. pallescens, F. solani, F. moniliforme, Aspergillus flavus, A. fumigatus, A. terreus, Chaetomium* sp., *Penicillium* sp.
AR 41	5.0	*Co. miyabeanus, F. solani, Cu. pallescens*
AR 725	0	-
AR 48	0	-
IR 28128-45-3-3-2	3.0	*N. oryzae, F. solani*
AR 47	4.0	*Co. miyabeanus, N. oryzae, F. solani*
AR 45	6.0	*Co. miyabeanus, N. oryzae, F. solani*
AR 10	1.0	*Co. miyabeanus, N. oryzae, Cu. pallescens , F. solani*
AR 52	2.0	*N. oryzae*
AR 49	2.0	*Co. miyabeanus*

[1] The score was computed from the mean percentage frequency of isolation from each variety.

3. Conclusion

In conclusion, brown spot, narrow brown leaf spot and grain discoloration were prevalent in all the rice-growing areas of Ghana. Other fungal diseases encountered were leaf blast, leaf scald and false smut. These foliar diseases occur together in many production systems. Variations in incidence and severity are also apparent in he different rice-growing locations. It is, therefore, important that programmes to develop cultivars with resistance to the major fungal diseases consider other factors including the environmental conditions and cultivation practices in the area of intended cultivation. The following research needs have been identified 1) undertake a survey of diseases and pests of upland/hydromorphic rice in northern Ghana and study the epidemiology of airborne spores of fungal pathogens of rice, 2)characterisation of key rice resistance screening sites in Ghana and developing varieties for pest and disease resistance and 3) development of integrated disease/pest management strategies and sustainable production of rice in northern Ghana.

4. References

Addison, E. A. (1971) Seedborne fungi of *Oryza sativa* in Ghana, *Ghana Journal of Agricultural Science*, **4**, 87-88.

Banniza, S. and Holderness, M. (1997) *Identification and Characterization of key screening sites for blast and scald resistance in West Africa* (X0357), Progress report (October 1996-August 1997).

Chipili, J., Sreenivasaprasad, S., Sere, Y. and Talbot, N. J. (1998) Genetic diversity of the rice blast pathogen population in West Africa. Abstract, *Proceedings of the 2nd International Rice Blast Conference*, Montpellier, France, August 4-8, 1998.

Dekuku, R. C., Oteng, J. W. and Papastaurou, A. L. (1986) Rice improvement in Ghana, Paper presented at *the Joint Rice Monitoring and Evaluation Workshop in Tanzania*, March 18-22, 1996.

Ministry of Food and Agriculture (MOFA) and Caisse Francaise de Developpment (CFD) (1997) *Feasibility study of a rice development project in the Northern Region in Ghana*, Draft Report.

Nutsugah, S. K. (1997a) Survey of rice diseases/pests in northern Ghana: Initiation of cooperative research programme for disease/pest management, in *Proceedings of the 6th annual WARDA IPM-TF meeting*, March 2-5, 1997, Bouaké, Cote d'Ivoire.

Nutsugah, S. K. (1997b) Survey of diseases/pests of upland/hydromorphic rice in northern Ghana, in *Proceedings of the SARI annual in-house review*, March 24-26, 1997, Nyankpala, Ghana.

Nyankpala Agricultural Experimental Station (1985) *Annual Report*.

Policy, Planning, Monitoring and Evaluation Department (PPMED) (1995) Statistics Division, Ministry of Food and Agriculture, Ghana.

Sreenivasaprasad, S. and Chipili, J. (1997) *Identification and characterization of key screening sites for blast and scald resistance in West Africa* (X0356), Progress report (October 1996-August 1997).

Sreenivasaprasad, S., Chipili, J. and Sere, Y. (1998) Molecular characterization of the blast pathogen *Magnaporthe grisea* from some West Africa screening sites, Abstract, *Proceedings of the 7th International Congress of Plant Pathology*, Edinburgh, UK, August 9-16, 1998.

Turner, H. C. (1997) Identification and characterization of key screening sites for blast and scald resistance in West Africa (A0578), Progress report (October 1996-August 1997).

Turner, H. C. (1998) Molecular variability of the rice leaf scald pathogen, *Monographella albescens*, Abstract, *Proceedings of the 7th International Congress of Plant Pathology*, Edinburgh, UK, August 9-16, 1998.

Twumasi, J. K. (1993) Studies on rice diseases in Southern Ghana, in *Proceedings of the CRI annual in-house review*, February 1993, Fumesua, Kumasi, Ghana.

Twumasi, J. K. (1995) The grain discoloration syndrome of rice in Ghana, in *Proceedings of the 4th annual WARDA IPM Task Force meeting*, March 15 – 17, 1995, Bouake, Cote d'Ivoire.

Twumasi, J. K. and Adu-tutu, K. O.(1995) Weeds and Fungal diseases of rice in selected in-land valleys in Ghana, in *Proceedings of the 4th annual WARDA IPM Task Force meeting*, March 15 – 17, 1995, Bouaké, Cote d'Ivoire.,

Twumasi, J. K. (1996) Survey on rice diseases in Southern Ghana, in *Proceedings of the 5th WARDA IPM Task Force meeting*, March 20 – 22, 1996, Bouaké, Cote d'Ivoire..

SEED-BORNE FUNGI OF RICE AND QUARANTINE SIGNIFICANCE

O. R. REDDY and N. SATHYANARAYANA
Plant Quarantine Station, Directorate of Plant Protection,
Ministry of Agriculture, Chennai- 600 001, India.

1. Introduction

Rice (*Oryza sativa* L) is the most widely grown cereal cultivated in more than 147 million hectares worldwide. Rice seed is known to be colonised by various types of fungi, among which many are identified as plant pathogens. The infected seed not only affects the quality of the grain but also spreads the seed-borne pathogens to different regions. Rice is known to be attacked by 56 fungal pathogens (Ou, 1985), of which 41 are reported to be seed-borne (Richardson, 1979,1981). Wei (1957) reported a number of fungal species on rice responsible for grain discoloration. Sevilla and Mamicpic (1988) reported 70 fungi as seed-borne on rice in the Philippines. A comprehensive list of fungi associated with rice seed is given in Table 1.

The seed-borne fungi have been grouped into different categories by many workers as field and storage fungi (Christensen, 1957), obligate parasites, facultative saprophytes and facultative parasites or saprophytes and parasites (Kulwant Singh and Mukerji, 1994). In this review the emphasis was given to fungi associated with rice seed. Most fungi identified on rice are either i) pathogenic fungi (causing characteristic disease symptoms) or ii) deleterious fungi (normally do not cause any disease but affect rice quality in storage, seed germination and cause seed rot).

Seeds are the best carriers of several pathogens which are responsible for most plant diseases leading to considerable loss of crop yield (Neergaard, 1977). Rice crop is known to be attacked by many seed-borne fungal diseases of major and minor importance. Some of these are widely distributed causing significant yield as well as economic loss particularly blast, brown spot, bakane diseases. A few have restricted distribution and are of minor importance, but becoming increasingly known in recent years (leaf scald, sheath rot, false smut and kernel smut).

2. Pathogenic fungi

Considerable progress has been made in recent years in furthering the knowledge on seed-borne fungal pathogens of rice. Most important diseases of this cereal crop are summarised.

S. Sreenivasaprasad and R. Johnson (eds.), Major Fungal diseases of Rice Recent Advances, 331-345.
Prof. K. Manibhushan Rao Festschrift

TABLE 1. Seed-borne fungi recorded from rice (*Oryza sativa* L.)

Name of the fungi	PF/DF	Reference
Acremoniella atra	PF	Wu and Dow (1993)
Alternaria sp.	DF	Sevilla and Mamicpic (1988)
A. alternata (A. tenuis)	PF	Koroleva *et al.* (1984)
A. longissima	PF	Richardson (1979, 1981)
A. padwickii (Trichoconiella padwickii)	PF	Godfrey (1920)
A. porri	DF	Sevilla and Mamicpic (1988)
Arthrinium phaeospermum	DF	Ramesh and Tiwari (1995)
Arthrobotrys sp.	DF	Sevilla and Mamicpic (1988)
Ascochyta oryzae	PF	Kanjanasoon (1962)
Aspergillus candidus	DF	Udagawa (1959)
A. flavus	DF	Hyun (1963)
A. fumigatus	DF	Koroleva *et al.* (1984)
A. glaucus	DF	Udagawa (1959)
A. niger	DF	Del Prado and Christensen (1952)
A. repens	DF	Hyun (1963)
A. restrictus	DF	Hyun (1963)
A. sydowii	DF	Mazen *et al.* (1993)
A. tamarii	DF	Mazen *et al.* (1993)
A. terreus	DF	Del Prado and Christensen (1952)
A. versicolor	DF	Udagawa (1959)
Bipolaris oryzae (Drechslera oryzae; C. miyabeanus)	PF	Breda de Haan (1900)
B. sorokiniana (Cochliobolus sativus)	PF	Nyvall *et al.* (1995)
Botryodiplodia palmarum	DF	Sevilla and Mamicpic (1988)
Brachysporium spp.	PF	Richardson (1979, 1981)
Cephalosporium sp.	DF	Sevilla and Mamicpic (1988)
C. gramineum (Hymenula cerealis)	DF	Kim (1992)
Cercospora sp.	PF	Sevilla and Mamicpic (1988)
C. janseana (C. oryzae; Sphaerulia oryzae)	PF	Miyake (1910)
Chaetomium sp.	DF	Sevilla and Mamicpic (1988)
C. globosum	DF	Khan *et al.* (1988)
Cladosporium sp.	DF	Sevilla and Mamicpic (1988)
C. cladosporoides	DF	Kim (1992)
C. herbarum	DF	Tullis (1936)
Colletotrichum sp.	DF	Sevilla and Mamicpic (1988)
C. dematium	DF	Sevilla and Mamicpic (1988)
C. gloeosporioides	DF	Sevilla and Mamicpic (1988)
C. lindemuthianum	DF	Sevilla and Mamicpic (1988)
Corynespora sp.	DF	Sevilla and Mamicpic (1988)
Curvularia spp.	PF	Rao and Salam (1954)
C. affinis	PF	Richardson (1979, 1981)
C. clavata	PF	Jin (1989)
C. cymbopogonis	PF	Sevilla and Mamicpic (1988)
C. eragrostidis (Cochliobolus eragrostidis)	PF	Jayaweera *et al.* (1988)
C. fallax	PF	Jin (1989)
C. geniculata (Cochliobolus geniculatus)	PF	Jin (1989)
C. inequalis	PF	Jin (1989)
C. intermedia	PF	Jin (1989)

C. lunata (Cochliobolus lunatus)	PF	Soave *et al.* (1983)
C. maculans	PF	Chevaugeon and Ravise (1957)
C. oryzae	PF	Sevilla and Mamicpic (1988)
C. pallescens (Cochliobolus pallescens)	PF	Jayaweera *et al.* (1988)
C. protuberata	PF	Sisterna *et al.* (1994)
C. trifolii	PF	Sevilla and Mamicpic (1988)
C. tuberculata	PF	Jain (1962)
C. verruculosa	PF	Jayaweera *et al.* (1988)
Dichotomophthoropsis nymphacearum	DF	Agarwal *et al.* (1989)
Diplodia oryzae	PF	Richardson (1979, 1981)
Drechslera sp.	PF	Sevilla and Mamicpic (1988)
D. australiensis	PF	Richardson (1979, 1981)
D. halodes	PF	Richardson (1979, 1981)
D. hawaiiensis	PF	Richardson (1979, 1981)
D. holmii	PF	Sharma and Siddiqui (1978b)
D. longirostrata	PF	Richardson (1979 & 1981)
D. maydis	PF	Mew *et al.* (1988)
D. neergardii	PF	Richardson (1979, 1981)
D. poae	PF	Wu and Dow (1993)
D. rostrata	PF	Mew *et al.* (1988)
D. sorokiniana	PF	Mew *et al.* 1988
D. specifera (Cochliobolus specier)	PF	Reddy (1984)
D. tetramera	PF	Sevilla and Mamicpic (1988)
Ephelis oryzae (E. pallida, Balansia oryzae)	PF	Sydow (1914)
Epicoccum spp.	PF	Richardson (1979, 1981)
E. hyalopes	PF	Kimura (1937)
E. nigrum	PF	Fisher and Petrini (1992)
E. purpurascens	PF	Soave *et al.* (1983)
Fusarium avenaceum	DF	Sevilla and Mamicpic (1988)
F. camptoceras	PF	Sharma and Siddiqui (1978b)
F. concolor	DF	Sevilla and Mamicpic (1988)
F. culmorum	DF	Sevilla and Mamicpic (1988)
F. equiseti	PF	Fisher and Petrini (1992)
F. graminearum (Gibberella zeae)	PF	Hemmi *et al.* (1931)
F. lateritium (Gibberella baccata)	DF	Wu and Dow (1993)
F. longipes	DF	Sevilla and Mamicpic (1988)
F. moniliforme (Gibberella fujikuroi)	PF	Ito and Kimura (1931)
F. nivale	DF	Chevaugeon and Ravise (1957)
F. oxysporum	PF	Koroleva *et al.* (1984)
F. pallidoroseum	PF	Bora *et al.* (1993)
F. poae	DF	Sevilla and Mamicpic (1988)
F. semitectum	PF	Reddy (1984)
F. solani	DF	Sevilla and Mamicpic (1988)
Ganobotrytis sp.	DF	Misra *et al.* (1991)
Helicoceras sp.	PF	Richardson (1979, 1981)
H. oryzae	PF	Padwick (1950)
Hendersonia sp.	PF	Richardson (1979, 1981)
Heterosporium echinulatum	DF	Agarwal *et al.* (1989)
Humicola lanuginosa	DF	Dewey *et al.* (1992)
Leptosphaeria sp.	DF	Mew *et al.* (1988)
Macrophoma sp.	DF	Sevilla and Mamicpic (1988)
Macrophomina phaseolina	PF	Parate and Lanjewar (1987)

Magnaporthe salvinii	PF	Richardson (1979, 1981)
Melonomma glumarum	PF	Richardson (1979, 1981)
Memnoniella echinata	DF	Mittal and Sharma (1978)
Microdochium oryzae (Gerlachia oryzae; Rhynchosporium oryzae; Monographella albescens)	PF	Hashioka and Ikegami (1955)
Monascus purpureus	PF	Uyeda (1901)
Mycogone nigra	DF	Mittal and Sharma (1978)
Mycosphaerella danubialis	PF	Richardson (1979, 1981)
M. shiraiana	PF	Richardson (1979, 1981)
Myrothecium roridum	PF	Khan et. al. (1988)
M. verrucaria	PF	Richardson (1979, 1981)
Nakataea sigmoidea	DF	Mew *et al.* (1988)
Nigrospora oryzae (Khuskia oryzae)	PF	Hudson (1963)
Oospora oryzetorum	PF	Padwick (1950)
Ophiobolus oryzinus	PF	Harris *et al.* (1962)
Penicillium adametzi	DF	Udagawa (1959)
P. chrysogenum	DF	Udagawa (1959)
P. citreoviride	DF	Tanaka and Inagaki (1957)
P. citrinum	DF	Tanaka and Inagaki (1957)
P. claviforme	DF	Koroleva *et al.* (1984)
P. cyclopium	DF	Udagawa (1959)
P. funiculosum	DF	Udagawa (1959)
P. implicatum	DF	Udagawa (1959)
P. puberulum	DF	Schroeder (1963)
P. regulosum	DF	Udagawa (1959)
Periconia sp.	DF	Sevilla and Mamicpic (1988)
Pestalotia sp.	PF	Vaid *et al.*(1994)
Phaeotrichoconis crotoloriae	PF	Richardson (1979, 1981)
Phoma glomerata	PF	Gajbe and Lanjewar (1991)
P. glumarum	PF	Reddy (1984)
P. sorghina (Phyllosticta glumarum)	PF	Inoue and Matsunaga (1952)
Pyrenochaeta oryzae	PF	Mew et. al. (1988)
P. terestris	PF	Jayaweera *et al.* (1988)
Pyricularia oryzae	PF	Goto (1955)
Rhizoctonia solani (Thanatephorus cucumeris)	PF	Miyake (1910)
Rhizopus nigricans (R. stolonifer)	PF	Chevaugeon and Ravise (1957)
Sarocladium oryzae (Acrocylindrium oryzae)	PF	Sawada (1922)
Sclerotium rolfsii	PF	Reinking (1918)
Sordaria fimicola	PF	Jayaweera *et al.* (1988)
Stachybotrys alternans (S. atra)	DF	Koroleva *et al.* (1984)
S. lobulata	DF	Mittal and Sharma (1978)
Stagnospora sp.	PF	Mendes and Ferreira (1994)
Stemphyllium botryosum (Pleospora herbarum)	DF	Mittal and Sharma (1978)
Tilletia barclayana (Neovossia horrida, N. barclayana, T. horrida)	PF	Takahashi (1896)
Trematosphaerella oryzae	PF	Richardson (1979, 1981)
Ulocladium spp.	DF	Bokhary (1991)
Ustilaginoidea virens (Ustilago virens, Tilletia oryzae)	PF	Takahashi (1896)
Wolkia decolorans	PF	Wolk (1913)

PF: Pathogenic fungi; DF: Deleterious fungi

2.1. *PYRICULARIA ORYZAE* CAV. (=*MAGNAPORTHE GRISEA* (HEBERT)BARR.)

Blast caused by *Pyricularia oryzae* Cav. is a serious and destructive disease of rice. It is reported to cause 75% grain loss in India (Padmanabhan, 1965), 40% loss in yield in Nigeria (Awoderu and Esuruoso, 1975) and Liberia (Carpenter, 1977) and major constraint for upland rice cultivation in Brazil (Prabhu and Morias, 1986). This pathogen is wide spread in most of the rice growing countries in Asia, Africa, Australia and Oceania, Europe, Central America, West Indies and North and South America. Extensive coverage on blast appeared from time to time (Padwick, 1950; Ou, 1985; Manibhushanrao, 1994).

Mathur (1981) reported that the fungus is present on the seed or in the tissues of embryo, endosperm, bran layers and glumes. In Korea the seeds collected from severely infected field showed 65% infection on the hulls, 25% on pericarp and 4% in embryo. Such infected seeds produced 7-8% seedlings with typical lesions (Chung and Lee, 1983). Further, Lamey (1970) observed that seeds having 40% surface infection produced 7-13% diseased seedlings. The fungus has divergent physiological races distributed around the world and has been reported to produce several phytotoxic compounds such as piricularin, α-picolinic acid (Tamari and Kaji, 1955), pyriculol and tenuazonic acid (Umetsu *et al.*, 1972). Pyriculol and tenuazonic acid are potent inhibitors of seed germination.

2. 2. *BIPOLARIS ORYZAE* (BREDA DE HAAN) SHOEM (=*DRECHSLERA ORYZAE* (BREDA DE HAAN) SUBRAM. & JAIN)

Brown spot of rice caused by *Bipolaris oryzae* is wide spread in all rice growing countries of Asia, America and Africa. In India the disease is particularly prevalent in West Bengal, Orissa, Andhra Pradesh and Tamil Nadu. This pathogen is one of the principal factors of the Great Bengal famine in 1942, causing crop loss ranging from 50-90% (Padmanabhan, 1973). Further, Chattopadhyay *et al.* (1975) recorded 65% loss in yield due to the disease in West Bengal, India. In Nigeria crop losses of 12- 43% were reported (Aluko, 1975), while loss in grain yield ranges from 12-30% depending upon the severity of the infection as well as susceptibility of the host.

The fungus is reported as seed-borne (Padmanabhan, 1953) and the infection noted in the seed coat, pericarp, glumes and endosperm (Fazli and Schroeder, 1966). The pathogen *B. oryzae* can survive in the infected seeds for varying lengths of time depending upon the storage condition. Infected seeds showed reduced germination compared to healthy seeds (Kulkarni *et al.*, 1980). Further, Singh *et al.* (1979) observed 11 - 29% reduction in seed of cvs. germination in T-36, C-8585 and PUSA-2-2. Hiremath and Hegde (1981) identified the fungus in 60-72.5% of the partially developed grains and the seedlings from such seeds died before maturity. The pathogen is known to produce a host specific toxin ophiobolin (Canonica,1972), which inhibits seed germination, growth of root and coleoptile (Janardhana, 1994).

2.3. *MICRODOCHIUM ORYZAE* (HASHIOKA & YOKOGI) SAMUELS & HALLETT (=*GERLACHIA ORYZAE* (HASHIOKA & YOKOGI) W. GAMS; *RHYNCHOSPORIUM ORYZAE* HASHIOKA & YOKOGI)

Leaf scald caused by *Microdochium oryzae* is widely prevalent in Africa, Asia, Australia, Europe, Central America, West Indies, North and South America. Earlier, Lamey and Williams (1972) reported that the disease could be some times as devastating as rice blast. It is known to cause 20-30% yield loss in high yielding varieties (Bakr and Miah 1975). The pathogen was recorded as seed-borne (Singh and Sen Gupta 1981) and located in the husk, endosperm and embryo (Mia *et al.*, 1986). Seed-borne infection also caused seed rot, seedling blight and brownish discoloration of the coleoptile (Kim *et al.*, 1984). The fungus survived in rice seed for more than 11 years when stored at 5°C (Mia *et al.*, 1985).

2.4. *RHIZOCTONIA SOLANI* KUHN

Sheath blight or banded blight/sclerotial disease is an economically important disease and is reported from Bangladesh, Brazil, China, India, Japan, Korea, Madagaskar, Nepal, Nigeria, Pakistan, Philippines, Sri Lanka, Surinam, Taiwan, USA and Venezuela. Initially the disease was identified as minor, but is now positioned next to blast in economic importance (Rush, 1971; Manibhushanrao, 1995).

Production losses of 0.8% - 5.2% were reported due to the disease from South East Asia and Japan (Anon, 1979). Further, Kozaka (1970) reported 30-40% yield loss in cases of severe infection of sheath and leaf blades and Lee and Rush (1983) reported yield loss up to 50% in the USA. The pathogen is reported as seed-borne and seeds collected in October-January had high infection rate (6.6%) than the seeds collected in February-May (5%) and June-September (4%) (Kannaiyan and Prasad, 1978).

2.5. *ACROCYLINDRIUM ORYZAE* SAWADA

Sheath rot of rice caused by *Acrocylindrium oryzae* was considered as a minor disease and its widespread occurrence was recorded from several rice growing countries of Asia, Kenya, Nigeria and Brazil. Gams and Hawksworth (1975) transferred *A. oryzae* to a new genus *Sarocladium* as *S. oryzae*. Subsequently, the relevance of another pathogenic species *S. attenuatum* as alternate or co-pathogen on rice was realised. The malady of the disease was reported by Sawada (1922) and later by Padwick (1950), Ou (1985) and in a recent monographic study by Manibhushanrao (1996).

This pathogen is known to cause 3-20% yield in Taiwan (Chen, 1957), sometimes as high as 52.8% and 85% in the Philippines (Estrada *et al.*, 1984)and 84% to almost total destruction in Thailand (Surin and Disthaporn, 1977). In India, yield losses up to 14.5% have been reported from different states (Chakravarthy and Biswas, 1978). The pathogen is seed borne and Milagrosa (1987) isolated it from the husks and kernels.

2.6. *ALTERNARIA PADWICKII* (GANGULY) M.B. ELLIS (=*TRICHOCONIS PADWICKII* GANGULY; *TRICHOCONIELLA PADWICKII* (GANGULY) JAIN)

Stack burn disease caused by *Alternaria padwickii* is seed-borne, wide spread in rice growing countries of Asia, Africa, Australia, North and South America. In India this disease is known to cause considerable loss in Kerala and West Bengal (Rangaswami, 1975). The seed infection was recorded upto 51-76% by Padmanabhan (1949), and 40-46% by Sharma and Siddiqui (1978a). Conidia of the pathogen are carried externally on the seed surface, however the resting mycelium was identified in the tissues of endosperm, embryo, bran layers and glumes (Ou, 1985). Again, the seed infection results in poor germination, rotting of seed, root and coleoptile (Kulik, 1977).

2.7. *FUSARIUM MONILIFORME* SHELD

Bakanae (bad seedling) disease or foot rot caused by *Fusarium moniliforme* is one of the oldest and most widespread diseases of rice, occurring in all the major rice growing areas with grain yield losses estimated between 4-20%. The pathogen is both soil and seed-borne. Agarwal and Singh (1974) readily isolated the fungus from the seeds. The fungus has been found to be harboured mainly in the embryo (Vidhyasekaran *et al.*, 1970). While Yu and Sun (1976) observed 100% infection of seeds collected from moderately infected bakanae rice fields and such seeds produced 30% diseased seedlings. Severely infected grains are discoloured and some times pink coloured due to the presence of conidial mass.

2.8. *USTILAGINOIDEA VIRENS*(COOKE) TAK. (*USTILAGO VIRENS* COOKE; *TILLETIA ORYZAE* PATOUILLARD)

This pathogen causes false-smut in rice and is known to be widely distributed in rice growing countries of Asia, Africa, Australia, North America, Central America, West Indies, South America and Italy. This pathogen is known to cause yield loss upto 20% in Colombia, 10% loss in Fiji (Morwood, 1966) and up to 44% in India (Singh and Dube, 1978). This pathogen is externally seed borne on the surface of seeds as contaminant. Earlier, Manibhushanrao, (1964) reported the role of environmental conditions on the incidence of false-smut in rice.

2.9. *TILLETIA BARCLYANA* (BREF.) SACC. & SYD. (=*NEOVOSSIA BARCLAYANA* BREF.; *TILLETIA HORRIDA* TAK.; *NEOVOSSIA HORRIDA* (TAK.) PADW. & KAHN)

Kernel smut or bunt disease is known to occur in rice as seed-borne in many rice growing countries of Asia, Africa, Australia, Europe, North America, South America and West Indies. It is considered as a minor disease and yield losses of 2-5% have been reported from Burma (Su, 1933). However, in India Tiagi and Sharma (1978) observed severe out break in Rajasthan and found loss in grain yield up to 36.48%. In Pakistan panicle infection up to 87% was recorded (Hassan, 1971). The pathogen is externally seed borne as contaminant.

2.10. *EPHELIS ORYZAE* SYD. (=*EPHELIS PALLIDA* PAT.)

Udbatta disease is reported from India, China, New Caledonia and West Africa. This disease is considered a scientific curiosity and the yield losses ranging from 1.8-3.7% were observed on different rice cvs. in India (Shivanandappa and Govindu, 1976). The pathogen is internally seed borne as mycelium is present in embryo and conidia are externally seed borne on the surface of the seed. Further, Ranganathaiah and Safeeulla (1994) observed 40,000 - 100,000 conidia /g of seeds collected from field infected rice plants and the viability of the conidia was 2 - 4%

3. Deleterious fungi

Fungi that do not normally cause any disease but under favourable conditions, their growth and metabolites may affect the seed quality, germination, viability, seedling vigour, growth of root and coleoptile are included as deleterious fungi. Seed borne fungi are responsible for deterioration in food reserves in seeds (Neergaard, 1977). A number of fungi present on grains are capable of producing mycotoxins, which include species of *Aspergillus, Penicillium, Alternaria, Curvularia, Fusarium* etc., and in many cases mycotoxin accumulation by seed borne fungi during storage may reach concentrations above the level of mammalian tolerance (Chenulu and DharamVir, 1979). Aflatoxins produced by *A. flavus* group are known to reduce seed germination, seedling elongation, and seedling vigour (Purushotham and Shetty, 1995) and also inhibit chlorophyll synthesis, inhibit various enzymes and degrannulate the endoplasmic reticulum (Dashak and Lewellyn, 1977). Further, Paul (1989) studied the aflatoxin contamination in paddy seeds and out of 119 seed samples tested, 32 yielded *A. flavus* and only 23 of these isolates produced aflatoxin ranging from 0.02-16.0 ppm. Subsequently, Vallejos and Mattos (1990) analysed 79 rice samples in Peru for seed borne fungi and observed that *A. candidus, A. flavus, A. fumigatus, A. versicolor, A. niger, Penicillium* sp., *Nigrospora oryzae* and *Alternaria padwickii* were predominant. Among the 79 samples tested 41 were contaminated with aflatoxin and 10% of the samples contained >20ppm concentration.

Many fungi, including those causing sheath/panicle infection cause seed discoloration of rice resulting in heavy losses to the crop (Das and Narain, 1988). The fungi associated with discoloured seeds may cause deterioration in quality and nutritional value of the grains due to physical, physiological and biochemical changes in the seeds (Duraiswamy and Mariappan, 1983). Further, Sachan and Agarwal (1995) observed that seeds infected with *A. alternata, S. oryzae/S. attenuatum* and *B. oryzae* revealed a decrease in starch content. Some fungi produce characteristic discoloration on seeds, *A. alternaria* (ash grey), *B. oryzae* (dark purple), *S. oryzae* (light brown) and *Fusarium* sp.(light pink) resulting in weight loss of the grains.

Misra and Dharam Vir (1991) analysed 66 rice cvs. and found that grains having <50% discoloration showed 31.2% loss in weight, while grains having >50%

discoloration showed up to 50.2% loss. The rice discoloration and total fungi count on seed was directly correlated with moisture content and storage period. At 92% and 100% RH, the viability of seed was completely lost when stored for 4 and 3 month periods respectively, and the predominant fungi recorded belong to *A. candidus, A. sydowii, A. tamarii, A. versicolor, A. flavus, P. chrysogenum, F. oxysporum* and *Cladosprium cladosporioides* (Mazen *et al.*, 1993).

4. Methods of detection

Standard blotter method is most commonly used for the detection of seed-borne fungi of rice (ISTA, 1985). However, slow growing fungi such as *Alternaria padwickii* (Syn. *Trichoconiella padwickii*) are observed only after a long period of incubation (12 - 15days) (Michail *et al.*, 1985). Water agar (Bacto Agar, Difco 1.5%) was described for inducing sporulation in seed-borne fungi (Srinivasan *et al.*, 1971). Detection of some fungal species such as *Gerlachia oryzae* was improved by the deep-freezing method (Kim *et al.*, 1984). Dry seed inspection is carried out for the detection of *Ephelis oryzae* (Syn. *Balansia oryzae*-sativae) (Agarwal *et al.*, 1989) and also for detection of sclerotia of *Rhizoctonia solani.* Kernel smut (*Tilletia barclayana*) infection in seeds with unruptured glumes can be detected by soaking the seeds in 0.2% solution of sodium hydroxide for 24 h at 18-25°C (Agarwal and Srivastava, 1981, 1985) and washing test is employed for the detection of surface borne spores of various fungi and for detecting *Ustilaginoidea virens* in rice seed samples (Agarwal *et al.*, 1989).

5. Quarantine significance

The international exchange of rice seed both by way of germplasm and commercial seed lots may pose considerable risk of introducing new strains or races of pathogens into areas previously known to be free. Earlier , Neergaard (1980) reviewed an example illustrative of the danger of introducing new pathogenic races of rice blast to upper volta rice growing areas of Africa . The introduction of new varieties from the International Rice Research Institute (IRRI) or other research centres enabled the new pathogenic races of rice blast to get introduced, spread and attacked the African rice which was hitherto bred for resistance to African races and thus caused devastating epidemics of blast in 1971. Therefore the importance of seed health testing in relation to exchange of germplasm has been greatly emphasised (Neergaard,1984).

There are specific quarantine restrictions for introduction of rice seed in many countries against the known major plant pathogens of rice . For example, the member countries of European Plant Protection Organisation (EPPO) prohibit the import of rice seeds infected with blast (*P. oryzae*), brown spot (*B. oryzae*) and *A. padwickii* and prescribe zero tolerances for bacterial leaf blight (*Xanthomonas oryzae*) and bacterial leaf streak (*X. oryzicola*) (Neergaard, 1981). The Australian quarantine regulations prohibit the entry of rice seeds infected with *A. padwickii, Tilletia barclayana* and *X. oryzae*. Further, Greece prohibits the rice seed infected with *P. oryzae* and *A. padwickii* while Japan with the later (Neergaard, 1981). The USDA has listed glume blotch (*Melanomma glumarum*) and bacterial blight (*X. oryzae*) as two of the most destructive foreign diseases of rice and advocate hot/water treatment of seeds at 56-57° C for 15 min before releasing the same to the importer (Cooper, 1988). Mendes and Ferreira (1994) intercepted *Bipolaris hawaiiensis, Stagnospora* sp. and *Pyrenochaeta* sp. on rice seeds introduced into Brazil and Paroda *et al.* (1988) intercepted *A. padwickii, B. oryzae, Tilletia barclayana* and *Ustiloginoides virens* on rice seed imported into India from Philippines and such infected germplasm lines were destroyed .

6. Germplasm preservation and exchange

A critical approach for safe exchange of rice germplasm is to determine the pest risk associated with introduction of seed-borne pathogens of rice. Philips *et al.*, (1992) evaluated potential pest and disease risks associated with paddy as a contaminant of milled rice imported into the Murrumbidgee Irrigation Area (MIA) Australia. It was estimated that the risk of disease infection occurring in MIA is of low order up to a tolerance level of 5 grains /Kg of milled rice. Singh (1988), while reviewing ASEAN plant quarantine systems proposed zero tolerance for *Gibberalla zeae* (Syn. *F. graminearum*), *P. oryzae, Balansia oryzae, Xanthomonas campestris* pv. *oryzae* (Syn. *X. oryzae*), *X.c.* pv. *oryzicola* (Syn *X. oryzicola*) and *Aphelenchoides besseyi.*

The exchange of large quantities of seed material by way of germplasm (about 70 to 80 thousand accessions) across the national boundaries of those countries involved in rice improvement programmes is becoming a formidable task and therefore , international cooperation has become all the more essential for safe conduct of germplasm. The following measures would greatly alleviate the risk of introducing potential seed-borne pathogens of rice into new areas:

The germplasm imports restricted to International Agricultural Research centres or other national research centres engaged in rice research, where facilities for proper seed-health testing of rice are available and subject to proper guidelines for exchange. As for as possible such germplasm exchanges shall be routed through national plant genetic resource centres.

The germplasm imports shall be accompanied by a phytosanitary certificate issued by the national plant protection services of the country of origin with appropriate additional declarations for freedom from quarantine pests as specified in the permit or the import quarantine regulations of the importing country.

The commercial exchange of seed shall be restricted to certified seed lots. To this end inspection of growing crop and adopting appropriate standards of seed-health testing is essential.

Application of appropriate pre and post-harvest chemical treatments to control seed-borne pathogens of rice. It is advised to carryout chemical or hot water treatment of seed just prior to shipment or before sowing the seed since extended storage of treated seed may adversely affect the viability of seed.

Last but not the least that each rice growing country should prepare an up to-date pest lists and list out those of potential economic importance for exclusion from the exchange of germplasm. It is needless to emphasise regular monitoring of pest status in rice growing regions of the country and global exchange of information about new pest records or occurrence of new strains or races of pathogen.

7. References

Agarwal, P.C., Carmen, N. M. and Mathur, S.B. (1989) Seed-borne diseases and seed health testing of rice, *Technical Bulletin No.*3, *Phytopathological Papers* No.**30,** pp106.

Agarwal, V.K. and Singh, O.V. (1974) Relative percentage of incidence of seed borne fungi associated with different varieties of rice seeds, *Seed Research* **2**, 23-25.

Agarwal, V.K. and Srivastava, A.K. (1981) A simpler technique for routine examination of rice seed lots for rice bunt, *Seed Technology News* **11,** 1-2.

Agarwal, V.K. and Srivastava, A.K. (1985) 'NaOH seed soak' method for routine examination of rice seed lots for rice bunt, *Seed Research* **13**, 159.

Aluko, M.O. (1975) Crop losses caused by the brown leaf spot disease of rice in Nigeria, *Plant Disease Reporter* **59**, 609-613.

Anonymous, (1979) International Rice Research Institute, *Annual Report*, pp. 52.

Awoderu, V.A. and Esuruoso, O.F. (1975) Reduction in grain yield of two rice varieties infected by the rice blast disease in Nigeria, *Nigerian Agricultural Journal* **11**, 170-173.

Bakr, M.A. and Miah, S.A. (1975) Leaf scald of rice, a new disease in Bangladesh, *Plant Disease Reporter* **59**, 909.

Bokhary, H.A. (1991) Seed-borne fungi of rice from Saudi Arabia, *Journal of Plant disease and Protection* **98**, 103.

Bora, L.C., Robin-Gogoi, Gogoi, R. (1993) Estimation of seed quality deterioration of deep water rice by seed-borne microflora in Assam, *Indian Journal of Mycology and Plant Pathology* **23**, 214-216.

Breda De Haan,J. (1900) Vorlaufilge Beschreibung von Pilzen, bei tropischen Kulturpflanzen beobachter, *Bull. Inst. Bot. Builtenzorg* **6**, 11-13.

Canonica, L. (1972) The biosynthesis of phytotoxins, in R. K. S. Wood, A. Grnite and A. Balio (eds.), *Phytotoxins in Plant Disease,* Academic Press London, pp. 157-173.

Carpenter, A.J. (1977) Proceedings of the crop losses affecting rice in Liberia, *Proceedings of the WARDA Seminar on Plant Protection for the rice crop,* pp. 160-173.

Chakravarty, D.K. and Biswas, S. (1978) Estimation of yield loss in rice affected by sheath rot, *Plant Disease Reporter* **62**, 226-227.

Chattopadhyay, S.B., Chakrabarti, N.K. and Ghose, A.K. (1975) Evaluation of loss in yield of rice due to infection of brown spot incited by *Helmithosporum oryzae, Indian Phytopathology* **14**, 88-92.

Chen, M.J. (1957) Studies on sheath rot of rice plant, *Journal of Agriculture, Taiwan* **84,** 102.

Chenulu, V.V. and Dharam Vir (1979) Seed microflora its role in grain spoilage and production of mycotoxins, *Bulletin of Grain Technology* **17**, 148-157.

Chevaugeon, J. and Ravise, A. (1957) Regime de l'eau et maladies parasitaires du riz en Afrique Occidentale. *Journal Agric. Trop. Bot. Appl* **4**, 143-151.

Christensen, C.M. (1957) *Botanical Review* **23**, 108-134.

Chung, H.S. and Lee, C.V. (1983) Deletion and transmission of *Pyricularia oryzae* in germinating seed. *Seed Science and Technology* **11**, 625-637.

Cooper, F.E., (1988) Plant quarantine system for rice in the United States, in *Rice seed health*, International Rice Research Institute, Philippines, pp. 57-60

Das, A.N. and Narain, A. (1988) Detection of grain discolouring fungal organisms of rice and production of disease free seeds, *Indian Journal of Mycology Plant Pathology* **18**, 24-30.

Dashak, W.V. and Lewellyn, G.C. (1977) Aflatoxin toxicity and mode of action in plant tissues, *Ann. Nutr Alim* **31**,841-856.

Del Prado,F.A. and Christensen,C.M. (1952) Grain storage studies. XII. The fungus flora of stored rice seed, *Cereal Chemistry* **29**, 456-462.

Dewey, F.M., Twiddy, D.R., Phillips, S.I., Grose, M.J. and Wareing, P.W. (1992) Development of a quantitative monoclonal antibody-based immunoassay for *Humilcola lanuginosa* on rice grains and comparison with conventional assays, *Food and Agricultural Immunology* **4**, 153-167.

Duraiswamy, V.S. and Mariappan, V. (1983) Biochemical properties of discoloured rice grains, *International Rice Research Newsletter* **9**, 9.

Estrada, B.A., Torris, C.Q. and Bonnman, J.M. (1984) Effect of sheath rot on some rice yield components, *International Rice Research Newsletter* **91**, 14.

Fazli, I. S.F. and Schroeder, H.W. (1966) Effect of kernel infection of rice by *Helmintrhosporium oryzae* on yield and quality, *Phytopathology* **56**, 1003-1005.

Fisher, P.J. and Petrini, O. (1992) Fungal saprobes and pathogens as endophytes of rice (*Oryza sativa* L.),. *New Phytologist* **120**, 137-143.

Gajbe, D.W. and Lanjewar, R.D. (1991) Antagonism of *Aspergillus niger* on some seed-borne fungi of two rice cultivars, *PKV Research Journal* **15**, 168-171.

Gams, A. and Hawksworth, D.L. (1975) The identity of *Acrocylindrium oryzae* Sawada and a similar fungus causing sheath rot of rice, *Kavaka* **3**, 57-61.

Godfrey, G.H. (1920) A seed borne sclerotium and its relation to a rice seedling disease, *Phytopathology* **10**, 342-343.

Goto, K. (1955) History of the blast disease and changes in methods of control, *Agric. Improv. Bur.*, Ministry of Agriculture, Japan **5**, 1-2

Harris, E., Pyne, C.T. and Fatmi, A.S. (1962) Plant Pathology. Rice. *Rep. Res. Spec. Serv.*, Ministry of Agriculture, Nigeria **62**, 16.

Hassan, S.F. (1971) *Fundamental studies on rusts and smuts of small grains in Pakistan*, Cereal Diseases Research Institute, Department of Plant Protection, Pakistan 1-186.

Hashioka, Y. And Ikegami, H. (1955) *The leaf scald of rice*, Jubilee Publ. Commem. 60th Birthday of Prof. Y. Tochinai and T. Fukushi, Sappro, Japan, pp. 46-51.

Hemmi, T., Seto, F. And Ikeya, J. (1931) Studies on the 'bakanae' disease of rice plant. II. On the infection of rice *by Lisea fujikuroi* Sawada and *Gibberella saubinetii* (Mont.)Sacc. in the flowering period. *Forschn. Geb. Pflankrankh.*, Kyoto **1**, 99-110.

Hiremath, P.C. and Hegde, R.K. (1981) Role of seed borne infection of *Drechslera oryzae* on the seedling vigour of rice, *Seed Research* **9**, 45-48.

Hudson, H.J. (1963) The perfect state of *Nigrospora oryzae*, *Transactions of the British Mycological Society* **46**, 355-360.

Hyun, J.S. (1963) Development of storage fungi in polished rice infected with rice weevil, *Sitophilus oryzae* L. *Seoul University Journal, Ser. B.* **13**, 77-86.

Inoue, Y. and Matsunaga, M. (1952) On the glume blight of the rice plant, *Kyushu Agricultural Research* **9**, 13-14.

International Seed Testing Association. (1985) International Rules for Seed Testing, Rules 1985. S.R. Draper (ed.), Zurich, Switzerland, pp. 520.

Ito, S. and Kimura, J. (1931) Studies on the 'bakanae' disease of the rice plant, *Report of the Hokkaido Agriculture Experimental Station* **27**, 95

Jain, B.L. (1962) Two new species of *Curvalaria, Transactions of the British Mycological Society* **45**, 539-545.

Janardhana, K. K. (1994) Phytotoxins produced by some seed borne fungi, in T. Singh and P. C. Trivedi (eds.), *Vistas in Seed Biology*, Printwell, Jaipur, India.

Jayaweera, K.P., Wijesundera, R.L.C. and Medis, S.A. (1988) Seed-borne fungi of *Oryza sativa, Indian Phytopathology* **41**, 355-358.

Jin, M.Z. (1989) Preliminary study of discoloured rice grains caused by *Curvularia, Acta Phytopathologica Sinica* **19**, 21-26.

Kanjanasoon, P. (1962) A note on a new leaf disease of rice in Thailand, *International Rice Comm. Newsletter* **11**, 22.

Kannaiyan, S. and Prasad, N.N. (1978) Seed-borne nature of sheath blight pathogen *Rhizoctonia solani* in rice, *International Rice Research Newsletter* **3**, 10.

Khan, S.A.J., Khanzada, A.K., Nasreen-Sultana, Aslam, M. and Sultana,N. (1988) Evaluation of seed health testing techniques for the assessment of seed-borne mycoflora of rice, *Pakistan Journal of Agricultural Research* **9**, 502-505.

Kim, C.H. (1992) Fungi associated with discoloured rice grains, *Research reports of the Rural development Administration,Crop protection* (Korea) **34**, 6-11.

Kim, W.G., Park, J.S. and Yu-S.H. (1984) Seed infection and damage to rice seeds and seedlings by seed borne *Gerlachia oryzae, Korean Journal of Plant Protection* **23**, 126-131.

Kimura, K. (1937) On the relation of fungi to discoloured rice seeds, *Forschn Geb. Pflkrankh* **3**, 209-233.

Koroleva, I.B., Gvozdyak, R.I. and Kirilenko, T.S. (1984) Infection of rice seeds by phytopathogenic micro organisms, *Microbiologicheskii-Zhurnal* **46**, 15-19.

Kozaka, T. (1970) *Pellicularia* sheath blight of rice plants and its control, *Japanese Agricultural Research* **5**, 12-16.

Kulik, M.M. (1977) Seed germinability tests for predicting field emergence of rice seeds infected with *Helminthosporium oryzae* and *Trichoconis padwickii, Phytopathology* **67**, 1303-1304.

Kulkarni, S., Ramakrishnan, K. and Hegde, R.K. (1980) Ecology, epidemiology and supervised control of rice brown leaf spot, *International Rice Research Newsletter* **5**, 13-14.

Kulwant Singh and Mukerjii, K.G. (1994) Recent trends in classification of seed-borne fungi, in T. Singh and P. C. Trivedi (eds.), *Vistas in seed biology*, Printwell Publishers, Jaipur, India, pp. 1-8.

Lamey, H.A. (1970) *Pyricularia oryzae* on rice seed in the United States, *Plant Disease Reporter* **54**, 931-935.

Lamey, H.A. and Williams, R.J. (1972) Leaf scald of rice in West Africa, *Plant Disease Reporter* **54**, 106-107.

Lee, F.N. and Rush, M.C. (1983) Rice sheath blight: a major rice disease, *Plant Disease* **67**, 829-832.

Manibhushanrao, K. (1964) Environmental conditions and false-smut incidence in rice, *Indian Phytopathology* **22**, 110-114.

Manibhushanrao, K. (1994) *Rice blast disease*, Daya Publishing House, Delhi pp. 180.

Manibhushanrao, K. (1995) *Sheath blight disease of rice*, Daya Publishing House, Delhi, pp. 180.

Manibhushanrao, K. (1996) Sheath rot disease of rice, Daya Publishing House, Delhi, pp. 71.

Martinez, E.C. (1953) Fungus disease in rice plantings in the Cauca Valley, *Notes Agron. Estac. Agric. Exp. Palmira* **5**, 1.

Mathur, S.C. (1981) Observations on diseases of dryland rice in Brazil, *International Rice Research Newsletter* **6**, 11-12.

Mazen, M.B., Abdel-Hafez, S.I., El-Kady, I.A. and Maghraby, O.M. (1993) Effect of level of relative humidity on fungi and germination capacity of paddy grains in Egypt, *Qatar University Science Journal* **13**, 81-84.

Mendes, M.A.S. and Ferreira, M.A.S.V. (1994) Pathogenic fungi detected in vegetable germplasm introduced in Brazil from 1990-1992, *Fitopatologia Brasileira* **19**, 449-454.

Mew, T.W., Bridge, J., Hibino, H., Bonman, J.M. and Merca, S.D. (1988) Rice pathogens of quarantine importance, *Proceedings of the International workshop on Rice Seed Health*, 16-20th March 1987 101-115.

Mia, M.A.T., Mathur, S.B. and Neergaard, P. (1985) *Gerlachia oryzae* in rice seed, *Translations of the British Mycological Society* **84**, 337-338.

Mia, M.A.T., Safeeulla, K.M. and Shetty, H.S. (1986) Seed-borne nature of *Gerlachia oryzae* the incitant of leaf scald of rice in Karnataka, *Indian Phytopathology* **39**, 92-93.

Michail, S.H., Mathur, S.B. and Neergaard, P. (1985) Infections of jute, maize and rice not revealed by the blotter method, *Journal of Phytopathology* **113**, 193-195.

Milagrosa, Sergia P. (1987) *Transmission of Sarocladium oryzae (Sawada) W.Gams and Hawksw through seed*, Ph.D. Thesis, University of the Philippines, Los Banos, Laguna, Philippines.

Misra, A.K., Dharam Vir and Vir, D. (1991) Wild rice and its crosses the alternative hosts of paddy grain discolouring fungi, *International Journal Tropical Plant Diseases* **9**, 123-125.

Mittal, R.K. and Sharma, M.R. (1978) Addition to the mycoflora associated with seeds of *Oryza sativa* L, *Proceedings of the National Academy of Sciences, India* **48**, 77-78.

Miyake, I. (1910) Studien uber die Pilze der Reispflanza in Japan, *Journal of the College of Agriculture*, University of Tokyo **2**, 237-276.
Morwood, R.B. (1966) Notes on plant diseases tested for Fiji, *Agricultural Journal* Fiji **27**, 83-86.
Neergaard, P. (1977) *Seed Pathology*, **Vol. I,** Macmillan Press, London.
Neergaard, P. (1980) A review on quarantine for seed, *National Academy of Sciences, India, Golden Jubilee Commemoration volume* 495-530.
Neergaard, P. (1981) Risks for the EPPO Region from seed-borne pathogens, *EPPO Bulletin* **11**, 207-212.
Neergaard, P. (1984) Seed health in relation to the exchange of germplasm, *International Board for Plant Genetic Resources* 1-18.
Nyvall, R.F., Percich, J.A., Porter, R.A. and Brantner, J.R. (1995) *Plant Disease* **79**, 249-250.
Ou, S.H. (1985) *Rice Diseases,* CAB International Mycological Institute, Kew, Surrey, UK.
Padmanabhan, S.Y. (1949) Occurrence of fungi inside rice kernels, *Current Science* **18**, 442-443.
Padmanabhan, S.Y. (1953) Specialization in pathogenicity of *Helminthosporium oryzae, Proceedings of the 40th Indian Science Congress* **Part 4**, Abstract 18.
Padmanabhan, S.Y. (1965) Estimating losses from rice blast in India, in *The Rice Blast Disease,* John Hopkins Press Baltimore, Maryland, pp. 203-221.
Padmanabhan, S.Y. (1973) The great Bengal famine, *Annual Review of Phytopathology* **11**, 11-26.
Padwick, G.W. (1950) *Manual of rice diseases*, Commonwealth Mycological Institute, Kew, pp. 198.
Parate, D.K. and Lanjewar, R.D. (1987) Studies on seed mycoflora of two rice cultivars grown in Rice tract of Vidharbha. *PKV - Res. J* **11**, 47-50.
Paroda, R.S., Ramnath, Reddy A.P.K. and Singh, B.P. (1988) Indian Plant Quarantine systems for rice, in: *Proceedings of the International workshop on Rice Seed Health*, International Rice Research Institute, Manila, Philippines, pp. 77-89.
Paul, M. (1989) *Studies on seed-borne diseases of paddy grown in tribal areas of Rajasthan*, Ph.D. Thesis, University of Rajasthan, Jaipur, India.
Phillips, D., Chandrashekar, M. and McLean, G. (1992) Evaluation of potential disease and pest risks associated with paddy as a contaminant of milled rice imported into Australia, *FAO Plant Protection Bulletin* **40**, 4-20.
Prabhu, A.S. and Morias, O.P. (1986) Blast disease management in upland rice in Brazil, in *Progress of Upland Rice Research*, Proceedings of the 1985 Jakarta Conference, pp. 383-392.
Purushotham, S.P. and Shetty, H.S. (1995) Effect of aflatoxin B1 on paddy seed germination, in *National Seminar on Seed Pathology* (Abstract), Department of studies on Applied Botany, University of Mysore, India, pp. 44.
Ramesh-Tiwari and Tiwari-R. (1995) Seasonal mycoflora analysis of four varieties of rice at Gorakhpur, *Journal of Living World* **2**, 63-67.
Ranganathaiah, K.G. and Safeeulla, K.M. (1994) Seed-borne nature and mode of infection of *Ephelis oryzae* incitant of udabatta disease of rice, *Mysore Journal of Agricultural Science* **28**, 60-63.
Rangaswami, G. (1975) *Diseases of crop plants in India,* Prentice Hall, New Delhi.
Rao, P.N. and Salam, M.A. (1954) *Curvularia* species from discoloured grains from Hyderabad, *Journal of the Indian Botanical Society* **33**, 268-271.
Reddy, O.R. (1984) Evaluation of triforine (Saprol) against seed-borne pathogens of rice (*Oryza sativa*), *International Journal of Tropical Plant Diseases* **2**, 133-138.
Reinking, O.A. (1918) Philippines economic plant diseases, *Philippines Journal of Science,* Sect. A, **13**, 225-228.
Richardson, M.J. (1979) *An annotated list of seed-borne diseases,* 3rd ed., Commonwealth Agricultural Bureaux, U.K.
Richardson, M.J. (1981) *Supplements to an annotated list of seed-borne diseases,* Commonwealth Agricultural Bureaux, U.K.
Rush, M.C. (1971) Factors contributing to yield losses in rice affected by the bordered leaf and sheath spot disease, *63rd Annual Progress Report of the Rice Experimental Station*, Growells, 221-225.
Sachan, I.P. and Agarwal, V.K. (1995) Seed discolouration of rice: Location of inoculum and influence on nutritional value, *Indian Phytopathology* **48**, 14-20.
Sawada, K. (1922) Descriptive catalogue of Formosan Fungi: II, Report of the Government Research Institute, Department of Agriculture, Formosa **2**, p. 135.
Schroeder, H.W. (1963) Orange stain, a storage disease of rice caused by *Penicillium puberulum, Phytopathology* **53**, 843-845.

Sevilla, E.P. and Mamicpic, N.G. (1988) Rice seed exchange in the Philippines, in *Rice Seed Health,* Proceedings of the International workshop on Rice Seed Health, 16-20th March, 1987, pp. 33-48.
Sharma, I.P. and Siddiqui, M.R. (1978a) Study on seed microflora of paddy from Assam and West Bengal, *Seed Research* **6**, 43-47.
Sharma, I.P. and Siddiqui, M.R. (1978b) *Drechslera holmii* and *Fusarium camptoceras* two new records on paddy seeds, *Indian Phytopathology* **31**, 508-509.
Shivanandappa, N. and Govindu, C.H. (1976) Direct and indirect effect of udbatta disease on total number of panicles per hill in different paddy cultures, *International Rice Research Newsletter* **1**, 12.
Singh, K.G. (1988) ASEAN plant quarantine systems, in *Rice seed health,* International Rice Research Institute, Philippines, pp. 49-55.
Singh, R.A. and Dube, K.S. (1978) Assessment of loss in seven rice cultivars due to false smut, *Indian Phytopathology* **31**, 186-188.
Singh, R.P., Singh, P.N. and Shukla, P. (1979) Note on assessment of losses in rice due to brown leaf spot disease, *Indian Journal of Agricultural Research* **13**, 57-58.
Singh, S.A. and Sen Gupta, P.K. (1981) Transmission of *Rhynchosporium oryzae* by seed, *International Rice Research Newsletter* **6**, 11.
Sisterna, M.N., Lori, G.A. and Marassi, J.J. (1994) Symptomology and fungi associated with rice grain spotting in cultivar Irga 409, *Revista-de-la-facultad-de-Agronomia-La-Planta* **70**, 13-21.
Soave, J., Azzini, L.E., Villela, O.V. and Gallo, P.B. (1983) Selection of irrigated rice cultivars for low incidence of spotted seeds, *Summa-Phytopathologica* **9**, 179-185.
Srinivasan, M.C., Chidambaram, P., Mathur, S.B. and Neergaard, P. (1971) A simple method for inducing sporulation in seed-borne fungi,. *Transactions of the British Mycological Society* **56**, 31-35.
Su, M.T. (1933) *Report of the Mycologist*, Burma, Mandalay for the year ending the 31st March, 1933, pp.**1**-12.
Surin, A. and Disthaporn, S. (1977) 'Rice abortion' or sheath rot- a serious rice disease in Thailand, *International Rice Research Newsletter* **2**, 16-17.
Sydow, H. (1914) Beitrage zur Kenntnis der Pilzflora des sudlichen Ostindiens II, *Ann. Mycol. Berl* **12**, 484-490.
Takahashi, Y. (1896) On Ustilago virens Cooke and a new species of *Tilletia* parasitic on rice plant. *Bot. Mag.*, Tokyo **10**, 16-20.
Tamari, K. and Kaji, J. (1955) On the biochemical studies on the mould (*Pyricularia oryzae*). The causative mold of the blast disease of rice plants. II. Studies on the physiological action of Pyricularin, a toxin produced by the blast mold on rice plants, *Journal of Agricultural Chemistry Society* Japan **29,** 85-190.
Tanaka, Y. and Inagaki. (1957) Studies on the technique for the isolation for the presence of rice grain fungi, *Bull. Natn. Hyg. Lab.* Tokyo **75,** 443-459.
Tiagi, R.N.S. and Sharma, N.D. (1978) Studies on losses in grain yield caused by the three smuts of rice in Kota region of Rajastan, *Indian Journal of Mycology and Plant Pathology* **8**, 227-228.
Tullis, E.C. (1936) Fungi isolated from discoloured rice kernels, *Technical Bulletin* U.S.D.A. **540**, pp. 12.
Udagawa, S. (1959) Taxonomic studies of fungi on stored rice grains. III. *Pencillium* group (*Penilcillia* and related genera) 2, *Journal of Agricultural Science* Tokyo **5**, 5-21.
Umetsu, N., Kaji, T. and Tamari, K. (1972) Investigations on the toxin acid as a novel toxin, *Agricultural Biological Chemistry* **30**, 859-866.
Uyeda, Y. (1901) Ueber den 'Benikoji-pilz' aus Formosa, *Bot. Mag.*, Tokyo **15**, 160-163.
Vaid, A., Sharma, O.P. and Kaushik, R.P. (1994) Effect of grain discolouration disease on some important quality parameters of rice, *Plant Disease Research* **9**, 190-192.
Vallejos, O.V. and Mattos, C.L. (1990) Mycoflora of *Oryza satilva* L. and its relationship with aflatoxins in Lambayeque, *Fitopatologia* **25**, 54-59.
Vidhyasekaran, P., Subramanyan, C.L. and Govindaswamy, C.V. (1970) Production of toxins by seed-borne fungi and its role in paddy field spoilage, *Indian Phyotpathology* **23**, 518-525.
Wei, C.T. (1957) *Manual of rice pathogens,* Science Press, Beijing, pp. 267.
Wolk, P.C.Van Der. (1913) *Protascus colorans,* a new genus and a new species of the Protascineae group, the source of yellow grains in rice, *Myc. Centralbl* **3**, 153-157.
Wu, W.S. and Dow, S.K. (1993) A survey of rice seed-borne fungi in Taiwan, *Plant Pathology Bulletin* **2,** 52-55.
Yu, K.S. and Sun, S.K. (1976) Ascospore liberation of *Gibberella fujikuroi* and its contamination of rice grains, *Plant Protection Bulletin* Taiwan **18**, 319-129.

INDUCED SYSTEMIC RESISTANCE FOR THE MANAGEMENT OF RICE FUNGAL DISEASES

P. VIDHYASEKARAN
Department of Plant Pathology
Tamilnadu Agricultural University
Coimbatore - 641 003, India.

1. Introduction

Plants are endowed with several defense genes. They are found in both susceptible and disease-resistant plants. They are involved in production of antimicrobial compounds like phenolics, phytoalexins and pathogenesis-related proteins or involved in reinforcement of plant cell wall by accumulating hydroxyproline-rich glycoproteins, callose, lignin and wall bound phenolics (Vidhyasekaran, 1993, 1997, 1988a, b). These defense genes are quiescent in healthy plants. Both biotic and abiotic signals can activate these genes (Baker *et al.*, 1997). Early and increased expression of these genes can induce systemic resistance in plants against various pathogens. This type of resistance is termed systemic acquired resistance (SAR) (Ward *et al.*, 1991). Technologies have been developed to manage various crop diseases exploiting SAR and the present paper critically analyses the possibility of managing rice diseases by exploiting SAR.

2. Elicitors

Elicitors are the signal molecules that elicit defense mechanisms in plants (Scheel and Parker, 1990). Elicitors have been isolated from fungal, bacterial, and viral pathogens. Elicitors have also been isolated from plants (Tepper and Anderson, 1990). Microbial elicitors are glycoproteins, glucans, chitins, chitosans, fatty acids and proteins (Mayama *et al.*, 1986 ; Schaffrath *et al.*, 1995) while oligogalacturonides have been identified as endogenous host plant elicitors (Lorenzo *et al.*, 1990). An elicitor has been isolated from *Pyricularia oryzae*, the rice blast pathogen (Schaffrath *et al.*, 1995). The elicitor is a glycoprotein with a molecular weight of 15.6 kDa. The elicitor activity was not lost when treated with pronase, but the activity was completely abolished when it was treated with sodium periodate indicating that carbohydrate is the active moiety. The carbohydrate consists mainly of mannose with some glucose and galactose (Schaffrath *et al.*, 1995). The elicitor isolated from *P.oryzae* induces 3-hydroxy-3-methylglutaryl

S. Sreenivasaprasad and R. Johnson (eds.), Major Fungal diseases of Rice Recent Advances, 347-358.
Prof. K. Manibhushan Rao Festschrift

coenzyme A reductase (HMGR), the key enzyme in the synthesis of the diterpene phytoalexins oryzalexins and momilactones in rice (Nelson *et al.*, 1994).

An elicitor has been isolated from *Rhizoctonia solani* (Vidhyasekaran and Velazhahan, 1996). It was also a glycoprotein and the elicitor activity was due to the carbohydrate moiety. It induced phenylalanine ammonia-lyase (PAL), the key enzyme in phenylpropanoid pathway and a 35 kDa chitinase (Vidhyasekaran *et al.*, 1996).

3. Signal transduction pathway

The elicitors bind to the receptor sites in host plasma membrane and signals are produced for activation of defense genes. Second messenger systems exist to transmit the elicitation signal intracellularly. The signal transduction system involves pathogen signals, calcium ions influx, protein phosphorylation, membrane potential, active oxygen species and salicylic acid. Calcium ion is known to transduct extracellular primary signals into intracellular events (Pitt and Kaile, 1990). Calcium passively crosses plasma membranes through `channel' - forming proteins embedded in the lipid bilayer of the membrane (Zocchi and Rabotti, 1993). Phosphorylation of proteins is an important component in the integration of external and internal stimuli. Protein phosphorylation reaction is catalysed by calcium- regulated protein kinases (Marme, 1989). Protein kinase C occurs as soluble or membrane bound enzyme in plants. A shuttle system between cytosol and membrane would allow the phosphorylation of either soluble or membrane - bound proteins and therefore would allow protein kinase C to participate in transmembrane signalling (Conrath *et al.*, 1991). Recently several disease - resistance genes have been cloned. The predicted peptide sequences of several cloned resistance genes indicate a function in signal transduction (Dixon *et al.*, 1996 ; Ronald, 1997). Tanscriptional activation of plant defense genes is modulated by phosphorylattion and can be blocked by inhibitors of protein kinases (Raz and Fluhr, 1993). It suggests that protein kinases may be involved in phosphorylation cascades leading to plant defense responses (Lamb, 1996).

Phospholipase D is a critical component in signal transduction. Phospholipases activate protein kinase C. Phospholipase D hydolyses phospholipids generating phosphatidic acid and a free head group, such as choline, which serve as second messengers. Phospholipases are calcium-dependent enzymes (Breviario *et al.*, 1995). Phospholipase C appears to be involved in regulating intracellular Ca^{2+} concentrations which act as second messenger in signal transduction system (Alexandre *et al.*, 1990). Phospholipase C catalyses hydrolysis of the cell membrane phosphoinositides to release diacylglycerol and inositol phosphates, predominantly inositol 1,4,5 - triphosphates (IP_3). Diacylglycerol activates protein phosphorylation and IP_3 releases Ca^{2+} from internal stores (Berridge and Irvine, 1989) and both of them function as second messangers. The role of phospholipase C in signal transduction pathway has been well established.

Transport of solutes across the plasma membrane is driven by H^+ - ATPase that produces an electric potential (Popham *et al.*, 1995). Elicitors induce a rapid membrane depolarisation (Popham *et al.*, 1995). The membrane potentials created or modulated by ATPases may be intermediates in the signal chain (Steffens *et al.*, 1989). A sudden burst in active oxygen production has been reported in plants treated with the elicitors (Apostol *et al.*, 1989). H_2O_2, O_2^{-1}, O_2 and OH^0 are the active oxygen species involved in signal transduction and among them H_2O_2 is the most important signal (Du and Klessig, 1997). Salicylic acid has been reported as one of the important signal molecules which act locally in intracellular signal transduction. Salicylic acid may enhance release of H_2O_2 and H_2O_2 - derived active oxygen species and induce activities of defense - related genes (Shirasu *et al.*, 1997). Salicylic acid binding protein has been detected in plants and it has been identified as catalase (Conrath *et al.*, 1995). Catalase suppresses the H_2O_2 - degrading activity of catalase (Conrath *et al.*, 1995). The other major H_2O_2 -scanvenging enzyme, ascorbate peroxidase, is also inhibited by salicylic acid (Durner and Klessig, 1995).

What may be the sequence of events in intracellular signal transduction system? In parsley cells treated with a fungal elicitor and incubated with the ion channel blocker anthracene-9-carboxylate, activation of the protein kinase was completely inhibited (Ligterink *et al.*, 1997). It suggests that protein phosphorylation may be in the downstream of pathogen signals and ion flux. The elicitor-induced responses such as active oxygen species production and membrane depolarisation are blocked by a Ca^{2+} channel blocker, and K25a, a protein kinase inhibitor (Popham *et al.*, 1995; Hoyos *et al.*, 1996). It suggests that both active oxygen species productin and membrane depolarisation may be in the downstream of Ca^{2+} influx and protein phosphorylation.

Diphenylene iodonium (DPI) inhibits production of active oxygen species. In elicitor-treated parsley cells DPI blocked the oxidative burst and defense gene activation and phytoalexin accumulation without affecting calcium ion fluxes (Ligterink *et al.*, 1997). It suggests that the oxidative burst may be in the down stream of ion channels in the elicitor signal transduction cascade. Elicitor activation of the protein kinase was not inhibited by DPI, indicating that this kinase may act upstream of the oxidative burst (Ligterink *et al.*, 1997). H_2O_2 accumulation by the elicitors in soybean cells could be blocked by the salicylic acid synthesis inhibitor α- aminooxy-β- phenylpropionic acid and this response could be rescued by exogenous application of salicylic acid (Shirasu *et al.*, 1997). It indicates that H_2O_2 accumulation may be in the downstream of salicylic acid accumulation.

The intracellular signals should target plant cell nucleus to activate defense genes. Most of the resistance genes cloned so far encode proteins with leucine-rich repeats (LRR), leucine zipper and nuclear binding site (NBS) domains (Baker *et al.*, 1997). LRRS may be involved in protein-protein interactions in a downstream step of the signal transduction (Kobe and Deisenhofer, 1993). Leucine zippers may directly interact with DNA and the NLS may direct transport of the signals to the plant nucleus (Yang and Gabriel, 1995).

The signals generated due to elicitors are translocated systemically and the plants synthesise several defense chemicals. Salicylic acid appears to be the most important systemic signal molecule (Ryals *et al.*, 1996). Salicylic acid induces systemic accumulation of several PR Proteins and induces resistance (Ward *et al.*, 1991). Methyl jasmonate and jasmonic acid also act as systemic signal molecules (Schweizer *et al.*, 1997a). Ethylene is known to regulate defense genes in plants (Xu *et al.*, 1994). A combination of signal molecules may be needed to induce a single defense gene and multiple signal transduction system may exist in plants (Schweizer *et al.*, 1997a).

The proposed sequence of events in the intracellular signal transduction pathway

Elicitor
↓
Calcium ion flux
↓
Protein phosphorylation
↓
Membrane depolarisation
↓
Salicylic acid
↓
Active oxygen species

4. Manipulation of signal transduction system to induce resistance in rice

Active oxygen species act both in upstream and downstream of salicylic acid in signal transduction pathway to induce systemic resistance. Salicylic acid is commonly present in all rice cultivars (Silverman *et al.*, 1995). Rice seedlings had the highest levels of salicylic acid among all plants tested for salicylic acid. Leaf salicylic acid levels in 28 rice cultivars showed a correlation with generalised blast disease (*Pyricularia oryzae*) resistance, suggesting that salicylic acid may play a role as a constitutive defense compound (Silverman *et al.*, 1995). Salicylic acid content did not increase after inoculation with the rice pathogens *P. oryzae* and *R. solani* (Silverman *et al.*, 1995). It suggests that to exploit salicylic acid as a signal molecule, we may have to activate biosynthesis of salicylic acid. H_2O_2 and H_2O_2 - generating chemicals may induce accumulation of salicylic acid (Leon *et al.*, 1995).

Lipid peroxidation by plant lipoxygenases (LOX) may enhance H_2O_2 production. In rice three LOXs have been isolated from leaves (Ohta *et al.*, 1990). The leaf LOXs produced preferentially 13-L-hydroperoxy-9,11 (Z-E)-octadecadienoic acid (13-HPODD) from linolenic acid (Ohta *et al.*, 1986). The LOX products of linolenic acid inhibit germination of conidia of *Pyricularia oryzae* (Shimura *et al.*, 1983). Three kinds of hydroperoxides were identified as a result of in vitro lipoxygenase enzyme reaction using the homogenate of blast-infected rice leaves (Namai *et al.*, 1991). These three metabolites correspond to a kind of active oxygen species and they inhibited germination and germ tube elongation of conidia of *P. oryzae* (Namai *et al.*, 1991). Allyl alcohols derived from hydroperoxides are also inhibitory to *P. oryzae* (Kato *et al.*, 1983). 9-Lineolate hydroperoxide [9-LOOH] showed high antifungal activity against *P. oryzae* at 50ppm, while the product of 9-LOOH metabolism, 9-LOH, inhibits conidial germination almost completely even at 2.5 ppm (Ohta *et al.*, 1990). 13-L-hydroperoxy-9,11 (Z,E) - octadecadienoic acid (13-HPODD) and 13-L-hydroxy-9,11 (Z,E) - octadecadienoic acid (13-HODD) are the major products of the LOX pathway in *P.oryzae* infected rice leaves (Ohta *et al.*, 1990). Thus several LOX products have been shown to be highly inhibitory to rice fungal pathogens.

LOX and its products can generate active oxygen species (Veldink and Vliegenthart, 1985). The LOX is capable of generating superoxide anion radicals (Lynch and Thompson, 1984) and singlet oxygen (Anderson, 1989). Superoxide anion undergoes dismutation to form H_2O_2 (Anderson, 1989). Superoxide anion is produced in rice leaves inoculated with *P.oryzae* (Haga *et al.*, 1986; Sekizawa *et al.*, 1985; 1987; Averyanov *et al.*, 1987; Averyanov and Lapikova, 1988).

LOX - derived fatty acid hydropeoxides are the precursors of jasmonic acid, an important systemic signal molecule. In resistant interactions, LOX activity rapidly increased in rice leaves infected with *P.oryzae*; but in susceptible interactions, LOX activity was suppressed (Ohta *et al.*, 1990). Hence in susceptible interactions no increase in salicylic acid or jasmonic acid was observed. Probably external application of inducers of LOX or salicylic acid or jasmonic acid may be needed to induce systemic resistance in a susceptible rice variety.

LOX activity was found to be induced by a synthetic compound, INA (2, 6-dichloro-isonicotinic acid) in rice . Enhanced LOX activity might result in the activation of the octadecanoid signalling pathway leading to synthesis of jasmonic acid (Sembdnes and Parthier, 1993), But Schweizer *et al.* (1997a,b) found that although the octadecanoid pathway is inducible by low concentrations (0.1 to 10ppm) of INA in rice, no elevated level of endogenous jasmonic acid occurred. However INA treatment at low concentrations induced systemic resistance against the rice blast fungus *P.oryzae* (Schweizer *et al.*, 1997a). It suggests that jasmonic acid alone may not be the signal molecule involved. INA enhances LOX which may enhance synthesis of H_2O_2 which is another systemic signal molecule. The role of octadecanoid pathway in induction of systemic induced resistance is not yet clear (Doares *et al.*, 1995; Kogel *et al.*, 1995).

INA appears to induce resistance in tobacco (Ward *et al.*, 1991), sugar beet (Nielsen *et al.*, 1994), cucumber (Ward *et al.*, 1991), barley (Kogel *et al.*, 1994), bean (Dann and Deverall, 1995) and rose (Hijwegen *et al.*, 1996) through activation of signal transduction system. INA treatment induced several pathogenicity - related proteins PR1, PR3, PR5, and PR9 in rice leaves (Schweizer *et al.*, 1997a). Two synthetic compounds structurally related to INA, N-phenylsulfonyl - 2 - chlorosonicotinamide (Yoshida *et al.*, 1990) and N-cynaomethyl - 2 - chloroisonicotinamide (Seguchi *et al.*, 1992) also induced resistance in rice against blast disease.

Spraying jasmonic acid on rice plants treated with drench application of INA, enhanced blast disease resistance was observed than that observed with jasmonic acid alone or INA alone treated plants (Schweizer *et al.*, 1997a). The combination treatments induced more accumulation of PR1 proteins and the jasmonic acid induced protein ASP31 in rice leaves (Schweizer *et al.*, 1997a). It suggests that combination of signals may enhance induced systemic resistance (Xu *et al.*, 1994). Jasmonic acid sprayed on rice leaves was quickly degraded (Schweizer *et al.*, 1997a). Repeated spray of jasmonic acid might control the blast disease.

Jasmonic acid-related compounds also induce resistance in rice. Amino acid conjugates of jasmonic acid are found to elicit production of the flavonoid phytoalexin, sakuranetin in rice leaves. The elicitation was shown to arise from induction of naringenin 7-0 - methyltransferase, a key enzyme of sakuranetin biosynthesis (Tamogami *et al.*, 1997).

Attempts were made to increase salicylic acid content of rice plants and induce systemic resistance. Phyenylalanine ammonia - lyase (PAL) is the key enzyme in inducing synthesis of salicylic acid. Rice genome contains a number of PAL genes. One of the genes, ZB8 has been found to be induced by the elicitor treatment in rice cells (Li *et al.*, 1993). The gene was cloned and transgenic rice plants which express the elicitor - inducible PAL have been developed (Zhu *et al.*, 1995). The transgenic rice plants expressing PAL ZB8 showed systemic resistance against rice pathogens (Lamb *et al.*, 1997).

Some saprophytic strains of *Pseudomonas fluorescens* have been shown to contain regions homologous to hypersensitive response and pathogenicity (*hrp*) genes which may encode secretion of elicitors. These genes are commonly found only in pathogens. *P.fluorescens* strain PfG32R conserves the regions homologous to hrpRS and hrpF of *Pseudomonas savastanoi* pv. *phaseolicola,* a pathogen of bean) and the regions homologous to hrpC and hrpD of *P. syringae* pv. *tabaci* (a pathogen of tobacco). This strain was able to multiply in tobacco or bean leaves (Mulya *et al.*, 1996). Presence of *hrp* cluster in a saprophytic strain has helped the bacteria to grow in plants and the *hrp* cluster could have played a role in inducing the signal transduction system. This strain induced phenylalanine ammonia - lyase (PAL) activity in tobacco and bean plants (Mulya *et al.*, 1996).

Several *P.fluorescens* strains have been shown to induce systemic resistance in plants. Fusarium wilt of radish is controlled by *P.fluorescens* strain WCS374 by inducing systemic resistance (Hoffland *et al.*, 1995; Raaijmakers *et al.*, 1995). Induced resistance in cucumber against *Pythium aphanidermatum* (Zhou and Paultiz, 1994), *Colletotrichum orbiculare* (Wei *et al.*, 1991), and cucumber mosaic virus (Liu *et al.*, 1992), in carnation against fusarium wilt (Van Peer *et al.*, 1991) and in tobacco against tobacco mosaic virus (Maurhofer *et al.*, 1994) has been reported.

Pseudomonas fluorescens strains that induce resistance in rice to blast (Vidhyasekaran *et al.*, 1997) and sheath blight (Rabindran and Vidhyasekaran, 1996) have been identified. When rice seeds were treated with *P. fluorescens* Pf1 strain, resistance in foliage was observed up to 45 days after sowing (Vidhyasekaran *et al.*, 1997). The induced resistance was observed 3 to 4 days after foliar spray and the resistance persisted up to 15 days in rice leaves (Vidhyasekaran *et al.*, 1997). Induced resistance against the rice bacterial blight by *Pseudomonas fluorescens* strains has also been reported (Ohno et al., 1992; Kamala Nayar and Vidhyasekaran, 1998).

P. fluorescens strains induce several defense genes encoding phenylalanine ammonia lyase (PAL), peroxidase, phenylalanine chalcone synthase, and induce accumulation of phenolics, callose, lignin, phytoalexins and chitinases (Van Peer *et al.*, 1991; Zdor and Anderson, 1992; Kloepper *et al.*, 1993; Benhamou *et al.*, 1996 a,b,c; M'Piga *et al.*, 1997). Induction of defense genes by *P. fluorescens* was observed only when the bacteria-treated plants were infected with pathogens; healthy plants treated with *P. fluorescens* do not show any induction of defense genes (Van Peer *et al.*, 1991; Zdor and Anderson, 1992; M'Piga *et al.*, 1997). It suggests that a signal, originating from the interaction between the plant and the pathogen, probably the release of an elicitor, is essential for the induction of defense genes in plants by *P. fluorescens*. Endophytic *P. fluorescens* may be involved in the signal transduction (M'Piga *et al.*, 1997).

5. Transgenic rice plants with enhanced disease resistance

Downstream of signal transduction system, several defense genes are induced. In an alternate approach, attempts have been made to overexpress these genes constitutively, without the signals to induce these genes, to control diseases. 3-Hydroxy-3--methyl-glutaryl-Coenzyme A reductase (HMGR) is an important enzyme in the synthesis of terpenoid phytoalexins in tomato. One of the HMGR genes in tomato, *HMG2* has been cloned and this gene has been transferred to rice (Lamb *et al.*, 1997). Constitutive expression of *HMG2* increased the synthesis of isoprenoids and accumulation of specific steroids. The transgenic plants showed high resistance to the blast disease (Lamb *et al.*, 1997).

Several chitinase genes have been cloned from rice (Kim *et al.*, 1994; Xu *et al.*, 1996). One of the genes, *RC24* accumulates in rice leaf tissues upon elicitor treatment (Xu *et al.*,1996). A 35-kDa chitinase has been purified from elicitor -treated rice cells. The 35-kDa chitinase inhibited the growth of the rice sheath blight pathogen *Rhizoctonia solani in vitro* (Vidhyasekaran *et al.*, 1996). A 1.1 kb rice genomic DNA fragment containing a chitinase gene was cloned into rice under the control of CaMV 35S promoter (Lin *et al.*,1995). Transgenic plants expressing the chitinase gene showed enhanced resistnace to the sheath blight pathogen *R. solani* (Lin *et al.*, 1995). Two PR-5 proteins with a molecular mass of 24 and 25 kDa have been detected in rice leaf sheaths infected with *R. solani*. They were absent in healthy rice plants (Velazhahan *et al.*, 1998). Transgenic rice plants constitutively expressing PR -5 proteins have been developed by fusing one of the cDNA clones, D34 (Datta *et al.*, 1997). The PR-5 expressing transgenic rice plants showed enhanced resistance to *R. solani* (unpublished data).

6. Summary

Plants are endowed with various defense genes; but they are quiescent in normal healthy plants. When these defense genes are activated, they induce resistance against diseases. Specific signals are needed to induce these defense genes. A well orchestrated signal transduction system operates in plants. Disease resistance genes may also form a component in the signal transduction system. By manipulating this signal transduction system, disease resistance can be induced. Abiotic inducers jasmonic acid, salicyclic acid, and related compounds have been used to induce resistance in rice plants. Biotic inducers like fluorescent pseudomonads have also been employed to manage rice fungal diseases. Constitutive expression of defense genes by genetic engineering techniques have also been found useful in the management of rice diseases. `Induced systemic resistance' appears to be a potential tool to manage rice diseases.

7. References

Alexandre, J., Lassalles, J.P. and Kado, R.T. (1990) Opening of Ca^{2+} channels in isolated red beet root vacuole membrane by inositol 1,4,5 trisphosphate, *Nature* **343**, 567-570.

Anderson, J.M.(1989) Membrane derived fatty acids as precursors to second messengers, in W.F. Boss and D.J. Morre (eds.) *Second Messengers in Plant Growth and Development*, Alan R. Liss, New york, pp. 182-212.

Apostol, I., Heinstein, P.F. and Low, P.S. (1989) Rapid stimulation of an oxidative burst during elicitation of cultured plant cells. Role in defense and signal transduction, *Plant Physiology* **90**, 109-116.

Averyanov, A.A. and Lapikova, V.P. (1988) Fungitoxicity determined by active forms of oxygen in excretions of rice leaves, *Fiziol. Rastenii* **35**, 1142 - 1151.

Averyanov, A.A. Lapikova, V.P. Umnov,A.M. and Dzhavakhiya, V.G. (1987) Generation of superoxide radical by rice leaves in relation to blast resistance, *Fiziol. Rastenii*, **34**, 373-379.

Baker, B., Zambryski, P., Staskawicz, B. and Dinesh-Kumar, S.P. (1997) Signaling in plant-microbe interactions, *Science* **276**, 726-733.

Benhamou, N., Belanger, R.R. and Paulity, T. (1996a) Ultrastructural and cytochemical aspects of the interaction between *Pseudomonas fluorescens* and Ri T-DNA transformed pea roots : host response to colonization by *Pythium ultimum* Trow, *Planta* **199**, 105-117.

Benhamou, N., Belanger, R.R. and Paulity, T. (1996b) Induction of differential host responses by *Pseudomonas fluorescens* in Ri-TDNA transformed pea roots upon challenge with *Fusarium oxysporum* f.sp. *pisi* and *Pythium ultimum*, *Phytopathology* **86**, 1174-1185.

Benhamou, N., Kloepper, J.W. Quadt-Hallmann, A. and Tuzun, S. (1996c) Induction of defense-related ultrastructural modifications in pea root tissues inoculated with endophytic bacteria, *Plant Physiology* **112**, 919-929.

Berridge, M.T. and Irvine, R.F. (1989) Inositol phosphates and cell signalling, *Nature* **341**, 197-205.

Breviario,D., Morello, L. and Gianai, S. (1995) Molecular cloning of two novel rice cDNA sequences encoding putative calcium -dependent protein kinases, *Plant Molecular Biology* **27**, 953-967.

Conrath, U., Jeblick, W. and Kauss, H. (1991) The protein kinase inhibitor, K-252a, decreases elicitor-induced Ca^{2+} uptake and K^{+} release, and increase coumarin synthesis in parsley cells, *FEBS Letters* **279**, 141-144.

Conrath, U., Chen, Z., Ricigliano, J.R. and Klessig, D. (1995) Two inducers of plant defense responses, 2-6-dichloroisonicotinic acid and salicylic acid, inhibit catalase activity in tobacco, *Proceedings of the National Academy of Sciences USA* **92,** 7143-7147.

Dann, E.K. and Deverall, B.J. (1995) Effectiveness of systemic resistance in bean against foliar and soil-borne pathogens as induced by biological and chemical means, *Plant Pathology* **44**, 458-466.

Datta, K., Velazhahan, R., Oliva, N., Vasquez, A., Alejar, M., Muthukrishnan, S. and Datta, S.K. (1997) Expression of PR-5 protein gene in rice to enhance sheath blight resistance. *Rockefeller Foundation International Programme Rice Biotechnology Meeting, Malacca, Malaysia,* p. 325 (abstract)

Dixon, M., Jones, D., Keddie, J., Thomas, C., Harrison, K. and Jones, J. (1996) The tomato Cf-2 disease resistance locus comprises two functional genes encoding leucine - rich repeat proteins, Cell **84,** 451-459.

Doares, S.H. Syrovets, T., Weiler, E.W. and Ryan, C.A. (1995) Oligogalacturonides and chitosan activate plant defensive genes through the octadecanoid pathway, *Proceedings of the National Academy of Sciences USA* **92**, 4095-4098.

Du, H. and Klessig, D.F. (1997) Role for salicylic acid in the activation of defense responses in catalase-deficient transgenic tobacco, *Molecular Plant Microbe Interactions* **10**, 922-925.

Durner, J. and Klessig, D.F. (1995) Inhibition of ascorbate peroxidase by salicylic acid and 2,6-dichloroisonicotinic acid, two inducers of plant defense responses, *Proceedings of the National Academy of Sciences USA* **92**, 11312-11316

Haga, M., Sekizawa, Y. Ichikawa, M. Hiramatsu, H., Hamamoto, A., Takino, Y. and Ameyama, M. (1986) Nitroblue tetrazolium reduction of rice leaf slices stimulated with blast fungus hyphal component, *Agricultural Biological Chemistry* **50**, 1427 - 1436.

Hijwegen, T., Verhaar, M.A. and Zadoks, J.C. (1996) Resistance to *Sphaerotheca pannosa* in roses induced by 2,6 - dichloroisonicotinic acid, *Plant Pathology* **45**, 631-635.

Hoffland, E., Pieterse, C.M.J., Bik, L. and Van Pelt, J.A. (1995) Induced systemic resistance is not associated with accumulation of pathogenesis-related proteins, *Physiological and Molecular Plant Pathology* **46**, 309-320.

Hoyos, M.E., Stanley, C.M. He, S.Y., Pike, S., Pu, X-A. and Novacky, A. (1996) The interaction of harpin Pss plant cell walls, *Molecular Plant Microbe Interactions* **9,** 608-616.

Kamala Nayar.and Vidhyasekaran, P. (1998) Development of a powder formulation of *Pseudomonas fluorescens P1* for management of rice bacterial blight, *Annals of Applied Biology* (in press).

Kato, T., Yamaguchi, Y., Uyechara, T., Yokoyama, T., Namai, T. and Yamanaka, S. (1983) Self defensive substances in rice plant against rice blast disease, *Tetrahedron Letters 24,* 4715-1418.

Kim, Y.K. Baek, J.M., Park, H.Y. Choi, Y.D. and Kim, S.I. (1994) Isolation and characterisation of cDNA clones encoding class I chitinase in suspension-cultured rice cells. *Biosci. Biotechnol. Biochem.* 58, 1164-1166.

Kloepper, J.W., Tuzun, S., Liu, L., and Wei, G. (1993) Plant-growth promoting rhizobacteria as inducers of systemic resistance, in R.D. Lumsden and J.L. Waughn (eds.) *Pest Management : Biologically Based Technologies,* American Chemical Society Press, pp. 156-165.

Kobe, B. and Deisenhofer, J. (1993) Crystal structture of porcine ribonuclease inhibitor, a protein with leucine-rich repeats, *Nature* **366**, 751-756.

Kogel, K-H., Beckhove, U., Dreschers, J., Munch, S. and Romme, Y. (1994) Acquired resistance in barley. The resistance mechanism induced by 2,6 - dichloroisonicotinic acid is a phenocopy of a genetically based mechanism governing race-specific powdery mildew resistance, *Plant Physiology* **106**, 1269-1277.

Kogel, K.H. Ortel, B., Jarosch, B., Atzorn, R., Schiffer, R., Wasternack, C. (1995) Resistance in barley against the powdery mildew fungus (*Erysiphe graminis f. sp. hordei*) is not associated with enhanced levels of endogenous jasmonates, *European Journal of Plant Pathology* **101,** 319-332.

Lamb, C. (1996) A ligand-receptor mechanism in plant-pathogen recognition, *Science 274,* 2038-2039.

Lamb, C., Zhu, Q., Dabi, T., Zhong, J. and Potnis, A. (1997) Emerging strategies for engineering enhanced disease resistance and yield in rice, *Proceedingd of the Rockfeller Foundation International Programme on Rice Biotechnology,* Malacca, Malaysia, pp. 166-167.

Leon, J., Lawton, M.A. and Raskin, I. (1995) Hydrogen peroxide stimulates salicylic acid biosynthesis in tobacco, *Plant Physiology* **108**, 1673-1678.

Li, L., Qu, R., Kochko, A., Fauquet, C. and Beachy, R.N. (1993) An improved rice transformation system using the biolistic method, *Plant Cell Reports* **12**, 250-255.
Ligterink, U., Kroj,T., Nieden, U.Z., Hirt, H. and Scheel, D. (1997) Receptor-mediated activation of a MAP kinase in pathogen defense of plants, *Science* **276**, 2054-2057.
Lin, W., Anuratha, C.S., Datta, K., Potrykus,I., Muthukrishnan, S. and Datta, S.K. (1995) Genetic engineering of rice for resistance to sheath blight, *Biotechnology* **13**, 686-691.
Liu, L., Kloepper, J.W. and Tuzun, S. (1992) Induction of systemic resistance against cucumber mosaic virus by seed inoculation with selected rhizobacterial strains, *Phytopathology* **82**, 1109.
Lorenzo, G. De., Ito,Y. D'Ovidio, R., Cervone, F., Albersheim, P. and Darvill, A.G. (1990) Host-pathogen interactions. XXXVII. Abilities of the polygalacturonase inhibiting proteins from three races of *Colletotrichum lindemuthianum, Physiological and Molecular Plant Pathology* 36, 421- 435.
Lynch, D.V. and Thompson, J.E. (1984) Lipoxygenase mediated production of superoxide anion in senescing plant tissue, *FEBS Letters* **173**, 251-254.
Marme, D.(1989) The role of calcium calmodulin in signal transduction. In W.F. Ross and D.J. Morre (eds.), *Second Messengers in Plant Growth and Development,* Alan R. Liss, NewYork, pp. 57-80.
Maurhofer, M., Hase, C., Meuwly, J.P. and Defago, G. (1994) Induction of systemic resistance of tobacco to tobacco necrosis virus by the root-colonizing *Pseudomonas fluorescens* strains CHAO : influence of the *gacA* gene and of pyoverdine production, *Phytopathology* **84**, 139-146.
Mayama, S., Tani, Tieno, T., Midland, S.L., Sims, J.J. and Keen, N.T. (1986) The purification of victorin and its phytoalexin elicitor activity in oat leaves, *Physiological and Molecular Plant Pathology* **29, 1-18**.
M'Piga, P., Belanger, R.R., Paulitz, T.C. and Benhamou, N. (1997) Increased resistance to *Fusarium oxysporum* f.sp. *radicis-lycopersici* in tomato plants treated with the endophytic bacterium *Pseudomonas fluorescens* strain 63-28, *Physiological and Molecular Plant Pathology* **50**, 301-320.
Mulya, K., Takikawa, Y. and Tsuyumu, S. (1996) The presence of regions homologous to *hrp* cluster in *Pseudomonas fluorescens* PBG32R, *Annals of the Phytopathol Society of Japan* **62**, 355-359.
Namai, T., Kato, T., Yamaguchi, Y. and Togashi, T. (1991) Cytotoxicity and anti-blast activity of hydroperoxide of linolenic acid produced by *in vitro* enzyme reaction using blast-infected rice leaves, *Annals of the Phytopathol Society of Japan* **57**, 339-344.
Nelson, A.J., Doerner, P.W., Zhu, Q. and Lamb, C.J. (1994) Isolation of a monocot 3-hydroxy-3-methylglutaryl coenzymeA reductase gene that is elicitor-inducible, *Plant Molecular Biology* **25**, 401-402.
Nielsen, K.K., Bojsen, K., Collinge, D.B. and Mikkelsen, J.D. (1994) Induced resistance in sugar beet against *Cercospora beticola* ; induction by dichloroisonicotinic acid is independent of chitinase and β-1,3-glucanase transcript accumulation, *Physiological and Molecular Plant Pathology* **45**, 89-99.
Ohno, Y., Okuda, S., Natsuaki, T. and Teranaka, M. (1992) Control of bacterial seedling blight of rice by fluorescent *Pseudomonas* spp, *Proceedings of the Kanto-Tosan Plant Protection Society* **39, 9-11**.
Ohta, H., Shida, K., Peng, Y.L., Furasawa, I., Shishiyama, J., Aibara,S. and Morita, Y. (1990) The occurrence of lipid hydroperoxide decomposing-enzyme activities in rice and the realtionship of such activities to the formation of antifungal substances, *Plant Cell Physiology* **31**, 1117-1122.
Ohta, H., Ida, S., Mikami, B. and Morita, Y.(1986) Purification and characterization of rice lipoxygenase component 3 from embryos, *Agricultural Biological Chemistry*, **50**, 3165-3171.
Ohta, H., Shida, K., Peng, Y.L., Furusawa, I., Shishiyama, J., Aibara, S. and Morita, Y. (1990) The occurrence of lipid hydroperoxide decomposing - enzyme activities in rice and the relationship of such activities to the formation of antifungal substances, *Plant Cell Physiology* **31**, 1117-1122.
Pitt, D. and Kaile, A. (1990) Transduction of the calcium signal witth special reference to Ca^{2+} - induced conidiation in *Penicillium notatum,* in J. Kuhn, A.P.J. Trinci, M.J. Jung, M.W. Goosey, and L.G. Copping (eds.), *Biochemistry of Cell Walls and Membranes in Fungi,* Springer-Verlag, Berlin, pp. 283-298.
Popham, P., Pike, S. and Novacky, A. (1995) The effect of harpin from *Erwinia amylovora* on the plasmalemma of suspension-cultured tobacco cells, *Physiological and Molecular Plant Pathology* **47**, 39-50.
Raaijmakers, J.M., Leeman, M., Van Oorschot, M.M.P., Van der Sluis, I., Schippers, B. and Bakker, P.A.H.M. (1995) Dose-response relationship in biological control of fusarium wilt of radish by *Pseudomonas* sp., *Phytopathology* **85**, 1075-0181.
Rabindran, R. and Vidhyasekaran, P. (1996) Development of a formulation of *Pseudomonas fluorescens* PfALR 2 of management of rice sheath blight, *Crop Protection* **15**, 715-721.

Raz, V. and Fluhr, R. (1993) Ethylene signal is transducted via protein phosphorylation events in plants, *The Plant Cell* **5**, 523-530.

Ronald, P.C. (1997) The molecular basis of disease resistance in rice, *Plant Molecular Biology* **35**, 179-186.

Ryals, J., Neuenschwander, U.H., Willits, M.G., Molina, A., Steiner, H.Y. and Hunt, M.D. (1996) Systemic acquired resistance, *The Plant Cell* **8**, 1809-1819.

Schaffrath, U., Scheinpflug, H. and Reisner, H. (1995) An elicitor from *Phricularia oryzae* induces resistance responses in rice : isolation, characterization and physiological properties, *Physiological and Molecular Plant Pathology* **46**, 293-307.

Scheel, D. and Parker, J.E., (1990) Elicitor recognition and signal transduction in plant defense gene activation, *Zeitschrift Naturforschung* **45C**, 569-575.,

Schweizer, P., Buchala, A. and Metraux, J.P. (1997a) Gene-expression patterns and levels of jasmonic acid in rice treated with the resistance inducer 2,6- dichloroisonicotinic acid, *Plant Physiology* **115**, 61-70.

Schweizer, P., Buchala A., Silverman, P., Seskar, M., Raskin, I. and Metraux, J.P. (1997b) Jasmonate-inducible genes are activated in rice by pathogen attack without a concomitant increase in endogenous jasmonic acid levels, *Plant Physiology* **114**, 79-88.

Seguchi, K., Kurotaki, M., Sekido, S. and Yamaguchi, I. (1992) Action mechanism of N-cyanomethyl-2-chloroisonicotinamide in controlling rice blast disease, *Journal of Pesticide Science* **17**, 107-113.

Sekizawa, Y., Haga, M., Hirabayashi, E., Takeuchi, N. and Takino, Y. (1987) Dynamic behaviour of superoxide generation in rice leaf tissue infected with blast fungus, and its regulation by some substances, *Agricultural Biological Chemistry* **51**, 763-770.

Sekizawa, Y., Haga, M., Iwata, M., Hamamoto, A., Chihara, C. and Takino, Y. (1985) Probenazole and burst of respiration in rice leaf tissue infected with blast fungus, *Journal of Pesticide Science* **10**, 225-231.

Sembdner, G. and Partier, B. (1993) The biochemistry and the physiological and molecular actions of jasmonates, *Annual Review of Plant Physiology and Plant Molecular Biology* **44**, 569-589.

Shimura, M., Mase, S. and Iwata, M. (1983) Anticonidial germination factor induced in the presence of probenazole in infected host leaves. III-Structural elucidation of substances A and C, *Agricultural Biological Chemistry* **47**, 1983-1989.

Shirasu, K., Nakajima, H., Rajasekhar, V.K., Dixon, R.A. and Lamb, C. (1997) Salicylic acid potentiates an agonist - dependent gain control that amplifies pathogen signals in the activation of defense mechanisms, *The Plant Cell* **9**, 261-270.

Silverman, P., Seskar, M., Kanter, D., Schweizer, P., Metraux, J.P. and Raskin, I. (1995) Salicylic acid in rice : biosynthesis, conjugation, and possible role, *Plant Physiology* **108**, 633-639.

Steffens, M., Ettl, F., Kranz, D. and Kindl, H. (1989) Vanadate mimicks effects of fungal cell wall in eliciting gene activation in plant cell cultures, *Planta* **177**, 160-168.

Tamogami, S., Randeep, R. and Kodama, O. (1997) Phytoalexin production by amino acid conjugates of jasmonic acid through induction of naringenin - 7- O-methyltransferase, a key enzyme on phytoalexin biosynthesis in rice (*Oryza sativa L.), FEBS Letters* **401**, 239-242.

Tepper, C.S. and Anderson, A.J. (1990) Interactions between pectic fragements and extracellular components from the fungal pathogen *Colletotrichum lindemuthianum, Physiological and Molecular Plant Pathology* **36**, 147-158.

Van Peer, R., Niemann, G.J. and Schippers, B. (1991) Induced resistance and phytoalexin accumulation in biological control of fusarium wilt of carnation by *Pseudomonas* sp. strain WCS417r, *Phytopathology* **81**, 728-734.

Velazhahan, R., Cole, K.C., Anuratha, C.S. and Muthukrishnan, S. (1998) Induction of thaumatin-like proteins (TLPs) in *Rhizoctonia solani* - infected rice and characterization of two new cDNA clones. *Physiologia Plantarum* **100** (in Press).

Veldink, G.A. and Vliegenthart, J.F.G. (1985) Lipoxygenase, nonheme iron-containing enzymes, *Advances in Inorganic Biochemistry* **6**, 139-161.

Vidhyasekaran, P. (1988a) *Physiology of Disease Resistance in Plants,* Vol. I, CRC Press, Florida, pp. 149.

Vidhyasekaran, P. (1988b) *Physiology of Disease Resistance in Plants*, Vol.II, CRC Press, Florida, pp. 127.

Vidhyasekaran, P. (1993) Defense genes for crop disease management, in P. Vidhyasekaran (ed.), *Genetic Engineering, Molecular Biology, and Tissue Culture for Crop Pest and Disease Management,* Daya Publishing House, Delhi, pp. 17-30.

Vidhyasekaran, P. (1997) Fungal Pathogenesis in *Plants and Crops : Molecular Biology and Host Defense Mechanisms,* Marcel Dekker, Newyork, pp. 553.

Vidhyasekaran, P. and Velazhahan, R. (1996) Elicitor induced defense responses in suspension cultured rice cells, in V.C. Chopra, R.P. Sharma, and M.S. Swaminathan (eds.) *Agricultural Biotechonology,* Oxford and IBH Publishing Co., New Delhi, pp. 249-255.
Vidhyasekaran, P., Velazhahan, R., Samiappan, R., Ruby Ponmalar, T. and Muthukrishnan, S. (1996). Isolation of a 35-kDa chitinase from suspension-cultured rice cells and its potential in the development of sheath blight-resistant transgenic rice plant, in G. khush (ed.), *Rice Genetics III,* International Rice Research Institute, the Philippines, pp. 868-872.
Vidhyasekaran, P., Rabindran, R., Muthamilan, M., Nayar, K., Rajappan, K., Subramanian, N. and Vasumathi, K. (1997) Development of a powder formulation of *Pseudomonas fluorescens* for control of rice blast, *Plant Pathology* **46**, 291-297.
Ward, E.B., Uknes, S.J., Willams, S.C., Dinchjer, S.S., Wiederhold, D.C., Alexander, D.C., Ahl-Goy, P., Metraux, J.P. and Ryals, J.A. (1991) Co-ordinate gene activity in response to agents that induce systemic acquired resistance, *The Plant Cell* **3**, 1085-1094.
Wei, G., Kloepper, J.W. and Tuzun, S. (1991) Induction of systemic resistance of cucumber to *Colletotrichum orbiculare* by select strains of plant growth-promoting rhizobacteria, *Phytopathology* **81**, 1508-1512.
Xu, Y., Chang, P.L., Liu, D., Narasimhan, M.L., Raghotthama, K.G., Hasegawa, P.M. and Bressan, R.A. (1994) Plant defence genes are synergistically induced by ethylene and methyl jasmonate, *The Plant Cell* 6, 1077-1085.
Xu, Y., Zhu, Q., Panbangred, W., Shirasu, K., and Lamb, C.,(1996) Regulation, expression and function of a new basic chitinase gene in rice (*Oryza sativa L.*), *Plant Molecular Biology* **30**, 387-401.
Yang, Yi. and Gabriel, D.W. (1995) Intragenic recombination of single plant pathogen gene provides a mechanism for the evolution of new host specificities, *Journal of Bacteriology* **177**, 4963-4968.
Yoshida, H., Konishi, K., Nakagawa, T., Sekido.S. and Yamaguchi, I. (1990) Characteristics of N-phenylsulfonyl-2-chloroisonicotinamide as an anti-rice blast agent, *Journal of Pesticide Science* **15**, 199-203.
Zdor, R.E. and Anderson, A.J. (1992) Influence of root colonizing bacteria on the defense responses of bean, *Plant and Soil* **140**, 99-107.
Zhou, T. and Paulitz, T.C. (1994) Induced resistance in the biocontrol of *Pythium aphanidermatum* by *Pseudomonas* spp. on cucumber, *Journal of Phytopathology* **142**, 51-63.
Zhu, Q, Chappell, J., Hedrick, S.A. and Lamb, C.J. (1995) Accurate *in vitro* transcription from circularized plasmid templates by plant whole cell extracts, *Plant Journal* **7**, 1021-1030.
Zocchi, G. and Rabotti, G. (1993) Calcium transport in membrane vesicles isolated from maize coleoptiles. Effect of indoleacetic acid and fusicoccin, *Plant Physiology* **101**, 135-139.

Key Words Index

www.ingramcontent.com/pod-product-compliance
Ingram Content Group UK Ltd.
Pitfield, Milton Keynes, MK11 3LW, UK
UKHW021454290926
475987UK00007B/76
* 9 7 8 1 4 0 2 0 0 0 5 0 8 *